Forest and Rangeland Soils of the United States Under Changing Conditions

Richard V. Pouyat • Deborah S. Page-Dumroese
Toral Patel-Weynand • Linda H. Geiser
Editors

Forest and Rangeland Soils of the United States Under Changing Conditions

A Comprehensive Science Synthesis

 Springer

Editors
Richard V. Pouyat
Northern Research Station
USDA Forest Service
Newark, DE, USA

Deborah S. Page-Dumroese
Rocky Mountain Research Station
USDA Forest Service
Moscow, ID, USA

Toral Patel-Weynand
Washington Office
USDA Forest Service
Washington, DC, USA

Linda H. Geiser
Washington Office
USDA Forest Service
Washington, DC, USA

ISBN 978-3-030-45218-6   ISBN 978-3-030-45216-2   (eBook)
https://doi.org/10.1007/978-3-030-45216-2

English Equivalents

When you know	Multiply by	To find
Millimeters (mm)	0.0394	Inches
Centimeters (cm)	0.394	Inches
Meters (m)	3.28	Feet
Kilometers (km)	0.621	Miles
Hectares (ha)	2.47	Acres
Square meters (m^2)	10.76	Square feet
Square kilometers (km^2)	0.386	Square miles
Kilograms (kg)	2.205	Pounds
Megagrams (Mg)	1	Tons
Petagrams (Pg)	1,102,311,311	Tons
Teragrams (Tg)	1,102,311.31	Tons
Degrees Celsius (°C)	1.8 °C + 32	Degrees Fahrenheit

Executive Summary

Overview and Purpose

Soils have life-giving and life-sustaining capabilities that directly or indirectly support all living organisms. Humanity's well-being and health are tied directly to the health of soils. Worldwide, soil processes contribute to an abundant supply of water, food, and fiber while also tempering a warming climate. More than 40% of global terrestrial carbon (C) is stored in the soils of forests, grasslands, and shrublands (Jackson et al. 2017). Forest and rangeland soils of the United States make a disproportionately large contribution to such ecosystem services compared to their geographic extent. Although forests and rangelands occupy only about one-third of the land area in the United States, they supply 80% of the surface freshwater (Sedell et al. 2000) and sequester 75% of the total C to a depth of 30 m stored in the nation (Liu et al. 2013). Ecosystem alterations associated with land-use change and other pressures may impair the ability of soils to fulfill their foundational role.

Today, a number of disturbances compound the vulnerability of forest and rangeland soils across the United States. Of greatest concern are various environmental changes, continued overgrazing, catastrophic wildfire, and invasive plant and animal species. Effects on soil health (soil functions mediated by living organisms) are expected to be more severe when two or more of these disturbances or stressors interact with each other. Changes in the climate are likely to magnify or accelerate all of these impacts on forest and rangeland soils. However, great uncertainty surrounds predictions of climate-induced impacts because overall effects will depend on the magnitude of temperature and precipitation changes and the frequency of extreme events.

The purpose of this report is to synthesize leading-edge science and management information about forest and rangeland soils of the United States, offer ways to better understand changing conditions and their impacts on soils, and explore directions that positively affect the future of forest and rangeland soil health. Other assessments provide similar information for agricultural soils, so agricultural soils are not included in this report. This report outlines soil processes and identifies the research needed to manage forest and rangeland soils in the United States. Chapter 1 provides an overview of the state of forest and rangeland soils research in the nation, including multidecadal studies. Chapters 2, 3, 4 and 5 summarize various human-caused and natural impacts and their effects on soil C, hydrology, biogeochemistry, and biological diversity. Chapters 6 and 7 consider the effects of changing conditions on forest soils in wetland and urban settings, respectively. Impacts include:

- *Climatic variability and change.* Shifts in precipitation patterns, temperature increases and variability, and increases in atmospheric carbon dioxide (CO_2) concentrations affect plant productivity, flooding, nutrient cycling, and biological populations. Changes in climate, coupled with an increase in the frequency and severity of extreme weather events, have had, and will continue to have, direct and cascading indirect effects on soil formation and associated productivity rates, as well as physical, chemical, and biological processes.
- *Severe wildfires.* Regardless of origin, wildfires are becoming larger, more frequent, more intense, and are lasting longer. Severe wildland fires impact the C and nitrogen (N) in soils,

alter the environment for various communities, and can change the trajectory of forest composition. Fire may also form hydrophobic (water repellent) layers in soil. These layers, coupled with loss of vegetation, can lead to accelerated erosion and nutrient leaching.

- *Invasive species, pests, and diseases.* The introduction of a wide range of invasive species, from microbes, macrofauna, and macroflora, has important impacts on soil processes which are exacerbated by climatic changes. Furthermore, the ranges of some invasive insects and pathogens are expanding. Invasive species contribute to tree stress, and lead to decline and mortality, decline of biodiversity, and soil changes in organic matter composition and nutrients. Similar to landscapes after wildland fires, problematic species can create soil conditions that enhance flash flooding, soil erosion, and sediment loading.

- *Pollution.* Air and water pollutants can dramatically affect soil characteristics and species composition. Acid deposition can cause nutrient leaching, while other pollutants such as some hydrocarbon compounds and various trace metals can cause other chemical changes in soils. Despite environmental policies designed to limit the release of pollutants, some continue to impact forest and range soils. Soil recovery from the impacts of these pollutants can take many decades.

- *Non-urban land uses* can potentially have an impact on forest and range soils. Compaction is considered one of the most critical issues on forest and rangeland soils because it can severely alter the movement and storage of air, water, and nutrients in the soil. These changes can slow tree growth and negatively affect microbial populations. In addition, mining has disturbed millions of hectares of forest and grasslands in the United States.

- *Urban land use and change.* Direct urban-induced effects that can impact soils include the introduction of built structures, landfills, stormwater facilities, impervious surfaces, and lawn management. Urban land use and change may also affect soils through indirect processes such as the urban heat island effect, emissions of various pollutants, and the spread of non-native species.

Chapter 8 considers approaches to maintaining or restoring forest and rangeland soil health in the face of these varied impacts. Chapter 9 discusses mapping, monitoring, and data sharing as ways to leverage scientific and human resources to address soil health at scales from the landscape to the individual parcel (monitoring networks, data-sharing Web sites, and educational soils-centered programs are tabulated in appendix B). Chapter 10 highlights opportunities for deepening our understanding of soils and for sustaining long-term ecosystem health (appendix C summarizes research needs). Nine summaries (appendix A) offer a more detailed regional look at forest and rangeland soils in the United States and its affiliates.

Key Messages

Key Benefits of Forest and Rangeland Soils

Carbon and Water

Soil organic carbon (SOC) is a key indicator and dynamic component of soil health. The functioning of a healthy ecosystem depends on the quality and quantity of SOC. For example, a high proportion of SOC in forest and rangeland soils results in greater water and nutrient retention, which in turn helps to buffer soil against drought and pollution effects. With climate factors being approximately equal, most SOC in the United States is in forests, wetlands, and rangelands that are not intensively managed. Very small changes to SOC can have profound effects on atmospheric CO_2 levels at a national or global scale. This sensitivity underscores the importance of gaining knowledge about the magnitude and extent of disturbance impacts on SOC. Some of the factors that directly affect SOC are climate change, overgrazing, overharvesting, invasive species, use of fertilizers, and catastrophic wildfire. Management can proactively help to limit the negative effects of these factors by managing fuel loads, fostering reforestation, controlling invasive weeds, and preventing further unnecessary site disturbances.

The scientific community is developing and beginning to adopt a new paradigm for understanding and predicting long-term accumulation of SOC. The new paradigm focuses on the complex relationships between microorganisms and minerals in the soil instead of the characteristics of the SOC itself. This analytical framework may allow for more accurate prediction of SOC than previous models and quantifies factors that are easier to measure than SOC molecular properties.

Soil organic carbon is also a key component of soil organic matter, which holds soil particles together through adhesion and also promotes fungi to form and stabilize soil aggregates. Soil organic matter helps to reduce erosion and facilitates plant uptake of nutrients and water. The loss of soil water holding capacity due to loss of SOC can result in catastrophic erosion, as has been the case with many wildfires in the western United States. Soil organic material also plays a key role in purifying drinking water and detoxifying pollutants in the soil. In fact, clean water from United States forested watersheds supplies approximately two-thirds of the drinking water for towns and cities.

Tree canopies usually prevent 2–30% of water from reaching the ground but also serve a regulating function in allowing the water to infiltrate into soils at a rate that prevents massive runoff events. The soil and its inherent characteristics affect how water moves on the surface and in the subsurface through infiltration and percolation. This in turn affects the quantity and quality of water from forests and rangelands. It is estimated that approximately 25% of precipitation is stored in soils and watersheds of forests. Less is known about rangelands due to wide swinging precipitation patterns (e.g., monsoons).

Many hydrologic models continue working towards fully capturing the dynamics of soil. Most models rely on climate and topography, but a growing body of research points to soil layering and its relationship with bedrock as strong influences on how water is redistributed across the landscape. Recent advances in data science, computing infrastructure, data availability, and new monitoring tools, create the opportunity to develop new frameworks for modeling that lead to an integrated understanding of water dynamics and biogeochemical cycles. The interaction of soil and water needs better quantification as current models continue working to improve on prediction of this interaction at watershed scales. However, the paired catchment studies of the twentieth century continue to provide a solid foundation for studying more recent hydrologic and soil changes.

Biodiversity and Indicators of Soil Health

Soils and their components result in complex ecosystems with multifaceted webs of flora, fauna, and other organisms that respond dynamically to external influences. Many of these organisms, especially bacteria and archaea, fungi, nematodes, and insects, are largely undescribed. Each gram of soil is estimated to contain 1×10^9 microorganisms and about 4000 species. Organisms that live in the soil also respond to myriad properties including textures and structures. These soil dwellers vary greatly in their size, function, mobility, and response to disturbances and environmental stress. The functional redundancy in soil is thought to increase its resistance and resilience (Turbé et al. 2010). Research on soil response to lesser disturbances suggests that short-term impacts can persist in soil for up to 5 years before soil conditions and communities return to the previous state. It is still unclear the impact that more intense disturbances (e.g., wildfire, invasive plants and animals, and climate change) and the resulting changes have on soil organisms over a longer period of time.

Biogeochemistry

Biogeochemistry, an area of study that emerged in the late twentieth century, explores how physical, chemical, biological, and geological processes interact and affect the natural environment. Soils have a prominent role in biogeochemical cycles in forests and rangelands. They are the major reservoir for plant nutrients as decomposition transforms organic matter into a continual supply of nutrients. While many ecosystem-level losses occur in forests and rangelands through multiple pathways, the biogeochemical cycles of forests have been found to be generally effective in maintaining pools of many essential nutrients such as N, phosphorus, sulfur, boron, and potassium.

Soil in Wetland and Urban Landscapes

Wetland and hydric soils are distinguished from "upland" or forested areas by their properties, composition, and biogeochemistry. Wetlands are subject to long periods of anoxia, or lack of oxygen, while upland soils are almost always oxic. Many wetlands are generally characterized by anaerobic processes, and the vegetation that grows in wetlands is specifically adapted to this environment. Wetlands also contribute to a wide variety of highly valued ecosystem services, such as water supply and quality, C sequestration, wildlife habitat, and recreational opportunities. For example, wetland soils can store and sequester C at much higher rates than upland areas.

Wetlands receive inputs from uplands through the soil by hydrologic forces. They discharge outputs to groundwater and adjacent waterways and uplands. Prior to discharging outputs that flow directly into streams and other bodies of water, wetlands absorb many pollutants including pesticides and fertilizers, heavy metals, hydrocarbons, and road salts. Wetlands can also be negatively impacted by pollutants that flow into them. Several regulatory control measures have resulted in declines in pollutants such as mercury and sulfur, but other kinds of disturbances continue to be a threat to wetlands. Higher temperatures, changes in precipitation, and an altering of species composition will all significantly affect wetland ecosystem processes.

Wetlands have not always been valued, and nearly 53% have been converted to other uses. These wetlands were drained and used for other functions, particularly agriculture, until the mid-1900s. Today, the primary threat to wetlands is urbanization (US EPA 2016).

By 2010 in the United States, almost 250 million people—about 81% of the total population—lived in urban areas (US Census 2010). The global population is projected to exceed 11 billion before the end of the twenty-first century with an estimated 70% living in urban areas (United Nations 2015). The relatively rapid expansion of urban populations in the United States and worldwide suggests the increasing importance of "urban soils."

Urban soil conditions vary based on the severity of disturbance and environmental changes that typically occur in our cities and towns. These impacts vary from soils associated with remnant forests or grasslands embedded in urban areas to drastically disturbed soils associated with human-created landfills; surfaces sealed with asphalt, concrete, or other impervious surfaces; and altered physical conditions such as those associated with residential yards.

Despite these impacts, urban soils can provide many ecosystem services and support an array of microorganisms and invertebrates. Some native and nonnative species survive and thrive in the face of urban environmental changes. Others are affected through various avenues such as management practices (e.g., irrigation, which supports particular biota) and landscaping practices (e.g., composting and mulching, removal of woody and leaf debris). Similarly, the use of green infrastructure such as green roofs and rain gardens provides habitats for many soil organisms.

Management of urban ecosystems requires a holistic approach that takes into account the many, and sometimes competing, services that the ecosystem can provide. These ecosystem services include C sequestration, improved water quality, food production, recreation, stable substrates for structures and underground utilities, and stormwater retention. The importance of services provided by urban ecosystems is magnified because of their close proximity to people.

In the United States, the classification and survey of soils in urban landscapes have advanced tremendously in the last 20 years. For example, modern soil surveys have been conducted in New York, NY; Chicago, IL; Los Angeles, CA; and Detroit, MI. These surveys have provided detailed information on physical, chemical, and mineralogical properties of human-altered and human-transported soils. The greater number of soil properties and the level of detail enable more reliable interpretations for stormwater management, revegetation and restoration efforts, urban agriculture, and resource inventories.

Degradation of Soil Health

Soils have finite qualities and take thousands of years to develop but can lose their ability to contribute ecosystem services in a fraction of that time due to human-induced and natural disturbances. Identifying key disturbances and understanding their effects are critical to developing sound management responses to sustain and restore soils. Key disturbances that impact soils include climate change, severe wildland fire, invasive species and pests, and pollution related to nonurban land uses and urban land uses.

Management

Management actions may cause soils to lose, retain, or improve their capacity to sustain life, maintain balance in hydrologic and nutrient cycling, and provide other ecosystem functions (Heneghan et al. 2008). Nutrient cycling is also susceptible to human-caused disturbances. Active management manipulates aboveground and belowground nutrient cycles to achieve various goals. These activities have direct and indirect impacts on the soil. Some effects are known, but others are not. Management has an important role in maintaining long-term soil health and thus ecosystem integrity. A proactive rather than reactive approach is needed to address negative impacts from disturbances. New management and restoration techniques merit exploration.

Managing, Restoring, and Addressing Soil Needs

Innovations in Soil Management

A proactive approach to strategic forest and rangeland management has been gradually gaining momentum, particularly over the last several decades. Since the 1980s, management has specifically focused on limiting soil loss on National Forest System lands and the restoration of soils. A recent revision of the Forest Service Manual (FSM) Chapter 2550 identifies six key functions of healthy soils: soil biology, soil hydrology, nutrient cycling, C storage, soil stability and support, and filtering and buffering (USDA FS 2010). Each function is an important area of further study that will be critical for land managers and scientists to consider. Other efforts, such as the interagency Ecological Site Handbook for Rangelands (Caudle et al. 2013), have included key foci for management of ecological sites, such as seeking to protect soils and minimize soil damage. New technology is under development to aid in minimizing impacts to soils, such as low-pressure tires and tracked vehicles to reduce compaction. Understanding how to balance and use disturbances for soil gain has also been gaining attention.

While wildland fire can be a major disturbance to soil health, it can also be beneficial when properly managed. Recently, prescribed burning has been used not only to reduce fuel loading to prevent catastrophic wildfire, but also to confer other ecosystem benefits, including improved soil health. When used effectively as a management tool, fire can assist in keeping the forest floor C/N ratios more optimal for N mineralization (DeLuca and Sala 2006), increase water availability to surviving trees, and improve overall nutrient cycling. When catastrophic fires do occur, proactive approaches such as using the USDA Forest Service's Burned Area Emergency Response (BAER) program help to mitigate losses and work to restore ecological function. Managing wildland fire involves a delicate balance that is still being refined; managers and scientists need to continue to expand understanding and application of ecological principles to ensure effective fire management.

Mining is another disturbance that has impacted soils. Approximately 161,000 mines on public lands have been abandoned. The cost of restoration is estimated in the billions of dollars. Efforts by the Department of the Interior's Bureau of Land Management (BLM) and the Forest Service have inspired state and federal legislation that guides mining processes and mine land reclamation. Although many lands have been reclaimed, many more need to be restored. The BLM and Forest Service embarked on mutual efforts to address Abandoned Mine Lands in 1997, and in 2016, the Forest Service adopted an approach to address maintenance and monitoring post-reclamation. Innovations now available for managers to restore soil and ecological health and functionality include biochar, seed coating technologies, and soil transplants.

The need is ongoing for continued innovative approaches that improve functionality of abiotic and biotic systems.

Monitoring Restored Systems

In order to understand the effectiveness of various practices, ecosystems must be monitored. Research has shown that this minimal amount of monitoring is not an accurate predictor of long-term success and can often give false indicators of success. Like all monitoring, restoration monitoring programs need to be well designed and standardized and have the support required for long-term monitoring. A restoration primer by the Society for Ecological Restoration (SER 2004) has put forward attributes of restored ecosystems (e.g., presence of indigenous species and functional groups required for long-term stability) that would be useful to incorporate into monitoring protocols.

Assessment, Mapping, and Measuring

High-quality soils information has been a growing demand since the development of the field of soil science in the nineteenth century. The US Government has long supported capturing data through soil surveys. Over the last 118 years, methods for mapping, storing, and delivering soil information have advanced. Mapping soils through surveys has been one way to extrapolate information and link it with other ecosystem health indicators. Early soil surveys were recorded through aerial photography and paper notes but are now available through digital means. Modern techniques have made the maps more accurate and their use more comprehensive in relaying data. Traditional and digital soil mapping can be used for operational activities to predict trajectories of a species or ecosystem or as a means for monitoring to understand the effects of a sequence of management actions. Other tools are available for more specialized applications. For example, the Forest Service's Forest Inventory and Analysis (FIA) program maintains nationwide monitoring that can be used at a variety of spatial scales. Regardless of the goals or tools chosen, effective monitoring and assessment are those that can be adapted to local and regional scales but follow protocols that can be used universally.

Agencies are working together with their different tool offerings to give users the best possible reference standards. Future soil mapping is likely to include more ecological variables and place greater emphasis on soil health and its response to disturbances. Using periodic assessments as a monitoring tool can help to quantify soil impacts related to disturbances and to determine to what degree, if at all, restoration is progressing.

Needs for the Future

One of the best ways to address future needs is through research opportunities. The quantity, quality, and accessibility of soil data have never been as good as they are now for scientists to make considerable headway in understanding soils, effects of disturbances on soils, and best practices for forest and rangeland soil management. Important areas for study include:

- *Carbon* and its interactions with other ecological components, disturbance, and recovery
- *Measuring and monitoring water dynamics of soils* and creating or improving hydrologic models that are applicable at watershed scales
- *Nutrient cycling* and refining ways to quantify mineral weathering and the relationship to forest and rangeland health
- *Biodiversity* and identifying and explaining ecosystem services provided by currently unexplored or unidentified species and organisms within soil and how they are affected by land management
- *Wetlands*, particularly soil and C sequestration in tidal freshwaters
- *Urban* soils, such as more basic understanding of how soils are formed and affected by human-made materials and land management, how urban soil characteristics vary spatially, and how to interpret these variations for their use in green infrastructure

Further research is essential for assessing how these listed factors interact with precipitation patterns, air temperatures, pollutant inputs, invasive species, management actions, and frequency and intensity of wildland fire.

Managers need information that they can apply to on-the-ground operations. Addressing the complex interactions in order to derive applicable information requires long-term field studies and interdisciplinary teams that are able to assess, understand, and accurately extrapolate information on changes in soil properties over time. The United States is a world leader in supporting long-term studies. These efforts need continued support and avenues for collaborating on knowledge generation and knowledge sharing across programs.

The current number of scientists who are able to do this work is dwindling as academic, government, and private industry jobs that have focused on soil health are being lost. Forest and rangeland soil scientists need to be supported so that there is a next generation of scientists who are equipped to continue and improve upon the work of today.

Much of these needs will be met by strategically planning and prioritizing resources to:

- Support and integrate existing soil networks
- Incorporate the best and most leading-edge methods for monitoring, assessing, and managing soils
- Invest in human capital through creating an array of opportunities for educating students and the public
- Link efforts made by public and private stakeholders

Considerable effort has been exerted for more than a century to build the understanding and application of effective soil management. The care taken to manage soil health contributes directly to the sustainable maintenance of ecosystem services such as forest products, water, food, fiber, and support of wildlife and plant health.

Soil health reflects, in part, the state of soil organic C stores, which are subject to natural and human-caused disturbances. Disturbance effects on aboveground and belowground C stores in turn impact nutrient cycles, which may impact economic, recreational, and conservation goals. Point and nonpoint source pollutants, climate change, wildfires and other extreme events, and distribution of invasive pests and species have significantly changed nutrient cycles through the centuries. In order to maintain sustainable nutrient cycles in the soils of forests and rangelands, balanced, intentional, and proactive management choices will need to be made for these ecosystems. In this time of more frequent and intense disturbances, accelerated population growth and urbanization, and a rapidly changing climate, soil health must be at the center of our management and restoration planning and actions for the nation's forests and rangelands.

Literature Cited

Caudle D, DiBenedetto J, Karl M et al. (2013) Interagency ecological site handbook for rangelands. U.S. Department of the Interior, Bureau of Land Management, Washington, DC, 109 p

DeLuca TH, Sala A (2006) Frequent fire alters nitrogen transformations in ponderosa pine stands of the inland Northwest. Ecology 87(10):2511–2522

Gartner T, Mehan GT III, Mulligan J et al. (2014) Protecting forested watersheds is smart economics for water utilities. J Am Water Works Ass. 106(9):54–64

Heneghan L, Miller SP, Baer S et al. (2008) Integrating soil ecological knowledge into restoration management. Restor Ecol 16(4):608–617

Jackson RB, Lajtha K, Crow SE et al. (2017) The ecology of soil carbon: pools, vulnerabilities, and biotic and abiotic controls. Ann Rev Ecol Evol Syst. 48(1):419–445

Liu S, Wei Y, Post WM et al. (2013) The Unified North American Soil Map and its implication on the soil organic carbon stock in North America. Biogeosciences. 10:2915

Sedell J, Sharpe M, Apple DD et al. (2000) Water and the Forest Service. FS-660. U.S. Department of Agriculture, Forest Service, Washington, DC, 26 p

Society for Ecological Restoration (2004) SER international primer on ecological restoration. Society for Ecological Restoration, International Science and Policy Working Group, Washington, DC

Turbé A, De Toni A, Benito P et al. (2010) Soil biodiversity: functions, threats and tools for policy makers. Bioemco, 00560420

U.S. Census Bureau (2010) Metropolitan and micropolitan change. Available at http://www.census.gov/programs-survey/metro-micro.html. Accessed 22 Apr 2019

United Nations (2015) World population prospects: the 2015 revision, key findings and advance tables. Working Paper ESA/P/WP.241. United Nations, Department of Economic and Social Affairs, Population Division

U.S. Department of Agriculture, Forest Service [USDA FS] (2010) Forest Service Manual, FSM 2500 Watershed and Air Management, Chapter 2550 Soil Management. WO Amendment 2500-2010-1. U.S. Department of Agriculture, National Headquarters, Washington, DC

U.S. Environmental Protection Agency (2016) National wetland condition assessment 2011: a collaborative survey of the nation's wetlands. EPA Report EPA-843-R-15-005. https://www.epa.gov/national-aquatic-resource-surveys/nwca

Disclaimer

The findings and conclusions in this publication are those of the authors and should not be construed to represent official USDA or US Government determination or policy.

Contents

About the Contributors

Mary Beth Adams is a research soil scientist, U.S. Department of Agriculture, Forest Service, Northern Research Station, Ecology and Management of Invasive Species and Forest Ecosystems, 180 Canfield Street, Morgantown, WV 26505.

Sarah M. Anderson is a presidential management fellow in forest management, Range Management and Vegetation Ecology, U.S. Department of Agriculture, Forest Service, Washington, DC 20250.

Vince A. Archer is a soil scientist, U.S. Department of Agriculture, Forest Service, Northern Region, 26 Fort Missoula Road, Missoula, MT 59804.

Elizabeth M. Bach is executive director of the Global Soil Biodiversity Initiative, housed at Colorado State University, School of Environmental Sustainability, Fort Collins, CO 80523.

Scott Bailey is a geologist, U.S. Department of Agriculture, Forest Service, Northern Research Station, Center for Research on Ecosystem Change, 234 Mirror Lake Road, North Woodstock, NH 03262.

Sheel Bansal is a research ecologist, U.S. Geological Survey, 8711 37th Street SE, Jamestown, ND 58401.

Erin Berryman is a quantitative ecologist at RedCastle Resources and contractor for U.S. Department of Agriculture, Forest Service, Forest Health Assessment and Applied Sciences Team, 2150 Centre Avenue, Building A, Suite 331, Fort Collins, CO 80526.

Dan Binkley is an adjunct professor, Northern Arizona University, School of Forestry, 200 E. Pine Knoll Drive, Flagstaff, AZ 86011.

Sally Brown is a research associate professor, University of Washington, School of Forest Resources, 203 Bloedel Hall, Box 352100, Seattle, WA 98195.

Jeff Bruggink is a soil scientist, U.S. Department of Agriculture, Forest Service, Intermountain Region, 324 25th Street, Ogden, UT 84401.

Matt Busse is a research soil scientist, U.S. Department of Agriculture, Forest Service, Pacific Southwest Research Station, 1731 Research Park Drive, Davis, CA 95618.

Mac A. Callaham Jr is a research ecologist, U.S. Department of Agriculture, Forest Service, Southern Research Station, Center for Forest Disturbance Science, 325 Green Street, Athens, GA 30602.

Steve Campbell is a soil scientist, U.S. Department of Agriculture, Natural Resources Conservation Service, West National Center, 1201 NE Lloyd Boulevard, Suite 801, Portland, OR 97232.

Chih-Han Chang is a postdoctoral scholar, University of Maryland, Department of Environmental Science and Technology, College Park, MD 20742, and Johns Hopkins University, Department of Earth and Planetary Sciences, Baltimore, MD 21210.

Rodney Chimner is a professor, Michigan Technological University, School of Forest Resources and Environmental Science, Noblet Building 114, 1400 Townsend Drive, Houghton, MI 49931.

Taniya Roy Chowdhury is a postdoctoral fellow, Pacific Northwest National Laboratory, Earth and Biological Sciences Division, Richland, WA 99352.

Stephanie J. Connolly is a forest soil scientist, U.S. Department of Agriculture, Forest Service, Supervisor's Office, Monongahela National Forest, Elkins, WV 26241.

David V. D'Amore is a research soil scientist, U.S. Department of Agriculture, Forest Service, Pacific Northwest Research Station, 11175 Auke Lake Way, Juneau, AK 99801.

Susan D. Day is an associate professor, Virginia Tech, Department of Forest Resources and Environmental Conservation, 307D Cheatham Hall, 310 West Campus Drive, Blacksburg, VA 24061.
Current address: Faculty of Forestry, University of British Columbia, Vancouver, BC, Canada.

Jonathan L. Deenik is a professor, University of Hawaii at Mānoa, Department of Tropical Plants and Soil Science, 3190 Maile Way, St. John 102, Honolulu, HI 96822.

David Diamond is executive coordinator of the Greater Yellowstone Coordinating Committee, 10 E. Babcock Street, Room 234, Bozeman, MT 59771.

Grant M. Domke is a research forester, U.S. Department of Agriculture, Forest Service, Northern Research Station, 1992 Folwell Avenue, St. Paul, MN 55108.

Cara L. Farr is a regional soil scientist, U.S. Department of Agriculture, Forest Service, Pacific Northwest Region, 1220 SW 3rd Avenue, Portland, OR 97204.

Brian Gardner is a major land resource area soil survey leader, U.S. Department of Agriculture, Natural Resources Conservation Service, 1848 S. Mountain View Road, Suite 3, Moscow, ID 83843.

Linda H. Geiser is Air and Soil Research program leader, U.S. Department of Agriculture, Forest Service, Research and Development, 201 14th Street SW, Washington, DC 20250.

Christian P. Giardina is a research ecologist, U.S. Department of Agriculture, Forest Service, Pacific Southwest Research Station, Institute of Pacific Islands Forestry, 60 Nowelo Street, Hilo, HI 96720.

Grizelle González is a research ecologist and project leader, U.S. Department of Agriculture, Forest Service, International Institute of Tropical Forestry, Jardín Botánico Sur, 1201 Ceiba Street, Río Piedras, PR 00926-1119.

James Gries is the assistant director of air, water, lands, soils, and minerals, U.S. Department of Agriculture, Forest Service, Eastern Region, 626 East Wisconsin Avenue, Milwaukee, WI 53202.

Jeffrey Hatten is an associate professor, Oregon State University, College of Forestry, Forest Engineering, Resources & Management, 210 Snell Hall, Corvallis, OR 97331.

Katherine A. Heckman is a research biological scientist, US Department of Agriculture, Forest Service, Northern Research Station, 410 MacInnes Drive, Houghton, MI 49931.

Scott M. Holub is a forest soil scientist, Weyerhaeuser, 785 N. 42nd Street, Springfield, OR 97477.

Mark J. Kimsey is a research assistant professor, University of Idaho, Department of Forest, Rangeland and Fire Sciences, 875 Perimeter Drive, MS 1133, Moscow, ID 83844.

Jennifer Knoepp is an emeritus scientist, U.S. Department of Agriculture, Forest Service, Southern Research Station, 3160 Coweeta Lab Road, Otto, NC 28763.

Randall K. Kolka is a team leader and research soil scientist, U.S. Department of Agriculture, Forest Service, Northern Research Station, Center for Research on Ecosystem Change, 1831 Highway 169 East, Grand Rapids, MN 55744.

Larry E. Laing is a Landscape Ecology program leader, U.S. Department of Agriculture, Forest Service, National Forest System, Washington, DC 20250.

Erik A. Lilleskov is a research ecologist and director's representative, U.S. Department of Agriculture, Forest Service, Northern Research Station, Climate, Fire and Carbon Cycle Sciences, 410 MacInnes Drive, Houghton, MI 49931.

Graeme Lockaby is Clinton McClure professor, director, and associate dean, Auburn University, School of Forestry and Wildlife Sciences, Auburn, AL 36849.

Richard A. MacKenzie is a research ecologist, U.S. Department of Agriculture, Forest Service, Pacific Southwest Research Station, Institute of Pacific Islands Forestry, 60 Nowelo Street, Hilo, Hawaii 96720.

Korena Mafune is a research assistant, University of Washington, School of Environmental and Forest Science, Box 352100, Seattle, WA 98195.

Erika Marín-Spiotta is an associate professor, University of Wisconsin-Madison, Department of Geography, 550 N. Park Street, Madison, WI 53706.

Anne S. Marsh is a national program lead, Bioclimatology and Climate Change Research, U.S. Department of Agriculture, Forest Service, Research and Development, 201 14th Street SW, Mailstop 1115, Washington, DC 20250.

Manuel Matos is a state soil scientist, U.S. Department of Agriculture, Natural Resources Conservation Service, 654 Muñoz Rivera Avenue, Suite 604, San Juan, PR 00918-4123.

Paul McDaniel is emeritus professor of pedology, University of Idaho, Department of Soil and Water Systems, Moscow, ID 83844.

Kevin McGuire is a professor, Virginia Tech, Virginia Water Resources Research Center and Department of Forest Resources and Environmental Conservation, 210-B Cheatham Hall, Blacksburg, VA 24061.

Patrick Megonigal is associate director for research and principal investigator, Smithsonian Institute, Smithsonian Environmental Research Center, 647 Contees Wharf Road, Edgewater, MD 21037.

Chelcy F. Miniat is a project leader, U.S. Department of Agriculture, Forest Service, Southern Research Station, 3160 Coweeta Lab Road, Otto, NC 28763.

Lucas E. Nave is an associate research scientist, University of Michigan Biological Station and Department of Ecology and Evolutionary Biology, 9133 Biological Road, Pellston, MI 49769.

Daniel G. Neary is a supervisory research soil scientist, U.S. Department of Agriculture, Forest Service, Rocky Mountain Research Station, 2500 S. Pine Knoll Drive, Flagstaff, AZ 86001.

Greg Nowacki is a regional ecologist and acting manager of the Regional Soils Program, U.S. Department of Agriculture, Forest Service, Eastern Region, 626 E. Wisconsin Avenue, Milwaukee, WI 53202.

Toby O'Geen is a soil resource specialist in Cooperative Extension, University of California, Davis, Department of Land, Air and Water Resources, 2152 Plant and Environmental Sciences Building, Davis, CA 95616.

Eunice Padley is a national forester, U.S. Department of Agriculture, Natural Resources Conservation Service, 1400 Independence Avenue, Room 6161-S, Washington, DC 20250.

Deborah S. Page-Dumroese is a research soil scientist, U.S. Department of Agriculture, Forest Service, Rocky Mountain Research Station, Moscow Forestry Sciences Laboratory, 1221 S. Main Street, Moscow, ID 83843.

Toral Patel-Weynand is a Director, Forest Management Science Research Staff, U.S. Department of Agriculture, Forest Service, Research and Development, 201 14th Street SW, Washington, DC 20250.

Charles H. (Hobie) Perry is a research soil scientist, U.S. Department of Agriculture, Forest Service, Northern Research Station, 1992 Folwell Avenue, St. Paul, MN 55108.

Richard V. Pouyat is an emeritus scientist, U.S. Department of Agriculture, Forest Service, Northern Research Station, Newark, DE 19711.

Jennifer Puttere is a soil scientist, U.S. Department of Interior, Bureau of Land Management, 3106 Pierce Parkway, Suite E, Springfield, OR 97477.

Daniel D. Richter is a professor of soils and forest ecology, Duke University, Nicholas School of the Environment, 9 Circuit Drive, Box 90328, Durham, NC 27708.

Peter R. Robichaud is a research engineer, U.S. Department of Agriculture, Forest Service, Rocky Mountain Research Station, 1221 S. Main Street, Moscow, ID 83843.

Lindsey E. Rustad is a research ecologist, U.S. Department of Agriculture, Forest Service, Northern Research Station, Center for Research on Ecosystem Change, 271 Mast Road, Durham, NH 03824.

Michael SanClements is a senior scientist, National Ecological Observatory Network, 1685 38th Street, Suite 100, Boulder, CO 80301.

Kirsten Schwarz is an associate professor, Departments of Urban Planning and Environmental Health Sciences, University of California, Los Angeles, CA.

D. Andrew Scott is a district ranger, U.S. Department of Agriculture, Forest Service, Bankhead National Forest, 1070 Highway 33, Double Springs, AL 35553.

Richard E. Shaw is a state soil scientist, U.S. Department of Agriculture, Natural Resources Conservation Service, New Jersey State Office, 2200 Davidson Avenue, 4th Floor, Somerset, NJ 08873.

Yvonne Shih is a resource assistant, U.S. Department of Agriculture, Forest Service, Research and Development, 1400 Independence Avenue, SW (Mailstop 1120), Washington, DC 20250.

Jane E. Smith is a research botanist/mycologist, U.S. Department of Agriculture, Forest Service, Pacific Northwest Research Station, Forestry Sciences Laboratory, 3200 S.W. Jefferson Way, Corvallis, OR 97331.

Marla J. Stelk is executive director of the Association of State Wetland Managers, 32 Tandberg Trail, Suite 2A, Windham, ME 04062.

Kyle Stephens is a soil data quality specialist, U.S. Department of Agriculture, Natural Resources Conservation Service, 1201 NE Lloyd Boulevard, Portland, OR. 97232.

Michael Strobel is director of the National Water and Climate Center, U.S. Department of Agriculture, Natural Resources Conservation Service, 1201 NE Lloyd Boulevard, Suite 802, Portland, OR 97232.

Chris W. Swanston is a project leader, Climate, Fire, and Carbon Cycle Sciences, U.S. Department of Agriculture, Forest Service, Northern Research Station, and director, U.S. Department of Agriculture, Forest Service, Northern Research Station, Northern Institute of Applied Climate Science, 410 MacInnes Drive, Houghton, MI 49931.

Katalin Szlavecz is a research professor, Johns Hopkins University, Morton K. Blaustein Department of Earth and Planetary Sciences, 226 Olin Hall, 3400 N. Charles Street, Baltimore, MD 21218.

Brian Tangen is an ecologist, U.S. Geological Survey, Northern Prairie Wildlife Research Center, 5231 S. 19th Street, Lincoln, NE 68512.

Melanie K. Taylor is an ecologist, U.S. Department of Agriculture, Forest Service, Southern Research Station, Center for Forest Disturbance Science, 325 Green Street, Athens, GA 30602.

Tara L. E. Trammell is John Bartram assistant professor of Urban Forestry, University of Delaware, Plant and Soil Sciences, 152 Townsend Hall, Newark, DE 19716.

Carl C. Trettin is a team leader and research soil scientist, U.S. Department of Agriculture, Forest Service, Southern Research Station, 3734 Highway 402, Cordesville, SC 29434.

Robert Vaughan is a senior specialist in landscape ecology, U.S. Department of Agriculture, Forest Service, Geospatial Technology and Applications Center, 2222 West 2300 South, Salt Lake City, UT 84119.

Daniel J. Vogt is an associate professor, University of Washington, Department of Environmental and Forest Science, Box 352100, Seattle, WA 98195.

Steven D. Warren is a research disturbance ecologist, U.S. Department of Agriculture, Forest Service, Rocky Mountain Research Station, 735 North 500 East, Provo, UT 84606.

Mary I. Williams is an ecologist, Nez Perce Tribe Wildlife Division, P.O. Box 365, Lapwai, ID 83540.

Stephanie A. Yarwood is an associate professor, University of Maryland, Department of Environmental Science and Technology, College Park, MD 20742.

Ian D. Yesilonis is a soil scientist, U.S. Department of Agriculture, Forest Service, Northern Research Station, Baltimore Field Station, 5523 Research Park Drive, Suite 350, Baltimore, MD 21228.

State of Forest and Rangeland Soils Research in the United States

1

Dan Binkley, Daniel D. Richter, Richard V. Pouyat, and Linda H. Geiser

Overview

Flying across the eastern United States at an altitude of 10,000 m, we see a landscape below that is a mosaic of forests, rivers, farm fields, towns, and cities. Almost all of the lands covered by forests today have undergone intensive harvest, and even regrowth and reharvests, following decades or centuries of cultivation-based agriculture and other land uses. The visible change in the boundaries of forests and fields is matched by similar, though less visible, patterns in the soils. Indeed, the soils that form the living surface of the Earth below may be as different on each side of the airplane as they are from one corner of the United States to another. Local differences in hillslopes and valley bottoms, in the types of bedrocks and sediments that sit below the living soil, and in the history of human land uses, may be greater than the differences driven by the climate of, for example, Virginia versus Oregon. The inhabitants of the land that became the United States survived on food and resources which were fundamentally derived from soils. The economic output of the colonies and the youthful United States flowed more from agriculture than from industry until at least the Civil War era (Gallman and Weiss 1969). Wood from the forests built the towns and cities, while energy from wood was the primary fuel of the American economy until the 1880s (U.S. Energy Information Administration 2011).

As we glide down to an altitude of 1000 m above the Sumter National Forest, near Union, SC, the tops of trees come into focus. These trees rise from soils formed over millions of years from the original granitic gneiss bedrock that underlies much of the Southeast. The soils were shaped by chemical and physical processes, largely mediated by the plants, animals, and microorganisms that form the biological engines of soil formation, weathering, and change. In the millennia before the arrival of the settlers, oaks (*Quercus* spp.), hickories (*Carya* spp.), and southern pines (*Pinus* spp., such as shortleaf [*P. echinata*]) shaped the soils that provisioned the first peoples—and further back in time, spruce (*Picea* spp.), fir (*Abies* spp.), and northern pines (e.g., eastern white pine [*Pinus strobus*]) graced the landscapes. After 1800, fields of cotton (*Gossypium hirsutum*) and other row crops and farm animals replaced almost all the forests, leading to soil degradation from massive erosion but also soil enrichment through liming and fertilization.

The economic depression of the 1930s brought changes to the soils of the Sumter National Forest, as agricultural abandonment was followed by reforestation with pine trees, either seeding in naturally or planted, to restore the fertility of soils. The Calhoun Experimental Forest was established in the 1940s to provide information that land managers would need to foster the regeneration of forests and forest soils across the entire southeastern region. The history of the soils of the Sumter National Forest entailed very large changes across decades, a century, or two centuries, against the backdrop of ongoing soil-forming processes on the timescale of millions of years. Some of the changes in the Sumter's soils may be unique. But if we glided down to any other landscape, we would find soils that, like the Sumter's, have been shaped by both natural and human factors. The unique histories of all forest soils across the United States share an imprint of

D. Binkley (✉)
School of Forestry, Northern Arizona University, Flagstaff, AZ, USA
e-mail: Dan.binkley@alumni.ubc.ca

D. D. Richter
Nicholas School of the Environment, Duke University, Durham, NC, USA

R. V. Pouyat
Northern Research Station, USDA Forest Service, Newark, DE, USA

L. H. Geiser
Washington Office, U.S. Department of Agriculture, Forest Service, Washington, DC, USA

© The Author(s) 2020
R. V. Pouyat et al. (eds.), *Forest and Rangeland Soils of the United States Under Changing Conditions*,
https://doi.org/10.1007/978-3-030-45216-2_1

1

changes driven by events and processes in recent times, overlying the product of long-term processes that shape the living soils of Earth's surface.

The Forest and Rangeland Soils of the United States

Forest soils are a vital component of most, if not all, of the United States. Although forests occupy only about one-third of the nation's land area, they provide 80% of the nation's surface freshwater (Sedell et al. 2000). Forest and rangeland soils were degraded across the United States at an alarming rate in the 1700s and 1800s, primarily due to land conversion to agriculture and unsustainable tree harvesting and grazing practices. Later, particularly in the East, many agricultural lands were abandoned and forests returned. The US forest land base has remained relatively stable at around 160 million ha since the 1920s, despite population growth. Accommodation of a growing population is expected to reduce cropland, pasture, range, and forest area in the future, largely as a result of urbanization and other land development (USDA FS 2012). Urban land area increased 44% between 1990 and 2010 (USDA FS 2016). The Southeast is expected to have the greatest loss of forest, ranging from 4.0 to 8.5 million ha between 2010 and 2060 or roughly 4–8% of the region's 2007 forest land base (USDA FS 2012). Appendix A explores forest and rangeland soils in greater depth by US region, state, territory, or affiliated island.

Today, forest and rangeland soils are vulnerable to degradation from several additional threats. Both natural and human-caused disturbances have degraded forest and range-land soils across the United States, with various environmental changes, overgrazing, overharvesting, severe wildfire, and invasive plant and animal species as the greatest concerns. Effects on soil health are expected to be more severe when two or more of these disturbances or stressors interact with each other. Additionally, in the Eastern United States, acid deposition remains an important concern for both soil acidification and nitrogen (N) enrichment, including potential changes in species composition and leaching of N into aquatic ecosystems. Even though the implementation of clean air laws and standards has dramatically decreased acid deposition, in many cases soil recovery has been slow (Likens et al. 1996). Any deleterious effects on forest and rangeland soils will be magnified or accelerated by changes in the climate. But there is great uncertainty about the extent and nature of these effects as impacts will depend on the magnitude of temperature and precipitation changes and the frequency of extreme events.

Changes in forest soils may enhance or degrade their ability to support trees and other life. For example, forest management may add substantial amounts of fertilizer to about 400,000 ha of pine forests each year in the Southeast, boosting both growth and profits from forest lands (Albaugh et al. 2018).

Soil Variability

The ability of soils to grow trees typically varies by twofold or more across local landscapes (Fig. 1.1) and across forested regions (Fig. 4.7). Returning to our local example, about one-quarter of the Sumter National Forest can grow

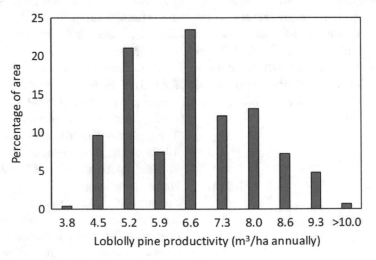

Fig. 1.1 Percentage of a 100,000 ha portion of the Sumter National Forest (including the Calhoun Experimental Forest) supporting various levels of loblolly pine annual productivity. The ability of soils to grow loblolly pine trees varies greatly across this area, owing to the differences in soil parent material, erosion history, landscape position, and the biogeochemical cycles that support tree nutrition (data from Soil Survey Staff n.d.). Silvicultural treatments and selected tree genotypes have more than doubled levels of inherent soil productivity for pine production in the Southeast

only 5 m³ ha⁻¹ annually or less, whereas another one-quarter can grow more than 8 m³ ha⁻¹ annually (Fig. 1.1). Most forest soils can grow trees even more rapidly if amended with fertilizers. The climate is relatively uniform across local areas, but differences in soil textures (especially the amount of clay), drainage, and slope position cause large differences in the ability of soils to retain water between storms. The ability of soils to supply nutrients for tree growth is probably more important than differences in water holding capacity. For example, the Southeast Tree Research and Education Site (SETRES, a long-standing collaboration between the Southern Research Station, North Carolina State University, Duke University, and the North Carolina State Forest Nutrition cooperative member companies) demonstrated that irrigating a loblolly pine (*Pinus taeda*) stand on a soil with low water holding capacity might increase growth by 25%, but fertilizing the stand would double growth (Albaugh et al. 2004).

The variation in soil productivity across landscapes is matched by the variation caused by changes over time and in response to management activities. Forest management can increase soil productivity or degrade it. Changes in the ability of soils to support tree growth are particularly important for national forests, because the Multiple-Use Sustained-Yield Act of 1960 requires "coordinated management of the various resources, each with the other, without impairment of the productivity of the land" and the National Forest Management Act of 1976 requires that plans "insure research on and (based on continuous monitoring and assessment) in the field evaluation of the effects of each management system to the end that it will not produce substantial and permanent impairment of the productivity of the land." Some forest management activities, such as planting genetically selected trees and controlling competing vegetation, can lead to siz-

able increases in wood production, though soils may not be greatly altered. Other activities, such as site preparation that entails soil compaction, removal of too much topsoil, or overly intense slash fires, may lower soil productivity.

Legacies of Forest Soils Research

Research in forest and rangeland soils, particularly research incorporating long-term measurements, provides important, fundamental insights into the processes that influence the ability of soils to support plant growth. The three investigations described next illustrate the kinds of insights that research can contribute to our foundational understanding of how trees and soils interact.

Calhoun Experimental Forest, Sumter National Forest, South Carolina

The highly eroded landscapes of the Sumter National Forest were restored to productive forests through intensive planting, natural regeneration, and conservation work by the Civilian Conservation Corps—all benefiting from a strong research component. Lou Metz was one of the early directors of the Calhoun Experimental Forest charged with restoring forest, land, and water resources following some of the most severely damaging agricultural impacts in the United States (Metz 1958). Carol Wells, a scientist with the USDA Forest Service (hereafter, Forest Service), developed a detailed soil sampling protocol in an old cotton field experimentally planted with loblolly pine in the Calhoun Experimental Forest (Fig. 1.2). Wells returned about every 5 years to examine how the trees drove changes in the soil.

Fig. 1.2 Shortly before this photo was taken, this former cotton field in the Sumter National Forest was planted with loblolly pine from seed collected across the entire Southeastern United States. (Photo credit: USDA Forest Service, circa 1958)

The timeframe of the development of forests and soils may extend beyond the career of individual scientists, and eventually Wells turned over his experiment to Dan Richter and colleagues. They continued this work and expanded it after Wells retired in about 1990. One dramatic long-term effect that the trees had on the soil was the restoration of the important O horizon, made up of fresh litter and decomposing organic matter (Richter and Markewitz 2001; Richter et al. 1999). The rate of carbon (C) accumulation in the O horizon was rapid, about one-quarter of the rate of accumulation in the trees themselves. In contrast, the A and B horizons of the mineral soils showed no overall change in the storage of C and were rapidly depleted of N, phosphorus, and calcium, nutrients that the regenerating trees required in large amounts. The lack of net change in organic C might indicate that little activity occurred in the mineral soil, but investigations using C isotopes showed that the mineral soil C was actually highly dynamic: High rates of C input were matched by high rates of decomposition and loss.

Sylvania Wilderness, Ottawa National Forest, Michigan

The long-term development of soils depends strongly on the influences of trees, but the Sylvania Wilderness demonstrates how soil formation can be very different under the influence of different species of trees. The old-growth forests of the Upper Peninsula of Michigan, where the Ottawa National Forest is located, often contain a mosaic of patches of eastern hemlock (*Tsuga canadensis*) and northern hardwoods, with vegetation changing from one type to the other at distances

of 50–100 m (Frelich et al. 1993). The boundaries between these conifer and hardwood patches in the Sylvania Wilderness have been remarkably stable at timeframes of 100 to more than 1000 years, especially considering the minimal differences in topography or soil drainage between patches. Major differences in the rates of nutrient cycling may be the most plausible explanations. Species such as sugar maple (*Acer saccharum*) foster higher rates of soil N turnover, whereas hemlock litter decomposes gradually and slows down N turnover. The hardwood species thrive with higher soil N supplies, while hemlock is adept at tolerating low soil N. After generations of trees have shaped the soils, the soils influence which trees will be most successful. Yet these long-term patterns are likely to change in the near future. Invasions of exotic earthworms (suborder Lumbricina) into the Sylvania Wilderness (and many other forests of North America) are drastically altering soils, N supplies, and the successful establishment of trees (Hendrix and Bohlen 2002). These effects show how soil-forming factors are dynamic—indeed, constantly changing—and, in our world today, most certainly include human-related interactions (Richter and Yaalon 2012).

Long-Term Soil Productivity Program, United States and Western Canada

One of the most valuable ongoing programs of wildland soils research is the Long-Term Soil Productivity (LTSP) program (Mushinski et al. 2017; Powers et al. 2005). This coordinated network of over 100 sites (Fig. 1.3) was initiated in 1991 in response to concerns that the removal of organic matter and

Fig. 1.3 The Long-Term Soil Productivity program has been following the effects of forest management on soils for three decades at more than 100 locations across North America. (Source: USDA FS 2018)

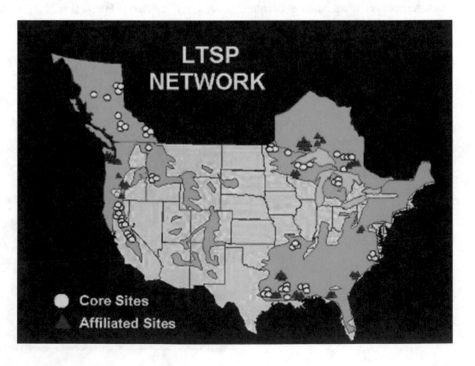

compaction of soils in forest harvests could alter soil fertility and site productivity. Two important insights from this network of experiments are as follows: (1) The effects of biomass removal on soil productivity depend strongly on the type of biomass removed. The removal of organic matter in the form of branches and logging slash generally did not affect the growth of regenerating forests, but removal of the organic matter from the O horizon frequently reduced site productivity. (2) The effects of harvesting vary substantially across sites, at both local and regional scales. Lessons learned from a single experimental forest site may have limited value for explaining forest and soil dynamics across the broader diversity of conditions on forested landscapes.

The LTSP approach is effective because it provides a framework for calculating average responses and variations around averages and for identifying the site factors that best explain the variation. As noted later, a large network such as the LTSP program takes great dedication from the scientists and forest managers who work on each site and sustained financial support from core institutions.

Monitoring to Detect Changes in Soil

Forests and the soils that support them are remarkably dynamic. Before colonial settlement, most forested sites in North America had been covered by forests for millennia. Any single forest might be young or old, depending on the time elapsed since stand-replacing events such as fires and hurricanes. A similar story applies to forest soils. Soils that developed through interactions with trees have fundamental characteristics (such as the presence of an O horizon above the A horizon) that persist across generations of trees. Long-term spatial patterns of soils are influenced by changes over shorter time periods, in response to changes in the species of trees, the time since stand-replacing events, land use history, and ongoing management activities. The turnover time for C in upper soil horizons averages about 15–300 years (Schmidt et al. 2011), not much different from the lifespan of trees.

We know that individual soils can change substantially over a period of years and decades, but we do not yet have the ability to describe the local changes that may be expected across landscapes and regions. Some countries have major programs devoted to determining rates of change in forest soils and identifying the factors that account for differences in rates of change. For example, Germany has a system of over 1800 plots that have been sampled every decade since the 1980s (Grüneberg et al. 2014). Sampling is 10 times more intensive in Sweden, where more than 20,000 locations are sampled every 10 years (Nilsson et al. 2015).

Considerable data are collected to document the state of soil resources in the United States. The primary source for soil information is the Soil Survey Geographic (SSURGO) database, which is accessible through the USDA Natural Resources Conservation Service (USDA NRCS) Web Soil Survey. This database contains hundreds of estimated properties for soil landscapes that cover more than 90% of the continental United States (1:24,000 spatial scale). The State Soil Geographic (STATSGO2) database, also distributed through the Web Soil Survey, provides a smaller set of estimated properties for the entire country at a 1:250,000 scale. The spatial resolution of the chemical data in SSURGO is sufficient for large, homogeneous landscapes, but in variable terrain with multiple soil parent materials (much of which is under forest and rangeland cover), this dataset is limited.

The National Cooperative Soil Survey (NCSS) Soil Characterization database, maintained by the USDA NRCS, contains measured data on over 1000 soil properties obtained from over 63,000 sites throughout the United States and the world. The NCSS also contains calculated data on many other soil properties. All of these datasets are based on consistent, well-documented standards and specifications.

The Forest Service's Forest Inventory Analysis (FIA) program produces a survey of the state of United States forests, including forest soils, and reports on issues such as land-cover change, C sequestration, and effects of pollutants and fires. The survey includes approximately 125,000 plots for core data collection, of which about 7800 are sampled intensively and include forest health and soil characteristics.

Several public-private collaborations aggregate and analyze large quantities of soil data. One example is the International Soil Carbon Network (ISCN), a platform working to develop a globally integrated database of soil C measurements. This network partners with several Federal programs, including the interagency United States Global Change Research Program (USGCRP) and the National Science Foundation-funded National Ecological Observatory Network (NEON). Many other datasets are hosted by Federal entities such as the United States Environmental Protection Agency and the Department of Energy. Despite all of these efforts, however, many existing datasets lack the requisite resolution for effective policy and soil management decisions. Many higher-resolution datasets are regional and lack integration into national databases. Thus, the United States lacks a single clearinghouse for soil data or infrastructure for intercomparison of heterogeneous datasets, especially those containing data collected via different protocols and with different objectives. Aggregation and intercomparison are inherently difficult due to the wide range of soil properties, the varying degree of importance of each property depending on the location and land use or land-cover type, scale, and the different research needs for different soil management goals.

Research Challenges

The most important challenges for research that will increase fundamental understanding of soils and sustain long-term ecosystem health are identified throughout this assessment and summarized in Chap. 10. These challenges include:

- Understanding how soils are affected by changes in precipitation patterns, air temperatures, carbon dioxide (CO_2) concentrations, atmospheric deposition of fertilizing and acidifying compounds, increasing frequency and intensity of wildfire, and urban land use change;
- Understanding the effects of invasive species including methods for tracking, mitigation, and adaptation;
- Research that can inform management practices to sustain and restore soil water holding capacity, organic matter, fertility, biotic diversity, and productivity under changing conditions; and,
- A long-term, consistent commitment of funding to support collection, analysis, and archiving of samples and the associated databases.

Physical and Human Resources for Knowledge Acquisition, Integration, Analysis, and Transfer

Finally, this report highlights the integral role of monitoring, modeling, mapping, digital databases, and human resources that underlie modern soils research. These resources are key to providing the science that can help to sustain agriculture, silviculture, grazing, abundant water, and other ecosystem services provided by forest and grassland soils. These essential resources will expand knowledge of how to protect and restore soil fertility and water holding capacity, prevent erosion, build and protect soil organic matter, and promote the soil invertebrates and microorganisms driving soil health-promoting chemical and physical processes. Some key existing networks and programs have already been mentioned. Additional resources are mentioned in Chaps. 2, 3, 4, 5, 6, 7, 8 and 9, and all are summarized in Appendix B. These resources and institutions provide the means and capacity to:

- Monitor, model, map, store, and access data networks supporting soils research and management;
- Provide training and opportunities for the next generation of scientists; and,
- Communicate findings and conduct outreach for the benefit of local communities; urban populations; rural landowners; growers; and local, regional, and Federal policymakers and decision-makers.

Key Findings

- Changes in forest soils across the United States are driven by events and processes in recent times, overlying the long-term processes that shape the living soils of Earth's surface. Long-term soil spatial patterns are influenced by short-term changes in tree species, time since stand-replacing events, land use history, and ongoing management activities.
- Today's forests occupy only about one-third of the nation's land area but provide 80% of the nation's surface freshwater.
- Historically over 50 million ha of eastern United States forests were converted to primarily agricultural uses. Much of this has reverted to forest since World War I, but about 400,000 forested ha annually are now being converted to urban or suburban purposes to accommodate urban and suburban growth.
- Research in forest and rangeland soils, particularly research networks that encompass a broad range of site conditions and long-term measurements, provides fundamental insights into the processes that influence the ability of soils to support plant growth.
- The turnover time for C in upper soil horizons averages about 15–300 years, not much different from the lifespan of trees. Removing soil organic matter significantly reduces soil C, whereas removing tree branches and logging slash has comparatively little effect.
- Key soils databases in the United States include SSURGO, STATSGO2, NCSS, LTSP, and FIA. Collaborations that aggregate and analyze large quantities of soils data include the ISCN, USGCRP, and NEON. (See also Appendix B.)

Key Information Needs

- A single national clearinghouse needs to be created and maintained for soil data or infrastructure for intercomparison of heterogeneous datasets, especially those containing data collected via different protocols and with different objectives and higher-resolution regional datasets. Data with higher resolution than is generally currently available are needed to more effectively inform policy and soil management decisions.
- The ability to describe expected local changes across landscapes and regions would require intensification of current soil monitoring networks as individual soils can change substantially over a period of years and decades. Sweden and Germany offer good models for systematic monitoring. The FIA soils indicator currently monitors about 7800 sites.

• Understanding how human-caused changes affect soil health and productivity will be of key importance to managing soils in the future. Specific information needs regarding soil water, nutrients, and biota, as well as wetland soils and urban soils, are detailed in Chaps. 3, 4, 5, 6 and 7, respectively.

Literature Cited

Albaugh TJ, Allen HL, Dougherty PM, Johnsen KH (2004) Long term responses of loblolly pine to optimal nutrient and water resource availability. For Ecol Manag 192(1):3–19

Albaugh TJ, Fox TR, Cook RL et al (2018) Forest fertilizer applications in the southeastern United States from 1969 to 2016. For Sci. https://doi.org/10.1093/forsci/fxy058

Frelich LE, Calcote RR, Davis MB, Pastor J (1993) Patch formation and maintenance in an old-growth hemlock-hardwood forest. Ecology 74(2):513–527

Gallman RE, Weiss TJ (1969) The service industries in the nineteenth century. In: Fuchs VR (ed) Production and productivity in the service industries. Columbia University Press, New York, pp 287–352

Grüneberg E, Ziche D, Wellbrook N (2014) Organic carbon stocks and sequestration rates of forest soils in Germany. Glob Chang Biol 20:2644–2662

Hendrix PF, Bohlen PJ (2002) Exotic earthworm invasions in North America: ecological and policy implications. Bioscience 52(9):801–811

Likens GE, Driscoll CT, Buso DC (1996) Long-term effects of acid rain: response and recovery of a forest ecosystem. Science 272(5259):244–246

Metz LJ (1958) The Calhoun Experimental Forest. U.S. Department of Agriculture, Forest Service, Southeastern Forest Experiment Station, Asheville, 24 p

Multiple-Use Sustained-Yield Act of 1960; Act of June 12, 1960; 16 U.S.C. 528 et seq

Mushinski RM, Boutton TW, Scott DA (2017) Decadal-scale changes in forest soil carbon and nitrogen storage are influenced by organic matter removal during timber harvest. J Geophys Res Biogeo 122(4):846–862

National Forest Management Act of 1976; Act of October 22, 1976; 16 U.S.C. 1600

Nilsson T, Stendahl J, Löfgren O (2015) Markförhållanden i svensk skogsmark—data från Markinventeringen 1993–2002. Rapport 19, Institutionen för mark och miljö, Sveriges lantbruksuniversitet, Uppsala. In Swedish only; for information in English, see: https://www.slu.se/en/Collaborative-Centres-and-Projects/Swedish-Forest-Soil-Inventory/

Powers RF, Scott DA, Sanchez FG et al (2005) The North American Long-Term Soil Productivity experiment: findings from the first decade of research. For Ecol Manag 220(1–3):31–50

Richter DD, Markewitz D (2001) Understanding soil change: soil sustainability over millennia, centuries, and decades. Cambridge University Press, New York, 255 p

deB Richter D, Yaalon DH (2012) "The changing model of soil" revisited. Soil Sci Soc Am J 76(3):766–778

Richter DD, Markewitz D, Trumbore SE, Wells CG (1999) Rapid accumulation and turnover of soil carbon in a re-establishing forest. Nature 400:56–58

Schmidt MW, Torn MS, Abiven S et al (2011) Persistence of soil organic matter as an ecosystem property. Nature 478:49–56

Sedell J, Sharpe M, Apple DD et al (2000) Water and the Forest Service. FS-660. U.S. Department of Agriculture, Forest Service, Policy Analysis, Washington, DC, 26 p

Soil Survey Staff (n.d.) Web soil survey. U.S. Department of Agriculture, Natural Resources Conservation Service. https://websoilsurvey.nrcs.usda.gov/app/WebSoilSurvey.aspx. Accessed November 20, 2017

USDA Forest Service [USDA FS] (2012) Future of America's Forest and Rangelands: Forest Service 2010 resources planning act assessment. General Technical Report WO-87. Washington, DC, 198 p

USDA Forest Service [USDA FS] (2016) Future of America's Forests and Rangelands: update to the 2010 Resources Planning Act Assessment. General Technical Report WO-GTR-94. Washington, DC, 250 p

USDA Forest Service [USDA FS] (2018) Research topics: Forest management. The Long-Term Soil Productivity Experiment. U.S. Department of Agriculture, Forest Service, Pacific Southwest Research Station, Albany. https://www.fs.fed.us/psw/topics/forest_mgmt/ltsp/. Accessed April 16, 2019

U.S. Energy Information Administration (2011) History of energy consumption in the United States, 1775–2009. https://www.eia.gov/todayinenergy/detail.php?id=10. Accessed November 20, 2017

Soil Carbon

Erin Berryman, Jeffrey Hatten, Deborah S. Page-Dumroese, Katherine A. Heckman, David V. D'Amore, Jennifer Puttere, Michael SanClements, Stephanie J. Connolly, Charles H. (Hobie) Perry, and Grant M. Domke

Introduction

Soil organic matter (OM) is a pervasive material composed of carbon (C) and other elements. It includes the O horizon (e.g., litter and duff), senesced plant materials within the mineral soil matrix, dead organisms (including macroorganisms and microorganisms), microbial and root exudates, and organic materials adhering to mineral surfaces. Soil organic carbon (SOC) is a very dynamic component of the soil; each year, the amount of SOC processed by microorganisms within the soil is roughly equal to the amount of inputs from plant detritus. The pervasive dynamic nature of SOC is key to the ecosystem services, or "the benefits people obtain from ecosystems" (Millennium Ecosystem Assessment 2003), that SOC provides.

Sidebar 2.1 Tool and research needs pertaining to soil organic carbon

- A mechanism to transfer knowledge about the new SOC paradigm to forest and rangeland managers and use that knowledge to develop best management practices for building up SOC
- Quantitative models of SOC stabilization and vulnerability designed for management applications (e.g., Forest Vegetation Simulator with SOC module)
- Improved linkage of Ecological Site Descriptions to management actions that impact SOC
- Synchronization of SOC data across multiple agencies and sampling initiatives
- Models that link forest health and drought resistance to changes in SOC

Soil organic carbon is an essential indicator of soil health. Soil health refers to the "ability of soil to function effectively as a component of a healthy ecosystem" (Schoenholtz et al. 2000, p. 335). The quantity and quality of SOC are linked to important soil functions including nutrient mineralization, aggregate stability, trafficability, permeability to air, water retention, infiltration, and flood control (Box 2.1). In turn, these soil functions are correlated with a wide range of ecosystem properties. For example, high SOC in mineral soils is usually associated with high plant productivity (Oldfield et al. 2017), with subsequent positive implications for wild-

E. Berryman (✉)
Forest Health Protection, State and Private Forestry, USDA Forest Service, Fort Collins, CO, USA

U.S. Department of Agriculture, Forest Service, Forest Health Assessment and Applied Sciences Team, Fort Collins, CO, USA
e-mail: Erin.berryman@usda.gov

J. Hatten
Oregon State University, College of Forestry, Forest Engineering, Resources & Management, Corvallis, OR, USA

D. S. Page-Dumroese
Rocky Mountain Research Station, USDA Forest Service, Moscow, ID, USA

K. A. Heckman
U.S. Department of Agriculture, Forest Service, Northern Research Station, Houghton, MI, USA

D. V. D'Amore
U.S. Department of Agriculture, Forest Service, Pacific Northwest Research Station, Juneau, AK, USA

J. Puttere
U.S. Department of Interior, Bureau of Land Management, Springfield, OR, USA

M. SanClements
National Ecological Observatory Network, Boulder, CO, USA

S. J. Connolly
U.S. Department of Agriculture, Forest Service, Northern Research Station, Newtown Square, PA, USA

C. H. (Hobie) Perry · G. M. Domke
U.S. Department of Agriculture, Forest Service, Northern Research Station, St. Paul, MN, USA

© The Author(s) 2020
R. V. Pouyat et al. (eds.), *Forest and Rangeland Soils of the United States Under Changing Conditions*,
https://doi.org/10.1007/978-3-030-45216-2_2

Box 2.1 Benefits of Soil Organic Matter and Soil Carbon

Soil carbon (C), a major component of soil organic matter (SOM), provides several benefits for the function of forests, rangelands, and other wildlands. In addition, soil C is correlated to many other properties that enhance ecosystem services and is thus a strong indicator of soil health. Many of these ecosystem services are interdependent: Promoting soil C buildup starts a chain reaction that ultimately improves many facets of ecosystem health. Among the benefits of soil C and soil organic matter are that they:

Boost nutrient storage—soils with high soil C tend to have high nutrient content, promoting growth of trees and forage. About 99% of soil nitrogen (N) is found within SOM. In addition, SOM provides much of the cation exchange capacity essential for making nutrients available to plant roots.

Enhance soil structure—SOM holds soil particles together through adhesion and entanglement, reducing erosion and allowing root movement and access to nutrients and water.

Act as a large biological carbon store—through stabilization mechanisms (see text), the formation of soil C feeds a major reservoir of global C. With the right management, increases in soil C storage can be large enough to offset a portion of anthropogenic greenhouse gas emissions.

Enhance plant carbon sequestration—by increasing nutrient and water availability, soils with high soil C and organic matter support increased growth of forests, rangelands, and wildlands, leading to increased uptake of atmospheric CO_2.

Increase ecosystem water storage—through enhancing soil structure and increasing soils' effective surface area, SOM increases the amount of water that can be retained in the soil for plant and downstream use, reducing evaporative and runoff losses.

Purify drinking water—the effects of SOM on water holding capacity and soil structure help to enhance soil's resilience to erosion. Soil organic matter also plays a role in reducing the bioavailability of pollutants. These functions contribute to SOM's strong role in purifying water for human uses.

Detoxify soil—by affecting nutrient availability and soil structure, and by serving as an energy source for microbes, SOM plays an important role in maintaining soil health. By reducing the content, bioavailability, and mobility of compounds, SOM supports soil's ability to detoxify pollutants that occur as a result of chemical spills or contamination.

life habitat, distribution, and abundance. Consequently, ecosystem services can be degraded when SOC is altered or lost from forest or rangeland sites. Measuring and monitoring SOC levels can lead to a more complete understanding of ecosystem and soil health at a particular site; indices of soil health incorporate measures of SOC and can be used to track changes in soil health over time and in response to management activities (Amacher et al. 2007; Chaer et al. 2009).

Soils account for the largest pool of terrestrial organic C globally, with an estimated $2.27-2.77 \times 10^{15}$ kg or 2270–2770 petagrams (Pg) of C in the top 2–3 m of soil (Jackson et al. 2017). This represents a pool that is two to three times larger than the atmospheric and biotic C pools. North American and US soils (all soil orders) store about 366 Pg C and 73.4 Pg C, respectively, in the top 1 m (Liu et al. 2013; Sundquist et al. 2009; USGCRP 2018). Most of the SOC stock in the United States is in nonintensively managed lands such as forests (Fig. 2.1), wetlands, and rangelands (Liu et al. 2012, 2014). Across land uses, most SOC is concentrated near the surface, where it may be vulnerable to loss; 74.5% of North America's SOC occurs in the top 30 cm of mineral soil (Batjes 2016; Scharlemann et al. 2014). Most assessments of SOC pools represent only mineral SOC and have omitted organic soil horizons that sit on top of the mineral soil, despite the importance of O horizons as a source of OM for building SOC. In this chapter we will refer to "O horizons" and "forest floors" when talking about organic soil horizons on top of the mineral soil, "mineral soil" when discussing mineral-dominated soil horizons, SOC as mineral soil organic C, and "soil" as everything from the O horizon and deeper. If included in the above estimates, O horizons would increase the global SOC pool estimates by about 43 Pg (Pan et al. 2011) or about 2% of the total SOC pool. However, O horizons are more important in forests than in rangelands. Domke and others (2016, 2017) found that forest floor O horizons accounted for about 12% of the SOC pool; forest SOC had a density of about 63 Mg ha^{-1} and litter represented roughly 8 Mg ha^{-1} across all USDA Forest Service (hereafter, Forest Service) Forest Inventory and Analysis (FIA) plots (Fig. 2.2; Box 2.2).

Accurate assessment and ongoing monitoring of national SOC stocks are a critical first step to understanding how management activities can impact this national resource. Because the SOC pool is large compared to other C pools (especially the atmosphere), a small change in SOC can produce a large change in atmospheric carbon dioxide (CO_2) levels. For example, a global decrease in SOC of 5% in the upper 3 m would result in 117 Pg of C released into the atmosphere, causing an increase in the atmospheric C pool (829 Pg in 2013) of 14%, i.e., from 400 ppm to 456 ppm CO_2. Conversely, sequestering a small percentage in this large C pool translates into a substantial increase that is globally relevant. Site-level studies suggest that reforestation and other land-use and management changes increase SOC by 0.1–0.4 Mg C ha^{-1} year^{-1}, and a national-scale (conterminous

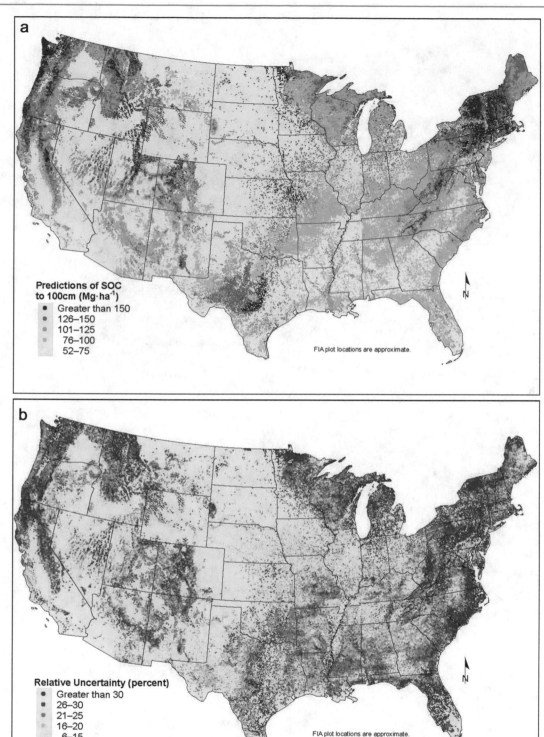

Fig. 2.1 (**a**) Map of soil C stores predicted at Forest Service Forest Inventory and Analysis plots. (**b**) Map of uncertainty in predictions of soil C stores at these plots. (Source: Domke et al. 2017)

United States) analysis suggests that reforesting topsoils are accumulating 13–21 × 10⁹ kg or 13–21 teragrams (Tg) C year.⁻¹ with the potential to sequester hundreds more teragrams C within a century (Nave et al. 2018).

In the past few decades, there have been several coordinated efforts to assess national-level SOC stocks in a statistically robust manner (Box 2.2). Such assessments provide a baseline for detecting future change in United States

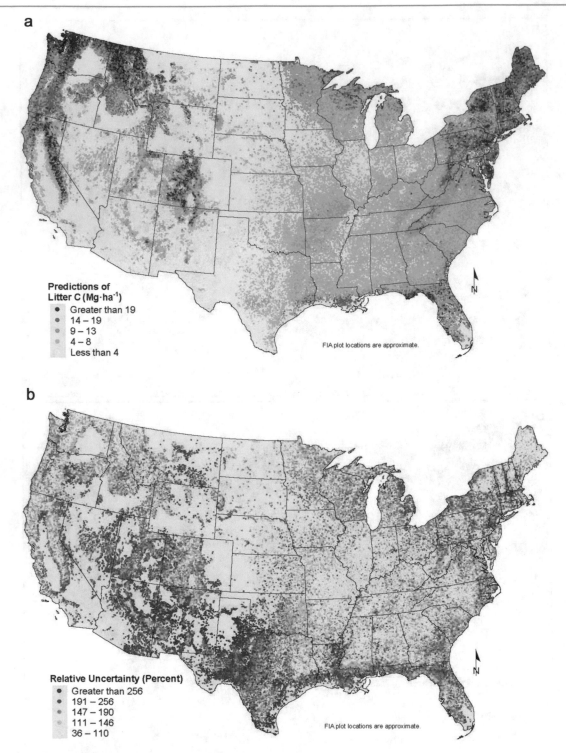

Fig. 2.2 (**a**) Map of litter (forest floor) C stores predicted at Forest Service Forest Inventory and Analysis plots. (**b**) Map of uncertainty in predictions of litter C stores at these plots. (Source: Domke et al. 2016)

SOC. Yet there are some limitations to national-scale assessments. One is varying depth of soil and, consequently, SOC stocks across the country. Most assessments do not consider soil deeper than 1 m, even though deeper soil can be an important reservoir of SOC (Harrison et al. 2011).

Furthermore, in any study of SOC on a specific site, it is difficult to detect SOC changes less than about 25% (e.g., Homann et al. 2001) due to soil heterogeneity (Fig. 2.1b) and sampling and measurement error. Though controlled experiments (e.g., laboratory incubations) demonstrate that some

Box 2.2 Assessing the Nation's Soil Organic Carbon Stores

There have been many efforts to characterize soil organic carbon (SOC) at a national scale. A major challenge to accurate accounting of SOC is the high spatial variability of SOC content, which results in large uncertainties and may preclude the detection of change over time. Each effort has slightly different goals and objectives, but they all emphasize free and open data availability.

The USDA Forest Service's Forest Inventory and Analysis (FIA) program (https://www.fia.fs.fed.us/ library/brochures/docs/Forest_Health_Indicators.pdf) reports on the status and trends of the nation's forest resource, across all ownerships. The field campaign of FIA collects information on the area of forest land and its location; the species, size, and health of trees; and total tree growth, mortality, and removals by harvest and land-use change. In 2000, the Forest Service greatly enhanced the FIA program in several ways. It changed from a periodic survey to an annual survey (the field crew returns to each plot every 5 or 10 years), and it expanded the scope of data collection to include soil, understory vegetation, tree crown conditions, coarse woody debris, and lichen community composition on a subsample of plots. It also increased our capacity to analyze and publish the underlying data as well as information and knowledge products derived from it. To facilitate forest SOC estimation, FIA collects data on litter thickness and mass, C content, mineral soil bulk density, and rock fraction at repeatable depth intervals (0–10 cm, 10–20 cm) (O'Neill et al. 2005). These measurements can be used in combination with other site attributes and ancillary information to generate robust, statistically sound estimates of forest floor and litter C (Domke et al. 2016) and SOC (Domke et al. 2017) at a national scale. An advantage of FIA over other assessment efforts is the repeated nature of the survey; eventually, change in soil properties over time may be possible to detect. However, the spatial density of the subsample of FIA plots where soil data are collected is very low (1 plot per 103,000 ha), so subnational assessments should consider additional soil data such as those collected by the national forests or Natural Resources Conservation Service (NRCS).

Rapid Carbon Assessment (RaCA; https://www. nrcs.usda.gov/wps/portal/nrcs/detail/soils/survey/?cid =nrcs142p2_054164) was initiated in 2010 as an organized, coordinated effort led by NRCS to systematically sample and measure SOC across the United States using a consistent and repeatable methodology. Data are intended to represent a snapshot in time of the national inventory of SOC under various land covers and differing agricultural management. This assessment sampled 6017 plots across the country and emphasized statistically reliable and defensible methods. Forthcoming analyses from RaCA data will represent a comprehensive accounting of SOC stocks under major land-use categories across the United States, regardless of ecosystem type or land ownership status.

National Cooperative Soil Survey (NCSS; https:// www.nrcs.usda.gov/wps/portal/nrcs/main/soils/survey/partnership/ncss) is a nationwide partnership of Federal, regional, state, and local agencies and private entities and institutions that was created to document soil taxonomy using consistent methods. Participants cooperate with each other to gather information about soils using common or shared procedures. Natural Resources Conservation Service runs soil surveys at the county level. The Forest Service conducts soil surveys in many of its national forests, as does the Department of the Interior's National Park Service in some national parks. An important product of the NCSS is the Gridded Soil Survey Geographic (gSSURGO) database, a spatial data layer (10 m resolution) of various soil properties derived from soil series that are mapped, delineated as "map units." Each map unit is linked to specific soil properties in a comprehensive database, which includes SOC. However, gSSURGO is not intended to be a statistically robust map of SOC. The spatial accuracy of specific soil attributes is not defined and may vary widely across the United States. In addition, gaps exist in gSSURGO in areas where soil survey data are not available, such as regions or counties that have not been mapped yet.

The *International Soil Carbon Network (ISCN; https://iscn.fluxdata.org)* is an ad hoc research coordination network that "facilitates data sharing, assembles databases, identifies gaps in data coverage, and enables spatially explicit assessments of soil C in context of landscape, climate, land use, and biotic variables" (ISCN n.d.). Data are derived from independent research projects, so they represent a wide range of geographic coverage, temporal resolutions, and methods.

perturbation will cause change (e.g., soil warming), it can be difficult to detect this change in the field. The inclusion of rocks larger than 2 mm in samples, changes in texture, and sensitivity of SOC stocks to bulk density all contribute to error in assessing SOC (Jurgensen et al. 2017; Page-

Dumroese et al. 1999). Improved measurement technology and statistical methods that account for different sources of uncertainty may help overcome these challenges and allow for the detection of more subtle changes in SOC.

While essential for national soil C accounting purposes, nationwide assessments like FIA may not be helpful for regional or local management challenges due to their coarse spatial resolution. In 1976, the National Forest Management Act was enacted and set forth three points that would necessitate soil monitoring and analysis on national forests to inform planning. The first point was that land management could not produce substantial or permanent impairment of site productivity. Second, trees could be harvested only where soil, slope, or watershed conditions would not be irreversibly damaged. Last, harvesting had to protect soil, watershed, fish, wildlife, recreational, and aesthetic resources. In support of national forest managers' decision-making, soil monitoring standards and guidelines were developed nationwide to determine baseline soil properties and identify changes associated with harvesting (Neary et al. 2010). In addition, many national forests have developed and maintained soil monitoring protocols unique to their needs. For example, the Monongahela National Forest in West Virginia has a long-running soil monitoring program (>20 years) that originated from the need to understand acid deposition impacts on soil and water health. It is important to recognize that National Forest System data may not include SOC nor is it incorporated into national-scale assessments. Many national forests lack funding and personnel to sample and analyze harvest-unit soil and vegetation changes, hindering the ability of local managers to consider SOC benefits and impacts when developing management plans. Remote sensing using light detection and ranging (LiDAR) or high-resolution satellite images can help quantify aboveground forest C (e.g., Gonzalez et al. 2010), and, where available, SOC can be predicted by using near infrared reflectance hyperspectral proximal data combined with remote sensing data (Gomez et al. 2008). In most cases, however, efforts to quantify changes in SOC must rely on archived soil samples combined with new sampling and analysis to determine changes in surface and subsurface pool sizes.

Mechanisms of Mineral Soil Organic Carbon Stability and Vulnerability: An Emerging Paradigm

For a long time, our understanding of SOC distribution and vulnerability was limited by the traditional SOC conceptual model in use for many decades. Arising from advances in technology that allow fine-scale molecular and microbial investigations of SOC interactions with mineral soil, a new conceptual framework of SOC stabilization and destabiliza-

tion is being developed that improves our ability to predict SOC behavior. Important advances have been made in our knowledge of the source and stabilization mechanisms of mineral-associated SOC.

Organic matter quality is used as a general descriptor for the combination of the chemical structure and elemental composition of OM that influences decomposition. Historically it was thought that the ability of an organism to effectively decompose OM was directly related to the material's molecular composition (such as lignin content) and concentration of nutrients (such as nitrogen (N)). These concepts are still useful when describing decomposition dynamics of organic soils or organic soil horizons, but measures of OM quality have been elusive. Furthermore, these concepts break down in attempts to describe the dynamics of SOC associated with horizons dominated by mineral materials.

The relative importance of aboveground sources (e.g., litterfall) and belowground sources (e.g., fine roots) of SOC is key to understanding impacts of disturbance and management on SOC. It is now understood that across many ecosystems most SOC is derived from root inputs and not aboveground inputs. In fact, root inputs may account for five times as much SOC as aboveground sources (Jackson et al. 2017). The intimate association of mineral soil and roots may be the primary cause of the disproportionate importance of roots on SOC. Historically, the focus has been on forest floor mass or litter layer depth; however, there is a growing recognition of the role of fine root production and turnover as OM inputs. This knowledge will have important implications for our ability to predict the response of SOC to disturbances that affect aboveground and belowground sources of OM.

The old paradigm suggested that OM entering the soil had three possible fates: (1) loss to the atmosphere as CO_2, (2) incorporation into microbial biomass, or (3) stabilization as humic substances (Schnitzer and Kodama 1977; Tate 1987). Humic substances were described as refractory, dark-colored, heterogeneous organic compounds of high molecular weight which could be separated into fractions based on their solubility in acidic or alkaline solutions (Sutton and Sposito 2005). Advances in analytical technology have revealed that SOC is largely made up of identifiable biopolymers, and the perceived existence of humic substances was an artifact of the procedures used to extract the material (Kelleher et al. 2006; Kleber and Johnson 2010; Lehman and Kleber 2015; Marschner et al. 2008).

Predicting the long-term behavior of SOC pools is difficult when using the old paradigm. Compounds thought to be chemically recalcitrant and resistant to decomposition (e.g., lignin) sometimes turned over rapidly, whereas compounds thought to be labile (e.g., sugars) were demonstrated to persist for decades (Grandy et al. 2007; Kleber and Johnson 2010; Schmidt et al. 2011). These inconsistencies uncovered key misconceptions of the old paradigm that prevented a predictive understanding of the vulnerability of SOC to change. As a result of such shortcomings, the conceptual model that soil scientists use to describe SOC and stabilization is undergoing a paradigm shift toward one that emphasizes the complex interactions between microorganisms and minerals in the soil.

This new paradigm for understanding SOC stability postulates that SOC exists across a continuum of microbial accessibility, ranging from free, unprotected particulate materials and dissolved OM to organic substances that are stabilized against biodegradation through association with mineral surfaces or occlusion within soil aggregates (or both) (Fig. 2.3) (Lehman and Kleber 2015). Under this paradigm, interactions of the microbial community and soil minerals, rather than characteristics inherent in the SOC itself, are the primary regulators of the pathways of OM stabilization and biodegradation. These factors may more accurately predict

the behavior of SOC pools and are also more easily measured than molecular properties of SOC, leading to new possibilities for management.

Sorption to mineral surfaces and occlusion within aggregates are the basic mechanisms of SOC stabilization under the emerging paradigm. Whether OM is sorbed to the surface of mineral soil particles or occluded depends partly on the chemical characteristics of SOC and whether microorganisms assimilate the SOC or use it as energy. Development of the concept of substrate use efficiency (SUE; the proportion of substrate assimilated versus mineralized or respired) has found that the ability of an organism to effectively decompose and transform the structural components of OM into stabilized SOC is related not only to the chemical characteristics of SOC, but also to the composition of the soil microbial community (Cotrufo et al. 2013). When microorganisms use the decomposition products of litter for energy (low

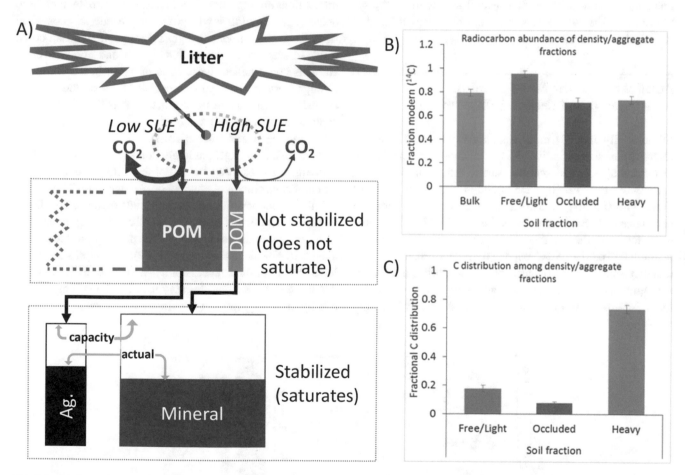

Fig. 2.3 (**a**) Conceptual model of emerging paradigm on soil organic carbon (SOC) stabilization. (1) SOC starts as litter that is deposited on the soil surface or belowground as senesced litter, roots, and other forms of organic matter (OM). (2) The microbial substrate use efficiency (SUE) determines whether the decomposition products of litter remain as particulate organic matter (POM) or are used to build microbial cells that eventually become dissolved organic matter (DOM). (3) POM is stabilized within aggregates (Ag.) while DOM is stabilized on

mineral surfaces. (**b**) Radiocarbon abundance by soil fraction. These different stabilization pathways lead to different stabilities as indicated by the ^{14}C abundance. The free/light fraction is not stabilized and has the youngest age, indicating that it is rapidly cycling. The occluded and heavy fractions are stabilized, cycle more slowly, and therefore have older ^{14}C ages. (**c**) Distribution of C by soil fraction. Most soil C in the mineral soil is found in the heavy fraction associated with mineral surfaces

SUE), that fraction remains as particulate OM stabilized within aggregates. Litter used for building microbial cells (high SUE) is stabilized on mineral surfaces and eventually becomes dissolved OM as a result of microbial exudation and death or lysis. Aggregate and mineral stabilized pools of C have a limited capacity and are said to saturate at a certain level. In contrast, free particulate matter is thought to have no upper threshold or a very high threshold.

Whether OM is stabilized within aggregates, through sorption on mineral surfaces, at depth, or in recalcitrant materials such as char, the presence of stabilized SOC is thought to be an ecosystem property: a property that arises as a result of an exchange of material or energy among different pools and their physical environment. Consequently, understanding the mechanisms that are important to the overall residence time of SOC as well as its response to a changing environment (Schmidt et al. 2011) will be valuable for monitoring and managing SOC. Further development and study of this paradigm are likely to find several interacting pathways to stabilize C in soils that involve microbial accessibility and chemical recalcitrance.

Application of the New Paradigm to Assessing Soil Carbon Vulnerability

Vulnerability of SOC (mineral soil and O horizons) to change refers to the susceptibility of SOC to change in the face of disturbance. Change could mean either increases or decreases, but usually the concern is with loss of SOC. Vulnerability of SOC can be described in terms of resistance and resilience (defined below). Soil C stores could resist losses as the result of a perturbation, or they could be resilient and recover SOC lost due to the perturbation. A system that is not affected by disturbance (scenario 1 in Fig. 2.4) is thought to be resistant to change. However, a system that loses SOC because of a disturbance, and regains lost SOC

post-disturbance, is resilient but not resistant (scenario 2 in Fig. 2.4). In scenario 3 (Fig. 2.4), the system has not resisted the disturbance and has not recovered, so it is neither resistant nor resilient.

These concepts are important because SOC loss and recovery can affect the C storage of landscapes over long timescales. Thus, the implications of SOC management and response to change need to be considered on a larger scale in both geographic extent and time. Due to heterogeneity of disturbance and time since disturbance on the landscape and over time, the system that resists change (scenario 1) will store the most SOC in the landscape over time, followed by 2 and then 3. Therefore, SOC in these systems increases in relative vulnerability from 3 to 1, with 3 being the most vulnerable to losses.

The vulnerability of SOC to change depends on the particular forcing (e.g., climate change, fire regime shift, invasive species, disturbance). Furthermore, there are usually interactions among forcing variables. For example, increased occurrence of wildfire will probably precede an ecosystem shift caused by climate change. This type of situation could result in scenario 3, where SOC is lost and not recovered. Humans could hasten these sorts of shifts in SOC content through certain management decisions, as in the case of intensive forest harvesting or grazing areas that may be susceptible to species shifts caused by climate change (Noss 2001; McSherry and Ritchie 2013).

With this new SOC paradigm comes a fresh approach to studying and predicting the response of SOC to disturbance, climate, environmental change, or management. The stability of any pool depends on the magnitude of, and controls on, its fluxes (inputs and outputs) (Fig. 2.5). The inputs are the quantity and quality of C fixed by the primary producers and altered by abiotic processes (e.g., fire); the outputs are regulated by microbial accessibility and microbial activity. Anything that changes the (1) quantity of OM inputs, (2) quality of OM inputs, (3) microbial accessibility, or (4)

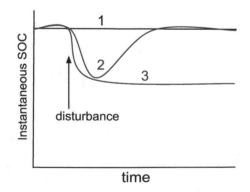
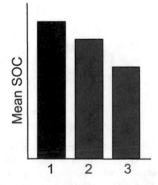

Fig. 2.4 Diagram of soil organic carbon (SOC) responses to disturbance. (**a**) The vulnerability of SOC to change depends on timescale, frequency and magnitude of change, and recovery. In scenario 1, soil C is stable through time, whereas in scenarios 2 and 3, a perturbation causes SOC to decrease. In scenario 2, SOC recovers to predisturbance levels; in scenario 3, it does not. (**b**) The mean SOC content over time would be highest for scenario 1, which never lost SOC, followed by 2 and 3

Fig. 2.5 Schematic showing the soil organic carbon (SOC) pool in balance with its inputs and outputs, which are regulated by primary production, quality of inputs, controls on microbial accessibility of SOC, and microbial activity

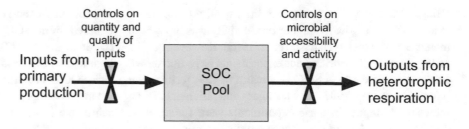

microbial activity will affect the magnitude and stability of the SOC pool. Examples of factors that could affect these inputs and outputs are listed next and will be referenced in the context of specific disturbances in the following paragraphs.

Changes affecting *quantity of OM inputs*—shifts in productivity; the removal or addition of biomass by fire, harvesting, or mulching

Changes affecting *quality of OM inputs*—change in species; changes in allocation of production (especially belowground versus aboveground production); transformation of biomass by pyrolysis

Changes affecting *microbial accessibility*—destruction of aggregates; destabilization of redox-active minerals; inputs of active minerals (e.g., ash deposition); changing the OC saturation state; changes to the quality and quantity of SOC inputs, which could affect priming (stimulation of decay of stabilized SOC); changes in the distribution of SOC with depth through erosion and deposition, leaching, bioturbation, and other influences

Changes affecting *microbial activity*—change in soil temperature and moisture, nutrient availability, freeze-thaw patterns, oxygen availability (i.e., redox), pH, or salinity; change in nutrient status from additions of substances such as herbicide or additions of N and sulfur from acid deposition

Soil Carbon Vulnerability Under Key Disturbances

Climate Change and Increasing Carbon Dioxide

As a primary factor in soil formation, climate has profound effects on SOC cycling (Jenny 1941). Quantity and quality of OM inputs will be impacted as warming temperatures and shifting precipitation regimes will lead to transitions in forest and rangeland plant communities (Clark et al. 2016). Microbial accessibility may be impacted as temperature and moisture changes alter rates of mineral complexation and leaching. Finally, microbial activity itself is sensitive to changes in temperature and moisture availability. In addition

to changes directly tied to climate, increases in CO_2 concentration will alter plant productivity, affecting the quantity of C inputs to soil, as well as the relative contributions of roots and shoots to SOC, potentially increasing root-derived OM inputs (Phillips et al. 2012). Some have also shown that litter quality will change or that species shifts could take place which change the quality of C inputs to soil (MacKenzie et al. 2004).

Impacts of climate change on SOC have been assessed by using manipulation experiments, ecosystem modeling, and field sampling along climate gradients (climosequences). Globally, studies of soil incubations report an increased loss of SOC from bulk soil under warmer temperatures (e.g., Sierra et al. 2015). Northern latitudes are expected to bear the brunt of this loss as permafrost thaws and decomposition is amplified, as shown by soil warming experiments (Schuur et al. 2015). Despite indications of increased mineralization of SOC, coupled Earth system-climate models suggest a change in global SOC pools of −72 to +253 Pg (USGCRP 2018). Projections of increasing SOC with warming are primarily a result of a modeled increase in SOC of northern latitude soils, driven by the effect of increased plant productivity on C inputs to soil (Genet et al. 2018). Recent field experiments show that shrub expansion in tundra, a phenomenon tied to climate change, may also promote stabilization of SOC (Lynch et al. 2018). This complex response of the soil-plant system to warming highlights the importance of a multifaceted approach to understanding climate change impacts on SOC stocks.

Climosequences are an approach that may better approximate a whole-ecosystem, longer-term response of SOC to climate change than short-term incubations. There is some indication of reduced SOC in warmer areas compared to cooler areas (Lybrand et al. 2017; Wagai et al. 2008). Along a tropical forest chronosequence, there was no trend in SOC storage or stabilization across a warming temperature gradient despite large increases in plant-driven inputs (Giardina et al. 2014). These results suggest that a temperature-driven reduction in stabilized SOC could offset increases in SOC inputs projected with climate change. Such projections emphasize the need for maintaining bulk density, aeration, and other soil properties that promote aggregation and mineral-associated stabilization under a warming climate.

Carbon dioxide fertilization effects on plant growth may lead to increases in plant productivity in some ecosystems

(Hickler et al. 2008) or changes in litter quality (e.g., Henry et al. 2005). Historically, higher productivity was thought to increase SOC stocks (Harrison et al. 1993). However, results of large-scale CO_2 enrichment experiments have not shown substantial increases in SOC as a result of CO_2 fertilization (Hungate et al. 1997; Schlesinger and Lichter 2001). Increased C inputs in these experiments were found to be disproportionately partitioned into rapidly cycling, nonstabilized SOC pools. Additionally, increases in root exudates have been shown to have a priming effect across many soils, stimulating the decay of SOC previously stabilized through mineral association (Sulman et al. 2014).

Precipitation changes may impact SOC cycling in ways independent of simply increasing or decreasing average soil moisture, with effects especially pronounced in arid and semi-arid ecosystems such as grasslands and deserts. Shifts in the timing of rainfall, rather than amount alone, have been shown to alter microbial activity via enzyme and nutrient dynamics (Ladwig et al. 2015). Precipitation event size and timing, which affect soil moisture levels between events, are a strong control over microbial mineralization of C in arid systems (Cable et al. 2008). In addition, moving from smaller, more frequent rainfall events to fewer, larger events can increase arid grassland productivity, thus increasing inputs of organic matter to soil (Thomey et al. 2011). Although more research is needed on the topic, especially in semiarid woodlands, shifts in microbial activity and plant production due to changing rainfall patterns are likely to affect SOC storage in water-limited ecosystems. Advances in understanding precipitation impacts on SOC are nonetheless overshadowed by the great uncertainty in projections of precipitation changes (IPCC 2014).

Fire

Fire can have many effects on SOC through changing the quantity and quality of C inputs to soil, such as the forest floor, and affecting conditions that control microbial activity and access to SOC. Fire, whether a prescribed burn or wildfire, has important first-order effects on SOC that are tied to fire intensity and duration of heating (Neary et al. 1999). Fire that mineralizes surface OM will reduce total C pools (SOC plus forest floor) and OM inputs to SOC. Fires in organic soils, such as peatland, are difficult to control, can persist for long periods of time, and combust large amounts of SOC (Reddy et al. 2015). In mineral soils, at least partial consumption of the O horizon is common in wildfires, especially in shrubby or forested sites with high fuel loading near the surface (Neary et al. 1999). Even controlled prescribed burns can generate enough heat to consume the forest floor and reduce O horizon and SOC storage (Boerner et al. 2008; Sackett and Haase 1992; Sánchez Meador et al. 2017).

Depending on soil bulk density and parent material, soil heating is usually strongly attenuated with depth in the mineral soil profile, and depths as shallow as 2.5 cm may be well buffered from SOC combustion during light surface fires (DeBano et al. 1977). Nonetheless, there is the potential for high loss of SOC via combustion during fire given high surface SOC concentrations. Soils start to lose significant SOC when soil temperatures exceed about 150 °C, temperatures achievable in surface soil during many burns and at depths greater than 2.5 cm during a moderate- or high-intensity burn (Araya et al. 2017; Neary et al. 1999). Previous research has quantified thresholds of soil heating for loss of different chemical fractions of SOC (González-Pérez et al. 2004); however, in light of the new SOC paradigm, loss of aggregates may be a more valuable indicator for overall SOC storage postfire. In a soil-heating laboratory experiment, soils exposed to temperatures that would be expected in high-intensity, high-severity fires had proportionately less SOC stored in macroaggregates than soils exposed to low temperatures (Araya et al. 2017). Even temperatures found in low- and moderate-severity burns, if long enough in duration, can degrade soil aggregates (Albalasmeh et al. 2013). Progress in this area is still inhibited by a lack of understanding of how common measures of burn severity, such as crown or duff consumption, relate to measurable soil effects (Kolka et al. 2014).

The first-order loss of SOC via combustion may be partially offset by creation of heat-altered C such as soot, charcoal, or biochar, collectively known as pyrogenic C (pyC). However, it may be difficult to detect significant increases in pyC from just one fire. Factors controlling pyC formation and accumulation are complex and likely to vary by soil type, climate, and ecosystem (Czimczik and Masiello 2007). In a fire-prone ponderosa pine (*Pinus ponderosa*) forest in Colorado, there was no difference in pyC content in the soil between recently burned and unburned areas, implying a role for erosion or legacy of past burns, or both, in present-day pyC content (Boot et al. 2015). In the same wildfire, postfire erosion and sedimentation were found to be an important control over spatial distribution of pyC (Cotrufo et al. 2016). DeLuca and Aplet (2008) estimate that charcoal may account for 15–20% of the total C in temperate, coniferous forest mineral soils and that some forest management activities (e.g., salvage logging, thinning) may reduce soil pyC content and long-term storage.

Second-order impacts of fire on the O horizon and SOC may take longer to become manifest. These effects result from fire's direct impact on soil microbial biomass, soil chemical characteristics, and plant productivity. After high-severity wildfire, the soil microbial community can shift in composition and size, which can impact microbial SOC transformations (Knelman et al. 2017; Prieto-Fernández et al. 1998). Fire-induced increases in soil pH and an initial increase in N availability can also affect microbial activity and mineralization of the O horizon and SOC (González-Pérez et al. 2004; Hanan et al. 2016; Kurth et al. 2014; Raison 1979). A darkening of the surface and decrease in surface albedo can raise soil temperatures, increasing SOC mineralization rates. Loss of vegetation can reduce the forest floor. The bare soil surface is left vulnerable to erosion, exposing deeper SOC for decomposition and loss as CO_2. However, it is unclear whether postfire erosion could increase SOC sedimentation enough to outweigh CO_2 losses (Cotrufo et al. 2016; Doetterl et al. 2016). A meta-analysis found that 10 years postfire, SOC increased across multiple forested sites, which could be attributed to a combination of secondary effects and pyC creation (Johnson and Curtis 2001). In areas that have repeated burning, these secondary fire effects accumulate over decades, with net effects on SOC that may vary by ecosystem type. A recent meta-analysis found an overall increase in SOC in frequently burned forests, yet a decrease in SOC in frequently burned grasslands (Pellegrini et al. 2017). These changes are thought to be largely tied to effects of fire on nutrient availability and plant productivity. Organic soils may lose exceptionally large amounts of C due to indirect effects of wildfire. In permafrost soils, wildfire increases the active layer depth, ultimately leading to increased C loss as CO_2 in the long term (Zhang et al. 2015a, b).

To understand the role that management can play in fire-SOC dynamics, we can consider how the reported effects are linked to fire behavior, which depends on fuel loading, weather, and topography. For example, slash piles can generate extremely high soil temperatures when burned. These temperatures, which are higher than typical broadcast burns, lead to chemical and microbial community transformations with potential feedbacks to SOC processing (Esquilín et al. 2007; Massman and Frank 2005). Slash piles in ponderosa pine forest in Arizona were found to have lower SOC in the top 15 cm of mineral soil 7 months after burning (Korb et al. 2004). However, high soil moisture and reduced bulk density near the surface can decrease surface heating (Frandsen and Ryan 1986).

Sidebar 2.2 Positive and negative effects of pyrogenic carbon on soils

Positive	Negative
Carbon sequestration	Toxicity to plants, microbes, soil animals, invertebrates
CO_2 capture	Carrier of contaminants
Improving aggregate stability and water holding capacity	Carbon fluxes
Stabilization of contaminants	Heavy metals
Sorption or removal of pollutants	Altering soil pH
Catalyst for microbes	Organic chemical release

Pyrogenic C (i.e., char, biochar, black carbon) may not exactly follow the emerging paradigm of SOC stabilization. Pyrogenic C is the product of incomplete combustion of OM and fossil fuels (see Chap. 7) and exists along a continuum of increasing alteration relative to its original OM from char to soot. It has been found that pyC can persist in soils and sediments for centuries to millennia and so is thought to be resistant to degradation. In addition, pyC affects many factors important for SOC stabilization. It increases cation exchange capacity (Liang et al. 2006), promotes water and nutrient retention, and reduces soil bulk density, encouraging microbial activity. Pyrogenic C is also highly utilized by ectomycorrhizal fungi (Harvey et al. 1976). The mechanism by which pyC resists degradation is thought to result from its complex molecular structure of condensed aromatic rings. However, pyC is often found in association with mineral surfaces and within soil aggregates (Brodowski et al. 2006; Wagai et al. 2009), suggesting that it may promote these stabilization processes, which in turn allow it to resist decomposition. We currently lack nationwide estimates of pyC in soil, although a recent global analysis estimated that pyC represents about 14% of total SOC (Reisser et al. 2016).

The first-order loss of SOC via combustion may be partially offset by creation of heat-altered C such as soot, charcoal, or biochar, collectively known as pyrogenic C (pyC). However, it may be difficult to detect significant increases in pyC from just one fire. Factors controlling pyC formation and accumulation are complex and likely to vary by soil type, climate, and ecosystem (Czimczik and Masiello 2007). In a fire-prone ponderosa pine (*Pinus ponderosa*) forest in Colorado, there was no difference in pyC content in the soil between recently burned and unburned areas, implying a role for erosion or legacy of past burns, or both, in present-day pyC content (Boot et al. 2015). In the same wildfire, postfire erosion and sedimentation were found to be an important control over spatial distribution of pyC (Cotrufo et al. 2016). DeLuca and Aplet (2008) estimate that charcoal may account for 15–20% of the total C in temperate, coniferous forest mineral soils and that some forest management activities (e.g., salvage logging, thinning) may reduce soil pyC content and long-term storage.

Harvesting and Thinning

Forest operations, such as harvesting and thinning, alter SOC by reducing C input quantity via forest floor and root OM inputs as the stand regenerates. In addition, microbial accessibility and activity are altered through the disturbance of the soil surface, which changes temperature and moisture regimes.

As a result of high heterogeneity in SOC, it can be exceedingly difficult to detect change as a result of forest harvesting in any specific study. The results of many individual experiments are synthesized in meta-analyses and can be used to detect changes that are broadly consistent across studies, even when heterogeneity obscures treatment effects within a single study. A few meta-analyses and review articles conclude that the net effect of harvest is a reduction in SOC, with forest and soil type determining the magnitude of C loss (Jandl et al. 2007; Johnson and Curtis 2001; Nave et al. 2010). Nave and others (2010) reported an 8% average reduction in SOC stocks after harvesting over all forest and soil types studied. Even whole-tree harvesting (but leaving roots in the soil) for biomass production may have little long-term effect on mineral SOC stocks if O horizons are left undisturbed (Jang et al. 2016; Powers et al. 2005). In general, postharvest reductions in SOC have been shown to occur as a result of soil disturbance during harvesting and site preparation (Achat et al. 2015a, b; James and Harrison 2016). However, advances in the understanding of how harvesting impacts belowground processes are difficult because most studies focus on the first 30 cm of the soil profile or even just the forest floor. Harrison and others (2011) report that for a variety of ecosystems and treatments, valid estimation of changes in ecosystem C was not even possible without sampling soil deeper than 20 cm. Therefore, these losses are primarily the result of a reduction in litter layer mass and organic matter inputs from growing trees; they may also reflect the sampling challenges of accurately tracking forest floor C over time (Federer 1993; Yanai et al. 2000).

Thinning of forest stands, as opposed to harvesting for timber, is a common practice to achieve various silvicultural objectives. Effects of thinning on inputs to SOC are variable, depending on residue management. Residues of the thinning process include the main parts of the cut trees: bole, branches, leaves, and the associated roots that have been severed. All of these components have a trajectory toward decomposition, which is, in general, accelerated due to the physical disturbance. Varying the timing (Schaedel et al. 2017) or intensity (D'Amore et al. 2015) of thinning may mitigate C losses. Forest thinning and competition control have a much smaller impact on soil characteristics and therefore affect SOC stocks less than forest biomass harvesting operations. Protection of natural forests through the use of intensively managed forests may also provide the benefit of C sequestration (Ouimet et al. 2007) or result in a release of C to the atmosphere (Harmon et al. 1990). Furthermore, herbicide application to improve seedling growth has been shown to have a positive impact on C storage aboveground but a negative impact on belowground C (Markewitz 2006).

Biomass harvesting and removal of woody residues by burning or for bioenergy are a concern in many forest ecosystems because of the potential adverse impacts on productivity (Janowiak and Webster 2010), ectomycorrhizae (Harvey et al. 1976), long-term nutrient cycling (Harmon et al. 1994), soil moisture content (Maser et al. 1988), N_2 fixation (Jurgensen et al. 1987), and regeneration success (Schreiner et al. 1996). Harvey and others (1981) noted that

harvesting has the potential to disturb soils and reduce the amount of woody residues, particularly in dry forest types. However, several studies have shown that coarse wood retention has very little effect on SOC or nutrients (Busse 1994), perhaps because all soils have been affected by coarse wood at some time (Spears et al. 2003). However, coarse wood functions as a C storage pool, creates wildlife forage areas, enhances fungal diversity, provides erosion control, and increases moisture retention. In addition, if soils have very low buffering capacity due to soil parent material chemistry and historical impacts of atmospheric deposition, biomass harvesting reduces the total amount of nutrients left on-site. Nutrient removal is particularly marked on sites that are extremely nutrient limited as a result of long-term anthropogenic acidification, overgrazing, wildfire, or excess OM piling and burning. Understanding inherent soil chemistry and composition, resilience to nutrient losses, and ecosystem dynamics dependent on nutrient cycling throughout a rotation or longer is necessary for assessing long-term sustainability (Jang et al. 2015).

Livestock Grazing

Rangelands, despite their lack of forest floor, can contain high amounts of SOC because grasses allocate a high percentage of biomass to roots. Rangeland SOC stocks are related to plant productivity, but management activities can have important effects on SOC stocks (Silver et al. 2010). Grazing by livestock can influence numerous factors that have control over SOC content with complex interactions that make it difficult to predict the net effect on SOC. Directly, grazing influences the quantity of OM that returns to the soil. Indirectly, grazing affects OM quality by altering plant physiology and ecological processes. Secondary feedbacks can occur if nutrient removal through grazing reduces grassland productivity.

Studies show that grazing rate, duration, and intensity can interact with wind erosion, site properties, and restoration activities to cause both increases and decreases in SOC (Piñeiro et al. 2010). Herbivores alter the quality of OM inputs by reducing C/N ratios of plant shoots and roots. Lower C/N ratios in plant litter increase decomposition rates and net N mineralization by reducing microbial demand for N; that is, N stocks are high enough to promote mineralization despite immobilization through microbial assimilation (Frank and Groffman 1998). These changes in decomposition rates suggest that microbial activity and substrate use efficiency are changed and that while decomposition rates may increase a return of CO_2 to the atmosphere, a portion of the C will return to the soil in dissolved forms that may be stabilized on mineral surfaces.

Grazing management techniques intended to increase forage production may also increase the quantity of inputs to

SOC, thus accumulating atmospheric C as a C sink (Conant et al. 2001). Nitrogen is typically the nutrient limiting primary production in grasslands and thus SOC content (Piñeiro et al. 2010). Maintenance of SOC is possible with grazing management systems that maintain N content and grassland productivity. High stocking rates tend to lead to decreased production (Conant et al. 2001), so systems such as slow rotation grazing with moderate stocking levels will increase vegetative heterogeneity and increase soil aggregate stability (Conant and Paustian 2002; Fuhlendorf and Engle 2001).

Soil stability is important for aggregation and microbial accessibility of SOC. Grazing can increase rates of erosion, which exports SOC from a site. Indeed, grazing in arid and semiarid systems can lead to a destabilization of soil surfaces that subsequently leads to losses of soil nutrients to wind and water erosion (Neff et al. 2005). Losses in soil nutrients can lower fertility, which can reduce plant productivity. As with postfire erosion, however, it is unclear whether this process increases sedimentation rates that outweigh other losses.

Local adaptations of grazing systems have been shown to increase net primary productivity, N storage, and, as a result of these pathways, SOC storage (Piñeiro et al. 2010). Abrupt changes in intensity in grazing systems ultimately reduce net C storage in soils by the alteration of plant communities through the direct action of grazing. When slow-growing native plants adapted to low disturbance are intensively grazed, SOC is lost through the alteration of plant roots, N availability, decomposition of plant litter, and the control of the soil microbial community (Klumpp et al. 2009).

The response of SOC to grazing depends on multiple factors: climate, soil properties, landscape position, plant community composition, and grazing management practices (Piñeiro et al. 2010; Reeder and Schuman 2002). Considering the sensitivity of rangelands to seasonal drought, for example, can help predict impacts of grazing on SOC stocks. Southwestern rangelands are particularly sensitive to drought; annual net C loss is a common occurrence due to low grassland productivity during drought years (Svejcar et al. 2008). In these instances, managers may reduce grazing intensity during drought periods to ensure recovery of grassland productivity the following year. The effects of reduced stocking rate on SOC in rangelands remain highly uncertain, however, and data are lacking in arid southwestern systems (Brown et al. 2010).

Nutrient Additions

The addition of fertilizing nutrients to mineral soils, through either nutrient management or N deposition (acid rain), can result in gains, losses, or no change in SOC stocks; the outcome depends on a large number of factors, not all of them known (Blagodatskaya and Kuzyakov 2008; Jandl et al. 2007; Janssens et al. 2010). Forest fertilization has been shown to increase or decrease SOC by increasing productivity, shifting production to aboveground vegetation components, increasing SOC mineralization rates, and depressing certain enzyme activity (Jandl et al. 2007; Van Miegroet and Jandl 2007). Effects of forest fertilization on SOC have been found to be site specific, but most studies show an increase in SOC stocks (Johnson and Curtis 2001).

Agroforestry may also benefit from planting N-fixing shrubs and trees as an economical N source for crops (Danso et al. 1992). Nitrogen-fixing tree species are associated with higher forest SOC; accumulation of SOC has been reported to be about 12–15 g of C for every gram of N fixed (Binkley 2005). The mechanisms leading to this SOC increase are incompletely understood but are thought to differ from the direct effects of N fertilization (Binkley et al. 2004; Forrester et al. 2013). Research suggests that OM derived from the N-fixing *Acacia* species is more protected from decomposition than litter from other trees in mixed stands (Forrester et al. 2013). Most research on the subject has been conducted outside of the United States; more studies are needed that focus on N-fixing trees, such as *Alnus* species, and forest management practices specific to the United States.

On a global scale, the largest source of nutrient additions to forest soils is atmospheric N deposition derived from both natural and anthropogenic sources. Anthropogenic N, from fossil fuel combustion and recirculated cropland fertilizers, accounts for about 60% of the approximately 130 Tg of N deposited globally each year (Kanakidou et al. 2016). Chronic N addition experiments consistently show that SOC increases under higher N availability (Frey et al. 2014). This result has been attributed to greater productivity due to the fertilization effect, as well as reductions in OM decomposition. Frey and others (2014) found that in addition to more C stored in tree biomass, there were significant shifts in SOC chemistry due to shifts in the microbial community (fewer fungi). Increases in SOC were attributed to a reduction in decomposition rate due to the lower abundance of fungi in the soil.

Tree Mortality

Carbon stocks in the forest floor and soil may be impacted when high levels (>50% canopy loss) of tree mortality result, such as that caused by drought and bark beetles (subfamily Scolytinae) in Western US pine and spruce (*Abies* spp.) forests and by invasive pests (detailed in the next section). Mass tree mortality effects on SOC are similar to effects of harvesting, but with the following important differences: (1) Mortality occurs more slowly than most harvesting operations; (2) mortality events do not usually kill as many trees as are harvested in a typical operation; and (3) dead trees are commonly in place throughout the mortality event, although some limited postmortality harvesting is conducted in high-use areas (e.g., national forest campgrounds).

Mortality events result in a reorganization of detritus over multiple years, impacting OM inputs to SOC formation (Edburg et al. 2012). Root OM inputs to the soil may increase as trees die but may later decline due to reduced live tree density; as a result, microbial activity in the rhizosphere is altered after tree mortality (Warnock et al. 2016). Litterfall is expected to increase in the first few years following mortality as dead trees drop their needles and fine branches; afterward, litterfall will decline, reducing forest floor mass in the longer term (Zhang et al. 2015b). Longer-term inputs to the soil are larger branches and boles, as wind topples standing dead trees throughout the next decade and beyond. Thus, mortality will change the rate and type of OM matter input to the soil. Shifts in nutrient dynamics could alter SOC mineralization rates, as soil-extractable N levels increase and SOC/N ratios decrease (Clow et al. 2011; Morehouse et al. 2008; Trahan et al. 2015). Finally, changes in microclimate postmortality, brought about by canopy loss and decreases in transpiration, could impact detrital C processing (Berryman et al. 2013). The duration of these effects depends on how fast the remaining living trees expand their canopies to compensate for the loss of the overstory.

As a result of this reorganization of detritus, some changes in SOC cycling have been detected following mortality events. Evidence from stable isotopes suggests shifts in C substrate type used for root and heterotrophic respiration starting in the first year after tree mortality (Maurer et al. 2016). Some studies have reported decreases in microbial biomass C (MBC) and increases in the aromaticity of dissolved organic C (DOC) in the soil, which may impact SOC stabilization (Brouillard et al. 2017; Kaňa et al. 2015; Trahan et al. 2015). Despite these changes in SOC substrates and cycling rates, changes in SOC stocks following tree mortality events are often undetectable and, on average, minor compared to impacts on soil respiration, DOC, and MBC (Morehouse et al. 2008; Zhang et al. 2015b). This suggests that though individual process rates may be affected by tree mortality, the balance between inputs to and outputs from the SOC pool may be constant enough to lead to undetectable changes in SOC.

Changes in SOC may be difficult to detect because they could be highly dependent on the amount of tree mortality. In bark beetle-impacted lodgepole pine (*Pinus contorta*) forests of Colorado, soil respiration 8 years after mortality depended on the relative amount of living versus dead trees (Brouillard et al. 2017). However, plot-level impacts of biotic disturbance may not scale predictably to the forest or watershed. Measured plot-level increases in extractable N and DOC following a bark beetle outbreak, for example, were not detected at the watershed scale in the Rocky Mountains in Colorado (Clow et al. 2011; Rhoades et al. 2017). This result suggests that landscape-scale patchiness in outbreaks could be important for buffering negative effects of mortality on SOC.

Invasive Species

Invasive species can alter nutrient and C cycling, as well as soil physical properties, all of which can affect SOC stocks. Exotic invasions impact factors important for OM quantity and quality and microbial activity, such as nutrient mineralization, N-fixation by soil bacteria, mycorrhizal inoculation, decomposition and aeration of soils by earthworms (suborder Lubricina), and aggregation of soils by fungi (Wolfe and Klironomos 2005).

The function of many ecosystems depends on regular disturbance, for example, to foster plant renewal and regeneration, but disturbance can also be detrimental by promoting invasion of nonnative and weedy plants (Hobbs and Huenneke 1992). Initial invasiveness is caused by chronic disturbance which disrupts the native nutrient and OM cycling that increases plant nutrient availability (Norton et al. 2007). In grasslands, disturbance and subsequent weed invasion can be caused by either temporary increases in nutrients or reduced competition from plant canopies and roots (Hobbs and Huenneke 1992), although the research is still unclear about the relative importance of these processes.

Several hypotheses have been offered to explain why exotic plants are so successful in disturbed ecosystems: (1) inherent properties of the invading plant (such as earlier colonization than native vegetation); (2) vegetation factors (such as species composition, richness, and heterogeneity); (3) soil microbial dynamics; and (4) climate factors such as rainfall amount and timing, aridity, and humidity (Blank and Sforza 2007). In the absence of disturbance, there are other factors such as plant-fungal interactions that may alter soil nutrient dynamics (Brundrett 2009) and contribute to understory plant invasion (Jo et al. 2018). Arbuscular mycorrhizal-dominant forests, which are characterized by thin litter layers and a low soil C/N ratio relative to ectomycorrhizal-dominant forests, are invaded by exotic plants to a greater extent (Jo et al. 2018). Other factors that may influence the invasion of exotic plants may be more indirect, for example, external factors such as deer (*Odocoileus* spp.) browsing or earthworm invasion (Nuzzo et al. 2009).

Plant invasion leads to a shift in plant species composition, which can influence ecosystem properties such as N accumulation and cycling, SOC storage, water availability and runoff, and disturbance regime (Mack et al. 2001). Nitrogen cycling may be a particularly sensitive indicator of changes in species composition, and changes to N cycling are prevalent during biological species invasions (Ehrenfeld 2003; Mack et al. 2001). The impacts of nonnative plant species on SOC are largely system- and plant-dependent. Scott and others (2001) and Ehrenfeld and others (2001) found increased SOC under invasive grasses and shrubs in a grassland and deciduous forest, respectively. These changes in SOC were attributed to higher inherent productivity of the invasive species relative to the native vegetation. On the other hand, Bradley and others (2006) found no change in SOC pools as a result of cheatgrass (*Bromus tectorum*) invasions in shrublands of the western United States. Invasive cogongrass (*Imperata cylindrica*) in the southeastern United States has higher litter decomposition rates than native vegetation, which has implications for SOC production and turnover (Holly et al. 2009).

Earthworms increase fragmentation and decomposition of litterfall and contribute to the formation of mull humus types, in which organic surface layers are mixed into the mineral soil. Therefore, they play an important role in C and N cycling in forest soils. However, exotic earthworms from Europe and Asia have invaded recently glaciated soils in northern temperate forests in North America (Alban and Berry 1994; Dymond et al. 1997; Scheu and Parkinson 1994). These novel soil engineers are affecting SOC and nutrient dynamics (Burtelow et al. 1998; Groffman and Bohlen 1999), including a loss of C from the soil profile (Alban and Berry 1994; Langmaid 1964). Invasive earthworms can eliminate the forest floor, decrease C storage in the upper soil mineral horizons, and reduce the soil C/N ratio (Bohlen et al. 2004).

Invasive forest pests and diseases alter ecosystem functioning of the forest, potentially leading to effects on SOC. Some exotic pests eventually cause forest mortality (see previous section), but many result in lesser disturbance than mortality events by affecting only a few trees or by causing defoliation without killing the tree. Recently, insects have expanded into previously climatically restricted geographic areas, and their activity induces a change in C sequestration (Kretchun et al. 2014).

Although invasive pests are relatively common in temperate forests, their effect on C pools is poorly understood. Studies on catastrophic disturbances from wildfire or clearcutting indicate a substantial loss of aboveground and belowground C (Amiro et al. 2006; Humphreys et al. 2006; Thornton et al. 2002). These results are similar to a study conducted in the Pine Barrens of New Jersey, where defoliation by the gypsy moth (*Lymantria dispar*) reduced C storage at the landscape scale (Clark et al. 2010). Modeled infestations of hemlock woolly adelgid (*Adelges tsugae*) indicate that, with inputs of dead wood and roots, mineral soil and forest floor C pools may remain static or increase in the long term (Krebs et al. 2017). In addition, modeled invasive gypsy moth defoliation episodes indicate a shift in overstory species, but not a decrease in SOC in the short term (Kretchun et al. 2014). Similarly, in New Hampshire, where emerald ash borers (*Agrilus planipennis*) attacked a mixed hardwood forest, there were no short-term changes in soil microclimate, respiration, or methane oxidation under the trees; these results were attributed to the sandy soil, which

diffused soil responses (Matthes et al. 2018). There are scant data on many plant, animal, insect, and pathogen invasive species, their impact on SOC, and how managers can effectively change land management to alter invasive species effects on ecosystem services. The impact of invasive pests on SOC is likely to be site- and species-specific. Soil scientists will play a key role in quantifying these impacts by measuring changes in SOC during land management or after insect outbreaks.

Managing for Soil Organic Carbon in Forests and Rangelands

As previously mentioned, SOC is critical for maintaining a host of ecosystem services. It is vulnerable to loss through events, either natural or human-induced, that remove large amounts of soil or OM inputs to soil, such as the forest floor or plant biomass (Table 2.1). Certain management actions

Table 2.1 Loss or gain in soil organic carbon (SOC) from key disturbances in forests, rangelands, and wildlands

Disturbance type	Direction of SOC effect	Magnitude of SOC effect	References
Biotic disturbance (insects, disease)	Loss	2% (+/−1%)	Zhang et al. (2015a, b)
Wildfire	Loss (<10 years) Gain (>10 years)	0–4% Long-term gain of 8%	Johnson and Curtis (2001) and Page-Dumroese and Jurgensen (2006)
Windthrow	Gain in pit or mound area	28%	Bormann et al. (1995)
Hurricane	Gain	9%	Sanford et al. (1991)
Prescribed fire	Loss	0–1%	Boerner et al. (2008) and Sánchez Meador et al. (2017)
Thinning and burning	Gain	2%	Sánchez Meador et al. (2017)
Warming	Loss	1–8%	Dieleman et al. (2012) and Lu et al. (2013)
Harvesting	Loss Gain	8% (+/−3%) 8% (+/− 8%)	Johnson and Curtis (2001) and Nave et al. (2010)
Grazing (conversion to improved)	Gain	1–7% per year	Chaplot et al. (2016) and Minasny et al. (2017)
Woody encroachment into grasslands	Gain	1% per year	Neff et al. (2009)

post-disturbance can counteract such losses. Application of organic amendments such as biochar or simply leaving harvest residues in place can nearly double existing SOC (Achat et al. 2015a, b; Liu et al. 2016; Page-Dumroese et al. 2018). Shifting land management priorities to place more value on C management practices may help encourage best practices for building up SOC pools on forest and rangeland sites.

Fuels management that aims to reduce the risk of high-severity wildfire could help mitigate SOC losses (Page-Dumroese and Jurgensen 2006; Neary et al. 1999). Prescribed fire is a common fuels reduction treatment on many federal lands. During prescribed fires, managers can control fire intensity and severity to reduce the amount of C that will be released into the atmosphere as CO_2 and increase the conversion of wood to pyC. One way to control burn intensity is by lowering fuel loadings on the soil surface before a prescribed burn. Slash pile size, season of burning, and soil texture and moisture will all determine the amount of SOC remaining in the soil after a prescribed burn (Busse et al. 2014).

Land managers are likely to have few options for sequestering SOC during a wildfire, but active management on forest lands can reduce the intensity and severity of fires. By using mechanical practices such as thinning, masticating, chipping, and mowing to reduce wildfire hazard, managers can restore healthy forests, limit damage from wildfire, and alter C storage. However, the loss of total site C during fuel treatments will negate any fire-prevention benefits if the site is unlikely to burn anyway. Therefore, risk for high-severity fire should be weighed against C removal benefits if maintaining ecosystem C over landscapes for the long term is the goal (Boerner et al. 2008; Krofcheck et al. 2017). Second-order negative impacts of high-severity fire on SOC can be mitigated by taking action to reduce postfire erosion and speed up the regrowth of vegetation postfire. Mulching treatments slow soil erosion and have the extra benefit of adding C to the mineral soil (Berryman et al. 2014; Robichaud et al. 2013).

In the short term, thinning forest stands reduces the amount of OM inputs and thus SOC storage (Jandl et al. 2007); it also changes the microclimate and therefore decomposition of OM inputs. Forest floor decomposition is temporarily stimulated because of warmer and wetter soil conditions, and SOC may decrease (Piene and Van Cleve 1978). Although forest thinnings can reduce C pools in the forest floor and mineral soil (Vesterdal et al. 1995), the overall effect on SOC may be rather small (Johnson and Curtis 2001). Whole-tree harvesting may have a greater impact on C reserves than thinnings or cut-to-length harvesting. However, 5 years after clear-cutting and OM removal on the Long-Term Soil Productivity study sites, there were no changes in SOC (Sanchez et al. 2006). These results highlight the resilience of some soils to harvest-induced losses of C and the importance of leaving roots and stumps on-site

after harvesting to help maintain SOC (Powers et al. 2005). This means that best management practices for harvesting should take into consideration soil texture, climatic regime, and inherent OM levels. In addition to selecting the most appropriate harvest methods, site preparation method selection can help improve SOC. Increasing the intensity of site preparation (e.g., soil mixing, stump pulling) increases SOC losses (Johannsson 1994; Nave et al. 2010; Post and Kwon 2000).

Management to reduce C flux from rangelands is important because rangelands are thought to have 30% of global terrestrial C stocks (Schuman et al. 2002). Many rangeland sites are dominated by near-surface C, so one method to reduce losses is to protect the surface mineral soil by increasing plant cover to lessen wind and water erosion (Booker et al. 2013). Grazing management can also prevent vegetative state transitions (e.g., increasing or decreasing woody plants), reduce soil disturbance, and limit bare soil exposure. For example, heavy grazing can contribute to a transition from grasses and forbs to invasion of woody shrubs, which choke out grasses (Russell and McBride 2003) or provide cover for wildlife (Laycock 1991). Different results on rangeland sites are due to varying soil moisture regimes (Booker et al. 2013); therefore, best management practices must be tailored to soil texture and climatic regime.

Compared to other disturbances like wildfire and harvesting, biotic disturbance effects on SOC storage appear to be minor. Site management may have a bigger impact on SOC storage than the disturbance event itself. Removing hazard trees or harvesting beetle-killed timber for wood products or bioenergy in a system already disturbed may lead to undesirable consequences for SOC storage. Avoiding soil compaction may promote conditions that favor SOC aggregation and stabilization.

Invasive species affect SOC storage and sequestration capacities differently from native species. For example, nonnative annual grasses that dominate in former sagebrush (Artemisia spp.) ecosystems have resulted in SOC losses (Prater et al. 2006), but when woody species replace grasses, SOC increases (Hughes et al. 2006). These differences in SOC sequestration are tied to local soil moisture regimes. All invasive species are difficult to manage because there is often more than one invader, and different species have different effects on SOC pools. Therefore, management is likely to entail identifying the species and then deciding whether action is needed to control the spread of the population (Hulme et al. 2008; Peltzer et al. 2010).

Other opportunities for increasing SOC during land management can be found with the use of soil amendments. Chipping or masticating nonmerchantable wood and using this material as a mulch (rather than burning it) can maintain surface SOC. In addition, biochar created from logging residues can be used to increase SOC on forest, rangeland, or mine sites (Page-Dumroese et al. 2017). Biochar additions to soil also have the benefit of increasing cation exchange capacity and soil moisture, further promoting SOC formation and stabilization. Biochar may have a positive effect on soil microbial populations as well (Steinbeiss et al. 2009). Soil amendments can also increase rangeland SOC. Long-term manure additions were found to increase SOC stocks by about 18 Mg ha^{-1} in California rangelands (Owen et al. 2015).

Links to Institutional Initiatives

There is growing awareness of soils as a key mitigating influence on global C cycles. Soils are emerging as a point of emphasis for maximizing the natural sinks available for C as offsets for increasing atmospheric greenhouse gas concentrations. Careful development of management plans that consider SOC goals within each of the disturbance scenarios outlined in this chapter could have a large multiplier effect on global C storage. Recognition of the impact of small SOC gains across large areas is the first step in impacting the overall global C budget.

Complementary initiatives in enhancing soil health provide a focal area for both increasing production and maintaining SOC gains or at least limiting or eliminating losses of SOC. As mentioned in the Introduction, SOC is an essential component of indices developed to indicate soil health. Widespread adoption of such indices would enhance our ability to track, maintain, and improve soil health across our nation's lands. Indeed, SOC turnover is critical to productive and healthy soils, and the new SOC paradigm presented at the beginning of this chapter can also serve as a rich framework for understanding the components of a healthy soil and valuing the maintenance of SOC. The potential for soils to store C and mitigate climate change has recently garnered significant attention with government agencies as well as the public (Barker and Polan 2015; Leslie 2017). Accordingly, several initiatives, briefly described in the following paragraphs, have been launched in an attempt to improve our appreciation and utilization of soils in the context of food security and climate change.

The "4 per 1000" Initiative (https://www.4p1000.org/) aims to increase SOC content through implementation of sustainable agricultural practices and thereby draw down atmospheric CO_2 concentrations. This program was launched by France in 2015 at the COP 21 (the 21st Conference of the Parties, held by the United Nations Framework Convention on Climate Change).

The National Soil Health Action Plan (https://soilhealthinstitute.org/) lays out strategies and best practices for safeguarding and enhancing the vitality and productivity of soils. This plan was drawn up for the United States by the

Soil Health Institute, an independent, multipartner organization. The plan was developed over 4 years with input from agronomists, government agency leaders, scientists, and nongovernmental organizations.

The Soil2026 Initiative[1] seeks to improve integration of soil survey and soil laboratory data streams while incorporating digital soil mapping. The Soil2026 campaign, launched by the USDA National Cooperative Soil Survey, will produce a complete soil inventory of the United States by 2026 along with establishing standards for digital soil mapping methods.

Key Findings

- Globally, SOC is the largest terrestrial pool of C. In the United States, most of this stock is found in forests and rangelands. It is an important provider and indicator of vital ecosystem services, such as nutrient and water cycling and C sequestration.

- Many scientific advances have been made in the past decade that have led to a better understanding of controls over SOC stabilization. Research shows that predicting vulnerability of SOC under disturbance can be aided by considering factors related to the quantity and quality of OM inputs and soil microbial accessibility and activity.

- Both natural and human-caused disturbances have degraded SOC stocks across the United States. Climate change, overgrazing, overharvesting, and catastrophic wildfire have emerged as the greatest concerns. Effects on SOC degradation are expected to be more severe when two or more of these disturbances interact with each other.

- Careful management can partially mitigate SOC degradation as well as promote rebuilding of SOC post-disturbance. In general, actions that promote retention and growth of native vegetation (fuels management, reforestation, and invasive weed control) and leave or add organic residues on-site will increase SOC stocks.

Key Information Needs

- Climate change effects on SOC remain highly uncertain. Specifically, research is needed to better understand how expected increases in precipitation and temperature variability—rather than changes in the mean—impact SOC vulnerability.

---

[1]Lindbo, D.L.; Thomas, P. 2016. Shifting paradigms – Soil Survey 2026. [Presentation and poster]. Resilience emerging from scarcity and abundance; meeting of the American Society of Agronomy–Crop Science Society of America–Soil Science Society of America; Nov. 6–9, 2016; Phoenix, AZ.

- Future research on SOC needs to measure both surface soil (0–30 cm) and soil deeper than 30 cm, preferably up to at least 1 m, including understanding the contribution of deeper roots to OM inputs to SOC.
- There are limited data on many plant, animal, insect, and pathogen invasive species, their impact on SOC, and how managers can effectively change land management to alter invasive species effects on ecosystem services.
- Estimates of changes in SOC over time are highly uncertain due to soil heterogeneity and rock content. Concentrated efforts to standardize technological and statistical solutions for dealing with uncertainty could help better constrain future SOC predictions.
- There is a lack of studies that examine how forest and rangeland management actions interact with global change phenomena (e.g., shifting plant communities, altered fire regimes) to influence SOC vulnerability.
- Most of our understanding of forest and rangeland SOC involves only a snapshot in time or very short-term (2–4 years) studies, leaving managers unable to predict longer-term impacts of decisions on SOC. Longer-term studies (longer than a rotation) of forest management effects (e.g., Adaptive Silviculture for Climate Change (Nagel et al. 2017)) are needed.
- There is a need to provide managers with the science needed to develop best management practices for enhancing SOC within the scope of the Forest Service 2012 Planning Rule (USDA FS 2012).

Literature Cited

Achat DL, Fortin M, Landmann G et al (2015a) Forest soil carbon is threatened by intensive biomass harvesting. Sci Rep 5(1):15991

Achat DL, Deleuze C, Landmann G et al (2015b) Quantifying consequences of removing harvesting residues on forest soils and tree growth—a meta-analysis. For Ecol Manag 348:124–141

Albalasmeh AA, Berli M, Shafer DS, Ghezzehei TA (2013) Degradation of moist soil aggregates by rapid temperature rise under low intensity fire. Plant Soil 362(1–2):335–344

Alban DH, Berry EC (1994) Effects of earthworm invasion on morphology, carbon, and nitrogen of a forest soil. Appl Soil Ecol 1(3):243–249

Amacher MC, O'Neill KP, Perry CH (2007) Soil vital signs: a new soil quality index (SQI) for assessing forest soil health. Research Paper RMRS-RP-65. U.S. Department of Agriculture, Forest Service, Rocky Mountain Research Station, Fort Collins, 12 p

Amiro BD, Barr AG, Black TA et al (2006) Carbon, energy and water fluxes at mature and disturbed forest sites, Saskatchewan, Canada. Agric For Meteorol 136:237–251

Araya SN, Fogel ML, Berhe AA (2017) Thermal alteration of soil organic matter properties: a systematic study to infer response of Sierra Nevada climosequence soils to forest fires. Soil 3:31–44

Barker, D.; Polan, M. 2015. A secret weapon to fight climate change: dirt. Wash Post, December 5

Batjes NH (2016) Harmonized soil property values for broad-scale modelling (WISE30sec) with estimates of global soil carbon stocks. Geoderma 269:61–68

Berryman E, Marshall JD, Rahn T et al (2013) Decreased carbon limitation of litter respiration in a mortality-affected piñon–juniper woodland. Biogeosciences 10:1625–1634

Berryman EM, Morgan P, Robichaud PR, Page-Dumroese D (2014) Post-fire erosion control mulches alter belowground processes and nitrate reductase activity of a perennial forb, heartleaf arnica (*Arnica cordifolia*). Research Note RMRS-RN-69. U.S. Department of Agriculture, Forest Service, Rocky Mountain Research Station, Fort Collins, 10 p

Binkley D (2005) How nitrogen-fixing trees change soil carbon. In: Binkley D, Menyailo O (eds) Tree species effects on soils: implications for global change. NATO Science Series IV: Earth and Environmental Sciences (NAIV, vol. 55). Springer, Dordrecht, pp 155–164

Binkley D, Kaye J, Barry M, Ryan MG (2004) First-rotation changes in soil carbon and nitrogen in a eucalyptus plantation in Hawaii. Soil Sci Soc Am J 68:1713–1719

Blagodatskaya E, Kuzyakov Y (2008) Mechanisms of real and apparent priming effects and their dependence on soil microbial biomass and community structure: critical review. Biol Fertil Soils 45(2):115–131

Blank RR, Sforza R (2007) Plant–soil relationships of the invasive annual grass *Taeniatherum caput-medusae*: a reciprocal transplant experiment. Plant Soil 298(1–2):7–19

Boerner REJ, Huang J, Hart SC (2008) Fire, thinning, and the carbon economy: effects of fire and fire surrogate treatments on estimated carbon storage and sequestration rate. For Ecol Manag 255:3081–3097

Bohlen PJ, Pelletier DM, Groffman PM et al (2004) Influence of earthworm invasion on redistribution and retention of soil carbon and nitrogen in northern temperate forests. Ecosystems 7(1):13–27

Booker K, Huntsinger L, Bartolome JW et al (2013) What can ecological science tell us about opportunities for carbon sequestration on arid rangelands in the United States? Glob Environ Chang 23(1):240–251

Boot CM, Haddix M, Paustian K, Cotrufo MF (2015) Distribution of black carbon in ponderosa pine forest floor and soils following the High Park wildfire. Biogeosciences 12(10):3029–3039

Bormann BT, Spaltenstein H, McClellan MH et al (1995) Rapid soil development after windthrow disturbance in pristine forests. J Ecol 83(5):747–757

Bradley BA, Houghton RA, Mustard JF, Hamburg SP (2006) Invasive grass reduces aboveground carbon stocks in shrublands of the western US. Glob Chang Biol 12(10):1815–1822

Brodowski S, John B, Flessa H, Amelung W (2006) Aggregate-occluded black carbon in soil. Eur J Soil Sci 57:539–546

Brouillard BM, Mikkelson KM, Bokman CM et al (2017) Extent of localized tree mortality influences soil biogeochemical response in a beetle-infested coniferous forest. Soil Biol Biochem 114:309–318

Brown J, Angerer J, Salley SW et al (2010) Improving estimates of rangeland carbon sequestration potential in the US Southwest. Rangel Ecol Manag 63(1):147–154

Brundrett MC (2009) Mycorrhizal associations and other means of nutrition of vascular plants: understanding the global diversity of host plants by resolving conflicting information and developing reliable means of diagnosis. Plant Soil 320:37–77

Burtelow AE, Bohlen PJ, Groffman PM (1998) Influence of exotic earthworm invasion on soil organic matter, microbial biomass and denitrification potential in forest soils of the northeastern United States. Appl Soil Ecol 9(1):197–202

Busse MD (1994) Downed bole-wood decomposition in lodgepole pine forests of Central Oregon. Soil Sci Soc Am J 58(1):221–227

Busse MD, Hubbert KR, Moghaddas EEY (2014) Fuel reduction practices and their effects on soil quality. General Technical Report PSW-GTR-241. U.S. Department of Agriculture, Forest Service, Pacific Southwest Research Station, Albany, 161 p

Cable JM, Ogle K, Williams DG et al (2008) Soil texture drives responses of soil respiration to precipitation pulses in the Sonoran Desert: implications for climate change. Ecosystems 11:961–979

Chaer GM, Myrold DD, Bottomley PJ (2009) A soil quality index based on the equilibrium between soil organic matter and biochemical properties of undisturbed coniferous forest soils of the Pacific Northwest. Soil Biol Biochem 41(4):822–830

Chaplot V, Dlamini P, Chivenge P (2016) Potential of grassland rehabilitation through high density-short duration grazing to sequester atmospheric carbon. Geoderma 271:10–17

Clark KL, Skowronski N, Hom J (2010) Invasive insects impact forest carbon dynamics. Glob Chang Biol 16:88–101

Clark JS, Iverson L, Woodall CW et al (2016) The impacts of increasing drought on forest dynamics, structure, and biodiversity in the United States. Glob Chang Biol 22(7):2329–2352

Clow DW, Rhoades C, Briggs J et al (2011) Responses of soil and water chemistry to mountain pine beetle induced tree mortality in Grand County, Colorado, USA. Appl Geochem J Int Assoc Geochem Cosmochem 26:S174–S178

Conant RT, Paustian K (2002) Spatial variability of soil organic carbon in grasslands: implications for detecting change at different scales. Environ Pollut 116(supplement 1):S127–S135

Conant RT, Paustian K, Elliott ET (2001) Grassland management and conversion into grassland: effects on soil carbon. Ecol Appl 11(2):343–355

Cotrufo MF, Wallenstein MD, Boot CM et al (2013) The Microbial Efficiency-Matrix Stabilization (MEMS) framework integrates plant litter decomposition with soil organic matter stabilization: do labile plant inputs form stable soil organic matter? Glob Chang Biol 19(4):988–995

Cotrufo MF, Boot CM, Kampf S et al (2016) Redistribution of pyrogenic carbon from hillslopes to stream corridors following a large montane wildfire. Glob Biogeochem Cycles 30(9):2016GB005467

Czimczik CI, Masiello CA (2007) Controls on black carbon storage in soils. Glob Biogeochem Cycles 21(3):GB3005

D'Amore DV, Oken KL, Herendeen PA et al (2015) Carbon accretion in unthinned and thinned young-growth forest stands of the Alaskan perhumid coastal temperate rainforest. Carbon Balance Manag 10(1):25

Danso SKA, Bowen GD, Sanginga N (1992) Biological nitrogen fixation in trees in agro-ecosystems. Plant Soil 141(1–2):177–196

DeBano LF, Dunn PH, Conrad CE (1977) Fire's effect on physical and chemical properties of chaparral soils. Research Paper PSW-RP-145. U.S. Department of Agriculture, Forest Service, Pacific Southwest Forest and Range Experiment Station, Berkeley, 21 p

DeLuca TH, Aplet GH (2008) Charcoal and carbon storage in forest soils of the Rocky Mountain west. Front Ecol Environ 6:18–24

Dieleman WIJ, Vicca S, Dijkstra FA et al (2012) Simple additive effects are rare: a quantitative review of plant biomass and soil process responses to combined manipulations of CO_2 and temperature. Glob Chang Biol 18(9):2681–2693

Doetterl S, Berhe AA, Nadeu E et al (2016) Erosion, deposition and soil carbon: a review of process-level controls, experimental tools and models to address C cycling in dynamic landscapes. Earth Sci Rev 154:102–122

Domke GM, Perry CH, Walters BF et al (2016) Estimating litter carbon stocks on forest land in the United States. Sci Total Environ 557–558:469–478

Domke GM, Perry CH, Walters BF et al (2017) Toward inventory-based estimates of soil organic carbon in forests of the United States. Ecol Appl 27:1223–1235

Dymond P, Scheu S, Parkinson D (1997) Density and distribution of Dendrobaena octaedra (Lumbricidae) in aspen and pine forests in the Canadian Rocky Mountains (Alberta). Soil Biol Biochem 29(3-4):265–273

Edburg SL, Hicke JA, Brooks PD et al (2012) Cascading impacts of bark beetle-caused tree mortality on coupled biogeophysical and biogeochemical processes. Front Ecol Environ 10(8):416–424

Ehrenfeld JG (2003) Effects of exotic plant invasions on soil nutrient cycling processes. Ecosystems 6(6):503–523

Ehrenfeld JG, Kourtev P, Huang W (2001) Changes in soil functions following invasions of exotic understory plants in deciduous forests. Ecol Appl 11(5):1287–1300

Esquilín AEJ, Stromberger ME, Massman WJ et al (2007) Microbial community structure and activity in a Colorado Rocky Mountain forest soil scarred by slash pile burning. Soil Biol Biochem 39:1111–1120

Federer CA (1993) The organic fraction-bulk density relationship and the expression of nutrient content in forest soils. Can J For Res 23:1026–1032

Forrester DI, Pares A, O'Hara C et al (2013) Soil organic carbon is increased in mixed-species plantations of Eucalyptus and nitrogen-fixing Acacia. Ecosystems 16(1):123–132

Frandsen W, Ryan K (1986) Soil moisture reduces belowground heat flux and soil temperatures under a burning fuel pile. Can J For Res 16:244–248

Frank DA, Groffman PM (1998) Ungulate vs. landscape control of soil C and N processes in grasslands of Yellowstone National Park. Ecology 79(7):2229–2241

Frey SD, Ollinger S, Nadelhoffer K et al (2014) Chronic nitrogen additions suppress decomposition and sequester soil carbon in temperate forests. Biogeochemistry 121(2):305–316

Fuhlendorf SD, Engle DM (2001) Restoring heterogeneity on rangelands: ecosystem management based on evolutionary grazing patterns. Bioscience 51(8):625–632

Genet H, He Y, Lyu Z et al (2018) The role of driving factors in historical and projected carbon dynamics of upland ecosystems in Alaska. Ecol Appl 28(1):5–27

Giardina CP, Litton CM, Crow SE, Asner GP (2014) Warming-related increases in soil CO_2 efflux are explained by increased belowground carbon flux. Nat Clim Chang 4:822–827

Gomez C, Rossel RAV, McBratney AB (2008) Soil organic carbon prediction by hyperspectral remote sensing and field Vis-NIR spectroscopy: an Australian case study. Geoderma 146(3-4):403–411

Gonzalez P, Asner GP, Battles JJ et al (2010) Forest carbon densities and uncertainties from Lidar, QuickBird, and field measurements in California. Remote Sens Environ 114(7):1561–1575

González-Pérez JA, González-Vila FJ, Almendros G, Knicker H (2004) The effect of fire on soil organic matter—a review. Environ Int 30(6):855–870

Grandy AS, Neff JC, Weintraub MN (2007) Carbon structure and enzyme activities in alpine and forest ecosystems. Soil Biol Biochem 39(11):2701–2711

Groffman PM, Bohlen PJ (1999) Soil and sediment biodiversity: cross-system comparisons and large-scale effects. Bioscience 49(2):139–148

Hanan EJ, Schimel JP, Dowdy K, D'Antonio CM (2016) Effects of substrate supply, pH, and char on net nitrogen mineralization and nitrification along a wildfire-structured age gradient in chaparral. Soil Biol Biochem 95:87–99

Harmon ME, Ferrell WK, Franklin JF (1990) Effects of carbon storage on conversion of old-growth forests to young forests. Science 247:699–702

Harmon ME, Sexton J, Caldwell BA, Carpenter SE (1994) Fungal sporocarp mediated losses of Ca, Fe, K, Mg, Mn, N, P, and Zn from conifer logs in the early stages of decomposition. Can J For Res 24(9):1883–1893

Harrison K, Broecker W, Bonani G (1993) A strategy for estimating the impact of CO_2 fertilization on soil carbon storage. Glob Biogeochem Cycles 7(1):69–80

Harrison RB, Footen PW, Strahm BD (2011) Deep soil horizons: contribution and importance to soil carbon pools and in assessing whole-ecosystem response to management and global change. For Sci 57(1):67–76

Harvey AE, Larsen MJ, Jurgensen MF (1976) Distribution of ectomycorrhizae in a mature Douglas-fir/larch forest soil in western Montana. For Sci 22(4):393–398

Harvey AE, Larsen MJ, Jurgensen MF (1981) Rate of woody residue incorporation into northern Rocky Mountain forest soils. Research Paper INT-RP-282. U.S. Department of Agriculture, Forest Service, Intermountain Forest and Range Experiment Station, Ogden, 5 p

Henry HA, Cleland EE, Field CB, Vitousek PM (2005) Interactive effects of elevated CO_2, N deposition and climate change on plant litter quality in a California annual grassland. Oecologia 142(3):465–473

Hickler T, Smith B, Prentice IC et al (2008) CO_2 fertilization in temperate FACE experiments not representative of boreal and tropical forests. Glob Chang Biol 14(7):1531–1542

Hobbs RJ, Huenneke LF (1992) Disturbance, diversity, and invasion: implications for conservation. Conserv Biol 6(3):324–337

Holly DC, Ervin GN, Jackson CR et al (2009) Effect of an invasive grass on ambient rates of decomposition and microbial community structure: a search for causality. Biol Invasions 11(8):1855–1868

Homann PS, Bormann BT, Boyle JR (2001) Detecting treatment differences in soil carbon and nitrogen resulting from forest manipulations. Soil Sci Soc Am J 65(2):463–469

Hughes RF, Archer SR, Asner GP et al (2006) Changes in aboveground primary production and carbon and nitrogen pools accompanying woody plant encroachment in a temperate savanna. Glob Chang Biol 12(9):1733–1747

Hulme PE, Bacher S, Kenis M et al (2008) Grasping at the routes of biological invasions: a framework for integrating pathways into policy. J Appl Ecol 45(2):403–414

Humphreys ER, Black TA, Morgenstern K et al (2006) Carbon dioxide fluxes in coastal Douglas-fir stands at different stages of development after clearcut harvesting. Agric For Meteorol 140:6–22

Hungate BA, Holland EA, Jackson RB et al (1997) The fate of carbon in grasslands under carbon dioxide enrichment. Nature 388(6642):576–579

Intergovernmental Panel on Climate Change [IPCC] (2014) Climate Change 2014: synthesis report. Contribution of Working Groups I, II and III to the Fifth Assessment Report of the Intergovernmental Panel on Climate Change (Core Writing Team, Pachauri RK, Meyer LA (eds)). Geneva, Switzerland, 151 p

International Soil Carbon Network (n.d.) Homepage https://iscn.fluxdata.org. Accessed March 27, 2019

Jackson RB, Lajtha K, Crow SE et al (2017) The ecology of soil carbon: pools, vulnerabilities, and biotic and abiotic controls. Annu Rev Ecol Evol Syst 48(1):419–445

James J, Harrison R (2016) The effect of harvest on forest soil carbon: a meta-analysis. Forests 7(12):308

Jandl R, Lindner M, Vesterdal L et al (2007) How strongly can forest management influence soil carbon sequestration? Geoderma 137(3):253–268

Jang W, Keyes CR, Page-Dumroese DS (2015) Long-term effects on distribution of forest biomass following different harvesting levels in the northern Rocky Mountains. For Ecol Manag 358:281–290

Jang W, Page-Dumroese DS, Keyes CR (2016) Long-term changes from forest harvesting and residue management in the northern Rocky Mountains. Soil Sci Soc Am J 80(3):727–741

Janowiak MK, Webster CR (2010) Promoting ecological sustainability in woody biomass harvesting. J For 108(1):16–23

Janssens IA, Dieleman W, Luyssaert S et al (2010) Reduction of forest soil respiration in response to nitrogen deposition. Nat Geosci 3(5):315–322

Jenny H (1941) Factors of soil formation: a system of quantitative pedology. McGraw-Hill, New York, 281 p

Jo I, Potter KM, Domke GM, Fei S (2018) Dominant forest tree mycorrhizal type mediates understory plant invasions. Ecol Lett 21:217–224

Johansson MB (1994) The influence of soil scarification on the turnover rate of slash needles and nutrient release. Scand J For Res 9(1-4):170–179

Johnson DW, Curtis PS (2001) Effects of forest management on soil C and N storage: meta analysis. For Ecol Manag 140(2-3):227–238

Jurgensen MF, Larsen MJ, Graham RT, Harvey AE (1987) Nitrogen fixation in woody residue of northern Rocky Mountain conifer forests. Can J For Res 17(10):1283–1288

Jurgensen MF, Page-Dumroese DS, Brown RE et al (2017) Estimating carbon and nitrogen pools in a forest soil: influence of soil bulk density methods and rock content. Soil Sci Soc Am J 81:1689–1696

Kanakidou M, Myriokefalitakis S, Daskalakis N et al (2016) Past, present, and future atmospheric nitrogen deposition. J Atmos Sci 73:2039–2047

Kaňa J, Tahovská K, Kopáček J, Šantrůčková H (2015) Excess of organic carbon in mountain spruce forest soils after bark beetle outbreak altered microbial N transformations and mitigated N-saturation. PLoS One 10(7):e0134165

Kelleher BP, Simpson MJ, Simpson AJ (2006) Assessing the fate and transformation of plant residues in the terrestrial environment using HR-MAS NMR spectroscopy. Geochim Cosmochim Acta 70(16):4080–4094

Kleber M, Johnson MG (2010) Advances in understanding the molecular structure of soil organic matter: implications for interactions in the environment. Adv Agron 106:77–142

Klumpp K, Fontaine S, Attard E et al (2009) Grazing triggers soil carbon loss by altering plant roots and their control on soil microbial community. J Ecol 97(5):876–885

Knelman JE, Graham EB, Ferrenberg S et al (2017) Rapid shifts in soil nutrients and decomposition enzyme activity in early succession following forest fire. Forests 8(9):347

Kolka R, Sturtevant B, Townsend P et al (2014) Post-fire comparisons of forest floor and soil carbon, nitrogen, and mercury pools with fire severity indices. Soil Sci Soc Am J 78:S58–S65

Korb JE, Johnson NC, Covington WW (2004) Slash pile burning effects on soil biotic and chemical properties and plant establishment: recommendations for amelioration. Restor Ecol 12(1):52–62

Krebs J, Pontius J, Schaberg PG (2017) Modeling the impacts of hemlock woolly adelgid infestation and presalvage harvesting on carbon stocks in northern hemlock forests. Can J For Res 47:727–734

Kretchun AC, Scheller RM, Lucash MS et al (2014) Predicted effects of gypsy moth defoliation and climate change on forest carbon dynamics in the New Jersey Pine Barrens. PLoS One 9(8):e102531

Krofcheck DJ, Hurteau MD, Scheller RM, Loudermilk EL (2017) Prioritizing forest fuels treatments based on the probability of high-severity fire restores adaptive capacity in Sierran forests. Glob Chang Biol 24(2):729–737

Kurth VJ, Hart SC, Ross CS et al (2014) Stand-replacing wildfires increase nitrification for decades in southwestern ponderosa pine forests. Oecologia 175:395–407

Ladwig LM, Sinsabaugh RL, Collins SL, Thomey ML (2015) Soil enzyme responses to varying rainfall regimes in Chihuahuan Desert soils. Ecosphere 6(3):1–10

Langmaid KK (1964) Some effects of earthworm invasion in virgin podzols. Can J Soil Sci 44(1):34–37

Laycock WA (1991) Stable states and thresholds of range condition on North American rangelands: a viewpoint. J Range Manag 44(5):427–433

Lehmann J, Kleber M (2015) The contentious nature of soil organic matter. Nature 528(7580):60–68

Leslie J (2017) Soil power! The dirty way to a green planet The New York Times, Opinion. December 2. https://www.nytimes.com/2017/12/02/opinion/sunday/soil-power-the-dirty-way-to-a-green-planet.html. Accessed June 21, 2018)

Liang B, Lehmann J, Solomon D et al (2006) Black carbon increases cation exchange capacity in soils. Soil Sci Soc Am J 70:1719–1730

Liu S, Liu J, Young CJ et al (2012) Chapter 5: Baseline carbon storage, carbon sequestration, and greenhouse-gas fluxes in terrestrial ecosystems of the western United States. In: Zhu Z, Reed BC (eds) Baseline and projected future carbon storage and greenhouse-gas fluxes in ecosystems of the western United States. Professional paper 1797. U.S. Department of the Interior, Geological Survey, Reston. Available at https://pubs.usgs.gov/pp/1797. Accessed March 21, 2019

Liu S, Wei Y, Post WM et al (2013) The Unified North American Soil Map and its implication on the soil organic carbon stock in North America. Biogeosciences 10:2915

Liu S, Liu J, Wu Y et al (2014) Chapter 7: Baseline and projected future carbon storage, carbon sequestration, and greenhouse-gas fluxes in terrestrial ecosystems of the eastern United States. In: Zhu Z, Reed BC (eds) Baseline and projected future carbon storage and greenhouse gas fluxes in ecosystems of the eastern United States, Professional paper 1804. U.S. Department of the Interior, Geological Survey, Reston, pp 115–156. Available at https://pubs.usgs.gov/pp/1804. Accessed March 21, 201

Liu S, Zhang Y, Zong Y et al (2016) Response of soil carbon dioxide fluxes, soil organic carbon and microbial biomass carbon to biochar amendment: a meta-analysis. GCB Bioenergy 8(2):392–406

Lu M, Zhou X, Yang Q et al (2013) Responses of ecosystem carbon cycle to experimental warming: a meta-analysis. Ecology 94(3):726–738

Lybrand RA, Heckman K, Rasmussen C (2017) Soil organic carbon partitioning and $\Delta^{14}C$ variation in desert and conifer ecosystems of southern Arizona. Biogeochemistry 134(3):261–277

Lynch LM, Machmuller MB, Cotrufo MF et al (2018) Tracking the fate of fresh carbon in the Arctic tundra: will shrub expansion alter responses of soil organic matter to warming? Soil Biol Biochem 120:134–144

Mack MC, D'Antonio CM, Ley RE (2001) Alteration of ecosystem nitrogen dynamics by exotic plants: a case study of C4 grasses in Hawaii. Ecol Appl 11(5):1323–1335

MacKenzie MD, DeLuca TH, Sala A (2004) Forest structure and organic horizon analysis along a fire chronosequence in the low elevation forests of western Montana. For Ecol Manag 203:331–343

Markewitz D (2006) Fossil fuel carbon emissions from silviculture: impacts on net carbon sequestration in forests. For Ecol Manag 236:153–161

Marschner B, Brodowski S, Dreves A et al (2008) How relevant is recalcitrance for the stabilization of organic matter in soils? J Plant Nutr Soil Sci 171(1):91–110

Maser C, Cline SP, Cromack K Jr et al (1988) What we know about large trees that fall to the forest floor. In: Maser C, Tarrant RF, Trappe JM, Franklin JF (eds) From the forest to the sea: a story of fallen trees. General Technical Report PNW-GTR-229. U.S. Department of Agriculture, Forest Service, Pacific Northwest Research Station; Portland, pp 25–46

Massman WJ, Frank JM (2005) Effect of a controlled burn on the thermophysical properties of a dry soil using a new model of soil heat flow and a new high temperature heat flux sensor. Int J Wildland Fire 13(4):427–442

Matthes JH, Lang AK, Jevon FV, Russell SJ (2018) Tree stress and mortality from emerald ash borer does not systematically alter short-term soil carbon flux in a mixed northeastern U.S. forest. Forests 9:37

Maurer GE, Chan AM, Trahan NA et al (2016) Carbon isotopic composition of forest soil respiration in the decade following bark beetle and stem girdling disturbances in the Rocky Mountains. Plant Cell Environ 39:1513–1523

McSherry ME, Ritchie ME (2013) Effects of grazing on grassland soil carbon: a global review. Glob Chang Biol 19(5):1347–1357

Millennium Ecosystem Assessment (2003) Ecosystem and human well-being: a framework for assessment. Island Press, Washington, DC

Minasny B, Malone BP, McBratney AB et al (2017) Soil carbon 4 per mille. Geoderma 292:59–86

Morehouse K, Johns T, Kaye J, Kaye M (2008) Carbon and nitrogen cycling immediately following bark beetle outbreaks in southwestern ponderosa pine forests. For Ecol Manag 255:2698–2708

Nagel LM, Palik BJ, Battaglia MA et al (2017) Adaptive silviculture for climate change: a national experiment in manager-scientist partnerships to apply an adaptation framework. J For 115(3):167–178

National Forest Management Act of 1976; Act of October 22, 1976; 16 U.S.C. 1600

Nave LE, Vance ED, Swanston CW, Curtis PS (2010) Harvest impacts on soil carbon storage in temperate forests. For Ecol Manag 259:857–866

Nave LE, Domke GM, Hofmeister KL et al (2018) Reforestation can sequester two petagrams of carbon in U.S. topsoils in a century. Proc Natl Acad Sci 115(11):2776–2781

Neary DG, Klopatek CC, DeBano LF, Ffolliott PF (1999) Fire effects on belowground sustainability: a review and synthesis. For Ecol Manag 122:51–71

Neary DG, Trettin CC, Page-Dumroese D (2010) Soil quality monitoring: examples of existing protocols. In: Page-Dumroese D, Neary D, Trettin C (eds) Scientific background for soil monitoring on national forests and rangelands: workshop proceedings. Proceedings RMRS-P-59. U.S. Department of Agriculture, Forest Service, Rocky Mountain Research Station, Fort Collins, pp 61–77

Neff JC, Reynolds RL, Belnap J, Lamothe P (2005) Multi-decadal impacts of grazing on soil physical and biogeochemical properties in southeast Utah. Ecol Appl 15(1):87–95

Neff JC, Barger NN, Baisden WT et al (2009) Soil carbon storage responses to expanding pinyon-juniper populations in southern Utah. Ecol Appl 19:1405–1416

Norton JB, Monaco TA, Norton U (2007) Mediterranean annual grasses in western North America: kids in a candy store. Plant Soil 298(1–2):1–5

Noss RF (2001) Beyond Kyoto: forest management in a time of rapid climate change. Conserv Biol 15(3):578–590

Nuzzo VA, Maerz JC, Blossey B (2009) Earthworm invasion as the driving force behind plant invasion and community change in northeastern North American forests. Conserv Biol 23(4):966–974

O'Neill KP, Amacher MC, Perry CH (2005) Soils as an indicator of forest health: a guide to the collection, analysis, and interpretation of soil indicator data in the Forest Inventory and Analysis program. General Technical Report NC-GTR-258. U.S. Department of Agriculture, Forest Service, North Central Research Station, St. Paul, 53 p

Oldfield EE, Wood SA, Bradford MA (2017) Direct effects of soil organic matter on productivity mirror those observed with organic amendments. Plant Soil 423(1–2):363–373

Ouimet R, Tremblay S, Périé C, Prégent G (2007) Ecosystem carbon accumulation following fallow farmland afforestation with red pine in southern Québec. Can J For Res 37(6):1118–1133

Owen JJ, Parton WJ, Silver WL (2015) Long-term impacts of manure amendments on carbon and greenhouse gas dynamics of range-lands. Glob Chang Biol 21(12):4533–4547

Page-Dumroese DS, Jurgensen MF (2006) Soil carbon and nitrogen pools in mid- to late-successional forest stands of the northwestern United States: potential impact of fire. Can J For Res 36:2270–2284

Page-Dumroese DS, Brown RE, Jurgensen MF, Mroz GD (1999) Comparison of methods for determining bulk densities of rocky forest soils. Soil Sci Soc Am J 63(2):379–383

Page-Dumroese DS, Coleman MD, Thomas SC (2017) Chapter 15: Opportunities and uses of biochar on forest sites in North America. In: Bruckman V, Varol A, Uzun BB, Liu J (eds) Biochar: a regional supply chain approach in view of mitigating climate change. Cambridge University Press, Cambridge, UK, pp 315–336

Page-Dumroese DS, Ott MR, Strawn DG, Tirocke JM (2018) Using organic amendments to restore soil physical and chemical properties of a mine site in northeastern Oregon, USA. Appl Eng Agric 34(1):43–55

Pan Y, Birdsey RA, Fang J et al (2011) A large and persistent carbon sink in the world's forests. Science 333(6045):988–993

Pellegrini AFA, Ahlström A, Hobbie SE et al (2017) Fire frequency drives decadal changes in soil carbon and nitrogen and ecosystem productivity. Nature 553:194–198

Peltzer DA, Allen RB, Lovett GM et al (2010) Effects of biological invasions on forest carbon sequestration. Glob Chang Biol 16(2):732–746

Phillips RP, Meier IC, Bernhardt ES et al (2012) Roots and fungi accelerate carbon and nitrogen cycling in forests exposed to elevated CO_2. Ecol Lett 15(9):1042–1049

Piene H, Van Cleve K (1978) Weight loss of litter and cellulose bags in a thinned white spruce forest in interior Alaska. Can J For Res 8(1):42–46

Piñeiro G, Paruelo JM, Oesterheld M, Jobbágy EG (2010) Pathways of grazing effects on soil organic carbon and nitrogen. Rangel Ecol Manag 63(1):109–119

Post WM, Kwon KC (2000) Soil carbon sequestration and land-use change: processes and potential. Glob Chang Biol 6(3):317–327

Powers RF, Scott DA, Sanchez FG (2005) The North American long-term soil productivity experiment: findings from the first decade of research. For Ecol Manag 220(1):31–50

Prater MR, Obrist D, Arnone JA III, DeLucia EH (2006) Net carbon exchange and evapotranspiration in postfire and intact sagebrush communities in the Great Basin. Oecologia 146(4):595–607

Prieto-Fernández A, Acea MJ, Carballas T (1998) Soil microbial and extractable C and N after wildfire. Biol Fertil Soils 27(2):132–142

Raison RJ (1979) Modification of the soil environment by vegetation fires, with particular reference to nitrogen transformations: a review. Plant Soil 51:73–108

Reddy AD, Hawbaker TJ, Wurster F et al (2015) Quantifying soil carbon loss and uncertainty from a peatland wildfire using multi-temporal LiDAR. Remote Sens Environ 170:306–316

Reeder JD, Schuman GE (2002) Influence of livestock grazing on C sequestration in semi-arid mixed-grass and short-grass rangelands. Environ Pollut 116(3):457–463

Reisser M, Purves RS, Schmidt MWI, Abiven S (2016) Pyrogenic carbon in soils: a literature-based inventory and a global estimation of its content in soil organic carbon and stocks. Front Earth Sci 4:80

Rhoades CC, Hubbard RM, Elder K (2017) A decade of streamwater nitrogen and forest dynamics after a mountain pine beetle outbreak at the Fraser experimental Forest, Colorado. Ecosystems 20(2):380–392

Robichaud PR, Lewis SA, Wagenbrenner JW et al (2013) Post-fire mulching for runoff and erosion mitigation: part I: effectiveness at reducing hillslope erosion rates. Catena 105:75–92

Russell WH, McBride JR (2003) Landscape scale vegetation-type conversion and fire hazard in the San Francisco bay area open spaces. Landsc Urban Plan 64(4):201–208

Sackett SS, Haase SM (1992) Measuring soil and tree temperatures during prescribed fires with thermocouple probes. General Technical Report PSW-GTR-131. U.S. Department of Agriculture, Forest Service, Pacific Southwest Research Station, Berkeley, 15 p

Sánchez Meador A, Springer JD, Huffman DW et al (2017) Soil functional responses to ecological restoration treatments in frequent-fire forests of the western United States: a systematic review. Restor Ecol 25(4):497–508

Sanchez FG, Tiarks AE, Kranabetter JM et al (2006) Effects of organic matter removal and soil compaction on fifth-year mineral soil carbon and nitrogen contents for sites across the United States and Canada. Can J For Res 36(3):565–576

Sanford RL Jr, Parton WJ, Ojima DS, Lodge DJ (1991) Hurricane effects on SOC dynamics and forest production in the Luquillo Experimental Forests, Puerto Rico: results of simulation modeling. Biotropica 24:364–372

Schaedel MS, Larson AJ, Affleck DLR et al (2017) Early forest thinning changes aboveground carbon distribution among pools, but not total amount. For Ecol Manag 389:187–198

Scharlemann JPW, Tanner EVJ, Hiederer R, Kapos V (2014) Global soil carbon: understanding and managing the largest terrestrial carbon pool. Carbon Manage 5:81–91

Scheu S, Parkinson D (1994) Effects of earthworms on nutrient dynamics, carbon turnover and microorganisms in soils from cool temperate forests of the Canadian Rocky Mountains—laboratory studies. Appl Soil Ecol 1(2):113–125

Schlesinger WH, Lichter J (2001) Limited carbon storage in soil and litter of experimental forest plots under increased atmospheric CO_2. Nature 411(6836):466–469

Schmidt MWI, Torn MS, Abiven S et al (2011) Persistence of soil organic matter as an ecosystem property. Nature 478:49–56

Schnitzer M, Kodama H (1977) Reactions of minerals with soil humic substances. In: Dixon JB, Weed SB (eds) Minerals and their roles in the soil environment. Soil Science Society of America, Madison, pp 741–770

Schoenholtz SH, Miegroet HV, Burger JA (2000) A review of chemical and physical properties as indicators of forest soil quality: challenges and opportunities. For Ecol Manag 138(1):335–356

Schreiner EG, Krueger KA, Houston DB, Happe PJ (1996) Understory patch dynamics and ungulate herbivory in old-growth forests of Olympic National Park, Washington. Can J For Res 26(2):255–265

Schuman GE, Janzen HH, Herrick JE (2002) Soil carbon dynamics and potential carbon sequestration by rangelands. Environ Pollut 116(3):391–396

Schuur EAG, McGuire AD, Schädel C et al (2015) Climate change and the permafrost carbon feedback. Nature 520(7546):171–179

Scott NA, Saggar S, McIntosh PD (2001) Biogeochemical impact of *Hieracium* invasion in New Zealand's grazed tussock grasslands: sustainability implications. Ecol Appl 11(5):1311–1322

Sierra CA, Trumbore SE, Davidson EA et al (2015) Sensitivity of decomposition rates of soil organic matter with respect to simultaneous changes in temperature and moisture. J Adv Model Earth Syst 7:335–356

Silver WL, Ryals R, Eviner V (2010) Soil carbon pools in California's annual grassland ecosystems. Rangel Ecol Manag 63(1):128–136

Spears JDH, Holub SM, Harmon ME, Lajtha K (2003) The influence of decomposing logs on soil biology and nutrient cycling in an old-growth mixed coniferous forest in Oregon, USA. Can J For Res 33(11):2193–2201

Steinbeiss S, Gleixner G, Antonietti M (2009) Effect of biochar amendment on soil carbon balance and soil microbial activity. Soil Biol Biochem 41(6):1301–1310

Sulman BN, Phillips RP, Oishi AC et al (2014) Microbe-driven turnover offsets mineral-mediated storage of soil carbon under elevated CO_2. Nat Clim Chang 4(12):1099–1102

Sundquist ET, Ackerman KV, Bliss NB et al (2009) Rapid assessment of U.S. forest and soil organic carbon storage and forest biomass carbon-sequestration capacity. Open-File Report 2009–1283. U.S. Department of the Interior, Geological Survey, Reston, 15 p. https://pubs.er.usgs.gov/publication/ofr20091283

Sutton R, Sposito G (2005) Molecular structure in soil humic substances: the new view. Environ Sci Technol 39(23):9009–9015

Svejcar T, Angell R, Bradford JA et al (2008) Carbon fluxes on North American rangelands. Rangel Ecol Manag 61:465–474

Tate RL III (1987) Soil organic matter: biological and ecological effects. Wiley Interscience, New York, 291 p

Thomey ML, Collins SL, Vargas R et al (2011) Effect of precipitation variability on net primary production and soil respiration in a Chihuahuan Desert grassland. Glob Chang Biol 17:1505–1515

Thornton PE, Law BE, Gholz HL et al (2002) Modeling and measuring the effects of disturbance history and climate on carbon and water budgets in evergreen needleleaf forests. Agric For Meteorol 113:185–222

Trahan NA, Dynes EL, Pugh E et al (2015) Changes in soil biogeochemistry following disturbance by girdling and mountain pine beetles in subalpine forests. Oecologia 177(4):981–995

U.S. Global Change Research Program [USGCRP] (2018) Second State of the Carbon Cycle Report (SOCCR2): a sustained assessment report (Cavallaro N, Shrestha G, Birdsey R et al (eds)). Washington, DC. 878 p

USDA Forest Service [USDA FS] (2012) National Forest System land management planning rule. 36 CFR part 219 RIN0596-AD2. Fed Regist Rules Regul 77(68):21162–21276

Van Miegroet H, Jandl R (2007) Are nitrogen-fertilized forest soils sinks or sources of carbon? Environ Monit Assess 128(1–3):121–131

Vesterdal L, Dalsgaard M, Felby C et al (1995) Effects of thinning and soil properties on accumulation of carbon, nitrogen and phosphorus in the forest floor of Norway spruce stands. For Ecol Manag 77(1–3):1–10

Wagai R, Mayer LM, Kitayama K, Knicker H (2008) Climate and parent material controls on organic matter storage in surface soils: a three-pool, density-separation approach. Geoderma 147(1–2):23–33

Wagai R, Mayer LM, Kitayama K (2009) Nature of the "occluded" low-density fraction in soil organic matter studies: a critical review. Soil Sci Plant Nutr 55(1):13–25

Warnock DD, Litvak ME, Morillas L, Sinsabaugh RL (2016) Drought-induced piñon mortality alters the seasonal dynamics of microbial activity in piñon–juniper woodland. Soil Biol Biochem 92:91–101

Wolfe BE, Klironomos JN (2005) Breaking new ground: soil communities and exotic plant invasion. Bioscience 55(6):477–487

Yanai RD, Arthur MA, Siccama TG, Federer CA (2000) Challenges of measuring forest floor organic matter dynamics. For Ecol Manag 138(90):273–283

Zhang Y, Wolfe SA, Morse PD et al (2015a) Spatiotemporal impacts of wildfire and climate warming on permafrost across a subarctic region, Canada: impacts of fire on permafrost. J Geophys Res Earth 120(11):2338–2356

Zhang B, Zhou X, Zhou L, Ju R (2015b) A global synthesis of belowground carbon responses to biotic disturbance: a meta-analysis. Glob Ecol Biogeogr 24(2):126–138

Soils and Water

3

Mary Beth Adams, Vince A. Archer, Scott Bailey,
Kevin McGuire, Chelcy F. Miniat, Daniel G. Neary,
Toby O'Geen, Peter R. Robichaud, and Michael Strobel

Introduction

Wildlands play a special role in providing a reliable supply
of high-quality water (Dissmeyer 2000), and, in particular,
we rely on forest and rangeland soils to ensure clean, abundant water. Soils retain water and make it available to support vegetation, facilitate drainage to soil and ultimately to
surface waters (streams and lakes), and recharge aquifers and
groundwater. Soils also help regulate water quality by filtering out pollutants and regulating sediments. In this chapter,
we explore the links between soil and water and evaluate
some potential threats to the ability of forest and rangeland
soils to provide clean, abundant water. We also identify
information gaps and research needs.

Forest and rangeland soils provide important ecosystem
services which can be difficult to quantify or describe in
terms of their economic value. However, there are examples
of the value of sound soil management for protecting water
quality. Almost two-thirds of drinking water in the United
States comes from forested watersheds and their soils, and
many towns and cities depend on water supplies from
national forest watersheds (Dissmeyer 2000; Gartner et al.
2014; NRC 2008).

One example of the value of forest, grassland, and other
wildland soils and the water they produce comes from
New York, NY. The water management bureau chose to
ensure drinking water quality for the millions of New York
City residents by protecting the upper Catskills watershed.
The bureau plans to maintain the watershed in forest land
and purchase conservation easements rather than build a filtration plant at an estimated cost of $10 billion, plus $100
million per year in operating costs (Hu 2018). Protecting forested watersheds is a sound economic choice for many water
utilities (Gartner et al. 2014).

The economic value of soil can also be described in terms
of the impacts of fires and postburn erosion. More than
765,000 m^3 of sediment entered Denver, CO's Strontia
Springs Reservoir following the 2002 Hayman Fire, which
burned 56,000 ha in Colorado (Robichaud et al. 2003).
Denver Water spent $27 million removing debris and sedi-
ment from the reservoir. Similarly, the Los Angeles (CA)
County Public Works estimated it would spend $190 million
dredging four reservoirs impacted by sediment from the
2009 Station Fire (Bland 2017). Keeping soil in place pro-
tects water quality and saves money.

M. B. Adams (✉)
U.S. Department of Agriculture, Forest Service, Northern Research
Station, Ecology and Management of Invasive Species and Forest
Ecosystems, Morgantown, WV, USA
e-mail: Mary.b.adams@usda.gov

V. A. Archer
U.S. Department of Agriculture, Forest Service, Northern Region,
Missoula, MT, USA

S. Bailey
Northern Research Station, Center for Research on Ecosystem
Change, U.S. Department of Agriculture, Forest Service,
North Woodstock, NH, USA

K. McGuire
Virginia Water Resources Research Center and Department of
Forest Resources & Environmental Conservation, Virginia Tech,
Blacksburg, VA, USA

C. F. Miniat
U.S. Department of Agriculture, Forest Service, Southern Research
Station, Otto, NC, USA

D. G. Neary
U.S. Department of Agriculture, Forest Service, Rocky Mountain
Research Station, Flagstaff, AZ, USA

T. O'Geen
Department of Land, Air and Water Resources, University of
California-Davis, Davis, CA, USA

P. R. Robichaud
U.S. Department of Agriculture, Forest Service, Rocky Mountain
Research Station, Moscow, ID, USA

M. Strobel
U.S. Department of Agriculture, Natural Resources Conservation
Service, National Water and Climate Center, Portland, OR, USA

© The Author(s) 2020
R. V. Pouyat et al. (eds.), *Forest and Rangeland Soils of the United States Under Changing Conditions*,
https://doi.org/10.1007/978-3-030-45216-2_3

Soils and the Water Cycle

Forest and rangeland soils regulate many important processes within the water cycle (Fig. 3.1). Not only does the soil strongly affect the vegetation, which intercepts, evaporates, and absorbs precipitation and transpires water back into the atmosphere, but the soil also affects surface and subsurface movement of water through infiltration and percolation. All of these processes can influence both the quantity and quality of water from forests and rangelands. Soil also serves as an essential water reservoir and can affect not only vegetation cover and type but also local and regional climate (Bonan 2008). Perturbations that affect soil through the important processes of infiltration, evaporation, surface runoff, and percolation are likely to affect watershed outputs.

To describe the amount of water coming from a watershed, processes and pools are captured in the water balance equation:

$$Q = P + ET + \Delta S,$$

where Q is the runoff or water yield from a watershed, P is the precipitation amount, ET is evapotranspiration, and ΔS is the change in soil storage. The water balance describes the water cycle for a watershed and is usually simplified to an annual basis.

The USDA Forest Service has conducted much long-term watershed research and has evaluated the water cycle in many forest ecosystems, by using paired watersheds and the water balance approach (Hornbeck et al. 1993; Lisle et al. 2010; Neary et al. 2012a; Swank and Crossley 1988; Verry 1997). In forests, streamflow generally increases with amount of precipitation over a year, although the timing and form of precipitation (rain vs. snow) affects that relationship. Interception of rain or snow by the canopy ranges from 2% to 13% in eastern hardwood catchments (Coweeta Hydrologic Laboratory in North Carolina) to 25% in second-growth hardwood forests (Caspar Creek Experimental Watershed in Northern California) to around 30% of precipitation (snow) lost to sublimated interception (Fraser Experimental Forest in Colorado) (Lisle et al. 2010). Water that reaches the ground ultimately infiltrates into the soil or runs off. Water that infiltrates is available for plant uptake and transpiration, and soil and groundwater recharge. Transpiration ranges from 25% of annual precipitation (Caspar Creek Experimental Watershed in California) to 51% (Fernow Experimental Forest in West Virginia) (Lisle et al. 2010). Water draining downward through soil or fractured bedrock ends up in streamflow or groundwater. Streamflow averages around 37–50% of annual precipitation at most long-term instrumented watersheds (Lisle et al. 2010). Therefore, by using the water balance equation, the estimated annual storage within forest watersheds, including soil storage, is about 25% of annual precipitation. Soil storage can vary significantly depending on season of the year, precipitation, type of vegetation, and soil depth.

The water balance of rangelands is perhaps not as well quantified as for forests. Streamflow that drains from rangeland can be difficult to quantify as these areas experience

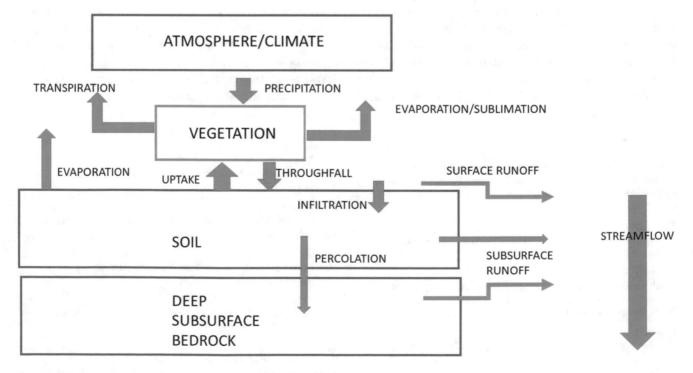

Fig. 3.1 Interactions among soil and water, as quantified in the water balance equation

wide swings in annual precipitation with seasonal pulses of rainfall from winter and summer monsoons. Furthermore, streamwater flowing from rangelands typically has high proportions of groundwater and may emanate from small parts of the watershed. However, evapotranspiration can account for 80–95% of water loss in semiarid rangelands, due to the evaporative draw from high temperatures (Renard 1970). At one research site at the Edwards Plateau in Texas, the annual interception losses ranged from 11% to 18% of precipitation for herbaceous vegetation to as high as 80% (45–80%) for woody vegetation (Wu et al. 2001), although these differences resulted in only subtle changes to overall evapotranspiration. However, soil water status varied with annual precipitation. At this Texas site, grasslands with deep soils sustained higher levels of evapotranspiration than nearby shallow soils with woody vegetation only during drought years, whereas differences were minimal during wet years (Heilman et al. 2014).

In these high-energy ecosystems, water status may relate also to the heterogeneity of the site, that is, the mixture of vegetation life-forms. As Breshears (2006) explained, the network of woody species patches and openings can have patch-scale differences in the amount of water available to plants. Shaded canopies can reduce evaporative losses while also providing organic matter for increasing soil water holding capacity. In degraded rangelands, the effects of that soil heterogeneity can be stark. For example, surface soil in degraded rangelands with honey mesquite (*Prosopis glandulosa*) had much higher water holding capacity and percolation rates under canopies compared to intercanopy spaces subject to wind abrasion and water erosion (Ravi et al. 2010).

Thus, forest and rangeland soils can interact substantially with the water balance of a watershed, mainly through the processes of infiltration, percolation, and soil storage and through relationships with vegetation amount and type. In forests, an intact forest floor (also known as duff, the O horizon, or organic horizon of the soil) is the most important factor affecting infiltration (Aust and Blinn 2004; Kochenderfer et al. 1997). When the forest floor is damaged or removed, infiltration decreases and surface runoff, which is rare in most forests, increases, with concomitant increases in erosion and transport of sediment. Changes in infiltration at the watershed scale will affect other parts of the water cycle, particularly surface runoff (overland flow).

Modeling Soils and the Water Cycle

Models have been developed that explain the basic hydrology of forests (e.g., Liang et al. 1994; Maneta and Silverman 2013; Tague and Band 2004), but they may not always incorporate soil processes well. These models may use simplified representations of soil properties and may not account for lateral redistribution of water through hydrologic flow paths. Furthermore, few models include the dynamics of soil-vegetation interactions (Maneta and Silverman 2013). In particular, issues of scale hinder our understanding of the hydrologic system (Blöschl and Sivapalan 1995). Questions surrounding the sustainability of water (supply and quality) need to be addressed at landscape scales, yet models often are limited by computational demands, data limitations, and the ability to adequately represent processes at that scale (Wood et al. 2002). Recent advances in data science, computing infrastructure, and data availability, as well as new monitoring tools, are providing the opportunity to build a new hydrologic modeling framework. With careful calibration, ecohydrology models can reproduce the dynamics of soil moisture and stream discharge and their interactions with vegetation (Kuppel et al. 2018; Nijzink et al. 2016). Results from these models can contribute to an integrated understanding of water dynamics and biogeochemical cycles.

Although most hydrologic models treat climate and topography as the primary drivers in watershed hydrology, an appreciable body of research has shown that redistribution of water across the landscape is strongly influenced by soil layering and relationship with bedrock (Bathke and Cassel 1991; Kienzler and Naef 2008; McDaniel and Falen 1994; McDaniel et al. 2001; Swarowsky et al. 2011, 2012; Tromp-Van Meerveld and McDonnell 2006a, b). The topography of the soil-bedrock interface is more important than surface topography in describing subsurface flow (Freer et al. 2002). In the same way, McNamara and others (2005) showed that subsurface lateral flow becomes an important hydrologic flowpath when hydrologic connectivity is established. The connectivity of saturated conditions (perched water tables), as controlled by subsurface claypans, was shown to influence streamflow characteristics in headwater catchments (Buttle and McDonald 2002; Detty and McGuire 2010; Newman et al. 1998; O'Geen et al. 2010; Swarowsky et al. 2011, 2012). Though the spatial characteristics of soils influence hydrology, documenting soil variability remains a challenge in terms of both adequately describing it and parameterizing models with these data.

Soil surveys are the standard tool for assessing soil-landscape relationships. However, the scale at which soil surveys are produced in rangelands and forests and the architecture of the data are not always applicable to hydrologic models or fully understood by hydrologic modelers (Gatzke et al. 2011; Terribile et al. 2011). The high degree of variability associated with soil prohibits the mapping at a 1:1 scale; thus, map units are used to depict patterns of soil distribution and hydrologic properties. Ultimately, most landscape- and watershed-scale decisions require soil property data at finer scales than what currently is available. Modelers have noted the need for basic statistical measures (e.g., mean, variance, confidence intervals) on reported soil properties as expressed in soil surveys and the Soil Survey Geographic (SSURGO) database (Brown and Huddleston 1991; Soil Survey Staff 2017), along with some measure of spatial variability (Brubaker and Hallmark 1991).

As with soil surveys, issues of scale pose challenges for hydrologic models (Blöschl and Sivapalan 1995). Process-based models that require a high degree of parameterization can be overwhelmed by data and processing time, limiting their applicability to the hillslope or small catchment scale. In contrast, large-scale (e.g., regional or continental) simulations performed with coarse-resolution data can reproduce discharge from very large basins with reasonable accuracy but are typically not able to resolve local hillslope-scale dynamics (Wenger et al. 2010). Watershed-scale models must aggregate and simplify soil-landscape relationships. Therefore, broad-scale hydrologic models minimize the complexity of soil and rely on primary drivers of the water budget such as evapotranspiration and precipitation. Important characteristics of watersheds such as soil storage, processes that give rise to lateral flow redistribution, and deep percolation are left poorly parameterized, if included at all. As a result of these challenges, no model yet exists to document soil water dynamics at watershed scales.

Threats to the Important Soil Function of Providing Clean, Abundant Water

Forest Harvesting

Forest harvesting effects on the water cycle have been extensively studied on experimental watersheds, many of them mountainous headwater catchments (Neary et al. 2012a). Results generally show that no effect on annual water yield is detectable until about 25% of a watershed is harvested (Hornbeck et al. 1993; Lisle et al. 2010). Above that level, the change in water yield (generally an increase) is related to the amount of the watershed that is harvested, although location of harvest area also matters (Boggs et al. 2016, Kochenderfer et al. 1997). The greatest change in annual streamflow from harvesting was observed in catchments with the highest annual precipitation (Bosche and Hewlett 1982). Recovery of annual water yield generally occurs within a relatively short time in wetter watersheds (eastern hardwoods and coastal Pacific Northwest) but takes longer in more arid ecosystems (Rocky Mountains and Southwest). Other parameters such as high flows can be affected for decades (Kelly et al. 2016).

Harvesting results in (generally) short-term decreases in transpiration and interception, due to lack of or decreased vegetative cover. In general, clear-cutting initially increases soil temperature and soil moisture after harvesting. The length of recovery to preharvesting state is related to the amount of time the soil is bare and the length of time to reach a "full" canopy (leaf area). This ranges from less than a decade in humid eastern deciduous forests to 50–60 years in some arid or Mediterranean western forests (Hornbeck et al. 1993; Lisle et al. 2010).

Changes in infiltration generally are minimal if the forest floor is mostly undisturbed and if care is taken to minimize soil compaction (Aust and Blinn 2004). The increased soil disturbance and water availability caused by timber harvesting can result in slight, but measurable, increases in stream sediment and nutrients. Generally, geologic (background) erosion losses are estimated at less than 500–1000 kg ha^{-1} year^{-1}. In most instances, the increased erosion rates associated with forest harvesting in the Eastern United States are comparable to or less than geologic erosion rates and well below the 2200–11,000 kg ha^{-1} year^{-1} erosion rates that are deemed acceptable and sustainable from agricultural lands (Aust and Blinn 2004). However, these numbers vary widely with forest type, climate, topography, and geology. For example, in coastal Oregon forests, the average sediment yield from unharvested reference watersheds was 180–250 kg ha^{-1} year^{-1} before harvesting (Grant and Wolff 1991). During the 30 years postharvest, however, sediment export increased by 4–12 times, with very large increases due to debris slides and flows associated with a single storm. The pattern of long-term sediment production reflected not just timber harvest but also mass movement history.

When site preparation increases the exposure of bare soil and removes more vegetation, it accentuates water quality problems, especially where sufficient slopes exist. Most water quality problems associated with forest harvesting do not arise from the loss of tree cover; instead, they are the result of poorly designed and constructed roads and skid trails, inadequate closure of roads and skid trails, stream crossings, excessive exposure of bare soil, or lack of adequate streamside management zones (Aust and Blinn 2004; Cristan et al. 2016; Neary 2014).

Grazing of Forests and Rangelands

Grazing most obviously impacts the vegetation biomass and exposes bare soil (Grudzinski et al. 2016; Kutt and Woinarski 2007; Teague et al. 2010) and the surface and near-surface soil. Thus, grazing mostly decreases the processes of interception, infiltration, runoff, and evapotranspiration and increases stream sediment concentrations (Bartley et al. 2010; Grudzinski et al. 2016; Olley and Wasson 2003; Vidon et al. 2008). Diminished riparian vegetation and access to streams by cattle (*Bos taurus*) can result in localized soil erosion from the streambanks greater than twofold that of vegetated streambanks (Beeson and Doyle 1995; Grudzinski et al. 2016; Zaimes et al. 2004). The amount of exposed bare soil and its spatial pattern in the landscape depend on the type of animal grazing. For example, bison (*Bison bison*) create more exposed bare soil at the watershed scale, and cattle create more exposed bare soil in riparian areas (Grudzinski et al. 2016). Excluding all livestock (e.g., cattle and sheep [*Ovis aries*]) and deer (*Odocoileus* spp.) from riparian areas has been shown to reduce soil erosion and decrease suspended sediment in streams (Line et al. 2016; Pilon et al. 2017). Grazed pastureland soils have higher bulk densities and lower infiltration rates and water holding capacities, compared to forest soils, all of which are attributable to compaction rather than differences in particle size distribution (Abdalla et al. 2018; Price et al. 2010).

Grazing may also have large legacy impacts on soil properties, but these impacts are poorly described. Areas that have been overgrazed in the past and are heavily eroded are likely to have altered infiltration over longer periods of time and larger scales (Renard 1970). Indeed, previous agricultural land uses have had long-term effects on surface hydrology in the southern Appalachian Mountains (Leigh 2010; Price et al. 2010), shifting the hydrologic response from soil infiltration and drainage input into streams toward overland flow (Schwartz et al. 2003; Trimble 1985). Nonmanaged systems that are grazed by herbivores may be more dynamic and resilient. Recent research on the global convergence of grassland and pasture structure has suggested that an interplay between bioturbation from roots and soil fauna and biocompaction from grazers creates a stable system of alternating mosaic patches (Howison et al. 2017). Researchers hypothesize that bare soil patches with low infiltration provide water via overland flow to nearby ungrazed patches and may subsequently recover due to bioturbation.

Active management may also ameliorate negative effects of grazing on soil properties. Grazing management treatments increased infiltration on average by about 60%, and the grazing management effect may be slightly larger in more humid environments (DeLonge and Basche 2018). Grazing can also alter infiltration in desert grasslands by damaging biological soil crusts (Belnap 2003). Finally, in semiarid grassland systems, excessive grazing can trigger wind erosion, which results in significant nutrient loss (Neff et al. 2005). The spatial variability of grazing effects points to the need for more research on the scaling of grazing impacts.

Fire and Related Activities

Projected climatic changes may increase drought and fire frequency, particularly in the western United States, and lead to greater stress for forest and rangeland watersheds (Vose et al. 2016). These projected increases in the frequency, size,

and severity of wildfire could double the rates of sedimentation in one-third of 471 large watersheds in the western United States by about 2040 (Neary et al. 2005), with sizable effects on stream channel characteristics, water quality for humans and wildlife, and management of extreme flows (Sankey et al. 2017). Fire can alter soil properties that influence infiltration, surface runoff, and erosion and that increase soil water repellency. Fire also affects nutrient cycling and the biological makeup of soils, which mediate nutrient cycling and water quality (DeBano et al. 1998; Neary et al. 2005). These effects on soils translate not only to changes in the amount of water coming from forest and rangeland soils but also to changes in the quality of water. When we think of wildfire effects on soils and water, the effects of wildfire come to mind most often, but activities associated with fire management, fire suppression, and prescribed fires can also have important impacts.

Wildfires usually are more severe than prescribed fires (DeBano et al. 1998), so they are more likely to produce significant effects on water. Fire severity is a qualitative term describing the amount of fuel consumed; fire intensity is a quantitative measure of the rate of heat released. Prescribed fires are designed to be less severe and are expected to have less effect on soils and water. The degree of fire severity is also related to the vegetation type. For example, in grasslands, the differences between prescribed fire and wildfire are small. In forests, the magnitude of the effects of fire on erosion and water quality will be much lower after a prescribed fire than after a wildfire because of the larger amount of fuel consumed in a wildfire. Canopy-consuming wildfires are the greatest concern to managers because of the loss of canopy coupled with the destruction of soil properties and function. These losses present the worst-case scenario for water quality. The differences between the effects of wildfires and prescribed fire in shrublands are intermediate between those in grass and forest environments.

The principal water quality concerns associated with wildland fires are (1) the introduction of sediment, (2) the potential of increasing nitrates in surface water and groundwater, (3) the possible introduction of heavy metals from soils and geologic sources within the burned area, and (4) the introduction of fire-retardant chemicals into streams that can reach levels toxic to aquatic organisms. The magnitude of the effects of fire on water quality is primarily driven by fire severity, rather than fire intensity. In other words, the more severe the fire, the greater the amount of fuel consumed and nutrients released and the more susceptible the site is to erosion of soil and nutrients into the stream, where they could potentially affect water quality.

Fire can produce significant changes in soil physical properties that in turn affect plants and other ecosystem components (Whelan 1995). The effect of fire on soil physical properties depends on the inherent stability of the soil property affected and the temperatures to which a soil is heated during a fire. At or near the soil surface, the sand, silt, and clay textural components have high temperature thresholds and are not usually affected by fire, unless they are subjected to extremely high temperatures (Fig. 3.2) (Lide 2001). Clay undergoes hydration and its lattice structure begins to collapse at 400 °C; internal clay structure is destroyed at 700–800 °C. Sand and silt, which are primarily quartz, melt and fuse at 1414 °C (Lide 2001, p. 81). When fusion occurs, soil texture becomes coarser and the soil more erodible. Temperatures deeper than 2–5 cm below the mineral soil surface are rarely high enough to alter clays, unless heated by smoldering roots. Fires can also affect other soil minerals, such as calcite. Calcite formation occurs at 300–500 °C, but the temperature threshold for formation is not consistent across soil or vegetation types (Iglesias et al. 1997).

Soil structure in the upper horizons can be dominated by organic matter and thus be readily affected by fire if it is directly exposed to heating during the combustion of aboveground fuels. The threshold value for irreversible changes in organic matter is low: 50–60 °C for living organisms and 200–400 °C for nonliving organic matter (DeBano 1990). Soil structure is related to productivity and water relations in wildland soils (DeBano et al. 1998). When soil heating destroys soil structure, both total porosity and pore size distribution are also affected (DeBano et al. 1998). Loss of macropores reduces infiltration rates and produces overland flow. Alteration of organic matter can also lead to hydrophobicity

(water repellent soil), further decreasing infiltration rates (see following section).

Soil chemical and nutrient changes during fires can also be dominated by organic matter and may be especially important in coarse-textured soils that have little remaining exchange capacity to capture the highly mobile cations released during the fire. Excessive leaching and loss can result, potentially reducing site fertility, particularly on nutrient-limited sandy soil.

Nitrogen (N) loss by volatilization during fires is of particular concern on low-fertility sites because N is replaced primarily by N-fixing organisms rather than by N mineralization (Hendricks and Boring 1999; Hiers et al. 2003; Knoepp and Swank 1993, 1994, 1998; White 1996). Forest disturbance in general frequently increases both soil inorganic N concentrations (due to reduced uptake by vegetation) and rates of potential N mineralization and nitrification (due to increases in soil moisture and temperature); however, total soil N typically declines with fire. The magnitude of decrease is related to fire severity. Total N, organic matter, and forest floor mass decrease as fire severity increases, whereas concentrations of N in the form of ammonium and exchangeable cations (calcium, magnesium, and potassium) in the soil increase with fire severity (Fig. 3.3). Soil pH generally increases following the loss of organic matter and its associated organic acids, which are replaced with an abundance of basic cations in the ash. In some systems, combustion of monoterpenes in the soil, which inhibit N mineralization, can also serve to increase soil inorganic N

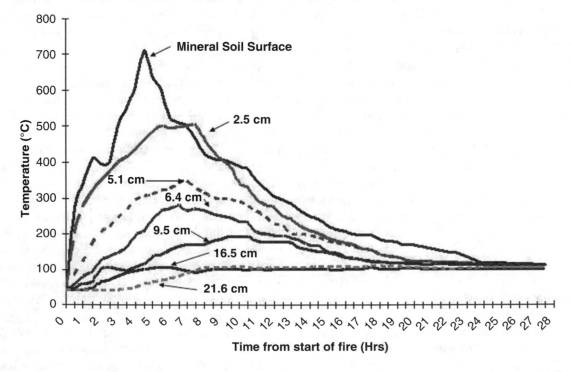

Fig. 3.2 Progression of soil temperature over time, by depth under windrowed logs. (Adapted from Neary et al. 2005; Roberts 1965)

Fig. 3.3 Generalized patterns of decreases in forest floor (duff) mass, total nitrogen (N), and organic matter (OM) and increases in soil pH, exchangeable cations, and N in the form of ammonium (NH_4^+) associated with increasing levels of fire severity. (USDA Forest Service, National Advanced Fire and Resource Institute; adapted from Neary et al. 2005)

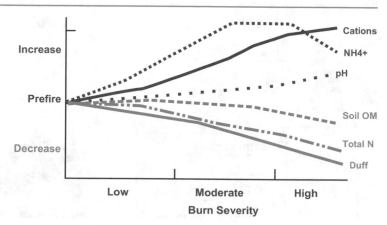

following fire (White 1991). The N increases resulting from a combination of changes in soil moisture and temperature, and the decreased plant uptake of N can make more N available for microbial populations in the soil.

Nitrogen is also an important concern relative to water quality. If soils are close to N saturation, it is possible to exceed maximum allowable contamination levels of N in the form of nitrate (10 mg L^{-1}) after a severe fire. On such areas, follow-up application of N-containing fertilizer is not recommended. Further, fire retardants typically contain large amounts of N, and they can cause water quality problems when dropped close to streams (Neary et al. 2005).

Soil microorganisms are complex (Borchers and Perry 1990), abundant, and most likely to experience fire (as they do not leave or go deeper in the soil). How microorganisms respond to fire will depend on numerous factors, including fire intensity and severity, site characteristics, and preburn community composition. In general, however, the effects of fire on microorganisms are greatest in the forest floor and decline rapidly with mineral soil depth, and mortality of microorganisms is greater in moist soil than in dry soil at high temperatures (DeBano et al. 1998; Neary et al. 2005). Most research has documented strong resilience by microbial communities to fire. However, while microorganisms are skilled at recolonizing disturbed forest soils, recovery of microbial and other biotic populations in the forest floor may not occur or may be slowed, particularly in dry systems with slow reaccumulation of organic material. Therefore, minimizing the loss of forest floor is important in prescribed fire. Finally, repeated burning of the forest floor may be detrimental to microbial biomass and activity, although the effects of repeated burning are not well documented.

Soil Water Repellency

Soil water repellency is a soil characteristic that prevents precipitation from wetting or infiltrating the soil. It has been documented in a wide range of vegetation types and climates (DeBano et al. 1998; Dekker and Ritsema 1994; Doerr et al. 2000, 2009). Water repellent conditions are of considerable interest to soil scientists, hydrologists, engineers, and land managers because of the implications for increasing runoff and erosion. Much of the research on soil water repellency has focused on the effects after wildfires, as soil water repellency caused by high burn severity is one of the primary factors in reduced infiltration rates postfire (Lewis et al. 2005). This reduction in infiltration is a primary cause of increased postfire runoff and erosion (DeBano 2000; Shakesby and Doerr 2006).

Burning induces or enhances natural soil water repellency by volatilizing the hydrophobic organic compounds in the litter and uppermost soil layers (Huffmann et al. 2001). Most of the compounds are lost to the atmosphere, but some are translocated downward in the soil profile by the thermal gradient (Fig. 3.4). The decline in soil temperature with depth means that these compounds will condense onto cooler soil particles below the soil surface (DeBano et al. 1976). Laboratory studies show that soil water repellency is intensified at soil temperatures of 175–270 °C but is destroyed at temperatures above 270–400 °C. The duration of heating can also affect the degree of soil water repellency; longer heating times influence the temperature at which these changes occur (e.g., DeBano et al. 1976; Doerr et al. 2004). The effect of wildfires depends primarily on the amount and type of organic matter consumed and the duration and amount of soil heating (DeBano and Krammes 1966; DeBano et al. 1998; Doerr et al. 2004, 2009; Robichaud and Hungerford 2000).

Several studies indicate a positive and significant relationship between soil water repellency and the amount of runoff from rainfall simulations (Benavides-Solorio and MacDonald 2001, 2005; Robichaud 2000), but few studies have rigorously isolated the effect of soil water repellency on infiltration and runoff (Leighton-Boyce et al. 2007). Only recently have these measurements been used to predict infiltration rates. Larson-Nash and others (2018) found significant correlations between minidisc infiltrometer measurements made

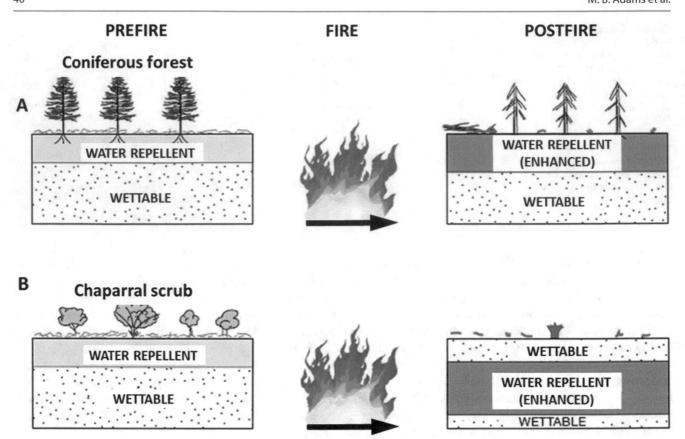

Fig. 3.4 Changes in soil water repellency following fire of moderate or high severity for (**a**) coniferous forest and (**b**) chaparral. Darker shading represents more severe repellency. (Adapted from Doerr et al. 2009)

at three depths to nonsteady-state total infiltration values taken at the 3-, 5-, and 10-min marks within 1-h rainfall simulations. Infiltration rates were reduced by 50–70% immediately after the fire and returned close to preburn conditions 5 years postfire (Larson-Nash et al. 2018).

Soil water repellency changes rapidly in response to changes in soil moisture. Both burned and unburned soils become less repellent or completely lose their water repellency as soil moisture increases. A water repellent soil can resist wetting for days or even months (e.g., Dekker and Ritsema 1994; King 1981), but the presence of macropores or other preferential flow paths means that water will eventually enter the soil. Over a period of months after the fire, soil water repellency decays toward prefire conditions. As a water repellent soil dries out, the soil water repellency is often reestablished. A series of wetting and drying cycles may eventually eliminate the soil water repellency induced by burning.

Temporal and spatial variability of soil water repellency is large, thus making it difficult to determine the single effect of soil water repellency on runoff rates at the watershed scale as compared to the point and plot scales (Woods et al. 2007). However, the greatest influence of soil water repellency on erosion is its potential for increasing overland flow. As the

amount of overland flow increases, so do its depth and velocity and hence the ability of the water to scour and transport particles by sheetwash and interrill erosion (Robichaud et al. 2016). The concentration of overland flow can initiate rill erosion (Benavides-Solorio and MacDonald 2005), and the topographic convergence of water at larger scales can result in gully, bank, and channel erosion (e.g., Martin and Moody 2001; Neary et al. 2012b).

Natural Gas Development

Recent advances in drilling technologies have dramatically increased the exploration for and extraction of natural gas in many areas of the United States. However, relatively little is known about the effects of natural gas development on the soils and water cycle processes, in either the eastern or western United States, due to a paucity of research. Other forms of energy development are also increasing but have been studied to some extent. Many of the processes involved in natural gas development—clearing vegetation and site preparation for well pad and pipeline construction—are similar to forest harvesting and land conversion; thus, some conclusions from that research may apply to disturbances during natural gas development. As noted earlier, forest harvesting research has shown that changes in streamflow often are not detectable until around 25% of a watershed is harvested (Hornbeck et al. 1993; Lisle et al. 2010). Thus, individual gas well pads may have little impact on hydrologic processes at larger watershed scales, depending on their location within a watershed. The cumulative effects of many well pads and roads in an area of intensive development (Fig. 3.5) may nevertheless be significant. More research is needed to document conditions that lead to impacts on the quantity of water in a variety of vegetation types and at various scales.

Further, relatively little research has been conducted on the effects of gas well development on surface water quality, although it is believed that surface water quality impacts are likely to be local and result from handling of the large amounts of water required for the hydrofracturing (much of which comes back up out of the well) and from accidental spills. Land application of fracking fluids is seldom permitted, and most fracking fluids are now recycled and reused rather than disposed of on land surfaces or down injection wells. However, localized effects on soils from accidental spills can be significant and may have implications for water quality and potentially water quantity. For example, surface soil concentrations of sodium and chloride increased 50-fold as a result of the land application of hydrofracturing fluids to a hardwood forest in West Virginia and declined over about

2 years to background levels (Adams 2011). The event resulted in major vegetation mortality through direct contact and uptake of soil solution by the trees. Significant impacts on soil chemistry persisted after removal of the flowback storage pond; vegetation failed to reestablish because the high levels of soil salts attracted deer (Adams et al. 2011). The effects on surface water were not evaluated, and such studies are lacking in the literature. Major effects on groundwater quality are not predicted from natural gas development, although concerns still exist, particularly relative to methane (Osborn et al. 2011). The concerns will vary with the type of well, its depth, and cementing technologies used.

Erosion from construction activities associated with the development of the well pad and associated infrastructure can be high and is related to length of time during which the ground-disturbing activities take place, precipitation, slope, best management practices (BMPs) used, and the rate of revegetation (Adams et al. 2011). Williams and others (2008) reported substantial sediment runoff from natural gas development in rangelands, which diminished as natural revegetation occurred. The estimated annual sediment loading from the natural gas field was almost 50 times greater than from typical undisturbed rangelands. Similarly, sediment concentrations and yields from a newly constructed natural gas pipeline were highest initially after completion, averaging about 1660 mg L^{-1} and 340 kg ha^{-1}, respectively, during the

Fig. 3.5 Natural gas development (well pads and roads) in Northwestern Pennsylvania. (Source: Google Earth)

first 3 months following completion of corridor reclamation. As revegetation of the right of way progressed, sediment concentrations and yields declined (Edwards et al. 2017). Early runoff and sediment movement occurred despite heavy straw mulch, suggesting that more than the usual BMPs may be necessary in forests and rangelands with steep slopes. Erosion from a pipeline installed in an existing skid road resulted in higher erosion rates, due to the negative effects of soil compaction on vegetation succession (Edwards et al. 2014). Therefore, during and after construction of gas well pads, reducing compaction to encourage infiltration and successful vegetation establishment is essential for controlling sediment losses.

Development for Recreational Activities

Recreational development and use have increased dramatically in recent decades in wildland areas (Hammett et al. 2015). Recreational use has intensified from primarily foot traffic to a wider variety of recreational activities including riding all-terrain vehicles and snowmobiles, mountain biking, snowboarding, and other activities.

The process of trampling by foot traffic has been recognized from the earliest studies as a major cause of adverse impacts on soil resources in recreation areas (Bates 1935), with implications for water quality and quantity (Fig. 3.6) (Manning 1979). These impacts can result from foot traffic by humans and horses; use of sports equipment such as mountain bikes, all-terrain vehicles, and skis; and management activities, such as grooming ski trails.

Soil erosion is the most widespread trail impact, because trails typically rely on, or result in, a bared soil surface. Sediment generation from trails will increase proportionally with recreation traffic since the shear stress frees soil particles available for transport, which may result in movement of soil from the surface to other areas (Olive and Marion 2009). In general, motorized and equestrian travel will displace trail soil surfaces more than human foot traffic (Cessford 1995; Newsome et al. 2004; Pickering et al. 2010). Careful trail planning can mitigate trail erosion by siting trails on side slopes where water bars and drainage-control features can limit erosion (Marion 2006). However, erosion along unauthorized routes can persist as travel on such routes may expose soil and damage or destroy vegetation that does not readily recover. Surface compaction from recreation activities on unauthorized routes can impede infiltration into soil. This compaction dries out soil along these tracks and paths (Settergren and Cole 1970), which can limit vegetative growth and further limit infiltration and percolation. Trail building helps to protect natural soil and vegetation communities by concentrating traffic and ensuring adequate storm drainage. Most research has evaluated localized impacts to water quality, because cumulative impacts from recreational activities are difficult to detect at larger watershed scales. Recreational activities affect water quality primarily via sediment delivery at trail crossings with streams and along the edges of water bodies (Hammett et al. 2015).

Winter sports resorts also affect soil and water resources, whether from creating and maintaining ski runs, grooming natural snow, or supplementing natural snowfall with artificial snow. The process of creating ski and snowboarding trails can alter temperatures at the soil surface, through compaction of snow (Rixen et al. 2003), and can lead to severe soil frost, with resultant impacts on vegetation, and a potential for increased erosion. At the highest altitudes, the processes of leveling to create ski trails can accelerate the thawing of permafrost (Haeberli 1992). Because of the

Fig. 3.6 An example of how trampling during recreational activity affects soil and water resources. (Source: Redrawn from Manning 1979)

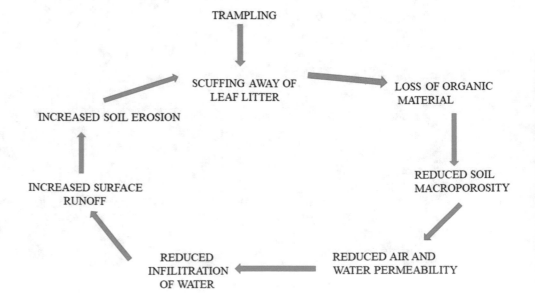

demand for longer ski seasons and increases in both the number and size of resorts to meet this demand, many ski resorts in the United States rely on snowmaking: In 2001, almost 90% of US resorts supplemented natural snowfall (Rixen et al. 2003). With increasing temperatures and climate change, this percentage is likely to rise. Making of artificial snow requires large amounts of water and creates a more granular snow. This artificial snow has a different chemical makeup than natural snowfall, which has implications for soil and water quality. Physically, adding artificial snow increases the depth of snow. Thus, the insulating value of snow is enhanced, decreasing the incidence of soil frost and possibly the erosion potential. However, the additional mass of artificial snow takes longer to melt, leading to prolonged snow cover in the spring, which also has implications for vegetative development (Rixen et al. 2003). The additional water (possibly up to five times the amount from natural snow) may increase water yield and increase the erosion potential (Wemple et al. 2007). Diverting water from local streams and lakes to manufacture snow also raises the likelihood of drying streams in the summer. The consequences of adding snow, whether using additives or only natural, locally available water, on water quality have not been widely studied.

Soil Pollution

Soil pollution is a global matter of concern, particularly as it relates to issues of food security and safety. It often is more a focus in developing nations than in more developed countries. Soil pollution is also often considered to be more of a problem in urban soils (see Chap. 7). However, soil pollution can be a local problem, as well as a regional, national, and international problem, with implications for wildland forest and rangeland soils and their ability to provide clean, abundant water. The pollutants of most concern are metals (such as lead, cadmium, and arsenic) and organic compounds (such as pesticide residues and by-products) (O'Connor et al. 2018). Though some of these substances occur naturally in the soil, they also originate from vehicle exhaust, waste disposal, untreated sewage disposal, industrial emissions, and, in the case of pesticides, direct application.

Important, but understudied, are soil contaminants carried in dust, including microbial organisms. Over the last two decades, dust emissions have increased by up to 400% in wildlands in the western United States (Brahney et al. 2013). Measurements of dust chemistry have shown that dust can transport appreciable amounts of phosphorus and carbonates to remote wildland areas (Lawrence and Neff 2009) and that heavy metals are also often transported in dust. The distribution of pollutants in soil is highly spatially variable, although generally contaminants are more concentrated in surface soil. Concentrations of contaminants tend to be greater at the surface because of deposition to the surface in wet or dry form and because organic matter content, which often binds metals and organic pollutants, is greater in surface horizons (Kaste et al. 2006). However, when the pollutant source is removed, the pollutant may "disappear" from the surface soil, only to be found deeper in the soil profile or ultimately in water bodies (Smith 1976). The pollutant may also be broken down or transformed into other compounds without moving through the soil (Kaste et al. 2006).

Soil contaminants can be taken up in vegetation and pose a health risk if the vegetation is consumed by humans or livestock. However, contaminants may also have effects on growth of plants and can affect water quantity and quality if the effects are sufficiently severe. For example, soil pollution from a smelter in Copper Hill, TN, which operated between 1843 and 1959, continues to suppress vegetative recovery (Raven et al. 2015). Research shows that soil acidification can increase the availability of heavy metals (Maiz et al. 2000), which has implications for soils still subject to the legacy effects of decades of acid deposition. As deposition of air pollutants decreases, the recovery of soil from acid precipitation is poorly understood and is a research need.

Pollutants can be moved from a site via surface runoff, groundwater flow transport (Mandal and Suzuki 2002), dust transport (Prospero 1999), and weathering (Nriagu 1989). Recent research has evaluated techniques for ameliorating soil pollution, and much of the focus has been on organic matter amendments, including the use of biochar (Ahmad et al. 2014). Only occasionally has biochar been evaluated in settings other than agricultural fields and developed beyond the research phase to operational application. Bioremediation, using plants or other organisms to remove pollutants from the soil, is another promising method and needs development beyond the research phase.

Priority Information Gaps

Linked Soil Climate Information

Soil moisture is a critical component in evaluating water budgets and assessing drought, yet monitoring of soil moisture on a national scale has its limitations. Assessment of soil moisture and soil temperature focuses on three sources of information: in situ sensors, remote sensing, and modeling. In situ measurements typically have sensors at various depths between the land surface and 1 m. These stations are valuable sources of information, such as direct measurements of soil moisture and temperature, but spatial distribution varies among states and regions; many areas have sparse or no coverage. Because each in situ network is installed and managed by different federal, state, local, tribal, university, and private industry groups, there is a lack of consistency in sensor depth, sensor type, transmission of data, data format, data availability (see Cosh et al. 2016 and the following discussion for details), and funding and network coordination. In addition, data are severely lacking in forested environments, where the tree canopy can limit data transmission and where root zones often exceed the 1 m depth typically used.

Remote sensing, both from fixed-wing flights and satellites, provides large spatial coverage of soil moisture but usually at a coarse grid resolution, and does not perform equally well everywhere. For example, the National Aeronautics and Space Administration's Soil Moisture Active Passive (SMAP) satellite provides soil moisture to 5 cm, but does not work well in forests because of the tree canopy (Panciera et al. 2014). In addition, there are temporal limitations due to flight schedules and satellite passes, with gaps of days to weeks as well as time needed to process data. Models, such as the North American Data Assimilation System (NLDAS-2) (Mitchell et al. 2004), offer spatial coverage across the country and use various sources of data, but need to be closely calibrated to in situ and satellite data, and generally have a coarse resolution.

A collaborative effort is underway to develop a National Soil Moisture Network (http://nationalsoilmoisture.com) that includes Federal, state, local, tribal, university, and private industry, as well as citizen science, data collection activities to develop a single dataset and map product. The goal is to combine in situ measurement, remote sensing, and modeling results into a single dataset that is used to develop a gridded map product of soil moisture with daily updates. The utility and application of such a product would benefit the United States Drought Monitor (https://droughtmonitor.unl.edu). This product would also serve as a tool for assessing hydrologic conditions related to agriculture, flooding, fire potential, and other applications at a sufficiently fine spatial resolution across the country.

Expanded Soil Moisture Monitoring

There are many areas where regular, direct measurements of soil moisture are sparse to nonexistent. On a national scale, the USDA operates the Soil Climate Analysis Network (SCAN), which consists of 218 stations in 40 states, Puerto Rico, and the United States Virgin Islands (https://www.wcc.nrcs.usda.gov/scan/). The National Oceanic and Atmospheric Administration operates the Climate Reference Network (CRN), which has 137 stations across the country (https://www.ncdc.noaa.gov/crn/). In addition, the USDA Snow Telemetry (SNOTEL) network in the Western United States has 446 stations that have soil moisture sensors as part of the data collection program (https://www.wcc.nrcs.usda.gov/snow/). These stations are especially important because they are in higher elevations, which are typically data-poor areas

for soil moisture, and generally in forested locations, which are also underrepresented in databases. The North American Soil Moisture Database consists of over 1800 in situ soil moisture stations in the United States, Canada, and Mexico (Quiring et al. 2016). These stations generally are concentrated in certain states or regions, so there are many areas with sparse data coverage. In addition, the Remote Automated Weather Stations (RAWS) network (https://raws.nifc.gov), an interagency effort of various wildland fire agencies that provides data to the National Interagency Fire Center, has about 2200 stations, some of which have soil moisture sensors (https://famit.nwcg.gov).

Although these national networks, in conjunction with regional, state, and local networks, provide soil moisture data from across the country, the data lack adequate density of spatial distribution, especially in forested regions. Moreover, many of the networks have a short period of record, typically less than 10 years, which makes it difficult to evaluate trends. In addition, comparing data among networks is hampered by different sensor types, depths, temporal data collection and transmission variations, data format, data type (volumetric water content vs. percentage), data access, and many other factors.

To improve on data collection efforts, a systematic approach to standards and specifications is needed. A more consistent approach would help ensure similar data quality and accessibility, adequate spatial coverage of data collection, and a single source of data storage and product generation to enable users to easily and effectively access and use data, tools, and products for assessing soil moisture.

Continued Support for Hydrologic Monitoring Networks

In the twentieth century, experimental forest catchment studies played a key role in understanding the processes contributing to high water quality (Neary 2016; Neary et al. 2012a). The hydrologic processes investigated on these catchments provided the science base for examining water quality responses to natural disturbances such as wildfire, insect outbreaks, and extreme hydrologic events and human-induced disturbances such as timber harvesting, site preparation, prescribed fires, fertilizer applications, pesticide usage, acidic deposition, and mining.

Another approach is the broad-scale landscape monitoring approach. The United States Geological Survey uses a landscape monitoring approach to acquire data on water resources from over 7200 gauging stations to report on the status and trends of water resources in the country (Neary 2016). It also uses data from cooperators to assemble information on 1.5 million sites in the United States. Landscape-level monitoring is important for discerning trends in national water resources.

Both methods need to be maintained in the light of a changing climate, and both approaches need to link more closely with available soils data and models. Some of the well-understood water-soil relationships could be altered by a more dynamic atmosphere and changing weather phenomena. There will need to be continued solid commitments from scientific organizations, government agencies, and private organizations and enterprises to achieve this goal.

Key Findings

- The key soil process with relevance to water quantity and quality from forests and rangelands is infiltration.
- Erosion and sedimentation remain major water quality concerns in forests and rangelands.
- Soil moisture is largely unexplored in most investigations and is often poorly quantified, particularly at depth.
- Current hydrologic models do not document or model soil water dynamics at watershed scales.
- Though some new threats to soil and water (e.g., natural gas development) can be understood in the context of traditional watershed research (e.g., harvesting impacts), this concept is less useful when processes other than infiltration are affected or when the spatial complexity is high.
- The legacy of twentieth-century paired catchment studies provides a solid framework for evaluating and predicting hydrologic and soil changes in the twenty-first century and beyond.

Key Information Needs

- Documenting soil variability remains a challenge in terms of adequately describing it and parameterizing models with these data.
- New tools should be explored to accurately quantify storage capacity and water dynamics at a variety of scales. The predictive capacity of terrain-based digital soil modeling needs to be further explored as a way to downscale soil surveys.
- Tools are needed to document the characteristics, storage capacity, and water utilization of deep soils and the rock materials overlying bedrock.
- Few studies have rigorously isolated the effect of soil water repellency on infiltration and runoff. The large temporal and spatial variability of soil water repellency makes it difficult to determine the single effect of soil water repellency on runoff rates at the watershed (catchment) scale, as compared to the point and plot scales.
- The effects of natural gas development and other energy and resource development on soils and the links with surface water quantity and quality in forest and rangeland soils need to be evaluated at the local and cumulative scales.
- Few studies have evaluated the cumulative impacts of recreational and trail development, particularly relative to water quality. Most studies have evaluated impacts of such development on a local scale.
- The consequences of adding snow, whether using additives or only locally available water, on water quality are not well described.
- More research on the scaling of grazing impacts on water movement and quality is needed.
- Soil pollution risks in forests and rangelands should be mapped and evaluated.
- The impacts of repeated fire and its effect on soil properties, including soil biota, should be evaluated in a spatially explicit way.
- Recent research to evaluate techniques for ameliorating soil pollution has focused on organic matter amendments. This research needs to be expanded and developed beyond the research phase to operational application. Consistent monitoring protocols and coordination among agencies are also needed.

Literature Cited

Abdalla M, Hastings A, Chadwick DR et al (2018) Critical review of the impacts of grazing intensity on soil organic carbon storage and other soil quality indicators in extensively managed grasslands. Agric Ecosyst Environ 253:62–81

Adams MB (2011) Land application of hydrofracturing fluids damages a deciduous forest stand in West Virginia. J Environ Qual 40:1340–1344

Adams MB, Edwards PJ, Ford WM et al (2011) Effects of development of a natural gas well and associated pipeline on the natural and scientific resources of the Fernow Experimental Forest. General Technical Report NRS-GTR-76. U.S. Department of Agriculture, Forest Service, Northern Research Station, Newtown Square, 24 p

Ahmad M, Rajapasha AU, Lim JE et al (2014) Biochar as a sorbent for contaminant management in soil and water: a review. Chemosphere 99:19–33

Aust WM, Blinn CR (2004) Forestry best management practices for timber harvesting and site preparation in the eastern United States: an overview of water quality and productivity research during the past 20 years (1982–2002). Water Air Soil Pollut Focus 4(1):5–36

Bartley R, Corfield JP, Abbott BN et al (2010) Impacts of improved grazing land management on sediment yields, part 1: Hillslope processes. J Hydrol 389(3–4):237–248

Bates GH (1935) The vegetation of footpaths, sidewalks, cart-tracks, and gateways. J Ecol 23(2):470–487

Bathke GR, Cassel DK (1991) Anisotropic variation of profile characteristics and saturated hydraulic conductivity in an Ultisol landscape. Soil Sci Soc Am J 55:333–339

Beeson CE, Doyle PF (1995) Comparison of bank erosion at vegetated and non-vegetated channel bends. Water Resour Bull 31(6):983–990

Belnap J (2003) The world at your feet: desert biological soil crusts. Front Ecol Environ 1:181–189

Benavides-Solorio J, MacDonald LH (2001) Post-fire runoff and erosion from simulated rainfall on small plots, Colorado Front Range. Hydrol Process 15:2931–2952

Benavides-Solorio JD, MacDonald LH (2005) Measurement and prediction of post-fire erosion at the hillslope scale, Colorado Front Range. Int J Wildland Fire 14:457–474

Bland A (2017) The West's wildfires are taking a toll on reservoirs. https://www.newsdeeply.com/water/articles/2017/09/20/the-wests-wildfires-are-taking-a-toll-on-reservoirs. Accessed 3 Mar 2019

Blöschl G, Sivapalan M (1995) Scale issues in hydrological modelling: a review. Hydrol Process 9(3–4):251–290

Boggs J, Sun G, McNulty S (2016) Effects of timber harvest on water quantity and quality in small watersheds in the Piedmont of North Carolina. J For 114(1):27–40

Bonan GB (2008) Forests and climate change: forcings, feedbacks, and the climate benefits of forests. Science 320(5882):1444–1449

Borchers JG, Perry DA (1990) Effects of prescribed fire on soil organisms. In: Walstad JD, Radosevich SR, Sandberg DV (eds) Natural and prescribed fire in Pacific Northwest forests. Oregon State University Press, Corvallis, pp 143–157

Bosche JM, Hewlett JD (1982) A review of catchment experiments to determine the effect of vegetation changes on water yield and evapotranspiration. J Hydrol 55(1–4):3–23

Brahney J, Ballantyne AP, Sievers C, Neff JC (2013) Increasing Ca^{2+} deposition in the western US: the role of mineral aerosols. Aeolian Res 10:77–87

Breshears DD (2006) The grassland-forest continuum: trends in ecosystem properties for woody plant mosaics. Front Ecol Environ 4(2):96–104

Brown RB, Huddleston JH (1991) Presentation of statistical data on map units to the user. In: Mausbach MJ, Wilding LP (eds) Spatial variabilities of soils and landforms. Proceedings of a symposium on spatial variability and map units for soil surveys. Special Publication No. 28. Soil Science Society of America, Madison, pp 127–148

Brubaker SC, Hallmark CT (1991) A comparison of statistical methods for evaluating map unit composition. In: Mausbach MJ, Wilding LP (eds) Spatial variabilities of soils and landforms. Proceedings of a symposium on spatial variability and map units for soil surveys. Special Publ. No. 28. Soil Science Society of America, Madison, pp 73–88

Buttle JM, McDonald DJ (2002) Coupled vertical and lateral preferential flow on a forested slope. Water Resour Res 38(5):18-1 to 18-6

Cessford GR (1995) Off-road impacts of mountain bikes: a review and discussion. Science & Research Series No. 92. Department of Conservation, Wellington, pp 42–70

Cosh MH, Ochsner TE, McKee L et al (2016) The soil moisture active passive Marena, Oklahoma, in situ sensor testbed (SMAP-MOISST): testbed design and evaluation of in situ sensors. Vadose Zone J 15(4)

Cristan R, Aust WM, Bolding MC et al (2016) Effectiveness of forestry best management practices in the United States: literature review. For Ecol Manag 360:133–151

DeBano LF (1990) Effects of fire on the soil resource in Arizona chaparral. In: Krammes JS (tech. coord) Effects of fire management of southwestern natural resources. General Technical Report RM-GTR-191. U.S. Department of Agriculture, Forest Service, Rocky Mountain Forest and Range Experiment Station, Fort Collins, pp 65–77

DeBano LF (2000) The role of fire and soil heating on water repellency in wildland environments: a review. J Hydrol 231–232:195–206

DeBano LF, Krammes JS (1966) Water repellent soils and their relation to wildfire temperatures. International Association of Hydrologic Sciences Bulletin, 11(2):14–19

DeBano LF, Savage SM, Hamilton DA (1976) The transfer of heat and hydrophobic substances during burning. Soil Sci Soc Am Proc 40:779–782

DeBano LF, Neary DG, Ffolliott PF (1998) Fire's effects on ecosystems. Wiley, New York, 333 p

Dekker LW, Ritsema CJ (1994) How water moves in a water-repellent sandy soil. 1. Potential and actual water-repellency. Water Resour Res 30:2507–2517

DeLonge M, Basche A (2018) Managing grazing lands to improve soils and promote climate change adaptation and mitigation: a global synthesis. Renew Agric Food Syst 33(03):267–278

Detty JM, McGuire KJ (2010) Topographic controls on shallow groundwater dynamics: implications of hydrologic connectivity between hillslopes and riparian zones in a till mantled catchment. Hydrol Process 24(16):2222–2236

Dissmeyer GE (2000) Drinking water from forests and grasslands: a synthesis of the scientific literature. General Technical Report SRS-GTR-39. U.S. Department of Agriculture, Forest Service, Southern Research Station, Asheville, 246 p

Doerr SH, Shakesby RA, Walsh RPD (2000) Soil water repellency, its characteristics, causes and hydro-geomorphological consequences. Earth Sci Rev 51:33–65

Doerr SH, Blake WH, Humphreys GS et al (2004) Heating effects on water repellency in Australian eucalypt forest soils and their value in estimating wildfire soil temperatures. Int J Wildland Fire 13:157–163

Doerr SH, Shakesby RA, MacDonald LH (2009) Soil water repellency: a key factor in post-fire erosion? In: Cerdà A, Robichaud PR (eds) Fire effects on soils and restoration strategies. Science Publishers, Enfield, pp 197–224

Edwards PJ, Harrison BM, Holz DJ et al (2014) Comparisons of sediment losses from a newly constructed cross-country natural gas pipeline and an existing in-road pipeline. In: Groninger JW, Holzmueller EJ, Nielsen CK, Dey DC (eds) Proceedings, 19th Central Hardwood Forest conference. General Technical Report NRS-P-142. U.S. Department of Agriculture, Forest Service, Northern Research Station, Newtown Square, pp 271–281. [CD-ROM]

Edwards PJ, Harrison BM, Williard KWJ, Schoonover JE (2017) Erosion from a cross-country natural gas pipeline corridor: the critical first year. Water Air Soil Pollut 228(7):232

Freer J, McDonnell JJ, Beven KJ et al (2002) The role of bedrock topography on subsurface storm flow. Water Resour Res 38(12):5-1 to 5-16

Gartner T, Mehan GT III, Mulligan J et al (2014) Protecting forested watersheds is smart economics for water utilities. J Am Water Works Assoc 106(9):54–64

Gatzke SE, Beaudette DE, Ficklin DL et al (2011) Aggregation strategies for SSURGO data: effects on SWAT soil inputs and hydrologic outputs. Soil Sci Soc Am J 75:1908–1921

Grant GE, Wolff AL (1991) Long-term patterns of sediment transport after timber harvest, Western Cascade Mountains, Oregon, USA. In: Peters NE, Walling DE (eds) Sediment and stream water quality in a changing environment: trends and explanation. Proceedings of a symposium held during the XX General Assembly of the International Union of Geodesy and Geophysics. IAHS Publication No. 203. International Association of Hydrological Sciences, Wallingford, pp 31–40

Grudzinski BP, Daniels MD, Anibas K, Spencer D (2016) Bison and cattle grazing management, bare ground coverage, and links to suspended sediment concentrations in grassland streams. J Am Water Resour Assoc 52(1):16–30

Haeberli W (1992) Construction, environmental problems and natural hazards in periglacial mountain belts. Permafr Periglac Process 3:111–124

Hammett WE, Cole DN, Monz CA (2015) Wildland recreation: ecology and management, 3rd edn. Wiley, Hoboken. 334 p

Heilman JL, Litvak ME, McInnes KJ et al (2014) Water-storage capacity controls energy partitioning and water use in karst ecosystems on the Edwards Plateau, Texas. Ecohydrology 17(1):127–138

Hendricks JJ, Boring LR (1999) N_2-fixation by native herbaceous legumes in burned pine ecosystems of the southeastern United States. For Ecol Manag 113(2–3):167–177

Hiers JK, Mitchell RJ, Boring LR et al (2003) Legumes native to long-leaf pine savannas exhibit capacity for high N_2-fixation rates and negligible impacts due to timing of fire. New Phytol 157(2):327–338

Hornbeck JW, Adams MB, Corbett ES et al (1993) Long-term impacts of forest treatments on water yield: a summary for northeastern USA. J Hydrol 150:323–344

Howison RA, Olff H, van de Koppel J, Smit C (2017) Biotically driven vegetation mosaics in grazing ecosystems: the battle between bioturbation and biocompaction. Ecol Monogr 87(3):363–378

Hu W (2018) A billion-dollar investment in New York's water. New York Times. January 18. https://www.nytimes.com/2018/01/18/nyregion/new-york-city-water-filtration.html

Huffmann FL, MacDonald LH, Stednick JD (2001) Strength and persistence of fire-induced hydrophobicity under ponderosa and lodgepole pine, Colorado Front Range. Hydrol Process 15:2877–2892

Iglesias T, Cala V, Gonzalez J (1997) Mineralogical and chemical modifications in soils affected by forest fire in the Mediterranean area. Sci Total Environ 204:89–96

Kaste JM, Bostick BC, Friedland AJ et al (2006) Fate and speciation of gasoline-derived lead in organic horizons of the northeastern U.S. Soil Sci Soc Am J 70:1688–1698

Kelly CN, McGuire KJ, Miniat CF, Vose JM (2016) Streamflow response to increasing precipitation extremes altered by forest management. Geophys Res Lett 43(8):3727–3736

Kienzler PM, Naef F (2008) Temporal variability of subsurface stormflow formation. Hydrol Earth Syst Sci 12:257–265

King PM (1981) Comparison of methods for measuring severity of water repellence of sandy soils and assessment of some factors that affect its measurement. Aust J Soil Res 19:275–285

Knoepp JD, Swank WT (1993) Effects of prescribed burning in the southern Appalachians on soil nitrogen. Can J For Res 23:2263–2270

Knoepp JD, Swank WT (1994) Long-term soil chemistry changes in aggrading forest ecosystems. Soil Sci Soc Am J 58:325–331

Knoepp JD, Swank WT (1998) Rates of nitrogen mineralization across an elevation and vegetation gradient in the southern Appalachians. Plant Soil 204:235–241

Kochenderfer JN, Edwards PJ, Wood F (1997) Hydrologic impacts of logging an Appalachian watershed using West Virginia's best management practices. North J Appl For 14:207–218

Kuppel S, Tetzlaff D, Maneta MP, Soulsby C (2018) What can we learn from multi-data calibration of a process-based ecohydrological model? Environ Model Softw 101:301–316

Kutt AS, Woinarski JCZ (2007) The effects of grazing and fire on vegetation and the vertebrate assemblage in a tropical savanna woodland in north-eastern Australia. J Trop Ecol 23(1):95–106

Larson-Nash SS, Robichaud PR, Pierson FB et al (2018) Recovery of small-scale infiltration and erosion after wildfires. J Hydrol Hydromech 66(3):261–270

Lawrence CR, Neff JC (2009) The contemporary physical and chemical flux of aeolian dust: a synthesis of direct measurements of dust deposition. Chem Geol 267:46–63

Leigh DS (2010) Morphology and channel evolution of small streams in the southern Blue Ridge Mountains of western North Carolina. Southeast Geogr 50(4):397–421

Leighton-Boyce G, Doerr SH, Shakesby RA, Walsh RDP (2007) Quantifying the impact of soil water repellency on overland flow generation and erosion: a new approach using rainfall simulation and wetting agents on in situ soils. Hydrol Process 21:2337–2345

Lewis SA, Wu JQ, Robichaud PR (2005) Assessing burn severity and comparing soil water repellency, Hayman Fire, Colorado. Hydrol Process 20(1):1–16

Liang X, Lettenmaier DP, Wood EF, Burges SJ (1994) A simple hydrologically based model of land surface water and energy fluxes for GSMs. J Geophys Res 99(D7):14415–14428

Lide DR (ed) (2001) CRC handbook of chemistry and physics, 82nd edn. CRC Press, New York. Chapter 4

Line DE, Osmond DL, Childres W (2016) Effectiveness of livestock exclusion in a pasture of central North Carolina. J Environ Qual 45(6):1926–1932

Lisle TE, Adams MB, Reid LM, Elder K (2010) Hydrologic influences of forest vegetation in a changing world: learning from Forest Service experimental forests, ranges, and watersheds. In: Adams MB, NcNeel J, Rodriguez-Franco C (eds) Meeting current and future conservation challenges through the synthesis of long-term silviculture and range management research. General Technical Report WO-GTR-84. U.S. Department of Agriculture, Forest Service, Washington Office, Washington, DC, pp 37–49

Maiz I, Arambarri I, Garcia R, Milan E (2000) Evaluation of heavy metal availability in polluted soils by two sequential extraction procedures using factor analysis. Environ Pollut 110(1):3–9

Mandal B, Suzuki K (2002) Arsenic round the world: a review. Talanta 58:201–235

Maneta MP, Silverman NL (2013) A spatially distributed model to simulate water, energy and vegetation dynamics using information from regional climate models. Earth Interact 17(2013):1–44

Manning RE (1979) Recreational impacts on soils. Am Water Resour Assoc Water Resour Bull 15(1):30–43

Marion JL (2006) Assessing and understanding trail degradation: results from Big South Fork National River and Recreational Area. U.S. Department of the Interior, US Geological Survey. 80 p

Martin DA, Moody JA (2001) Comparison of soil infiltration rates in burned and unburned mountainous watersheds. Hydrol Process 15:2893–2903

McDaniel PA, Falen AL (1994) Temporal and spatial patterns of episaturation in a Fragixeralf landscape. Soil Sci Soc Am J 58:1451–1457

McDaniel PA, Gabehart RW, Falen AL et al (2001) Perched water tables on Argixeroll and Fragixeralf hillslopes. Soil Sci Soc Am J 65(3):805–810

McNamara JP, Chandler DG, Seyfried M, Achet S (2005) Soil moisture states, lateral flow, and streamflow generation in a semi-arid, snowmelt-driven catchment. Hydrol Process 19:4023–4038

Mitchell KE, Lohman D, Houser PR et al (2004) The multi-institution North American Land Data Assimilation System (NLDAS): utilizing multiple GCIP products and partners in a continental distributed hydrological modeling system. J Geophys Res 109:(D7)

National Research Council, Water Science and Technology Board [NRC] (2008) Hydrologic effects of a changing forest landscape. The National Academy Press, Washington, DC. 167 p

Neary DG (2014) Best management practices for bioenergy feedstock production. International Energy Agency, Bioenergy Task 43 Special Publication. Chalmers University, Gothenburg. 125 p

Neary DG (2016) Long-term forest paired catchment studies: what do they tell us that landscape-level monitoring does not? MDPI For Special Issue 7:164

Neary DG, Ryan KC, DeBano LF (eds) (2005) (revised 2008) Fire effects on soil and water. General Technical Report RMRS-GTR-42-vol.4. U.S. Department of Agriculture, Forest Service, Rocky Mountain Research Station, Ogden, 250 p

Neary DG, Hayes D, Rustad L et al (2012a) The U.S. Forest Service experimental forests and ranges network: a continental research platform for catchment scale research in the United States. In: Webb AA, Bonell M, Bren L et al (eds) Revisiting experimental catchment studies in forest hydrology. Proceedings of a workshop held during the XXV General Assembly of the International Union of Geodesy and Geophysics. IAHS Publication No. 353. International Association of Hydrological Sciences, Wallingford, pp 49–57

Neary DG, Koestner KA, Youberg A, Koestner PE (2012b) Post-fire rill and gully formation, Schultz Fire 2010, Arizona, USA. Geoderma 191:97–104

Neff JC, Reynolds RL, Belnap J, Lamothe P (2005) Multi-decadal impacts of grazing on soil physical and biogeochemical properties in southeast Utah. Ecol Appl 15:87–95

Newman BD, Campbell AR, Wilcox BP (1998) Lateral subsurface flow pathways in a semiarid ponderosa pine hillslope. Water Resour Res 34:3485–3496

Newsome D, Cole DN, Marion JL (2004) Chapter 5: Environmental impacts associated with recreational horse-riding. In: Buckley R (ed) Environmental impacts of ecotourism. CABI Publishing, Wallingford, UK, pp 61–83

Nijzink R, Hutton C, Pechlivanidis I et al (2016) The evolution of root zone moisture capacities after deforestation: a step towards hydrological predictions under change? Hydrol Earth Syst Sci 20(12):4775

Nriagu JO (1989) A global assessment of natural sources of atmospheric trace metals. Nature 338:47–49

O'Connor D, Peng T, Zhang J et al (2018) Biochar application for the remediation of heavy metal polluted land: a review of in situ field trials. Sci Total Environ 619–620:815–826

O'Geen AT, Dahlgren RA, Swarowsky A et al (2010) Research connects soil hydrology and stream water chemistry in California oak woodlands. Calif Agric 64(2):78–84

Olive ND, Marion JL (2009) The influence of use-related, environmental, and managerial factors on soil loss from recreational trails. J Environ Manag 90(3):1483–1493

Olley JM, Wasson RJ (2003) Changes in the flux of sediment in the Upper Murrumbidgee catchment, southeastern Australia, since European settlement. Hydrol Process 17(16):3307

Osborn SG, Vengosh A, Warner NR, Jackson RB (2011) Methane contamination of drinking water accompanying gas-well drilling and hydraulic fracturing. Proc Natl Acad Sci 108(20):8172–8176

Panciera R, Walker JP, Jackson TJ et al (2014) The Soil Moisture Active Passive Experiments (SMAPEx): toward soil moisture retrieval from the SMAP Mission. IEEE Trans Geosci Remote Sens 52(1):490–507

Pickering CM, Hill W, Newson D, Leung Y-F (2010) Comparing hiking, mountain biking and horse riding impacts on vegetation and soils in Australia and the United States of America. J Environ Manag 91:551–562

Pilon C, Moore PA, Pote DH, Pennington JH, Martin JW, Brauer DK, Raper RL, Dabney SM, Lee J (2017) Long-term Effects of Grazing Management and Buffer Strips on Soil Erosion from Pastures. J Environ Qual 46(2):364-372

Price K, Jackson CR, Parker AJ (2010) Variation of surficial soil hydraulic properties across land uses in the southern Blue Ridge Mountains, North Carolina, USA. J Hydrol 383:256–268

Prospero JM (1999) Long-range transport of mineral dust in the global atmosphere: impact of African dust on the environment of the southeastern United States. Proc Natl Acad Sci 96(7):3396–3403

Quiring SM, Ford TW, Wang JK et al (2016) The North American Soil Moisture Database: development and applications. Bull Am Meteorol Soc 97(8):1441–1459

Raven PH, Hassenzahl DM, Hager MC et al (2015) Environment, 9th edn. Wiley, New York. 528 p

Ravi S, Breshears DD, Huxman TE, D'Odorico P (2010) Land degradation of drylands: interactions among hydrologic-aeolian erosion and vegetation dynamics. Geomorphology 116:236–245

Renard KG (1970) The hydrology of semiarid rangeland watersheds. Agricultural Research Service Publication No. 41-162. U.S. Department of Agriculture, Agricultural Research Service. 27 p

Rixen C, Stoeckli V, Ammann W (2003) Does artificial snow production affect soil and vegetation of ski pistes? Perspectives in plant ecology. Evol Syst 5(4):219–230

Roberts WB (1965) Soil temperatures under a pile of burning eucalyptus logs. Aust For Res 1(3):21–25

Robichaud PR (2000) Fire effects on infiltration rates after prescribed fire in northern Rocky Mountain forests, USA. J Hydrol 231–232:220–229

Robichaud PR, Hungerford RD (2000) Water repellency by laboratory burning of four northern Rocky Mountain forest soils. J Hydrol 231–232:207–219

Robichaud P, MacDonald L, Freehouf J, Neary DG (2003) Hayman Fire case study analysis: post-fire rehabilitation. In: Graham RT (ed) Interim Hayman Fire case study analysis. General Technical Report RMRS-GTR-114. U.S. Department of Agriculture, Forest Service, Rocky Mountain Research Station, Fort Collins, pp 293–314

Robichaud PR, Wagenbrenner JW, Pierson FB et al (2016) Infiltration and interrill erosion rates after a wildfire in western Montana, USA. Catena 142:77–88

Sankey JB, Kreitler J, Hawbaker TJ et al (2017) Climate, wildfire, and erosion ensemble foretells more sediment in western USA watersheds. Geophys Res Lett 44:8884–8892

Schwartz RC, Evett SR, Unger PW (2003) Soil hydraulic properties of cropland compared with reestablished and native grassland. Geoderma 116:47–60

Settergren CD, Cole DM (1970) Recreation effects on soil and vegetation in the Missouri Ozarks. J For 68:231–233

Shakesby RA, Doerr SH (2006) Wildfire as a hydrological and geomorphological agent. Earth Sci Rev 74:269–307

Smith WH (1976) Lead contamination of the roadside ecosystem. J Air Pollut Control Assoc 26(8):753–766

Soil Survey Staff (2017) Web soil survey. U.S. Department of Agriculture, Natural Resources Conservation Service. Available at https://websoilsurvey.nrcs.usda.gov. Accessed 6 Mar 2019

Swank WT, Crossley DA Jr (eds) (1988) Forest hydrology and ecology at Coweeta. Springer, New York

Swarowsky A, Dahlgren RA, Tate KW et al (2011) Catchment-scale soil water dynamics in a Mediterranean-type oak woodland. Vadose Zone J 10:1–16

Swarowsky A, Dahlgren RA, O'Geen AT (2012) Linking subsurface lateral flowpath activity with stream flow characteristics in a semiarid headwater catchment. Soil Sci Soc Am J 76:532–547

Tague CL, Band LE (2004) RHESSys: Regional Hydro-Ecologic Simulation System—an object-oriented approach to spatially distributed modeling of carbon, water, and nutrient cycling. Earth Interact 8(19):1–42

Teague WR, Dowhower SL, Baker SA et al (2010) Soil and herbaceous plant responses to summer patch burns under continuous and rotational grazing. Agric Ecosyst Environ 137:113–123

Terribile F, Coppola A, Langella G et al (2011) Potential and limitations of using soil mapping information to understand landscape hydrology. Hydrol Earth Syst Sci 15:3895–3933

Trimble SW (1985) Perspectives on the history of soil erosion control in the eastern United States. Agric Hist 59(2):162–180

Tromp-Van Meerveld HJ, McDonnell JJ (2006a) Threshold relations in subsurface stormflow: 1. A 147-storm analysis of the Panola hillslope. Water Resour Res 42:W02410

Tromp-Van Meerveld HJ, McDonnell JJ (2006b) Threshold relations in subsurface stormflow: 2. The fill and spill hypothesis. Water Resour Res 42:W02411

Verry ES (1997) Chapter 13: Hydrological processes of northern forested wetlands. In: Trettin CC, Jurgensen MF, Grigal DF et al (eds) Northern forested wetlands: ecology and management. CRC Press, Boca Raton, pp 163–188

Vidon P, Campbell MA, Gray M (2008) Unrestricted cattle access to streams and water quality in till landscape of the Midwest. Agric Water Manag 95(3):322–330

Vose JM, Clark JS, Luce CH, Patel-Weynand T (eds) (2016) Effects of drought on Forests and Rangelands in the United States: a comprehensive science synthesis. General Technical Report WO-GTR-93b.

U.S. Department of Agriculture, Forest Service, Washington Office, Washington, DC, 299 p

Wemple B, Shanley J, Denner J et al (2007) Hydrology and water quality in two mountain basins of the northeastern US: assessing baseline conditions and effects of ski area development. Hydrol Process 21:1639–1650

Wenger SJ, Luce CH, Hamlet AF et al (2010) Macroscale hydrologic modeling of ecologically relevant flow metrics. Water Resour Res 46(9):W09513

Whelan RJ (1995) The ecology of fire. Cambridge University Press, Cambridge. 346 p

White CS (1991) The role of monoterpenes in soil nitrogen cycling processes in ponderosa pine. Biogeochemistry 12:43–68

White CS (1996) The effects of fire on nitrogen cycling processes within Bandelier National Monument, NM. In: Allen CD (tech. ed) Fire effects in southwestern forests. Proceedings of the 2nd La Mesa fire symposium. General Technical Report RM-GTR-286. U.S. Department of Agriculture, Forest Service,

Rocky Mountain Forest and Range Experiment Station, Fort Collins, pp 123–139

Williams HFL, Havens DL, Banks KE, Wachal DJ (2008) Field-based monitoring of sediment runoff from natural gas well sites in Denton, County, Texas, USA. Environ Geol 55(7):1463–1471

Wood AW, Maurer EP, Kumar A, Lettenmaier DP (2002) Long-range experimental hydrologic forecasting for the eastern United States. J Geophys Res 107(D20):4429

Woods SW, Birkas A, Ahl R (2007) Spatial variability of soil hydrophobicity after wildfires in Montana and Colorado. Geomorphology 86(304):465–479

Wu XB, Redeker EJ, Thurow TL (2001) Vegetation and water yield dynamics in an Edwards Plateau watershed. J Range Manag 54(2):98–105

Zaimes GN, Schultz RC, Isenhart TM (2004) Stream bank erosion adjacent to riparian forest buffers, row-crop fields, and continuously-grazed pastures along Bear Creek in central Iowa. J Soil Water Conserv 59(1):19–27

Biogeochemical Cycling in Forest and Rangeland Soils of the United States

4

Lindsey E. Rustad, Jennifer Knoepp, Daniel D. Richter, and D. Andrew Scott

Introduction

In the Sand County Almanac (Leopold 1949), Aldo Leopold writes of the odyssey of element X and thus of the circulation of all nutrient elements as they cycle through the Earth's forests, rangelands, lakes, and oceans:

> The break came when a bur-oak root nosed down a crack and began prying and sucking. In the flush of a century the rock decayed, and X was pulled out and up into the world of living things. He helped build a flower, which became an acorn, which fattened a deer, which fed an Indian, all in a single year. From his berth in the Indian's bones, X joined again in chase and flight, feast and famine, hope and fear. He felt these things as changes in the little chemical pushes and pulls that tug timelessly at every atom. When the Indian took his leave of the prairie, X moldered briefly underground, only to embark on a second trip through the bloodstream of the land.

Also in mid-century, G.E. Hutchinson, while always one to praise aesthetic values, commented disparagingly about the quantitative science of element cycling, specifically that ecosystem carbon (C) data were "wretchedly inadequate" (Hutchinson 1954). Hutchinson's comment was not only a complaint but also a challenge to all ecosystem scientists who followed to quantify the Earth's biogeochemical cycles, for he understood that the resilience and functioning of ecosystems was entirely dependent on how plants, animals, and decomposers used and reused the chemical elements that we call nutrients.

In the late twentieth century, the science of biogeochemistry proliferated internationally to become a major interdisciplinary science, and enormous amounts of nutrient cycling data have since been collected and synthesized. Scientists recognized that the fundamental unit in ecology is the ecosystem, a dynamic, indivisible system of biota and the abiotic environment, and that elements cycle into, through, and out of these systems in generally predictable ways (Duvigneaud and Danaeyer-de Smet 1970; Likens and Bormann 1995; Ovington 1962). Many scientists followed Hutchinson's lead and were excited that forest and rangeland ecosystems conserve nutrients and recycle large fractions of the nutrients taken up each year by plant roots (Cole and Rapp 1981; Stone 1975; Switzer and Nelson 1972; Vitousek 1982; Wells et al. 1972).

Through this early body of work, it was quickly learned that individual nutrients cycle differently through ecosystems. Atmospheric inputs of the mineral elements nitrogen (N) and sulfur (S) were found to be substantial, and these elements accumulated in the system over time. Mineralization of soil organic matter is the major immediate source of N, S, and boron (B) and, in some soils, is also the major source of phosphorus (P). The soil's cation exchange capacity and the weathering of primary and secondary soil minerals are the most important sources for calcium (Ca), potassium (K), magnesium (Mg), and many trace elements such as iron (Fe), copper (Cu), and zinc (Zn). Overall, soil proved to be the major reservoir of nutrients for plant uptake and for the decomposition system that mineralizes and transformes organic matter, ensuring continued supplies of bioavailable nutrients. Ecosystems were known to lose nutrients through leaching losses to surface waters, gaseous losses to the atmosphere, and erosional losses. However, despite these losses, the biogeochemical cycles of forests were found to be generally conservative, such that the annual uptake of many nutrients, including N, P, S, B, and K (i.e., those used to drive photosynthesis in plants), was rapidly returned to the soil in aboveground and belowground

L. E. Rustad (✉)
U.S. Department of Agriculture, Forest Service, Northern Research Station, Center for Research on Ecosystem Change,
Durham, NH, USA
e-mail: Linsey.rustad@usda.gov

J. Knoepp
U.S. Department of Agriculture, Forest Service, Southern Research Station, Otto, NC, USA

D. D. Richter
Nicholas School of the Environment, Duke University,
Durham, NC, USA

D. A. Scott
U.S. Department of Agriculture, Forest Service, Bankhead National Forest, Double Springs, AL, USA

© The Author(s) 2020
R. V. Pouyat et al. (eds.), *Forest and Rangeland Soils of the United States Under Changing Conditions*,
https://doi.org/10.1007/978-3-030-45216-2_4

51

litter inputs (Gosz et al. 1972; Prescott 2002; Vogt et al. 1986), thereby remaining within the ecosystem. The woody biomass of trees accumulates only a small fraction of most nutrients taken up from soil (Switzer and Nelson 1972). Examples of biogeochemical pools and fluxes for N, P, and Ca for representative ecosystems are provided in Fig. 4.1, which illustrates the range in biogeochemical cycling of these elements in forests and grasslands of the United States.

While the soil is a reservoir of nutrients, this store is often unable to supply nutrients at rates that meet the biological potential of the vegetation. In other words, the productivity of many of the forests and rangelands of the United States is nutrient-limited (Binkley and Fisher 2012) or may become nutrient-limited if soils are degraded by land uses (Richter Jr. and Markewitz 2001). Understanding these limitations is important when assessing the uses of forests and rangelands and how they function to provide food and fiber, clean and plentiful water, C sequestration, wildlife habitat, recreation, and reservoirs for biodiversity. Many millions of hectares of industrial forest in the United States are intentionally fertilized, mainly with N and P, to boost productivity to its full potential (Fox et al. 2007). Many more millions of hectares also receive high amounts of point and nonpoint source pollutants, which unintentionally alter inputs or removals of nutrients at rates that are biologically significant and can impact forest and rangeland function (Buol et al. 2011; Richter Jr. and Markewitz 2001).

Soils and soil nutrient cycles vary greatly in space and are dynamic through time. Variations in space are better quantified in the literature than variations in time. Explanations for local- to continental-scale spatial variations are well studied and can be attributed to multiple soil forming factors, including human actions (Jenny 1980; Richter and Yaalon 2012). These factors include interactions of climate, biota, geomorphology, substrates, and human impacts, as they all influence soils over time. Spatial variations in soil properties can be extreme at local scales (due to drainage classes), regional scales (due to soil series), and continental scales (due to soil order) (Fig. 4.2).

Soil physical, chemical, and biological processes play out over timescales that range from milliseconds to millennia (Fig. 4.3). Most soils are now recognized to be polygenetic, which means that soils have long enough residence times to have been exposed to and influenced by varying soil forming processes, including changing climate, vegetation, and human management (Richter and Yaalon 2012). Human imprints on forest and rangeland soils and nutrient cycles, including impacts from agriculture, forestry, industrialization, and urbanization, have been increasing in intensity and extent during the last century and are shaping soils in potentially novel ways (Richter 2007).

To investigate soil forming processes across time, many studies have substituted space for time in a classic chronosequence approach that can tell investigators the general direction of temporal soil change (Hotchkiss et al. 2000). Perhaps the most well-known of these is the 2-My soil chronosequence on the Mendocino Staircase in California, where soils evolve from Entisols and Mollisols to Alfisols and their more acidic relatives the Ultisols and, finally, to extremely acidic Spodisols (Jenny 1980). At shorter timescales, local- and landscape-scale heterogeneity limits the precision and accuracy of such space-for-time approaches (Buol et al. 2011; Richter Jr. and Markewitz 2001), and researchers rely instead on a remarkably few long-term field studies that directly observe time-dependent soil changes by repeated soil sampling. In an inventory of well over 200 long-term field studies worldwide, about 15% are studies of forest and rangeland soils (Richter and Yaalon 2012). These field studies demonstrate that soil is highly responsive to management and that the soil system is highly dynamic on decadal timescales (Mobley et al. 2015).

Human Impacts on Forest and Rangeland Biogeochemical Cycling in the United States

Nutrient cycling in forest and rangeland soils is highly malleable and can be impacted positively or negatively by human-caused disturbances. Active forest and rangeland management, by definition, manipulates aboveground and belowground C and nutrient cycles to achieve economic, recreational, or conservation goals or combinations of these goals. Timber harvesting, grazing, changes in species composition, fertilization, and prescribed fire all have direct and indirect impacts on nutrient pools and cycles. Modern civilization and associated industrialization and urbanization impose further alterations on these processes. Examples include point and nonpoint source pollutants; anthropogenic-driven changes in climate and the frequency and severity of extreme weather events; changes in distribution of invasive species, pests, and pathogens; and extreme disturbances such as wildfires, fracking, mining, and urbanization.

Harvest and Grazing

Forest harvesting and rangeland grazing can potentially affect nutrient cycling in many ways, including the direct removal of aboveground nutrients in the harvested or grazed material, redistribution of nutrient-rich material from aboveground vegetation to the soil surface, disruption of the hydrologic cycle due to reduction in evapotranspiration, reductions in plant nutrient uptake, increased leaching loss of nutrients to surface waters, and adverse impacts of soil compaction and erosion. The impacts of forest harvest on soil C and N have been examined in detail across the United States and elsewhere. Nave and others (2010) synthesized the results from

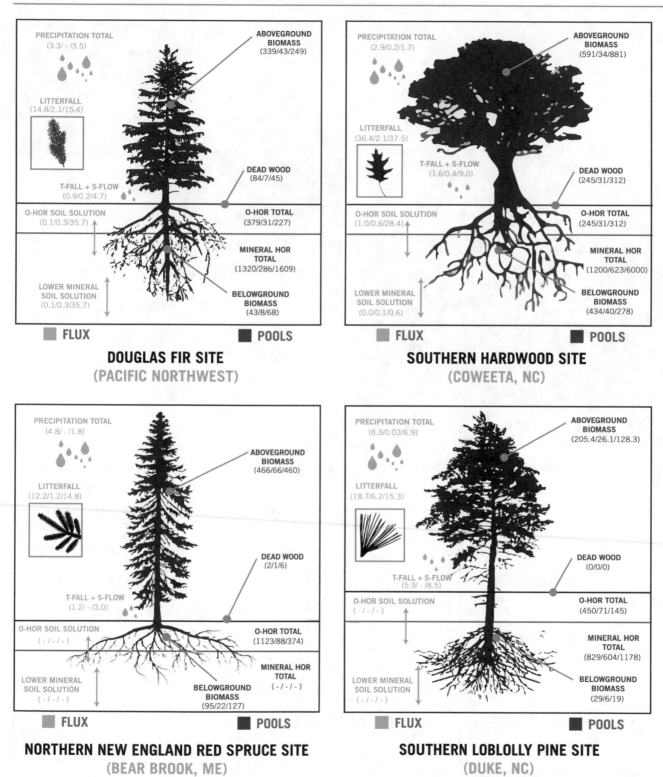

Fig. 4.1 Element pools and cycling of nitrogen (N), phosphorus (P), and calcium (Ca) in representative forest ecosystems in the United States. Values are given as N/P/Ca and are expressed as kg ha⁻¹ for flux and kg ha⁻¹ for pools (source: Johnson and Lindberg 1992)

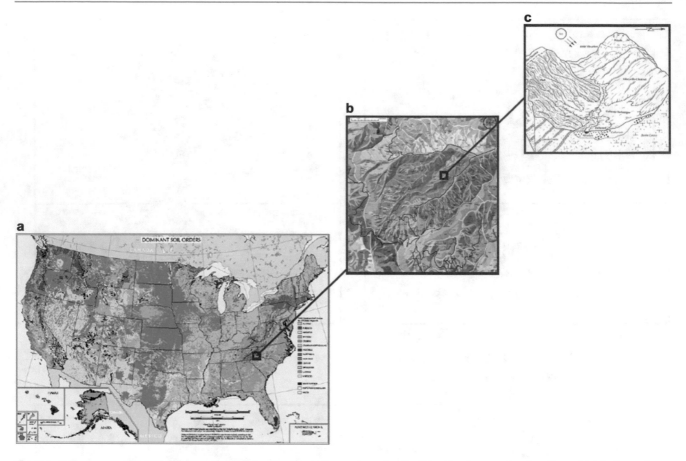

Fig. 4.2 Soils show high spatial variability across (a) soil orders at the continental scale, (b) soil series at regional scales, and (c) drainage classes at local hillslope scales

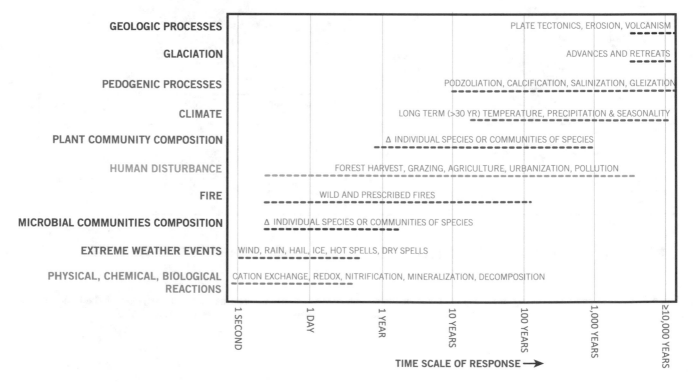

Fig. 4.3 Examples of timescale response for major drivers of soil change

many of these studies in a comprehensive meta-analysis, which shows that traditional bolewood harvesting reduces soil organic C by an average of 8%, with most of this reduction occurring in the organic horizons. Removing even greater quantities of biomass for bioenergy products was shown to have relatively little added effect (Johnson and Curtis 2001; Nave et al. 2010). In contrast, traditional forest harvests in which wood residues are left on site have been shown to increase soil organic C (Johnson and Curtis 2001). In addition, site preparation for plantation forestry, which often involves physical soil manipulations such as bedding, tillage, or ripping, can also reduce soil organic C (Nave et al. 2010).

In addition to the effects of biomass removals and redistributions on N, harvesting is known to increase rates of soil nitrification and mineralization in some cases (Likens 1989). When coupled with reduced plant uptake due to aboveground biomass removal and belowground root disturbance, these effects can result in transient losses of N from soils to surface waters (and thus removal from the ecosystem) or to the atmosphere as a gas (Jerabkova et al. 2011; Likens 1989; Vitousek and Melillo 1979). Additional disturbance, such as site preparation, can magnify these processes (Burger and Pritchett 1984). This loss of N to surface waters following harvesting is well demonstrated at Coweeta Hydrologic Laboratory in North Carolina. Research conducted at this site showed that intensive forest management, such as plantation establishment (Adams et al. 2014) and clear-cutting, followed by site preparation (Webster et al. 2016) shifts forest N cycling and stream N export from a biologically controlled process in undisturbed reference streams to a hydrologically controlled process in highly disturbed watersheds. This overall "regime shift," which persists to this day, is also accompanied by a shift in seasonal patterns of N export. Reference stream N export is greatest in the summer, when soil N transformation rates are greatest, and least during the fall, when the addition of litter fall immobilizes N; the disturbed watersheds only show a decline in N export in the fall.

Differences in the retention of N following forest harvest is a function of differences in N pool capacity (i.e., sinks in soils and vegetation) and kinetics of N processing (plant uptake and soil sorption) (Lovett and Goodale 2011). For example, Adams and others (2014) compared long-term stream data from watersheds with similar management in the Central (Fernow Experimental Forest in West Virginia) and Southern (Coweeta Hydrologic Laboratory in North Carolina) Appalachian Mountains. In both locations, reference watersheds retained more than 95% of incoming inorganic N deposition, and experimental clear-cutting shifted watersheds from net N retention to net loss. The net N loss at the Fernow Experimental Forest, however, was equal to approximately 150% of input compared to a net loss of only about 20% at the Coweeta Hydrologic Laboratory, reflecting the different pools and processes at these two sites (Box 4.1).

The connection between forest harvest and potential nutrient losses of soil base cations and P, and consequent reductions in productivity for subsequent rotations, has caused concerns in parts of the United States where these nutrients are most limiting. For base cations, many studies have documented short-term and longer-term losses of base cations after the removal of biomass by harvest or by leach-

Box 4.1

Watershed (WS) response to experimental clear-cut logging (L) and conversion to conifers (C) compared to a reference (R) in Coweeta Hydrologic Laboratory in North Carolina and the Fernow Experimental Forest in West Virginia shows how forest management impacts stream nitrogen (Box Fig. 4.1). Additional research at Coweeta indicates a regime shift in the seasonal patterns of N concentrations over time (Box Fig. 4.2), and data shows the relationship that exists between N export and stream discharge following experimental clear-cutting (Box Fig. 4.3).

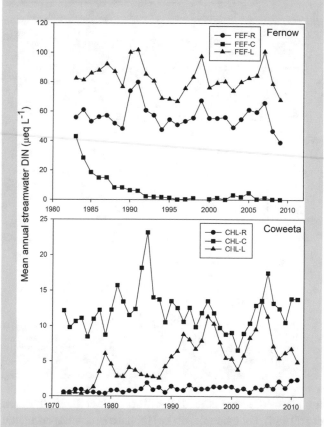

Box Fig. 4.1 Mean annual dissolved inorganic N (DIN) concentration in streams draining the six watershed at the Fernow Experimental Forest (top panel) and Coweeta Hydrologic Laboratory (bottom panel). For DIN, 1 ueg L^{-1} is 1 umol L^{-1}. (Source Adams 2014, reprinted with permission)

(continued)

Box 4.1 (continued)

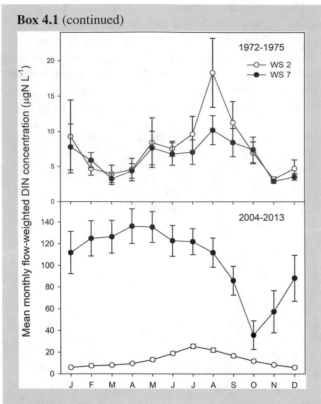

Box Fig. 4.2 Average (+/- SE) flow-weighted monthly disso-love inorganic nitrogen (DIN) concentrations for WS2 and WS7 before clearcutting (1972-1975, top panel) and for the most recent 10 years (2004-2013, bottom panel). In the bottom panel most error bars for WS2 are smaller than the symbols

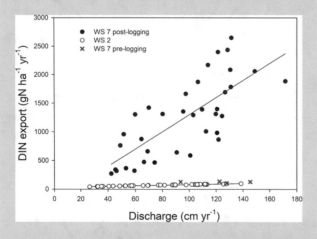

Box Fig. 4.3 Relationship between nitrogen (N) export and stream discharge following clearcutting

ing losses (Federer et al. 1989; Hornbeck et al. 1990; Mann et al. 1988). For example, at the Hubbard Brook Experimental Forest in New Hampshire, commercial whole-tree and strip-cut harvests resulted in a 4.8 and 1.9 times, respectively, greater stream export of Ca^{2+} over the first 8 years following cutting in the harvested watersheds compared to the undisturbed reference watershed (Likens et al. 1998). Longer-term

studies found continued elevated Ca^{2+} export in streams draining the cut watersheds compared to the controls even after 32 years (Bailey et al. 2003). In another long-term study, Johnson and others (2016) reviewed base cation response for over 30 years following stem-only and whole-tree harvest in a mixed oak (*Quercus* spp.) forest in Tennessee; they found that base cations were reduced in proportion to the amounts removed in harvest.

In the glaciated portion of the northeastern United States, potential base cation losses due to forest harvest, compounded by observed losses due to acidic precipitation, have caused concerns that soils of this region are becoming depleted of base cations, a condition referred to as "soil base cation depletion" (Adams et al. 2000; Federer et al. 1989; Huntington 2000). Soil base cation depletion has been documented by a variety of approaches, including repeated soil sampling (Bailey et al. 2005), mass balance models (Likens et al. 1996), and watershed scale acidification experiments (Fernandez et al. 2003). Soil base cation depletion, particularly Ca^{2+} depletion, has been implicated in regional forest declines, including sugar maple (*Acer saccharum*) decline (Horsley et al. 2002) and winter injury in red spruce (*Picea rubens*) (DeHayes et al. 1999). More research is needed to measure and model how base cations, input through the weathering of soil minerals or atmospheric deposition of dust, can replenish soil base cation stocks following harvest or during recovery from acid rain inputs (Yanai et al. 2005).

Phosphorus is a critical nutrient for tree growth and is often a limiting or colimiting nutrient in production forestry, particularly at sites on older, more highly weathered soils (Fox et al. 2007; Vitousek and Farrington 1997; Vitousek et al. 2010). Harvest-induced reductions in biologically available forms of soil organic P have been shown to reduce productivity at some sites, especially sites where P is already limiting (Scott and Dean 2006). Other studies have shown that forests may have strong internal controls on P loss. For example, Yanai (1998) reported that increases in P in surface water, following a whole-tree harvest at the Hubbard Brook Experimental Forest, reduced rates of soil P mineralization due to feedbacks with phosphatase production and microbial immobilization, thereby reducing overall P losses from the system.

Change in Species Composition

Forest and rangeland management practices that alter plant species composition by changing the distribution of existing species or by introducing new species may fundamentally alter nutrient cycling dynamics (Hooper and Vitousek 1998; Tilman et al. 1997). These altered dynamics can be attributed to differences in plant life histories, aboveground and belowground vegetative structure, phenology, chemical composition of plant tissues, photosynthetic pathways, N fixation properties, and mycorrhizal associations. An often cited example is the difference in nutrient cycling between evergreen gymnosperms and deciduous angiosperms. In a comprehensive meta-analysis, Augusto and others (2014) compared nutrient cycling under these two forest types and showed that across 200 species representing 100 genera, soil pH, foliar and litter nutrient concentrations, decomposition rates, and net N mineralization and nitrification were generally lower. They also found that accumulation of soil organic matter in soils, weathering of soil minerals, and interception of nutrients in precipitation were generally higher in ecosystems dominated by gymnosperms compared to angiosperms. Forest management practices that shift stands from one of these forest types to the other, or that introduce one or more of these species, can thus have profound impacts on aboveground and belowground nutrient cycling. However, it should be noted that differences between individual species within these forest types can be as great as between forest types. Augusto and others (2014), for example, found that larch (*Larix*) species had higher pH litter and faster N cycling than other coniferous species, Binkley and Valentine (1991) found that N cycling was faster under white pine (*Pinus strobus*) than under green ash (*Fraxinus pennsylvanica*) or Norway spruce (*Picea abies*), and Binkley and others (1986) found higher nitrate (NO$_3^-$) leaching under white pine compared to hardwoods. In more arid environments, both woody plant expansion into grasslands and the invasion of annual grasses into shrublands have been shown to alter soil C and nutrient cycling, largely due to differences in plant life form, phenology, and tissue chemistry (Gill and Burke 1999).

Forest Fertilization

The growth of most forests is limited by current supplies of one or more nutrients. The most common limiting nutrients include N and P (Tamm 1991; Vitousek et al. 2010), but the base cations Ca^{2+}, Mg^{2+}, and K$^+$, or even micronutrients such as B, Mn, Zn, or Si, can be limiting as well (Heiberg et al. 1964; Rashid and Ryan 2004; Thiffault et al. 2011). Forest managers have long recognized the benefits of forest soil fertilization to remedy these limitations and boost forest productivity. Fox and others (2007), for example, demonstrated the value of fertilization in managing the extensive loblolly pine (*Pinus taeda*) plantations in the southeastern United States. Their research showed that over an 8-year period, (1) the addition of 112 kg N ha^{-1} increased forest growth by about 7 m^3 ha^{-1} annually, (2) the addition of 56 kg P ha^{-1} yielded virtually the same growth response, and (3) the addition of the same amount of both N and P almost tripled the growth response to 20 m^3 ha^{-1} annually.

Alvarez (2012) further demonstrated this point in a modeling study that showed that soil nutrients limit forest growth across large swaths of forest lands of the southeastern United States, leading to large regional differences in the growth of loblolly pine plantations (Fig. 4.4). If soil nutrient limitations were alleviated, forest growth could be increased by 10–15%. Similar nutrient limitations to productivity have been reported elsewhere across the country. In the north central United States, for example, Reich and others (1997) showed that forest plantations with an annual soil N supply of 45 kg ha^{-1} produced about 14 m^3 ha^{-1} of wood annually, compared with forest plantations with an annual soil N supply of 90 kg ha^{-1}, which produced well over 21 m^3 ha^{-1} year^{-1}.

Prescribed Fire

Prescribed fire is the intentional use of low to moderate intensity fire as a management tool for a variety of purposes, including restoring or maintaining natural fire ecosystems, increasing soil pH and available nutrients, promoting new plant growth, controlling certain pathogens, creating wildlife habitat, and reducing the risk or severity of future wildfires (Elliott et al. 1999; McCullough et al. 1998; Neary et al. 1999; Vose and Elliott 2016; Vose et al. 1999). Prescribed fires affect forest and rangeland nutrient cycling by altering the distribution of nutrients within biomass pools and altering the physical, chemical, and biological properties of soils (Certini 2005; Knoepp et al. 2004; Richter et al. 1982). Depending on the intensity of the prescribed fire, aboveground and O-horizon nutrients can be oxidized and lost to the atmosphere, converted to ash or charcoal, or remain in incompletely burned vegetation or detritus (Vose and Swank 1993; Vose et al. 1999). Subsequent cycling of these nutrients follows similar pathways (Bodí ct al. 2014; Boerner 1982; Certini 2005; Giovannini et al. 1988). Knoepp and others (2005) summarized variability in the soil nutrient response with fire severity. They showed that while total soil N and organic matter decline with increasing fire severity, available soil N increases after burning and remains higher for the first year, before returning to preburning levels. This increase was evident in stream N export (Knoepp et al. 2004). High severity fires typically result in a significant loss of ecosystem N and soil organic matter (SOM) (Swank and Vose 1993). The loss of SOM, in turn, results in decreased cation exchange capacity, especially in surface soils in which a greater proportion of exchange capacity is attributed to organic matter, not clay minerals.

Ash and charcoal, the materials that remain after partial or near-complete combustion of organic matter, consist of mineral elements and charred organic components (Bodí et al.

Fig. 4.4 3PG model simulations for forest productivity across the southeastern United States with and without nutrient limitation (from Alvarez 2012). (**a**) Current growth of loblolly pine plantations ranges from about 300 to 600 m³ ha⁻¹ of wood volume at age 25. (**b**) If soil fertility did not limit growth, many locations would show much higher productivity. (**c**) Current limits on soil fertility reduce forest growth by about 60–90 m³ ha⁻¹, a reduction of 15–20%. (Source: Graphs based on illustrative simulations using the 3PG model, provided by Jose Alvarez-Munoz)

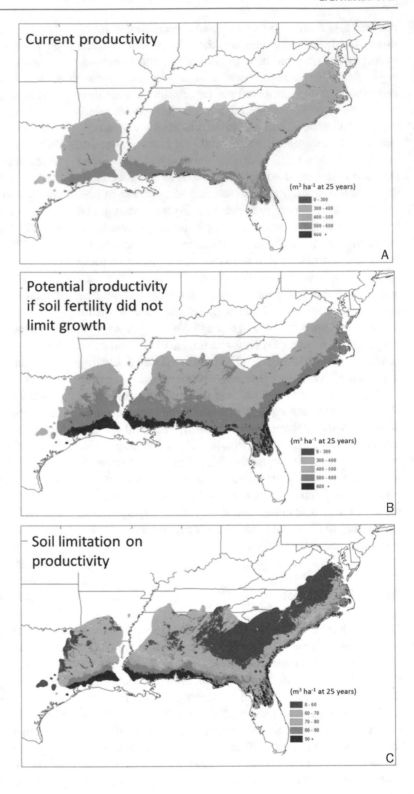

2014). The transformation of living biomass, fresh detritus, and especially SOM in the surface O-horizon to ash has direct and indirect effects on soil nutrient cycling. Base cations and small amounts of P, S, and N are released during organic matter combustion and can contribute to increased soil pH and plant nutrient availability. However, under con-

ditions of severe burning, large layers of ash may remain on site. This material represents a range of pyrolysis compounds that tend to be resistant to chemical and biological degradation, thereby producing long-lasting effects on soil organic matter following fire (González-Pérez et al. 2004).

Prescribed fire can also leave behind charcoal. Charcoal is not readily decomposed and can persist in the soil for many decades and may be effective in sequestering C in soils (Wardle et al. 2008) Additionally, charcoal can increase soil cation exchange capacity, nutrient availability, and N transformations (DeLuca et al. 2006) A review of the impacts of charcoal found widespread evidence for increased soil microbial biomass in addition to increased soil nutrients, including K, P, N, and total C. However, the addition of charcoal did not significantly impact plant productivity (Biederman and Harpole 2013).

Environmental Pollutants

Environmental pollutants include an array of point and non-point source contaminants that are introduced into the natural environment and cause adverse change. Pollutants that impact forest and rangeland nutrient cycles include atmospheric deposition of S, N, and mercury (Hg); toxic and priority pollutants; and contaminants of emerging concern, such as pharmaceuticals and healthcare products.

Atmospheric Deposition of Sulfur, Nitrogen, and Mercury

A Short History of Regulation and Research

Anthropogenic emissions of sulfur dioxide (SO_2) and nitrogen oxides (NO_x) began with the Industrial Revolution around the late 1700s. These pollutants mix with water in the atmosphere and form the acidic compounds sulfuric (H_2SO_4) and nitric (HNO_3) acids, which are returned to the Earth at potentially long distances from their source in what is generally known as atmospheric deposition or, colloquially, "acid rain." The impacts of atmospheric deposition on forested ecosystems as a subject of research began in the mid-twentieth century in Europe (Odén 1968), the northeastern United States (Likens and Bormann 1974; Likens et al. 1972). By the early 1980s, national programs were implemented to study the effects of atmospheric deposition on ecosystems and to examine the impacts of deposition on vegetation, soils, and water. The result was a series of major studies, including the Integrated Forest Study (IFS), an international effort with 15 sites in the United States and 1 each in Canada and Sweden, designed to understand the impacts of atmospheric deposition on forested ecosystems (EPRI 1972), and the National Acid Precipitation Assessment Program (NAPAP 1980). The final output of the IFS was the publication of a synthesis document and the development of the Nutrient Cycling Model known as NuCM (Johnson and Lindberg 1992; Liu et al. 1991), which utilizes detailed vegetation, soil, and water data to predict ecosystem responses to deposition scenarios and to increases, decreases, and

changes in atmospheric chemistry. These comprehensive research programs contributed the scientific underpinnings for the passage of the Clean Air Act Amendment (1990) and additional emission regulations, which have resulted in significant declines in industrial SO_2 and NO_x emissions and H_2SO_4 and HNO_3 concentrations in rainfall (Fig. 4.5). The US Environmental Protection Agency (US EPA) reported that by 2011, S deposition across the United States had decreased by over 50% (US EPA 2015) (Fig. 4.5). The IFS ecosystem-level measurements and research efforts continue in many locations (e.g., USDA Forest Service Experimental Forests and Rangelands and National Science Foundation Long Term Ecological Research sites) and are now focusing on recovery processes that are occurring due to deposition reductions that resulted from the successful implementation of national and regional air quality improvement efforts. The story of acid rain, including the identification of the problem, the history of scientific research, the communication of this research to stakeholders and policymakers, the implementation of pollution controls, and the resulting decline in emissions and recovery of ecosystems, is a model for the successful integration of research, management, and policymaking.

Mercury, which comes from both anthropogenic and natural sources, is another atmospheric pollutant of concern. An estimated 48% of atmospheric Hg emitted in 2015 was from coal- and oil-fired power plants (US EPA 2018). As with S and N, industrial Hg emissions began increasing in the 1800s (Swain et al. 1992; Yin et al. 2010). However, as a result of clean air legislation, shifts from coal to other fuel sources, and use of new technology, Hg emissions declined by nearly 50% between 2005 and 2015 (Fig. 4.5). Mercury has known deleterious biotic effects and is of particular concern to human health. While the original Clean Air Act and its Amendments did not include Hg standards, the US EPA finalized Hg emissions regulations in February 2015 and completed reporting protocols in April 2017 that went into effect in June 2018.

Impacts of Atmospheric Sulfur, Nitrogen, and Mercury Deposition on Forest and Rangeland Biogeochemical Cycling

Research on the effects of atmospheric deposition has focused on plant, microbial, soil, and surface water processes. Both N and S are essential nutrients and are taken up by plants and microbes. This uptake is mediated by seasonal patterns of plant and microbial growth driven by changes in temperature and water availability. Ecosystem N uptake is dominated by biological processes. In some forests, added atmospheric N can have a fertilizing effect; in others, especially where N deposition is high, the added atmospheric N can exceed the capacity of the forest to take up the added N. This can result in N saturation (Aber et al. 1989, 1998),

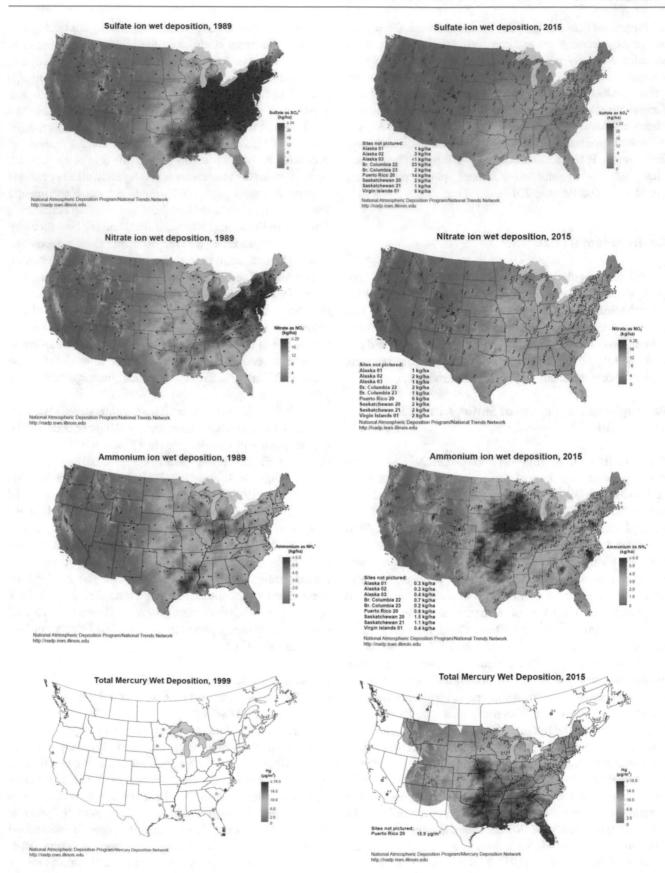

Fig. 4.5 Total wet deposition of sulfate, nitrate, and ammonium ions and mercury in 1999 and 2015. (Source: US EPA 2015)

where excess N appears, sometime deleteriously, in soil and surface waters. Ecosystem S cycling, in contrast, is dominated by chemical and physical processes (Johnson 1984; Swank et al. 1984). Sulfate anions (and to a lesser extent NO_3^- anions) are retained within soil by pH-dependent, variable-charge reactions with soil clay particles and organic matter (Strahm and Harrison 2007). The strength and capacity for these reactions is dependent on the soil mineralogy and organic matter chemistry with a sorption affinity for SO_4^{2-} that is greater than NO_3^-. Inputs of atmospheric SO_4^{2-} (and NO_3^-) anions that exceed the soils capacity for retention can result in the removal of soil base cations such as Ca^{2+} (Bailey et al. 2005; Johnson 1984) and potentially mobilize aluminum (Al^{3+}) (Lawrence et al. 2007), which can negatively impact plant roots (Shortle and Smith 1988) as well as surface water quality and aquatic flora and fauna (Lawrence et al. 2007).

In contrast to N and S, Hg is not a nutrient, and it can be highly toxic to terrestrial and aquatic organisms, particularly as it accumulates and becomes concentrated in the food chain in the form of methylmercury (Giller et al. 1998; Matida et al. 1971; Patra and Sharma 2000). This toxicity to plants and microorganisms in the soil will directly or indirectly alter soil biogeochemical cycles. Mercury can be taken up in gaseous form by plants, or it can be deposited as particles on vegetation surfaces; in both cases, Hg reaches the soil in litterfall (Jiskra et al. 2015). A portion of the deposited Hg can quickly revolatilize (Frossard et al. 2017). As the litter in the O horizon decomposes, Hg is incorporated into the soil organic matter and ultimately moves into the mineral soil where a large proportion is retained (Allan and Heyes 1998). Formation of methylmercury compounds occurs through the reduction of organically bound Hg by soil microbial activity. Methylmercury has a particular affinity for sulfhydryl compounds, which increases its toxicity. Under saturated conditions, methylmercury can be re-emitted to the atmosphere. Emissions of Hg from soil surfaces increase following disturbance from forest harvesting (Carpi et al. 2014; Mazur et al. 2014) or burning (Carpi et al. 2014). Although our understanding of the inputs and cycling of Hg through the air-plant-soil continuums is improving, more research is needed on the impacts of this toxic pollutant on the biogeochemical cycling of other elements in forest and rangeland ecosystems. See Chap. 6 for more details about Hg in wetlands.

Ecosystem Response to Declining Atmospheric Pollutants of Sulfur, Nitrogen, and Mercury

Implementation of clean air legislation, shifts to cleaner fuel sources, and use of new technology have resulted in declining emissions of the atmospheric pollutants S, N, and Hg. Studies have linked declining industrial S emissions to declining atmospheric inputs of SO_4^{2-}, increases in surface water pH and base cations (largely Ca^{2+}), and declines in stream water Al^{3+} (Lawrence et al. 2012) and SO_4^{2-} concentrations (Kahl et al. 2004; Likens et al. 2002; Rice et al. 2014). Despite these signs of "recovery" in response to reduced SO_4^{2-} inputs, many studies also report lags in recovery processes. These lags in recovery have been largely attributed to SO_4^{2-} that had been adsorbed onto soil exchange sites or stored in soil organic matter that is released back into soil and surface waters (Mitchell and Likens 2011). The shift from soil SO_4^{2-} retention that occurred under high SO_4^{2-} deposition scenarios to the release of soil SO_4^{2-} under lower deposition scenarios is predicted to occur over the next two decades in unglaciated soils of the eastern United States, with the dates advancing from North to South (Rice et al. 2014). Lags in recovery have been reported for soil and stream acidity because the base cations (Ca, Mg, K, and Na), which neutralize acidity and were removed from soils by decades of acidification, are only slowly replenished through atmospheric inputs and mineral weathering processes. To better understand the soil processes involved in ecosystem recovery from acidic deposition and to improve the temporal estimates of the recovery process (Lawrence et al. 2015), more research is needed that focuses on long-term soil resampling efforts, using previously sampled sites with archived soil samples.

Nitrogen dynamics present a different set of research challenges. Nitrate deposition across the continental United States is declining, while ammonium (NH_4^+) deposition is increasing (Li et al. 2016) (Fig. 4.5). Although both forms of N are taken up by the biota, the principles of chemical neutrality require that the uptake of NO_3^- is accompanied by the release of a negatively charged ion, typically an organic compound, whereas the uptake of NH_4^+ is accompanied by the release of a hydrogen ion (H^+), an acidifying element. The impacts of the change in the dominant form of atmospheric N deposition on forest and rangeland biogeochemical cycling is currently under investigation.

Compounding the uncertainties about recovery from atmospheric N deposition is the observation that nutrient pools and processes change over time due to changes in species composition, N saturation, climate, or other factors. Argerich and others (2013), for example, examined changes in stream NO_3^- concentrations at Forest Service experimental forests across the continental United States and found different patterns of long-term change. Stream NO_3^- concentrations increased over time at Coweeta Hydrologic Laboratory and Fernow Experimental Forest; declined at Hubbard Brook Experimental Forest, H.J. Andrews Experimental Forest in Oregon, and Luquillo Experimental Forest in Puerto Rico; and did not change at Marcell Experimental Forest in Minnesota. These patterns of change vary depending on the length of record and may be due, in part, to shifts from NO_3^- to NH_4^+ deposition or changes in forest productivity. Additional research is needed to address this issue.

The body of work on the response of forest soils to increases and decreases in atmospheric mercury deposition is sparse. For more information on this topic, see Chap. 6.

Critical Loads of Atmospheric Sulfur and Nitrogen

The concept of critical loads has been used extensively in Europe and more recently in the United States to help inform air pollutant control policies (Nilsson 1988; Pardo et al. 2011). A critical load has been defined as "the quantitative exposure to one or more pollutants below which significant harmful effects on sensitive elements of the environment do not occur, according to present knowledge" (Nilsson and Grennfelt 1988). Exceedances of critical loads of atmospheric inputs of SO_4^{2-} and NO_3^- can result in soil and surface water acidification, N saturation, and changes in biotic community composition and activity, resulting in shifts in biogeochemical cycling of a range of nutrients (Burns et al. 2008; Pardo et al. 2011). In a continental study, McNulty and others (2007) estimated that about 15% of forest soils in the conterminous United States were in exceedance of their critical acid load. These exceedances were most common in New England and West Virginia, with only rare occurrences in the western United States. Although critical loads are a useful tool for expressing chronic ecosystem vulnerability to acidic S and N inputs, additional research on mineral weathering rates and biologically relevant thresholds of response are needed to better constrain these short- and longer-term estimates.

The Clean Air Act and its Amendments have resulted in improved air quality and declining inputs of SO_4^{2-} and NO_3^-, and current models predict that SO_4^{2-} and NO_3^- deposition will soon be below soil critical load values for much of the country. However, the effect of increases in the atmospheric deposition of NH_4^+ remains unknown and warrants further study.

Toxic and Priority Pollutants

The Clean Water Act and its Amendments identify 56 "toxic pollutants" and 126 "priority pollutants" that are now regulated as part of the US national standards for wastewater discharges to surface water (Copeland and Library of Congress, Congressional Research 1993). These pollutants are made up of a wide range of compounds including lead (Pb), Hg, organic chemicals, and pesticides that are commonly found in municipal and industrial effluent and for which there are approved analytical testing methods. These substances can enter soil systems and be incorporated into forest and rangeland nutrient cycles via diffuse pollution pathways including subsurface migration from treatment plants; stormwater from streets; spreading of waste materials or sludge on agricultural or forest lands; and runoff from agricultural, urban, exurban, and industrial sites. Many of these pollutants are directly toxic to trees and soil microorganisms and macroorganisms and thus directly or indirectly affect forest nutrient cycles. Although this is largely an issue for urban or suburban soils, the harmful impacts are likely to become more widespread with increased urbanization, industrialization, and fragmentation of forest and rangelands across the United States (see also Chap. 7).

Contaminants of Emerging Concern

In addition to the toxic and priority pollutants, novel classes of pollutants, collectively known as "contaminants of emerging concern" by the US EPA, are increasingly being released into the environment. These consist of a range of pharmaceuticals and personal care products (collectively called PPCPs) and endocrine-disrupting agents (Caliman and Gavrilescu 2009; Doerr-MacEwen and Haight 2006). Pharmaceuticals include a wide array of steroidal and nonsteroidal anti-inflammatory drugs, antidepressants, anticonvulsants, and antibacterial agents; personal care products that contain chemicals such as phthalates, parabens, and formaldehyde; endocrine-disrupting agents including diethylstilbestrol (the synthetic estrogen DES), dioxin, and dioxin-like compounds; and polychlorinated biphenyls (PCBs), DDT, and some other pesticides. These chemicals have been engineered for specific human and animal therapeutic and hygienic functions, but may cause adverse environmental toxicological effects in soils and surface waters, and thus impact biogeochemical cycles in these systems. These compounds are excreted or disposed of in home, farm, and municipal sewage systems, where they are either degraded or released into the environment in wastewater or sludge. They are now commonly detected near population and agricultural centers and are increasingly being detected in relatively remote environments (Kallenborn et al. 2017). Bernhardt and others (2017) caution that the rate of increase in the production and variety of pharmaceuticals and other synthetic chemicals over the past four decades outpaces the increase in other drivers of global change, including rising atmospheric carbon dioxide (CO_2) concentrations, nutrient pollution, habitat destruction, and biodiversity loss. More information is urgently needed on the unintended impacts of these substances on organisms, their biochemical degradation, persistence in the environment, and bioaccumulation in the food chain. The release of these substances into the environment is undoubtedly having an impact on forest and rangeland nutrient cycles, especially those near urban and suburban areas. Effects on nutrient cycles will likely continue into the future as urbanization increases and the race for new medications and healthcare products continues.

Climate Change, Climate Variability, and Extreme Weather Events

The climate and weather of the United States are undergoing a period of rapid change, characterized by continental-scale increases in temperature; longer growing seasons; alterations in the amount, distribution, and intensity of precipitation; decreases in snow and ice cover; and an increase in the frequency and severity of extreme weather events (Box 4.2). Coupled atmosphere-ocean general circulation models suggest that this pat-

> **Box 4.2 Climate Change in the United States**
> The climate of the continental United States has changed over geologic, millennial, and decadal timescales:
>
> **Climate Change over Geologic and Millennial Timescales**
>
> On a geologic timescale, or during the Quaternary period from 2.58 million years ago to the present, North America has experienced a series of glacial and interglacial periods that have shaped the landscape, soils, and biota of the country. The last ice sheet, which covered much of the northeastern corner and northern tier of the contiguous United States, retreated approximately 11,700 years ago. Based on temperature reconstructions from ice cores, the estimated mean annual temperatures have fluctuated approximately 9 °C during these glacial-interglacial cycles. Roughly 6000 years ago, the Earth (especially the Northern Hemisphere) experienced a warmer period called variously the Mid-Holocene warm period, Hypsithermal, or the Climatic Optimum. This warmer period was characterized by temperatures as much as 4 °C warmer than present and was caused by changes in the Earth's orbit. Over the past millennium, the climate has fluctuated between the cooler period during the Little Ice Age, which lasted from about 1300 to the mid-1800s, and the warming period characterizing the past 100 years. Estimated mean annual temperatures, based on a variety of direct measurements and temperature reconstructions from ice cores, pollen, and tree ring records (IPCC 2014), have fluctuated approximately 1.4 °C during the last 1000 years.
>
> **Recent Climate Change**
>
> Since approximately the end of the Little Ice Age, which coincided with the beginning of the Industrial Revolution, the global climate has changed more rapidly than at any time in the preceding 800,000 years (IPCC 2014). The overwhelming scientific consensus is that this rapid change is in large part due to increased anthropogenic emissions of heat-trapping greenhouse gases, including carbon dioxide, methane, and nitrous oxide (IPCC 2014), resulting from the burning of fossil fuels, cement production, and global shifts in land use, such as the loss of significant portions of the world's C-rich tropical forests. In the United States, the average annual air temperature has increased by an average of 0.6–1.1 °C for the period from 1895 to 2014, with the greatest warming observed in the more northern states and the least across the Southeast. The average annual precipitation across the United States has increased by about 5% during this same period, with increased precipitation in the Northeast, Midwest, and southern Great Plains and decreased precipitation in the Southeast and Southwest. Other changes include an increase in the amount of rain falling in large events (especially in the Northeast), an increase in the length of the frost-free period, and a general intensification of the hydrologic cycle (Hayhoe et al. 2007; Huntington 2006; Melillo et al. 2014).
>
> **Projected Future Climate Change**
>
> Coupled atmosphere-ocean general circulation models predict more change to come, with the amount of change dependent on future greenhouse gas emissions (Melillo et al. 2014). Under a low emissions scenario, which assumes substantial reductions in emissions, mean annual temperatures in the United States are projected to increase by 1.7–2.8 °C; under a higher emissions scenario, the temperature increase could be as much as 2.8–5.6 °C, with the largest increases expected for the upper Midwest and Alaska. Projected future precipitation changes are multidirectional (i.e., some regions receive more precipitation; some receive less) and vary by region and season. The overall trend is for more precipitation and wetter future conditions in the North and less precipitation and dryer conditions in the South (Melillo et al. 2014). Summers are projected to be dryer in most areas of the continental United States, especially in the Northwest and South-Central region. The Southwest is projected to be particularly dry in the winter and spring, while Alaska is projected to be wetter in all seasons.

tern of change will continue into the future (IPCC 2014; Melillo et al. 2014). These continental-scale changes in climate, together with an increasing frequency and severity of extreme weather events, have had and will continue to have direct and cascading indirect impacts on the distribution of terrestrial ecosystems across the country, the productivity of these ecosystems, and the rates of internal physical, chemical, and biological processes, all of which affect nutrient cycling dynamics.

Temperature and Precipitation as Major Factors of Biome Distribution

Temperature and precipitation have long been recognized as major dynamic factors that determine the distribution of terrestrial biomes across the landscape, with desert biomes occupying the hottest and driest regions, temperate forests and grasslands occupying regions characterized by moderate temperatures and rainfall, and tropical rainforest occupying the hottest and wettest regions (Holdridge 1947; Whittaker 1975). Temperature and precipitation also influence the change in biomes with elevation, including the often sharp delineation between boreal forests and alpine tundra that characterizes the tree line. Across the United States, the observed and projected changes in climate are expected to result in the northward expansion of the deserts of the Southwest, the Mediterranean shrublands of the West Coast, the subtropical forests of the Southeast, and the temperate and boreal forests and grasslands of the central and northern United States. Shrub encroachment on arid and semiarid grasslands (D'Odorico et al. 2010; Knapp et al. 2008), invasion of the tundra ecosystems of Alaska by shrubs and trees (Suarez et al. 1999), and the retraction of broad expanses of permafrost soils (Camill 2005; Payette et al. 2004) have already been documented. Increasing temperature may also move biome ecotones upward in elevation (Parmesan and Yohe 2003; Walther 2003). The increasing frequency of extreme weather events, such as hurricanes, droughts, floods, and ice storms, will also cause extensive damage leading to the exceedance of local ecological or physiological thresholds, resulting in transient to permanent state changes, such as conversions of forests to shrublands or shrublands to grasslands (Smith 2011). Because different biomes, as well as different assemblages of species within a biome, are characterized by different intrinsic rates of nutrient cycling, major or even subtle climate shifts or extreme weather-driven shifts in the distribution of these biomes or species assemblages will impact landscape-scale nutrient cycling dynamics. Warming-induced increases in decomposition and the associated C and nutrient loss along with declining permafrost are a stark example of consequences of a climate-induced biome shift.

Temperature and Precipitation as Determinants of Ecosystem Productivity

Temperature and precipitation, along with N and P availability, are primary factors limiting terrestrial ecosystem productivity at local, regional, and global scales. Ecosystem productivity generally increases along geographic gradients of increasing temperature and precipitation (Kang et al. 2006; Raich et al. 2006; Wu et al. 2011). These changes in ecosystem productivity influence the amount and distribution of aboveground and belowground plant biomass and alter the flow of C and nutrients as these elements flux between the soils, microbiota, and vegetation. Soil C, for example, has been shown to increase along the same geographic gradients of increasing temperature and precipitation as plant productivity (Callesen et al. 2003). An increase in plant uptake associated with increased productivity can reduce leaching of nutrients to surface waters or gaseous loss of N compounds (N_2O and N_2); conversely, a decrease in productivity can make nutrients vulnerable to loss through these same pathways.

Temperature and Precipitation as Drivers of Physical, Chemical, and Biological Reactions

At the molecular level, temperature and moisture are fundamental drivers of virtually all biological, chemical, and physical reactions, and any change in temperature and precipitation associated with changing climate or weather will have both direct and indirect effects on the cycling of elements in soils (Rustad and Norby 2002). Biological reactions, such as those associated with plant and microbial metabolic processes, are particularly sensitive to changes in both temperature and moisture.

Rates of reactions typically increase exponentially with an increase in temperature, up to a temperature optima, after which they decline precipitously. The decline in reaction rates at temperatures beyond the optimum is often attributed to the denaturation of biological enzymes that occurs at high temperatures (>40 °C). Although these high temperatures are typically found outside the thermal regime of most soils in the United States, they do occur in hot, arid ecoregions, and they may increasingly occur during hot spells associated with a warming climate (Box 4.2) or for short periods of time following site disturbance (Waide et al. 1988). If not limited by other factors such as nutrient supply, water, or toxins, the rates of biological soil nutrient cycling processes will generally be enhanced as temperature increases. This has been corroborated by long-term field research studies as well as experimental manipulations of soil and air temperatures (Rustad et al. 2001; Wu et al. 2011).

In temperate and boreal ecosystems, increases in temperatures are also advancing the dates of bud break and canopy leaf out, delaying dates of leaf senescence and leaf fall, and overall expanding the length of the growing season. All of these increase the amount of time that plants and microbes are actively cycling soil nutrients. In snow-dominated regions, the number of days with snow cover, the maximum snow depth, and the snow water equivalent are all declining (Box 4.2). Although there is increasing recognition that biological activity occurs under the snowpack (Brooks et al. 1995; Edwards et al. 2007), most of soil biological activity occurs at temperatures above approximately 4 °C, and thus the longer that soil temperatures are above 4 °C, the greater the biological activity and greater the rates of overall nutrient cycling.

The decline or loss of snowpack also has a surprising contradictory impact. Because snow acts as an effective insulator of soils, warmer winters with less snow may ironically result in colder soils with a greater frost depth, leading to colder soils in a warmer world (Groffman et al. 2001a). Greater soil freezing can result in damage to fine roots, decreases in plant uptake, and increased leaching of N and nutrients to surface waters (Comerford et al. 2013; Groffman et al. 2001b). In soils with permafrost (defined as soils with subsurface material that remains below 0 °C for at least 2 consecutive years), the increase in temperature can cause melting of these historically frozen soils, which will potentially release large amounts of stored C (as the greenhouse gases CO_2 and CH_4 and as dissolved organic C), N (as the greenhouse gases N_2O, NO, and N_2 and as dissolved organic and inorganic N), and other nutrients (Schuur et al. 2015). The warming-induced releases of greenhouse gases from historically frozen soils are a cause of great concern because this mechanism provides a powerful, positive feedback to climate change (Schuur et al. 2015). Hypothesized changes in plant growth and nutrient cycling associated with lengthening of growing seasons, declines in extent and duration of snowpacks, and melting of permafrost have been corroborated by long-term field studies as well as experimental manipulations of air and soil temperatures and of the snowpack (Arft et al. 1999; Comerford et al. 2013; Groffman et al. 2001b; Loik et al. 2013).

Water plays several critical roles in soil nutrient cycling. Water sustains plant, microbial, and animal life; controls soil aeration by occupying pore spaces in soils; transports soil nutrients within the soil matrix to plant roots and microsites (via diffusion and mass flow); and transports soil nutrients out of the soil and ecosystem via erosion and leaching. In soils where oxygen is not limiting, soil nutrients typically become less available under drought conditions, as microbially driven aerobic processes regulating soil nutrient cycling (e.g., decomposition, ammonification, nitrification, nitrous oxide production, and aerobic respiration) typically decline with declining moisture, particularly as soil moisture falls below critical thresholds. These same processes also decline under saturated conditions, when water fills soil pores and oxygen becomes limiting (Arnold et al. 1999; Burton et al. 1998; Davidson et al. 2008; Emmett et al. 2004; Pilbeam et al. 1993; Rey et al. 2002; Rustad et al. 2000; Schlesinger 2013; Stark and Firestone 1995; Tate et al.

1988). In hydric soils, such as those found in wetlands, soils are permanently or seasonally saturated by water, and without vigorous oxygen diffusion aided by plant aerenchyma, oxygen is limiting and anaerobic processes dominate (Brady and Weil 2013). Under these conditions, drought can actually induce more favorable conditions for aerobic microbial processes by increasing the oxygen status of the soils (Emmett et al. 2004). Concurrently, rates of anaerobic processes, such as methanogenesis and denitrification, may decline. Increased precipitation may have little impact on these already water-saturated soils.

Overall, even though changes in the timing, intensity, frequency, and type of precipitation have been documented at regional and continental scales, these types of changes receive less attention than changes in the total precipitation amount (Laseter et al. 2012; Melillo et al. 2014). However, as the seasonal distribution of precipitation continues to change (e.g., the monsoon season in the Southwest (Petrie et al. 2014) and the summer versus fall precipitation in the Southeast (Laseter et al. 2012)), the intensity of precipitation continues to increase (as observed in the Northeast (Melillo et al. 2014) and the high rainfall areas of the southern Appalachians (Aber et al. 1993; Laseter et al. 2012)), the frequency of precipitation changes (e.g., projected longer periods between larger rain events for the Northeast (Melillo et al. 2014)), and more precipitation falls as rain or mixed precipitation than snow (as observed in the long-term record at the Hubbard Brook Experimental Forest (Likens 2013)), understanding these impacts on forest and rangeland nutrient cycling dynamics will become increasingly important.

In summary, the short-term changes in weather and longer-term changes in climate that have been observed in the United States over the past century (Melillo et al. 2014) have affected soil nutrients, and these changes along with changes that are projected for the future (IPCC 2014) will continue to have profound effects on forest and rangeland nutrient cycles.

Extreme Disturbance

Extreme natural and anthropogenic ecosystem disturbances, such as wildland fires, drought, mining of minerals, extraction of fossil fuels, and urbanization, are affecting an increasingly larger portion of the landscape. These events can lead to catastrophic disruptions of nutrient cycling within the forests and rangelands. Potential disturbances and their impacts are discussed in greater detail in Chaps. 2 and 3.

Invasive Species, Insect Pests, and Pathogens

A discussion of human-induced impacts on forest and rangeland biogeochemical cycling would be incomplete without a discussion of invasive species, insect pests, and pathogens. The impacts of these organisms occur both aboveground and below-ground. Aboveground impacts can cause defoliation, tree stress, tree decline and mortality, and plant species extirpation; belowground impacts can cause organic and mineral soil disturbance, changes in amount and chemistry of litter inputs, fine root mortality, changes in bulk density, and changes in amount and composition of organic matter (Ayres and Lombardero 2000; Dukes et al. 2009; Lovett et al. 2010). Escalation of impacts from these combined "nuisance" species is expected to continue under future population, land use, and climate scenarios, so these disturbance factors remain an area of concern for land managers and the public (Dukes et al. 2009).

Invasive Species

An invasive species has been defined as "a species that is non-native to the ecosystem under consideration and whose introduction causes or is likely to cause economic or environmental harm or harm to human health" (Beck et al. 2008). Invasive species that impact forest and rangeland nutrient cycles include invasive plants, insects, annelids, and other animals.

Invasive nonnative plant species impact native vegetation through competition for space, water, and nutrients. In some instances, an infestation by invasive plant species can result in the complete (or nearly complete) mortality of a foundational species (Ellison et al. 2005). In all instances, new species assemblages will alter element cycles, although the specific response varies with the invasive species and the ecosystem. A few examples of invasive plant species include kudzu (*Pueraria montana*), Oriental bittersweet (*Celastrus orbiculatus*), English ivy (*Hedera helix*), Canada thistle (*Cirsium arvense*), and sweet clover (*Melilotus officinalis*). Kudzu, introduced to the United States in the 1870s to reduce soil erosion, is a particularly virulent invasive plant species. Since its introduction, it has spread to over three million ha (Forseth and Innis 2004) and continues to spread at a rate of 50,000 ha per year (Hickman et al. 2010). Kudzu is an invasive vine that climbs on existing forest vegetation, adding a structural component to growth suppression (Fig. 4.6). It

Fig 4.6 Kudzu, a fast-growing invasive plant species, can overgrow and shade out native vegetation (Photo credit: Jerry Asher, USDI Bureau of Land Management, via Bugwood.org)

also emits isoprenes, organic molecules that inhibit plant growth. Kudzu spreads rapidly due to the rooting potential of stems, high photosynthetic rate, high water use capacity, and its ability to fix atmospheric N. This increases N inputs to the soil, which can result in elevated N leaching to streams and increased N_2O (a potent greenhouse gas) emissions from the soil to the atmosphere (Hickman et al. 2010). Kudzu's positive growth response to temperature suggests that its range will continue to expand in the future.

Introduced, nonnative insects and annelids, such as beetles, termites, and earthworms (suborder Lumbrica), can have profound impacts on soil nutrient cycling through bioturbation and transformations of aboveground vegetation and soil organic matter. A few examples of these species include Asian long-horned beetle (*Anoplophora glabripennis*), Formosan subterranean termite (*Coptotermes formosanus*), red imported fire ant (*Solenopsis invicta*), and earthworms (*Lumbricus terrestris*).

Earthworms play a particularly important role in soil formation processes and C cycling (Edwards 2004). Fahey and others (2013), for example, showed that in forests of the Northeast and Central United States, the introduction of nonnative earthworms reduced soil C storage in the upper 20 cm of the soil profile by 37%, leading to dramatic reductions in O-horizon organic matter and redistributions of nutrients within the soil profile.

Introduced, nonnative animals can alter both aboveground and belowground element cycling. Wild hogs (*Sus scrofa*), for example, are a common invasive species in the United States, representing a mix of European wild boar released for hunting in the 1940s and feral domestic hogs. Wild hogs reproduce at a high rate and have considerable negative impact on native plants and rates of decomposition and nutrient cycling. Research in the Great Smoky Mountains National Park (located in North Carolina and Tennessee) found that hogs caused considerable damage to the herbaceous layer plants, but not overstory species. Their activities increased turnover and rates of decomposition of the O horizon and increased leaching of N into streams adjacent to areas disturbed by hog activity (Bratton 1975).

Insect Pests

Both the introduction of new insect pests and changes in range or virulence of native insect pests can also have dramatic impacts on forest and rangeland biogeochemical cycling. Examples of insect pests include forest tent caterpillar (*Malacosoma disstria*), gypsy moth (*Lymantria dispar*), spruce budworm (*Choristoneura fumiferana*), southern pine beetle (*Dendroctonus frontalis*), Asian long-horned beetle,

and emerald ash borer (*Agrilus planipennis*). These insects are increasing in abundance and virulence, as ranges expand in response to increasing minimum temperatures and to increased host sensitivity due to larger areas of forests and grasslands being under stress from other factors such as heat, drought, and air pollution. An example on the East Coast is the hemlock woolly adelgid (*Adelges tsugae*), which was introduced into the United States in the 1950s. After a slow initial expansion, the adelgid has moved rapidly northward in response to an increase in minimum winter temperature and now occupies most of the range of eastern Hemlock (*Tsuga canadensis*) (Knoepp et al. 2011). Impacts of the adelgid on pure stands of hemlock common in the northeastern United States resulted in increased soil N (Orwig et al. 2008), while in the mixed stands of the Southern Appalachians, there were no initial changes in N cycling (Knoepp et al. 2011) due to the dominance of a native evergreen shrub in the understory (Elliott et al. 2016; Ford et al. 2012). Forests without the evergreen shrub showed an increase in both N and P cycling (Block et al. 2012, 2013).

An example of an insect pest that has expanded its range on the West Coast is the mountain pine beetle (*Dendroctonus ponderosae*). The range of this native insect of pine (*Pinus* spp.) forests in western North America has historically extended from Mexico to Canada and at elevations from sea level to 3353 m. With an increase in winter low temperatures and increased climate-change-related stress to host trees, the mountain pine beetle has expanded its range to the Northeast and upwards in elevation. Western coniferous forests, mostly in Wyoming, Montana, Colorado, and Idaho, have been ravaged by this insect pest.

Pathogens

Pathogens are also increasing in virulence, particularly as conditions become warmer and wetter and the abundance of stressed trees, forests, and grasslands increases (Sturrock et al. 2011). A few examples of pathogens include chestnut blight (*Cryphonectria parasitica*), beech bark disease (*Cryptococcus fagisuga/Neonectria* spp. complex), sudden oak death (*Phytophthora ramorum*), and root rot fungus (*Armillaria mellea*). An example of the devastating impact of a forest pathogen is the infection of the American chestnut (*Castanea dentata*) by chestnut blight in the 1920s and 1930s, which reshaped American forests across the eastern United States (Elliott and Swank 2008). The loss of chestnut resulted in changes in tree species composition, and while there was certainly a shift in forest biogeochemical cycling, the effects are unknown due to lack of research at that time. More recently, the impact of beech bark disease on forests in the Catskill Mountains of the northeastern

United States was examined by Lovett and others (2010). This disease resulted in the decline of American beech (*Fagus grandifolia*) followed by an increase in sugar maple. The result was an increase in rates of litter decomposition and increased soil N and N leaching (Lovett et al. 2010).

Key Findings

- Maintaining forest and rangeland nutrient pools and cycles is essential to supporting healthy and productive ecosystems. We have learned that intentional actions from forest and rangeland management; unintended consequences of point and nonpoint source pollutants; changes in climate and weather; extreme events such as wildfire, mining, and urbanization; and increases in the distribution and abundance of pests, pathogens, and invasive species have significantly altered forest and rangeland nutrient cycles across the sweep of American history. We anticipate more changes to come. Some of these future changes are predictable; other changes are unpredictable or as yet unknown. In the face of these changes, it is important to be proactive rather than reactive in maintaining forest and rangeland nutrient cycles. We must act before soils are degraded or altered in new ways, as the cost of remediation far exceeds the cost of proactive management.

Key Information Needs

- *The nature and dynamics of soil organic matter*—Our view on the nature and dynamics of soil organic matter continues to evolve. A few decades back, most soil scientists were comfortable in their explanations for why soils contain organic matter. The theory was that some of the material in plant tissues was recalcitrant to decomposition and that some of the by-products of decomposition included complex organic molecules resistant to degradation (i.e., humic substances). The current state of knowledge, however, is shifting as it acknowledges that short-term decomposition of plant materials (over several years or a decade) gives little insight into the processes that regulate the longer-term accumulation and loss of soil organic matter. More startling is the recognition that complex chains of organic molecules may not be common in soils and the apparent dominance of humic compounds may in part be an artifact of the chemicals used in processing soils (Schmidt et al. 2011). The current view of soil organic matter accumulation and loss from soils no longer focuses on molecules that are unusually resistant to decay. The focus is now on complex interactions of soil

structure, which can physically protect molecules from decay (Six et al. 2006; Torn et al. 1997), and interactions with minerals, many of which are redox sensitive and whose availability and oxidation state may change with a changing climate. A great deal of work will be needed to challenge or support the emerging views of soil organic matter (Schmidt et al. 2011). Forest soil scientists will need to pay special attention to soil horizons because the dynamics of O horizons at the top of soil profiles may be very different from the dynamics that characterize mineral soil horizons, where most of the organic matter resides. Understanding these dynamics will take a combination of long-term field experiments with periodic sampling and archiving of samples. These kinds of studies are not common, but there are several successful examples of long-term soil-ecosystem experiments with periodic sampling that are ongoing (Lawrence et al. 2013; Richter et al. 2007).

- *Mineral weathering*—The history of the science on mineral weathering is somewhat different than soil organic matter in that the science of mineral weathering has long existed in a state of uncertainty. While scientists have been certain in their understanding that mineral weathering is a major source of nutrients entering nutrient cycles (as was so elegantly stated by Leopold (1949) in our introduction), we have not had the ability to quantitatively estimate rates of mineral weathering. New conceptual models and interdisciplinary research teams (e.g., teams that have gathered for critical zone science), new and transformative instrumentation, and new uses of long-term forest research sites will be needed to provide future quantitative estimations of mineral weathering (Richter and Billings 2015).

- *Changes in soil biogeochemistry over time*—Better understanding of soil organic matter and mineral weathering will complement some of the most important lessons from soil science over the past 200 years—that soils vary greatly across landscapes and over time. We have a very good understanding of the changes in soils across landscapes, including valuable surveys produced at the county level by the USDA Natural Resources Conservation Service. Knowledge about changes over time is not nearly as strong, and we need to know much more about how forest and rangeland soils change over time, including how the rates of change vary among nutrients and among types of soils, and how management activities influence those rates. A powerful, two-pronged approach for filling in these gaps in knowledge is to study forest and rangeland nutrient cycles intensively at long-term research sites while also making periodic observations of soil change.

- *Investment in long-term research and monitoring*—The United States is a world leader in supporting long-term field studies, with large multiagency, multigenerational networks of research sites such as the USDA Forest Service's Experimental Forest and Ranges; USDA Agricultural Research Service's Long-Term Agroecosystem Research sites; and National Science Foundation's Long-Term Ecological Research sites, National Ecological Observatory Network, and Critical Zone Observatory sites. These research efforts need to be sustained, along with being augmented by new opportunities for integrating knowledge across programs. In addition, it will be important to maintain or develop less-intensive monitoring of soil conditions across many sites because the number of intensive sites does not provide the statistical power needed to extrapolate to actual forest and rangeland soils across landscapes. In particular, developing systematic approaches to soil resampling programs across gradients of natural and human-impacted landscapes will likely be an essential tool for an effective environmental monitoring and assessment (Lawrence et al. 2013). New ways of supporting these long-term research sites are needed, perhaps in new public-private partnerships.

- *Investment in human capital*—Finally, in order to continue to investigate past, current, and future changes in forest and rangeland nutrient cycles, it is critical to invest in human capital by educating forest and rangeland soil scientists. We are losing academic departments, industries, and federal, state, and private jobs that have focused on the science of soils and nutrient cycling. Rather than only building up what we had in the past, we need to channel new support to successful departments and positions, as well as enhance networking opportunities among sites, scientists, and agencies with the goal to form a more cohesive national interorganizational task force on understanding and protecting forest and rangeland nutrient pools and cycles.

Acknowledgments We gratefully acknowledge the contributions from Dan Binkley on the early versions of the manuscript, the astute editing assistance of Linda Geiser and Casey Johnson, and graphics from Mary Zambello.

Literature Cited

Aber JD, Nadelhoffer KJ, Steudler P, Melillo JM (1989) Nitrogen saturation in northern forest ecosystems. Bioscience 39(6):378–386

Aber JD, Magill A, Boone R et al (1993) Plant and soil responses to chronic nitrogen additions at the Harvard Forest. Massachusetts. Ecol Appl 3:156–166

Aber J, McDowell W, Nadelhoffer K et al (1998) Nitrogen saturation in temperate forest ecosystems. Bioscience 48:921–934

Adams MB, Burger JA, Jenkins AB, Zelazny L (2000) Impact of harvesting and atmospheric pollution on nutrient depletion of eastern US hardwood forests. For Ecol Manag 138:301–319

Adams MB, Knoepp JD, Webster JR (2014) Inorganic nitrogen retention by watersheds at Fernow Experimental Forest and Coweeta Hydrologic Laboratory. Soil Sci Soc Am J 78:S84–S94

Allan C, Heyes A (1998) A preliminary assessment of wet deposition and episodic transport of total and methyl mercury from low order Blue Ridge Watersheds, S.E. U.S.A. Water Air Soil Pollut 105:573–592

Alvarez J (2012) Modeling potential productivity of loblolly pine for the Southeastern US Forest Productivity Cooperative. North Carolina State University. Ph.D. dissertation, Raleigh

Arft AM, Walker MD, Gurevitch J et al (1999) Responses of tundra plants to experimental warming: meta-analysis of the international tundra experiment. Ecol Monogr 69:491–511

Argerich A, Johnson S, Sebestyen S et al (2013) Trends in stream nitrogen concentrations for forested reference catchments across the USA. Environ Res Lett 8:014039

Arnold S, Fernandez I, Rustad L, Zibilske L (1999) Microbial response of an acid forest soil to experimental soil warming. Biol Fertil Soils 30:239–244

Augusto L, De Schrijver A, Vesterdal L, [et al.]. (2014) Influences of evergreen gymnosperm and deciduous angiosperm tree species on the functioning of temperate and boreal forests: spermatophytes and forest functioning. Biol Rev 90(2):444–466

Ayres MP, Lombardero MJ (2000) Assessing the consequences of global change for forest disturbance from herbivores and pathogens. Sci Total Environ 262:263–286

Bailey SW, Buso DC, Likens GE (2003) Implications of sodium mass balance for interpreting the calcium cycle of a forested ecosystem. Ecology 84:471–484

Bailey SW, Horsley SB, Long RR (2005) Thirty years of change in forest soils of the Allegheny Plateau, Pennsylvania. Soil Sci Soc Am J 69(3):681–690

Beck K, Zimmerman K, Schardt J et al (2008) Invasive species defined in a policy context: recommendations from the Federal Invasive Species Advisory Committee. Invasive Plant Sci Manag 1(4):414–421. https://doi.org/10.1614/IPSM-08-089.1

Bernhardt ES, Rosi EJ, Gessner MO (2017) Synthetic chemicals as agents of global change. Front Ecol Environ 15:84–90

Biederman LA, Harpole WS (2013) Biochar and its effects on plant productivity and nutrient cycling: a meta-analysis. GCB Bioenergy 5(2):202–214

Binkley D, Fisher RF (2012) Ecology and management of forest soils, 4th edn. Wiley, New York, 362 p

Binkley D, Valentine D (1991) Fifty-year biogeochemical effects of green ash, white pine, and Norway spruce in a replicated experiment. For Ecol Manag 40:13–25

Binkley D, Aber J, Pastor J, Nadelhoffer K (1986) Nitrogen availability in some Wisconsin forests: comparisons of resin bags and on-site incubations. Biol Fertil Soils 2:77–82

Block CE, Knoepp JD, Elliott KJ, Fraterrigo JM (2012) Impacts of hemlock loss on nitrogen retention vary with soil nitrogen availability in the southern Appalachian Mountains. Ecosystems 15:1108–1120

Block CE, Knoepp JD, Fraterrigo JM (2013) Interactive effects of disturbance and nitrogen availability on phosphorus dynamics of southern Appalachian forests. Biogeochemistry 112:329–342

Bodí MB, Martin DA, Balfour VN et al (2014) Wildland fire ash: production, composition and eco-hydro-geomorphic effects. Earth Sci Rev 130:103–127

Boerner RE (1982) Fire and nutrient cycling in temperate ecosystems. Bioscience 32:187–192

Brady NC, Weil R (2013) The nature and properties of soils. Pearson new international edition. Pearson Education, Hoboken

Bratton SP (1975) The effect of the European wild boar, *Sus scrofa*, on gray beech forest in the Great Smoky Mountains. Ecology 56(6):1356–1366

Brooks PD, Williams MW, Walker D, Schmidt SK (1995) The Niwot Ridge snowfence experiment: biogeochemical responses to changes in the seasonal snowpack. In: Tonnessen KA, Williams MW, Tranter M (eds) Biogeochemistry of seasonally snow covered basins, IAHS–AIHS Publ. 228. International Association of Hydrological Sciences, Wallingford, pp 293–302

Buol SW, Southard RJ, Graham RC, McDaniel PA (2011) Soil genesis and classification, 6th edn. Wiley, New York, 544 p

Burger J, Pritchett W (1984) Effects of clearfelling and site preparation on nitrogen mineralization in a southern pine stand 1. Soil Sci Soc Am J 48:1432–1437

Burns DA, Blett T, Haeuber R, Pardo LH (2008) Critical loads as a policy tool for protecting ecosystems from the effects of air pollutants. Front Ecol Environ 6:156–159

Burton AJ, Pregitzer KS, Zogg GP, Zak DR (1998) Drought reduces root respiration in sugar maple forests. Ecol Appl 8:771–778

Caliman FA, Gavrilescu M (2009) Pharmaceuticals, personal care products and endocrine disrupting agents in the environment—a review. Clean Soil Air Water 37(4–5):277–303

Callesen I, Liski J, Raulund-Rasmussen K et al (2003) Soil carbon stores in Nordic well-drained forest soils—relationships with climate and texture class. Glob Chang Biol 9:358–370

Camill P (2005) Permafrost thaw accelerates in boreal peatlands during late-20th century climate warming. Clim Chang 68:135–152

Carpi A, Fostier AH, Orta OR et al (2014) Gaseous mercury emissions from soil following forest loss and land use changes: field experiments in the United States and Brazil. Atmos Environ 96:423–429

Certini G (2005) Effects of fire on properties of forest soils: a review. Oecologia 143:1–10

Clean Air Act of 1970, as amended November 1990; 42 U.S.C. s/s 7401 et seq.; Public Law 101–549. November 15, 1990

Cole D, Rapp M (1981) Elemental cycling in forest ecosystems. Dyn Properties For Ecosyst 23:341–409

Comerford DP, Schaberg PG, Templer PH et al (2013) Influence of experimental snow removal on root and canopy physiology of sugar maple trees in a northern hardwood forest. Oecologia 171:261–269

Copeland C, Library of Congress, Congressional Research Service (1993) Clean Water Act reauthorization. Library of Congress, Congressional Research Service, Washington, DC

Davidson EA, Nepstad DC, Ishida FY, Brando PM (2008) Effects of an experimental drought and recovery on soil emissions of carbon dioxide, methane, nitrous oxide, and nitric oxide in a moist tropical forest. Glob Chang Biol 14:2582–2590

DeHayes DH, Schaberg G, Hawley GJ, Strimbeck GR (1999) Acid rain impacts on calcium nutrition and forest health. Bioscience 49(10):789–800

DeLuca T, MacKenzie M, Gundale M, Holben W (2006) Wildfire-produced charcoal directly influences nitrogen cycling in ponderosa pine forests. Soil Sci Soc Am J 70:448–453

D'Odorico P, Fuentes JD, Pockman WT et al (2010) Positive feedback between microclimate and shrub encroachment in the northern Chihuahuan desert. Ecosphere 1:1–11

Doerr-MacEwen NA, Haight ME (2006) Expert stakeholders' views on the management of human pharmaceuticals in the environment. Environ Manag 38:853–866

Dukes JS, Pontius J, Orwig D et al (2009) Responses of insect pests, pathogens, and invasive plant species to climate change in the forests of northeastern North America: what can we predict? Can J For Res 39:231–248

Duvigneaud P, Danaeyer-de Smet S (1970) Biological cycling of minerals in temperate zone forests. Analysis of temperate forest ecosystems. Springer-Verlag, New York, pp 119–205

Edwards CA (2004) Earthworm ecology, 2nd edn. CRC Press, Boca Raton, 456 p

Edwards AC, Scalenghe R, Freppaz M (2007) Changes in seasonal snow cover of alpine regions and its effect on soil processes: a review. Quat Int 162–163:172–181

Elliott KJ, Swank WT (2008) Long-term changes in forest composition and diversity following early logging (1919–1923) and the decline of American chestnut (Castanea dentata). Plant Ecol 197(2):155–172

Elliott KJ, Hendrick RL, Major AE et al (1999) Vegetation dynamics after a prescribed fire in the southern Appalachians. For Ecol Manag 114:199–213

Elliott K, Miniat CF, Knoepp J et al (2016) Restoration of southern Appalachian riparian forest affected by eastern hemlock mortality. In: Stringer CE, Krauss KW, Latimer JS (eds) Headwaters to estuaries: advances in watershed science and management—proceedings of the fifth interagency conference on research in the watersheds, e-Gen. Tech. Rep. SRS-GTR-211. U.S. Department of Agriculture, Forest Service, Southern Research Station, Asheville, p 285

Ellison AM, Bank MS, Clinton BD et al (2005) Loss of foundation species: consequences for the structure and dynamics of forested ecosystems. Front Ecol Environ 3:479–486

Emmett BA, Beier C, Estiarte M et al (2004) The response of soil processes to climate change: results from manipulation studies of shrublands across an environmental gradient. Ecosystems 7:625–637

Fahey TJ, Yavitt JB, Sherman RE et al (2013) Earthworms, litter and soil carbon in a northern hardwood forest. Biogeochemistry 114:269–280

Federer CA, Hornbeck JW, Tritton LM et al (1989) Long-term depletion of calcium and other nutrients in eastern U.S. forests. Environ Manag 13:593

Fernandez IJ, Rustad LE, Norton SA et al (2003) Experimental acidification causes soil base-cation depletion at the Bear Brook Watershed in Maine. Soil Sci Soc Am J 67:1909–1919

Ford CR, Elliott KJ, Clinton BD, [et al]. (2012) Forest dynamics following eastern hemlock mortality in the southern Appalachians. Oikos 121:523–536

Forseth IN, Innis AF (2004) Kudzu (Pueraria montana): history, physiology, and ecology combine to make a major ecosystem threat. Crit Rev. Plant Sci 23:401–413

Fox TR, Allen HL, Albaugh TJ et al (2007) Tree nutrition and forest fertilization of pine plantations in the southern United States. South J Appl For 31:5–11

Frossard A, Hartmann M, Frey B (2017) Tolerance of the forest soil microbiome to increasing mercury concentrations. Soil Biol Biochem 105:162–176

Gill RA, Burke IC (1999) Ecosystem consequences of plant life form changes at three sites in the semiarid United States. Oecologia 121:551–563

Giller KE, Witter E, Mcgrath SP (1998) Toxicity of heavy metals to microorganisms and microbial processes in agricultural soils: a review. Soil Biol Biochem 30(10–11):1389–1414

Giovannini G, Lucchesi S, Giachetti M (1988) Effect of heating on some physical and chemical parameters related to soil aggregation and erodibility. Soil Sci 146:255–261

González-Pérez JA, González-Vila FJ, Almendros G, Knicker H (2004) The effect of fire on soil organic matter—a review. Environ Int 30:855–870

Gosz JR, Likens GE, Bormann FH (1972) Nutrient content of litter fall on the Hubbard Brook Experimental Forest, New Hampshire. Ecology 53(5):769–784

Groffman PM, Driscoll T, Fahey J et al (2001a) Colder soils in a warmer world: a snow manipulation study in a northern hardwood forest ecosystem. Biogeochemistry 56:135–150

Groffman PM, Driscoll CT, Fahey TJ et al (2001b) Effects of mild winter freezing on soil nitrogen and carbon dynamics in a northern hardwood forest. Biogeochemistry 56:191–213

Hayhoe K, Wake CP, Huntington TG et al (2007) Past and future changes in climate and hydrological indicators in the US Northeast. Clim Dyn 28:381–407

Heiberg SO, Madgwick H, Leaf AL (1964) Some long-time effects of fertilization on red pine plantations. For Sci 10:17–23

Hickman JE, Wu S, Mickley LJ, Lerdau MT (2010) Kudzu (*Pueraria montana*) invasion doubles emissions of nitric oxide and increases ozone pollution. Proc Natl Acad Sci 107:10115–10119

Holdridge LR (1947) Determination of world plant formations from simple climatic data. Science 105:367–368

Hooper DU, Vitousek PM (1998) Effects of plant composition and diversity on nutrient cycling. Ecol Monogr 68:121–149

Hornbeck JW, Smith CT, Martin CW et al (1990) Effects of intensive harvesting on nutrient capitals of three forest types in New England. For Ecol Manag 30(1–4):55–64

Horsley SB, Long RP, Bailey SW et al (2002) Health of eastern North American sugar maple forests and factors affecting decline. North J Appl For 19:34–44

Hotchkiss S, Vitousek PM, Chadwick OA, Price J (2000) Climate cycles, geomorphological change, and the interpretation of soil and ecosystem development. Ecosystems 3:522–533

Huntington TG (2000) The potential for calcium depletion in forest ecosystems of southeastern United States: review and analysis. Glob Biogeochem Cycles 14:623–638

Huntington TG (2006) Evidence for intensification of the global water cycle: review and synthesis. J Hydrol 319:83–95

Hutchinson G (1954) The biochemistry of the terrestrial atmosphere. In: Kuiper GP (ed) The Earth as a planet. University of Chicago Press, Chicago. Chapter 8

Intergovernmental Panel on Climate Change [IPCC] (2014). Climate Change 2014: synthesis report. Contribution of Working Groups I, II and III to the Fifth Assessment Report of the Intergovernmental Panel on Climate Change (Core Writing Team, Pachauri RK, Meyer LA eds). Geneva, Switzerland. 151 p

Jenny H (1980) The soil resource, Ecological Studies 37. Springer, New York

Jerabkova L, Prescott CE, Titus GD et al (2011) A meta-analysis of the effects of clearcut and variable-retention harvesting on soil nitrogen fluxes in boreal and temperate forests. Can J For Res 41:1852–1870

Jiskra M, Wiederhold JG, Skyllberg U et al (2015) Mercury deposition and re-emission pathways in boreal forest soils investigated with hg isotope signatures. Environ Sci Technol 49:7188–7196

Johnson DW (1984) Sulfur cycling in forests. Biogeochemistry 1:29–43

Johnson DW, Curtis PS (2001) Effects of forest management on soil C and N storage: meta analysis. For Ecol Manag 140:227–238

Johnson DW, Lindberg SE (eds) (1992) Atmospheric deposition and forest nutrient cycling: a synthesis of the integrated forest study. Springer-Verlag, New York

Johnson D, Trettin C, Todd D Jr (2016) Changes in forest floor and soil nutrients in a mixed oak forest 33 years after stem only and whole-tree harvest. For Ecol Manag 361:56–68

Kahl JS, Stoddard JL, Haeuber R et al (2004) Have U.S. surface waters responded to the 1990 Clean Air Act Amendments? Environ Sci Technol 38:484A–490A

Kallenborn R, Brorström-Lundén E, Reiersen L-O, Wilson S (2017) Pharmaceuticals and personal care products (PPCPs) in Arctic environments: indicator contaminants for assessing local and remote anthropogenic sources in a pristine ecosystem in change. Environ Sci Pollut Res 25:33001. https://doi.org/10.1007/s11356-017-9726-6

Kang S, Kimball JS, Running SW (2006) Simulating effects of fire disturbance and climate change on boreal forest productivity and evapotranspiration. Sci Total Environ 362:85–102

Knapp AK, Beier C, Briske DD et al (2008) Consequences of more extreme precipitation regimes for terrestrial ecosystems. Bioscience 58:811–821

Knoepp JD, Vose JM, Swank WT (2004) Long-term soil responses to site preparation burning in the southern Appalachians. For Sci 50(4):540–550

Knoepp JD, DeBano LF, Neary DG (2005) Soil chemistry. In: Neary DG, Ryan KC, DeBano LF (eds) Wildland fire in eco-systems: effects of fire on soils and water, General Technical Report RMRS-GTR-42-vol.4. U.S. Department of Agriculture, Forest Service, Rocky Mountain Research Station, Ogden, pp 53–72

Knoepp JD, Vose JM, Clinton BD, Hunter MD (2011) Hemlock infestation and mortality: impacts on nutrient pools and cycling in Appalachian forests. Soil Sci Soc Am J 75:1935–1945

Laseter SH, Ford CR, Vose JM, Swift LW (2012) Long-term temperature and precipitation trends at the Coweeta Hydrologic Laboratory, Otto, North Carolina, USA. Hydrol Res 43:890–901

Lawrence GB, Sutherland CW, Boylen SW et al (2007) Acid rain effects on aluminum mobilization clarified by inclusion of strong organic acids. Environ Sci Technol 41:93–98

Lawrence GB, Shortle WC, David MB et al (2012) Early indications of soil recovery from acidic deposition in US red spruce forests. Soil Sci Soc Am J 76:1407–1417

Lawrence GB, Fernandez IJ, Richter DD et al (2013) Measuring environmental change in forest ecosystems by repeated soil sampling: a North American perspective. J Environ Qual 42:623–639

Lawrence GB, Hazlett PW, Fernandez IJ et al (2015) Declining acidic deposition begins reversal of forest-soil acidification in the northeastern U.S. and eastern Canada. Environ Sci Technol 49:13103–13111

Leopold A (1949) A Sand County almanac. Oxford University Press, New York

Li Y, Schichtel BA, Walker JT et al (2016) Increasing importance of deposition of reduced nitrogen in the United States. Proc Natl Acad Sci 113:5874–5879

Likens GE (1989) Some aspects of air pollutant effects on terrestrial ecosystems and prospects for the future. Ambio 18:172–178

Likens GE (2013) Biogeochemistry of a forested ecosystem, 3rd edn. Springer, New York, 208 p

Likens GE, Bormann FH (1974) Acid rain: a serious regional environmental problem. Science 184:1176–1179

Likens GE, Bormann FH (1995) Biogeochemistry of a forested ecosystem, 2nd edn. Springer-Verlag, New York, 159 p

Likens GE, Bormann FH, Johnson NM (1972) Acid rain. Environ Sci Policy Sustain Dev 14:33–40

Likens GE, Driscoll CT, Buso DC (1996) Long-term effects of acid rain: response and recovery of a forest ecosystem. Science 272:244–246

Likens GE, Driscoll CT, Buso DC et al (1998) The biogeochemistry of calcium at Hubbard Brook. Biogeochemistry 41:89–173

Likens GE, Driscoll CT, Buso DC et al (2002) The biogeochemistry of sulfur at Hubbard Brook. Biogeochemistry 60:235–316

Liu S, Munson R, Johnson D et al (1991) Application of a nutrient cycling model (NuCM) to a northern mixed hardwood and a southern coniferous forest. Tree Physiol 9:173–184

Loik ME, Griffith AB, Alpert H (2013) Impacts of long-term snow climate change on a high-elevation cold desert shrubland, California, USA. Plant Ecol 214:255–266

Lovett GM, Goodale CL (2011) A new conceptual model of nitrogen saturation based on experimental nitrogen addition to an oak forest. Ecosystems 14:615–631

Lovett GM, Arthur MA, Weathers KC, Griffin JM (2010) Long-term changes in forest carbon and nitrogen cycling caused by an introduced pest/pathogen complex. Ecosystems 13:1188–1200

Mann LK, Johnson DW, West DC et al (1988) Effects of whole-tree and stem-only clearcutting on postharvest hydrologic losses, nutrient capital, and regrowth. For Sci 34:412–428

Matida Y, Kumada H, Kimura S et al (1971) Toxicity of mercury compounds to aquatic organisms and accumulation of the compounds by the organisms. Bull Freshwat Fish Res Lab 21:197–227

Mazur M, Mitchell C, Eckley C et al (2014) Gaseous mercury fluxes from forest soils in response to forest harvesting intensity: a field manipulation experiment. Sci Total Environ 496:678–687

McCullough DG, Werner RA, Neumann D (1998) Fire and insects in northern and boreal forest ecosystems of North America. Annu Rev Entomol 43:107–127

McNulty SG, Cohen EC, Moore Myers JA et al (2007) Estimates of critical acid loads and exceedances for forest soils across the conterminous United States. Environ Pollut 149:281–292

Melillo JM, Richmond TC, Yohe GW (eds) (2014) Climate change impacts in the United States: the third national climate assessment. U.S. Global Change Research Program, Washington, DC, 841 p

Mitchell MJ, Likens GE (2011) Watershed sulfur biogeochemistry: shift from atmospheric deposition dominance to climatic regulation. Environ Sci Technol 45(12):5267–5271

Mobley ML, Lajtha K, Kramer MG et al (2015) Surficial gains and subsoil losses of soil carbon and nitrogen during secondary forest development. Glob Chang Biol 21:986–996

Nave L, Vance E, Swanston C, Curtis P (2010) Harvest impacts on soil carbon storage in temperate forests. For Ecol Manag 259(5):857e866

Neary DG, Klopatek CC, DeBano LF, Ffolliott PF (1999) Fire effects on belowground sustainability: a review and synthesis. For Ecol Manag 122:51–71

Nilsson J (1988) Critical loads for Sulphur and nitrogen. Air pollution and ecosystems. Springer, New York, pp 85–91

Nilsson J, Grennfelt P (eds) (1988) Critical loads for Sulphur and nitrogen. NORD 97. Copenhagen, Nordic Council of Ministers, 418 p

Odén, S. (1968) The acidification of air and precipitation and its consequences in the natural environment. Swedish National Science Research Council, Ecology Committee. Bull. 1. (In Swedish). 68 p

Orwig DA, Cobb RC, D'Amato AW et al (2008) Multi-year ecosystem response to hemlock woolly adelgid infestation in southern New England forests. Can J For Res 38:834–843

Ovington J (1962) Quantitative ecology and the woodland ecosystem concept. Adv Ecol Res 1:103–192

Pardo LH, Fenn ME, Goodale L et al (2011) Effects of nitrogen deposition and empirical nitrogen critical loads for ecoregions of the United States. Ecol Appl 21:3049–3082

Parmesan C, Yohe G (2003) A globally coherent fingerprint of climate change impacts across natural systems. Nature 421:37

Patra M, Sharma A (2000) Mercury toxicity in plants. Bot Rev 66:379–422

Payette S, Delwaide A, Caccianiga M, Beauchemin M (2004) Accelerated thawing of subarctic peatland permafrost over the last 50 years. Geophys Res Lett 31:L18208

Petrie MD, Collins SL, Gutzler DS, Moore DM (2014) Regional trends and local variability in monsoon precipitation in the northern Chihuahuan Desert. J Arid Environ 103:63–70

Pilbeam C, Mahapatra B, Wood M (1993) Soil matric potential effects on gross rates of nitrogen mineralization in an Orthic Ferralsol from Kenya. Soil Biol Biochem 25:1409–1413

Prescott CE (2002) The influence of the forest canopy on nutrient cycling. Tree Physiol 22:1193–1200

Raich JW, Russell AE, Kitayama K et al (2006) Temperature influences carbon accumulation in moist tropical forests. Ecology 87:76–87

Rashid A, Ryan J (2004) Micronutrient constraints to crop production in soils with Mediterranean-type characteristics: a review. J Plant Nutr 27:959–975

Reich PB, Grigal DF, Aber JD, Gower ST (1997) Nitrogen mineralization and productivity in 50 hardwood and conifer stands on diverse soils. Ecology 78:335–347

Rey A, Pegoraro E, Tedeschi V et al (2002) Annual variation in soil respiration and its components in a coppice oak forest in Central Italy. Glob Chang Biol 8:851–866

Rice KC, Scanlon TM, Lynch JA, Cosby BJ (2014) Decreased atmospheric sulfur deposition across the southeastern US: when will watersheds release stored sulfate? Environ Sci Technol 48:10071–10078

Richter DD (2007) Humanity's transformation of Earth's soil: pedology's new frontier. Soil Sci 172(12):957–967

Richter DD, Billings SA (2015) 'One physical system': Tansley's ecosystem as Earth's critical zone. New Phytol 206:900–912

Richter DD Jr, Markewitz D (2001) Understanding soil change: soil sustainability over millennia, centuries, and decades. Cambridge University Press, New York, 255 p

Richter DD, Yaalon DH (2012) "The changing model of soil" revisited. Soil Sci Soc Am J 76:766–778

Richter DD, Ralston CW, Harms WR (1982) Prescribed fire: effects on water quality and forest nutrient cycling. Science 215:661–663

Richter DD, Hofmockel M, Callaham MA et al (2007) Long-term soil experiments: keys to managing Earth's rapidly changing ecosystems. Soil Sci Soc Am J 71:266–279

Rustad L, Norby R (2002) Temperature increase: effects on terrestrial ecosystems. Encyc Glob Environ Chang 2:575–581

Rustad LE, Huntington TG, Boone RD (2000) Controls on soil respiration: implications for climate change. Biogeochemistry 48:1–6

Rustad LE, Campbell JL, Marion GM et al (2001) A meta-analysis of the response of soil respiration, net nitrogen mineralization, and aboveground plant growth to experimental warming. Oecologia 126(4):543–562

Schlesinger WH (2013) An estimate of the global sink for nitrous oxide in soils. Glob Chang Biol 19:2929–2931

Schmidt MW, Torn MS, Abiven S et al (2011) Persistence of soil organic matter as an ecosystem property. Nature 478:49

Schuur EAG, McGuire AD, Schadel C et al (2015) Climate change and the permafrost carbon feedback. Nature 520:171–179

Scott DA, Dean TJ (2006) Energy trade–offs between intensive biomass utilization, site productivity loss, and ameliorative treatments in loblolly pine plantations. Biomass Bioenergy 30:1001–1010

Shortle WC, Smith KT (1988) Aluminum-induced calcium deficiency syndrome in declining red spruce. Science 240:1017–1018

Six J, Frey SD, Thiet RK, Batten KM (2006) Bacterial and fungal contributions to carbon sequestration in agroecosystems. Soil Sci Soc Am J 70:555

Smith MD (2011) An ecological perspective on extreme climatic events: a synthetic definition and framework to guide future research. J Ecol 99:656–663

Stark JM, Firestone MK (1995) Mechanisms for soil moisture effects on activity of nitrifying bacteria. Appl Environ Microbiol 61:218–221

Stone E (1975) Effects of species on nutrient cycles and soil change. Philos Trans R Soc Lond B 271:149–162

Strahm BD, Harrison RB (2007) Mineral and organic matter controls on the sorption of macronutrient anions in variable-charge soils. Soil Sci Soc Am J 71:1926–1933

Sturrock RN, Frankel SJ, Brown AV et al (2011) Climate change and forest diseases. Plant Pathol 60:133–149

Suarez F, Binkley D, Kaye MW, Stottlemyer R (1999) Expansion of forest stands into tundra in the Noatak National Preserve, Northwest Alaska. Ecoscience 6:465–470

Swain EB, Engstrom DR, Brigham ME et al (1992) Increasing rates of atmospheric mercury deposition in midcontinental North America. Science 257:784–787

Swank WT, Vose JM (1993) Site preparation burning to improve southern Appalachian pine-hardwood stands: aboveground biomass, forest floor mass, and nitrogen and carbon pools. Can J For Res 23:2255–2262

Swank WT, Fitzgerald JW, Ash JT (1984) Microbial transformation of sulfate in forest soils. Science 223:182–184

Switzer G, Nelson L (1972) Nutrient accumulation and cycling in loblolly pine (Pinus taeda L.) plantation ecosystems: the first twenty years. Soil Sci Soc Am J 36:143–147

Tamm CO (1991) Nitrogen in terrestrial ecosystems. Springer-Verlag, Berlin, 116 p

Tate K, Ross D, Feltham C (1988) A direct extraction method to estimate soil microbial C: effects of experimental variables and some different calibration procedures. Soil Biol Biochem 20:329–335

Thiffault E, Hannam D, Paré D et al (2011) Effects of forest biomass harvesting on soil productivity in boreal and temperate forests—a review. Environ Rev 19:278–309

Tilman D, Knops J, Wedin D et al (1997) The influence of functional diversity and composition on ecosystem processes. Science 277:1300–1302

Torn MS, Trumbore SE, Chadwick OA et al (1997) Mineral control of soil organic carbon storage and turnover. Nature 389:170

Vitousek P (1982) Nutrient cycling and nutrient use efficiency. Am Nat 119:553–572

Vitousek PM, Farrington H (1997) Nutrient limitation and soil development: experimental test of a biogeochemical theory. Biogeochemistry 37:63–75

Vitousek PM, Melillo JM (1979) Nitrate losses from disturbed forests: patterns and mechanisms. For Sci 25:605–619

Vitousek PM, Porder S, Houlton BZ, Chadwick OA (2010) Terrestrial phosphorus limitation: mechanisms, implications, and nitrogen–phosphorus interactions. Ecol Appl 20:5–15

Vogt KA, Grier CC, Vogt D (1986) Production, turnover, and nutrient dynamics of above-and belowground detritus of world forests. Adv Ecol Res 15:303–377

Vose JM, Elliott KJ (2016) Oak, fire, and global change in the eastern USA: what might the future hold? Fire Ecology 12(2):160–179

Vose JM, Swank WT (1993) Site preparation burning to improve southern Appalachian pine-hardwood stands: aboveground biomass, forest floor mass, and nitrogen and carbon pools. Can J For Res 23:2255–2262

Vose JM, Swank WT, Clinton JD et al (1999) Using stand replacement fires to restore southern Appalachian pine–hardwood ecosystems: effects on mass, carbon, and nutrient pools. For Ecol Manag 114:215–226

Waide J, Caskey W, Todd R, Boring L (1988) Changes in soil nitrogen pools and transformations following forest clearcutting. In: Swank WT, Crossley DA Jr (eds) Forest hydrology and ecology at Coweeta. Springer-Verlag, New York, pp 221–232

Walther GR (2003) Plants in a warmer world. Perspect Plant Ecol Evol Syst 6(3):169–185

Wardle DA, Nilsson M-C, Zackrisson O (2008) Fire-derived charcoal causes loss of forest humus. Science 320:629–629

Webster J, Knoepp J, Swank W, Miniat C (2016) Evidence for a regime shift in nitrogen export from a forested watershed. Ecosystems 19(5):881–895

Wells C, Whigham D, Lieth H (1972) Investigation of mineral nutrient cycling in upland Piedmont forest. J Elisha Mitchell Sci Soc 88:66–78

Whittaker RH (1975) Communities and ecosystems, 2nd edn. Macmillan Publishing Co, New York/London

Wu Z, Dijkstra P, Koch GW et al (2011) Responses of terrestrial ecosystems to temperature and precipitation change: a meta-analysis of experimental manipulation. Global Change Biology 17:927–942

Yanai RD (1998) The effect of whole-tree harvest on phosphorus cycling in a northern hardwood forest. For Ecol Manag 104:281–295

Yanai RD, Blum JD, Hamburg SP et al (2005) New insights into calcium depletion in northeastern forests. J For 103:14–20

Yin R, Feng X, Shi W (2010) Application of the stable-isotope system to the study of sources and fate of Hg in the environment: a review. Appl Geochem 25:1467–1477

Forest and Rangeland Soil Biodiversity

Stephanie A. Yarwood, Elizabeth M. Bach, Matt Busse,
Jane E. Smith, Mac A. Callaham Jr, Chih-Han Chang,
Taniya Roy Chowdhury, and Steven D. Warren

Introduction

Regardless of how soil is defined, soils are the most diverse of all ecosystems. It is estimated that 25–30% of all species on Earth live in soils for all or part of their lives (Decaëns et al. 2006). A single gram of soil is estimated to contain 1×10^9 microorganisms, roughly the same population size as the number of humans in Africa (Microbiology by Numbers 2011). That same gram of soil likely contains 4000 species. They are only one part of a larger food web, however, that includes roundworms (phylum Nematoda), springtails (order Collembola), and other fauna (Fig. 5.1). The soil fauna has equally astounding numbers (e.g., 40,000 springtails in 1 m²). Soil organisms, ranging from microbes to moles (family Talpidae), promote crop growth and livestock production (Barrios 2007; Kibblewhite et al. 2008), produce antibiotics (Wall et al. 2015), control nutrient loads in surface soils and groundwater (De Vries et al. 2011), and regulate greenhouse gas emissions (Singh et al. 2010).

Understanding the processes governing soil community size and composition and the functional implications of biodiversity is challenging. Many microorganisms and soil animals utilize the same carbon (C) sources and mineralize at least a portion of that C to carbon dioxide (CO_2), but within any soil ecosystem, specialized organisms break down lignin, transform nitrogen (N)-containing molecules, and produce methane (CH_4). Few studies have systematically tested the role of biodiversity in maintaining both general and specialized functions, however. One study reported a loss of N cycling functions when microbial biomass was experimentally decreased using heat (Philippot et al. 2013b), but less drastic changes may also result in functional shifts undetectable with some of our current measurement methods, such as soil enzyme assays (Burns et al. 2013) or gas fluxes. Disturbance, development, and climate change may impact the functional capacity of forest and rangeland soils, but currently we have little information on the resistance and resilience of soil communities under these scenarios (Allison and Martiny 2008; Coyle et al. 2017) (Box 5.1).

Several studies have reported changes in soil biodiversity following disturbance. For example, bacterial diversity has been observed to increase in intermediate-aged soils created by glacial retreat (Sigler and Zeyer 2004) and land uplift (Yarwood and Högberg 2017), and mature soils typi-

S. A. Yarwood (✉)
Department of Environmental Science and Technology, University of Maryland, College Park, MD, USA
e-mail: syarwood@umd.edu

E. M. Bach
Global Soil Biodiversity Initiative, School of Environmental Sustainability, Colorado State University, Fort Collins, CO, USA

M. Busse
U.S. Department of Agriculture, Forest Service, Pacific Southwest Research Station, Davis, CA, USA

J. E. Smith
Forestry Sciences Laboratory, U.S. Department of Agriculture, Forest Service, Pacific Northwest Research Station, Corvallis, OR, USA

M. A. Callaham Jr
U.S. Department of Agriculture, Forest Service, Southern Research Station, Center for Forest Disturbance, Athens, GA, USA

C.-H. Chang
Department of Environmental Science and Technology, University of Maryland, College Park, MD, USA

Department of Earth and Planetary Sciences, Johns Hopkins University, Baltimore, MD, USA

T. R. Chowdhury
Earth and Biological Sciences Division, Pacific Northwest National Laboratory, Richland, WA, USA

S. D. Warren
U.S. Department of Agriculture, Forest Service, Rocky Mountain Research Station, Provo, UT, USA

R. V. Pouyat et al. (eds.), *Forest and Rangeland Soils of the United States Under Changing Conditions*,
https://doi.org/10.1007/978-3-030-45216-2_5

Fig. 5.1 Soil organisms are linked together in a multi-trophic food web. (Illustration by S. Yarwood, University of Maryland)

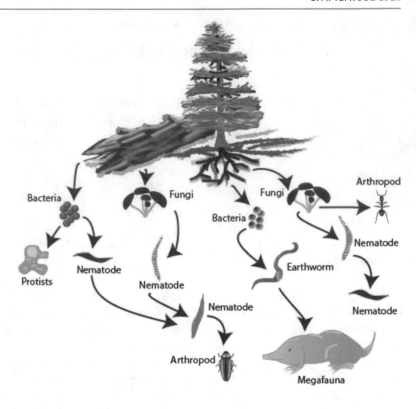

Predicting how the community will react to disturbance requires information on both the resistance and resilience of the community. Resistance refers to the level of disturbance that a community can withstand before it is altered by the disturbance. In this case, predisturbance and postdisturbance measurements along a gradient, such as a fire intensity gradient, can be used to develop the metric (Cowan et al. 2016). A number of studies have been conducted on soil resistance (reviewed in Griffiths and Philippot 2013). Measuring resilience, however, is more complex. Resilience is defined as the rate at which a community returns to its original state following disturbance (Allison and Martiny 2008). Whereas resistance can be measured between two time points, resilience requires monitoring beyond the disturbance event for an often-unspecified timeframe. A disturbance may also lead to a new stable state that may even be functionally similar to the predisturbance condition, but this too is difficult to determine (Shade et al. 2012).

The large amount of functional redundancy (high biodiversity) of soils is thought to contribute to both resistance and resilience, but there are a number of studies that have observed sensitivity of the soil community to disturbance (Shade et al. 2012). Whether community changes lead to a change in function is

largely unknown, but one model suggests that understanding a loss of biodiversity requires not only understanding the functions of the ecosystem under the environmental conditions that existed at the time of disturbance but also understanding how the community functions under different environmental conditions (Fetzer et al. 2015). In other words, a loss of biodiversity may not have immediate impacts on function, but changes may alter the functional capacity of the soil and affect the response when environmental conditions change.

cally contain more soil fauna, a phenomenon explained by the intermediate disturbance hypothesis (Connell 1978; Huston 2014). Soil detritivores also follow a successional pattern as plant litter decomposes, and such fine-scale temporal succession likely contributes to the high diversity found in soils (Bastow 2012). Both conditions highlight the importance of site legacy and temporal change on determining species richness. This begs the question: How do we interpret the majority of studies that have only examined the soil biota at a single time point?

To add another level of complexity, methods used to examine the soil community have changed, and new methods have been adopted for several decades. Following a long history of culture-based and microscopic approaches, the use of biomarkers such as lipids and nucleic acids has now become commonplace. The use of high-throughput sequencing technology to sufficiently capture soil biodiversity has only existed for the last decade, however, and easily accessible computational tools to analyze these data are still being developed. Trying to reconcile observations made using different methodologies continues to be challenging but will hopefully become easier as the soil ecology community pushes for increased consistency through the Earth Microbiome Project, the Global Litter Invertebrate Decomposition Experiment (GLIDE) (Wall et al. 2008), and other initiatives.

Many emerging studies of soil biodiversity, particularly for microorganisms and microfauna, target DNA (deoxyribonucleic acid) to capture the full community. However, this approach can also be problematic because of dormancy. At any time, 25–50% of all detected cells in soils may be dormant (Lennon and Jones 2011), and DNA includes both active and inactive communities. Aside from dormancy, there is also a growing recognition that soils contain extracellular relic DNA (Carini et al. 2016; Dlott et al. 2015). Relic DNA complicates our understanding of the current communities because it simultaneously reflects current conditions as well as legacy conditions. In the case of relic DNA, treating soils with DNAse before extractions may eliminate this inactive pool (Lennon et al. 2017), but this is not commonly done. It should also be noted that relic DNA can be used to study soil animals such as earthworms (Ficetola et al. 2015). It is likely that relic DNA and inactive cells have hampered efforts to connect the biological community to targeted functions. At the end of the chapter, we will discuss other emerging methods that may provide better insights into structure-function relationships.

Before discussing new methods, this chapter will introduce the various groups that compose the soil community, with emphasis on the known trends for each group of organisms in the forest and rangeland soils of the United States. These patterns will be discussed while acknowledging this caveat: All methods used have limitations and underlying assumptions that might not be met, and it is difficult to compare between studies that use different methods. After introducing the groups, we consider properties of the soil habitat that impact soil organisms, such as soil texture and plant communities. We will also discuss the likely impact that periodic disturbances (e.g., from fire or invasive species) and long-term climate changes have on soil biology. Finally, we will consider the future of forest and rangeland soil biology, highlighting emerging methods and long-term studies that promise to deepen our knowledge of soil biodiversity.

Major Groups of Soil Organisms

Viruses

Viral particles vary greatly in abundance across ecosystems; only a few hundred viruses were counted in hot desert soils, but forests and wetlands can contain 10^9 viruses per gram

Table 5.1 Estimated diversity and abundance of soil organisms. These estimates should be considered preliminary, as most soil species have not been described

Taxon	Diversity per amount of soil or area	Abundance (estimated)
Prokaryotes	100–9000	$4–20 \times 10^9$ cm^{-3}
	Genome equivalents cm^{-3}	
Fungi	200–235	100 mg^{-1}
	Operational taxonomic units g^{-1}	
Arbuscular mycorrhizal fungi AMF (species)	10–20 m^{-2}	81–111 m cm^{-3}
Protists	600–4800 sequences g^{-1}	$10^4–10^7$ m^{-2}
Nematodes (genera)	10–100 m^{-2}	$2–90 \times 10^5$ m^{-2}
Enchytraeids	1–15 ha^{-1}	12,000–311,000 m^{-2}
Collembola	20 m^{-2}	$1–5 \times 10^4$ m^{-2}
Mites (Orbatida)	100–150 m^{-2}	$1–10 \times 10^4$ m^{-2}
Isopoda	10–100 m^{-2}	10 m^{-2}
Diplopoda	10–2500 m^{-2}	110 m^{-2}
Earthworms (Oligochaeta)	10–15 ha^{-1}	300 m^{-2}

Modified from Bardgett and van der Putten (2014), which includes additional references

dry weight (Kimura et al. 2008). Current knowledge of soil viruses is scarce, with only a handful of studies enumerating and characterizing them. Most of the literature describes soil viruses as bacteriophages, in part because bacteria are numerous in soils and because bacteriophages have been the target of most studies. Currently there are only about 2000 described soil viruses, a number too low to generate confidence in any prediction of diversity (Williamson et al. 2017). In aquatic systems, 10–50% of bacterial cell lysis is due to viruses, but their impact on bacteria and other soil populations is uncertain and needs more study (Williamson et al. 2017).

Bacteria and Archaea

Prokaryotes are the most abundant cellular organisms in soil and represent much of the DNA diversity on Earth (Table 5.1). Bacteria substantially outnumber archaea in terrestrial ecosystems, including forest soils, where archaea only account for 2% of the community (Bates et al. 2011). In both forests and grasslands, the dominant archaeal groups match putatively identified ammonia oxidizers (Bates et al. 2011), suggesting that their contribution to soil function may be narrowly defined. In contrast, bacteria include a wide range of functional groups: chemoheterotrophs, chemoautotrophs, and photosynthetic cyanobacteria (Fig. 5.2f). Particularly in arid rangelands, cyanobacteria are important in biological soil crusts (Briske 2017) (Box

Fig. 5.2 Soil is home to more than 25% of the Earth's total biodiversity, including (**a**) slime molds (protists), (**b**) earthworms (Oligochaeta), (**c**) termites (Blattodea), (**d**) nematodes (Nematoda), (**e**) springtails (Collembola, *Dicyrtoma fusca* var. *rubrocula*), (**f**) bacteria, and (**g**) millipedes (Myriapoda) and mushrooms (fungi, genus *Chlorophyllum*).

(Photo Credits: (**a**) Creative Commons, Stu's Images; (**b**) Creative Commons slappytheseal; (**c**) M. Bertone; (**d**) D. Robson; (**e**) V. Gutekunst; (**f**) P. Turconi/Fondazione, Istituto Insubrico di Ricerca per la Vita; and (**g**) A. Harrington. All images used with permission)

5.2, Box figs. 5.1 and 5.2). Although major bacterial phyla such as *Acidobacteria*, *Proteobacteria*, and *Verrucomicrobia* are widely distributed in soils, there are also numerous candidate phyla with no cultured representatives (Youssef et al. 2015) and unknown function. Many studies have focused

on comparing compositional differences among soil types. In the case of both archaea and bacteria, soil pH has been found to correlate to community composition across different types of ecosystems (Fierer et al. 2009). Although soil characteristics such as pH, mineralogy, and texture appear

Box 5.2 Biological Soil Crusts

Biological soil crusts develop where various combinations of diminutive bacteria, cyanobacteria, algae, non-lichenized fungi, lichens, bryophytes, and similar microorganisms occupy the surface and the upper few millimeters of the soil (Box Figs. 5.1 and 5.2). Historically, they have been called cryptobiotic, cryptogamic, microbiotic, microfloral, microphytic, or organogenic crusts. Biological soil crusts and the organisms that are part of the crust can be present individually or as consortia in a wide range of ecological, successional, and climatic conditions when and where disturbance or aridity, or both, has limited vascular plant cover and resulted in opportunities for colonization. However, they are most prevalent in arid and semiarid ecosystems where vascular plant cover and diversity are characteristically low, leaving large areas available for colonization by some combination of the organismal groups previously mentioned.

Crust organisms are distributed and dispersed globally in the atmosphere, and they are precipitated wherever and whenever climatic and atmospheric conditions allow (Warren et al. 2017). They are found in all ecosystems but are less represented in dense forests, grasslands, glaciers, and icecaps, where they seldom contact the mineral soil. The ecological roles of biological soil crusts are many and varied and include nutrient cycling, hydrology, and soil stabilization (Belnap and Lange 2001; Warren 1995; Weber et al. 2016). Soil crusts also serve as an essential food source for protozoans, nematodes, tardigrades, rotifers, mites, Collembola, and even larger arthropods and mollusks (Weber et al. 2016). Given the thousands of species involved and their variety, abundance, diversity, and ecological roles, biological soil crust organisms play

Box Fig. 5.2 Surface of a cyanobacterial crust in a fine-textured soil. (Photo credit: Steve Warren, USDA Forest Service.)

an essential role in the soil biodiversity of rangelands. Biological soil crusts and their ecological functions can be disturbed by a variety of factors, including livestock trampling (Warren and Eldridge 2001), off-road vehicles (Wilshire 1983), and fire (Johansen 2001), all of which are common on rangelands.

Box Fig. 5.1 Piece of soil crust with cyanobacteria dangling underneath. (Photo credit: Steve Warren, USDA Forest Service.)

to be most important in influencing bacterial composition, shifts in species composition of vegetation communities have been correlated with shifts in microbial communities in some forest soil comparisons (Urbanová et al. 2015; Uroz et al. 2016).

Fungi

Across all soil types, fungi include both unicellular yeasts belonging to several taxonomic groups and filamentous species, the largest of which span 9.6 km^2. Because of visible fruiting bodies, forest fungi have been studied for centuries, but the links between their aboveground and belowground structures have only been investigated for a few decades. Both fruiting body collections and more recent DNA analyses have helped to highlight the importance of plant communities in shaping fungal communities. An estimated 25% of all C in boreal forest soils is due to fungal biomass (Högberg et al. 2011), most of which belongs to mycorrhizal fungi. The types of mycorrhizal fungi vary between ecosystems, however, with boreal and tundra systems dominated by ectomycorrhizal (EM) and ericaceous mycorrhizae, respectively. The coniferous forests of the Pacific Northwest are also dominated by EM fungi, some of which produce hyphal mats that can cover 40% of the forest floor (Kluber et al. 2010). Unlike plant biodiversity that increases with decreasing latitude, the opposite trend is true for EM (Tedersoo and Nara 2010). Hardwood forests and ecosystems dominated by herbaceous plants tend to host arbuscular mycorrhizal fungi (AMF). Unlike EM fungi, which are represented by numerous species of basidiomycetes and ascomycetes, arbuscular mycorrhizal fungi are relatively closely related, and all belong to the glomeromycetes.

Aside from symbiotic and mutualistic fungi, both saprotrophic and parasitic fungi shape plant communities and provide important ecosystem services. Lignolytic basidiomycetes are unique in their ability to degrade lignin and are visible in forests as white rot. Additionally, basidiomycetes and ascomycetes degrade cellulose, causing brown rot in fallen wood. Fungal pathogens can dramatically alter forests, as evidenced by the death of 100 million elm trees (*Ulmus* spp.) in the United Kingdom and United States due to Dutch elm disease and the death of 3.5 billion chestnut trees (*Castanea dentata*) in the United States due to blight (Fisher et al. 2012). In addition to these historic outbreaks, numerous other pathogenic fungi are considered to be the cause of emerging infectious diseases. Since the early 2000s, spread of the laurel wilt fungus (*Raffaelea lauricola*) by the invasive redbay ambrosia beetle (*Xyleborus glabratus*) has led to widespread mortality of redbay (*Persea borbonia*) and other native trees in the Southeastern United States. Molecular techniques are an increasingly important tool for studying fungi, including both their mutualistic relationships and for pathogen tracking. Soil fungi are not well represented in public sequence databases, however, and it is typical to have one-third of fungal DNA sequences in a soil library match to unknown species (Smith and Peay 2014).

Protists

Protists are divided into seven key taxonomic groups: red and green algae (Archaeplastida), Amoebozoa, Opisthokonta, Stramenopiles, Alveolata, Rhizaria, and Excavata (Orgiazzi et al. 2016). Within the Amoebozoa, slime molds are one of the better studied groups and have primarily been described in temperate forests (Fig. 5.2a). These organisms have complex life cycles that involve a great deal of social interaction. During reproduction, they create complex structures that are multinucleated and visible to the naked eye (Stephenson 2011). Although many protozoa feed on bacteria, there is evidence for selective feeding. The activities of protists are not commonly included in examinations of microbial community composition (Bonkowski et al. 2009; Geisen 2016). Although larger bacteriovores such as nematodes eat patches of bacteria within the soil, protists often feed on single cells. Protists affect plant growth both directly and indirectly:

directly by promoting N mineralization and indirectly by selective predation within the rhizosphere (Ekelund et al. 2009). Using stable isotope labeling, Crotty and others (2012) demonstrated the central role that amoebae have in the soil food web. When the researchers added labeled amoebae to both grassland and woodland soils, they were able to track that label into a wide variety of soil microfauna and mesofauna species.

Microfauna

Animals within the soils are divided according to size rather than on functional or taxonomic attributes (Coleman et al. 2018). The first grouping, microfauna, are less than 0.1 mm in size and include tardigrades (phylum Tardigrada), rotifers (phylum Rotifera), and nematodes (phylum Nematoda) (Fig. 5.2d). Nematodes, also known as roundworms, are ubiquitous and diverse with over 14,000 described species (Kergunteuil et al. 2016) (Fig. 5.2d). Soil nematodes specialize on many food sources. Eight different nematode feeding strategies have been described and include feeding on bacteria, fungi, plants, and other nematodes (Orgiazzi et al. 2016). These feeding strategies mean that nematodes are often found in numerous trophic levels within the soil food web (Fig. 5.1), are important to C flow, and are a link between microbes and fauna. Nematode diversity has been observed to be greater in forests compared with other ecosystems (Ettema and Yeates 2003), and forest age has a major effect on nematode communities and feeding channels (Zhang et al. 2015). Nematode community structure differs among tree species (Keith et al. 2009) and is affected by forest fire (Butenko et al. 2017) and grazing (Wang et al. 2006). In turn, nematodes can alter the fungal and bacterial communities (Blanc et al. 2006) and can enhance plant growth through increased N and phosphorus (P) mineralization (Gebremikael et al. 2016). Studies of nematode communities in North America have largely focused on agricultural and grassland sites, but more research on nematodes in forests is needed to fill an important knowledge gap.

Mesofauna

Mesofauna in the intermediate size range of 2–20 mm includes pot worms or enchytraeids (family Enchytraeidae), mites (subclass Acari), springtails (order Collembola), coneheads (order Protura), two-pronged bristletails (order Diplura), and false scorpions (order Pseudoscorpionida). Mesofauna feed on bacteria, fungi, plant detritus, and microfauna including nematodes, tardigrades, and rotifers. Enchytraeids are small segmented worms that feed primarily on bacteria, fungi, and, to a lesser extent, dead plant material. They are smaller than earthworms and are important members of soil communities in cold, wet ecosystems (Orgiazzi et al. 2016). Springtails are small arthropods with six legs and a short tube on their backside (cellophore) that aids in balancing fluids and electrolytes (Orgiazzi et al. 2016) (Fig. 5.2e). Springtails have been shown to preferentially feed on fungi and plants (Ruess et al. 2007), including both saprotrophic and mycorrhizal fungi. Feeding on arbuscular mycorrhizae would presumably hamper plant growth, but a recent study suggests Collembola enhance plant growth by discouraging arbuscular mycorrhizal sporulation, which in essence maintains the fungi in an active growth state (Ngosong et al. 2014). Of the approximately 7000 species of springtails (Deharveng 2004), some are found in a multitude of ecosystems and others are specialists. Across forest types, springtail communities vary by tree species, with the largest differences observed between conifers and hardwoods (Sławska et al. 2017). There are many mite species that prey on other microarthropods in forests, including springtails and coneheads that can be used as biocontrol agents for some disease-causing microorganisms and microfauna (Schneider and Maraun 2009).

Macrofauna

The macrofauna group includes ants (family Formicidae), termites (order Blattodea), pill bugs or wood lice (order Isopoda), centipedes, millipedes, pauropods, symphylans (order Myriapoda), earthworms, beetles (order Coleoptera), and numerous insect larvae (Fig. 5.2b, c, and g). To belong to this group, animals must be larger than 2 mm.

Ants are ubiquitous and are valuable ecosystem engineers, aerating soils and increasing drainage via their underground galleries (Nemec 2014). Ant activity can also lead to an altered distribution of soil organic matter as some species bring subsoil material to the surface and bury organic matter, while other species transport large quantities of surface organic matter and fresh leaves into mineral soils (Orgiazzi et al. 2016). There are approximately 1000 species of ants native to North America (Miravete et al. 2014), and it is likely there are a few hundred introduced ants that are mostly found in urban environments. Although most ant introductions appear to have little effect on forest and rangeland ecosystem, there are notable exceptions. The introduction of red fire ants (Solenopsis spp.) from South America into the Southern United States has decreased other invertebrate populations and hindered some bird populations, and their sting can cause blindness or death in livestock (Belnap et al. 2012). Fire ants have continued to spread across the Southern United States since their introduction in the 1930s, and this invasion is estimated to cost $1 billion annually (Pimentel 2014).

In a survey of eastern hardwood forests from Connecticut to Florida, 52% of all macroinvertebrates were ants and 45% were termites. Termites are the most numerous organisms in wood (King et al. 2013) (Fig. 5.2c). Like ants, termites are often thought of as a pest species, but termites are also the primary wood degraders in southern forests where they can consume 15–20% of newly deposited wood over 3 years (Ulyshen et al. 2014). Not only do they directly impact the decomposition rates of woody debris, but they also alter the bacterial and fungal populations (Ulyshen et al. 2014). Both ants and termites produce antimicrobial compounds that can suppress microbial activity in decaying wood. Ulyshen and others (2014) observed that wood decomposition proceeded at the same rate in wood where macroinvertebrates were present or absent, most likely due to the suppression of the microbial population. Little research has been done that looks at the link between macroinvertebrates and microorganisms in forest and rangeland systems.

Earthworms are annelids (segmented worms) with the capacity to physically modify soils. For this reason, they are called ecosystem engineers. Earthworms have three different behavioral traits, depending on their preferred food source. Epigeic species live in leaf litter and consume fresh organic matter. Endogeic species live in the mineral soil and consume microbial biomass and other forms of organic matter within the soil matrix. Anecic species build permanent burrows and mix leaf litter with mineral soil by pulling leaves from the surface into their burrows and by casting mineral soil around these leaves. Most earthworm species, to a greater or lesser extent, are responsible for bioturbation of surface soil layers, resulting in the mixing of surface-deposited materials with mineral soil (Orgiazzi et al. 2016). Although many ecosystems in the unglaciated portions of North America contain native earthworm populations, the invasion of earthworms has dramatically altered some US hardwood forests (see Invasive Organisms Shape Soil Biodiversity section).

Megafauna

Soil megafauna include mammals, reptiles, and amphibians. Moles (e.g., Scalopus aquaticus), voles (subfamily Arvicolinae, tribe Arvicolini), mice (Mus spp.), gophers (family Geomyidae), prairie dogs (Cynomys spp.), ferrets (Mustela putorius furo), and badgers (e.g., Taxidea taxus) are some of the most commonly known soil dwellers. These mammals dig in soil, altering the air and water flow through the soil. They may also pull food belowground, where other soil organisms may use it. Amphibians such as caecilians (order Gymnophiona) and salamanders (order Caudata) can live in or on soils and can have a major impact on dead plant litter decomposition.

Reptiles including turtles (order Testudines), tortoises (family Testudinidae), snakes (suborder Serpentes), and lizards (suborder Sauria) often lay eggs in soil, even if the adults primarily live aboveground (Orgiazzi et al. 2016). Several species of snakes live in leaf litter or are weakly fossorial. These snakes prey on insects and other invertebrates, and the larger species prey on soil-dwelling mammals such as voles, moles, and shrews. Some of the fossorial mammals (e.g., gophers and prairie dogs) are important belowground herbivores and can have significant impacts on plant productivity and plant community dynamics in ecosystems where they are abundant. In some cases, these animals can have a significant effect on local topography by building mounds that can persist for centuries. Moles are important soil-dwelling mammals that influence the aeration of soil and also prey on multiple soil invertebrate groups (including large numbers of earthworms in some ecosystems). Their mound building activities are another important source of bioturbation when they are abundant.

The Soil Habitat

Texture and Aggregation

At the heart of the soil habitat lies a complex of structures, niches, and labyrinth-like networks of pores that vary across all spatial scales, from aggregates to landscapes, which helps to explain the unparalleled diversity of soil life (Young and Crawford 2004). One of the most fascinating and challenging realities of studying soils is their heterogeneity. For example, aerobic and anaerobic processes can occur in the same soil profile and can even take place simultaneously. Saturation can change within a few

millimeters, and water can penetrate soil aggregates and create micro-anaerobic spaces in an otherwise well-drained soil. Not only are redox variations caused by water movement, but biological hot spots can quickly deplete oxygen in micropores, leading to anaerobiosis (Raynaud and Nunan 2014). Soil ecologists must constantly integrate the biological community with abiotic soil conditions to understand what an organism is likely to experience in situ.

Soil texture and structure are foundational components of the soil habitat (Fig. 5.3). Texture refers to the particle size distribution of soil minerals (sand, silt, clay), and structure is the three-dimensional arrangement of minerals, organic matter, and pore spaces into aggregates of varying size (Wall et al. 2012). Aggregate formation depends on soil organic matter, fungal hyphae, exudates, earthworm casts, roots, clay, and ionic bridging from metal ions (Bronick and Lal 2005). Categorizing microaggregates and macroaggregates (Six et al. 2000) provides a framework for understanding how the physical habitat affects soil biota. Microaggregates (20–50 μm) protect soil C from decomposition and microbial biomass from predation (Li et al. 2016), and they support greater microbial diversity (Bach et al. 2018). Microaggregates are also physically stable due to strong binding and cementing properties, which make them resistant to disruption from natural and anthropogenic disturbance. Macroaggregates (>250 μm) are more loosely arranged and offer organisms less protection. They are also more easily disrupted by forest and rangeland disturbances (Li et al. 2016).

Pore spaces are the primary habitat for soil organisms (Fig. 5.3). Like aggregates, pore spaces differ in their size, function, and dominant organism types. Micropores (<0.15 μm) exclude nearly all organisms and retain water at tensions unavailable for root uptake. Mesopores (0.15–

Fig. 5.3 A comparison of sizes of structures within soil

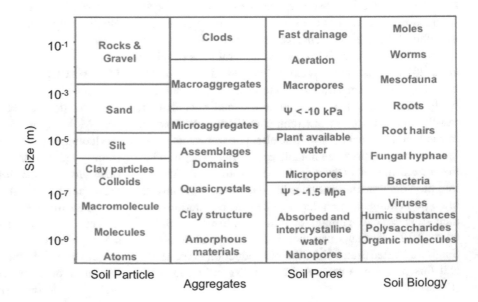

30 μm) offer habitat for bacteria, fungi, and microfauna (e.g., protozoa, nematodes) and contain plant available water; macropores (>30 μm) are critical for gravitational water flow and provide habitat for fungal hyphae and meso-fauna. Larger soil organisms (e.g., earthworms, ants, termites) serve as soil engineers because their movement requires pushing aside and mixing soil particles. Pore size distribution regulates how water, gases, nutrients, heat, and organisms move—and thrive—in soil (Wall et al. 2012). Soil heterogeneity encourages hot spots of biological life and supports high functional redundancy (Wall et al. 2012; Wilhelm et al. 2017a). Organisms may be physically isolated from adjacent predators or resource competitors, sequestered in the safety of small pores, or, alternatively, may be left to experience cosmopolitan life and death where sufficient pore size continuity exists. The arrangement and total volume of pore space moderates the microclimate for soil organisms (Jury et al. 1991). Consequently, temperature fluctuations are greatest near the soil surface and are quelled with soil depth by the insulating effect of soil pores (Busse et al. 2010).

Generalizations about the soil physical habitat and its influence on soil biodiversity can be made. For example, coarse-textured sandy soils support organisms that are best adapted to moisture- and nutrient-limiting stresses. Finer-textured soils provide a greater variety of structural habitat, leading to comparatively high diversity of soil organisms. Generalizations are not always helpful, however, because of unpredictable interactions of the physical habitat with climate, topography, plant communities, time, and site disturbances. Consequently, inferences about habitat conditions and their relationship to forest and rangeland soil biodiversity are still unfolding and are best viewed on a site-by-site basis.

Soil Chemistry

As primary minerals weather into clays, charged surfaces and cation exchange capacity increase. Soil organic matter sticks to mineral surfaces, and the accumulation of both clays and organic matter leads to increased hydrogen ions (H^+) in soil solution. Numerous studies report that microbial composition correlates to soil pH (Fierer and Jackson 2006; Fierer et al. 2009). Soil pH is determined by the many biochemical reactions and mineralogy of the soil (Fernández-Calviño et al. 2011). The relationship between the soil biota and mineralogy likely changes with ecosystem age. It has been hypothesized that biology shapes mineralogy in early ecosystem development. For example, lichens secrete organic acids that solubilize minerals for nutrient acquisition. As the ecosystem ages, however, some

nutrients are lost from the system, and such deficiencies restrict the biota (Brantley et al. 2011).

Minerals differ in their molecular composition and can be preferentially weathered by soil biota. The best-known examples of this weathering involve the actions of mycorrhizal fungi, which are capable of weathering P-containing minerals (Quirk et al. 2012). Bacteria can also colonize specific minerals. For example, bioleaching takes advantage of sulfate reducers that colonize minerals that contain iron (Fe) and sulfur (S) as a means to recover precious metals (Hutchens 2009). Although acquisition of nutrients in acidic forest soils has primarily been ascribed to mycorrhizal associations, aerobic bacteria colonize minerals such as biotite, making Fe and P more plant available (Uroz et al. 2009). Bioturbation of soil by animals can result in subsoils being brought to the surface where they are exposed to surficial weathering processes. Furthermore, the passage of soil through the earthworm gut results in a significant increase in mineral weathering rates (Carpenter et al. 2007; Resner et al. 2011).

Heavy metals can also impact forest and rangelands. In forests they can alter the understory (Stefanowicz et al. 2016) and the soil biota (Tyler et al. 1989). There are about 50,000 US locations that have metal concentrations higher than the normal range of 1–6.5 ppm (Bothe and Słomka 2017). The United States spends $6–8 billion annually remediating metal-contaminated areas (Gall et al. 2015). The patchy nature of effects of heavy metal concentrations and bioavailability make it difficult to track toxicity and quantify ecosystem effects. Laboratory-based studies have assessed bacterial toxicity levels, but they do not correspond to in situ measurements (Giller et al. 2009). Impacts of heavy metals on the soil microflora usually include a decrease in microbial biomass and a shift in community composition (Gall et al. 2015), and although metals can be distributed across the soil food web, heavy metal tolerance varies greatly. For example, earthworms and oribatid mites appear sensitive to heavy metals, but springtails are more tolerant (Tyler et al. 1989).

The Rhizosphere

The rhizosphere, the soil that surrounds and is influenced by a plant root, hosts a large number of biogeochemical processes. Due to the inherent complexity and diversity of plant root systems, the rhizosphere is not a region of definable size or shape but rather includes gradients of chemical, biological, and physical properties that change both radially and longitudinally along the root (Fig. 5.4). Roots can release 10–250 mg C g^{-1} annually or about 10–40% of their total photosynthetically fixed C (Jones et al. 2009).

Fig. 5.4 A cross-sectional view of the association of microorganisms with the rhizosphere or root zone of the *Arabidopsis* plant. (Photo credit: Scanning electron microscope image captured at the U.S. Department of Energy's Environmental Molecular Sciences Laboratory, Richland, WA, and colorized by Alice Dohnalkova)

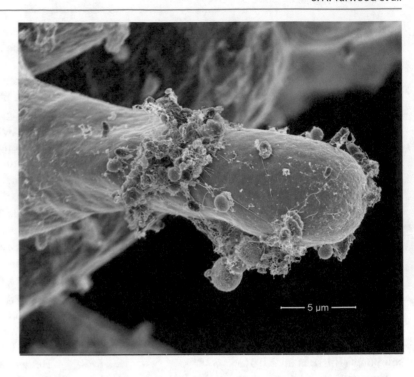

The composition and quantity of released compounds can be influenced by plant species, canopy position, climatic conditions, herbivory, nutrient deficiency or toxicity, and properties of the surrounding soil. Secreted compounds include organic acids, amino acids, proteins, sugar, phenolics, and other secondary metabolites that are readily available substrates for soil microorganisms. These compounds influence rhizosphere processes, including nutrient acquisition (e.g., acquisition of Fe and P), allelopathy, chemotaxis (e.g., between rhizobia and legumes), and the promotion of colonization of beneficial microbes on root surfaces (e.g., *Bacillus subtilis*, *Pseudomonas fluorescens*) (Bais et al. 2004; Park et al. 2004).

The root system architecture is determined by plant species as well as the biotic and abiotic soil conditions. The distribution of nutrients in soils is heterogeneous, and plants can modify their morphology when they sense the presence of nutrients. They are able to allocate more resources to the root system and direct root growth. For example, rangeland grasses such as the invasive cheatgrass (*Bromus tectorum*) have great root plasticity, allowing it to quickly take advantage of nutrient-rich patches (Arredondo and Johnson 1999). Nutrient acquisition from calcareous soils involves rhizosphere processes, such as the exudation of phosphate-mobilizing carboxylates (Hinsinger 2001) or the release of Fe-chelating phytosiderophores (Ma et al. 2003; Robin et al. 2008).

A growing body of research points to the influence the plant rhizosphere has on shaping the legacy of the soil microbial community (Philippot et al. 2013a) and nematode diversity (Keith et al. 2009). There is evidence for a long-term, evolutionary dimension to the interactions between plant roots, microorganisms, and soil in the rhizosphere (Lambers et al. 2009). Aside from the role that these interactions have in cropping systems, there has been a recent interest in how rhizosphere dynamics may influence restoration success (Kardol and Wardle 2010; Philippot et al. 2013a). Reestablishing mutualistic relationships and limiting pathogen load are now acknowledged components of restoration efforts.

The Impact of Disturbance on Soil Biodiversity

Compaction

Maintaining adequate soil porosity is one of the most important objectives of forest soil management (Powers 2006), with long-standing guidelines for disturbance thresholds established for many forests. Soil compaction and rutting during harvesting are of concern, as they often alter total porosity, pore size distribution, and pore continuity and can lead to detrimental changes in soil quality and plant productivity (Powers et al. 2005). It is common to find compaction-caused declines in macroporosity following harvesting, with concomitant increases in mesoporosity and microporosity as large pores are reduced in size (Shestak and Busse 2005). Because of reduced macroporosity, slower rates of water infiltration and related effects on erosion potential, waterlogging in clay soils, and indirect changes in soil life may result. In addition, the collapse of macropores represents a

Fig. 5.5 (a) In 2003, the B&B Fire burned over 36,422 ha in the Deschutes and Willamette National Forests, Mount Jefferson Wilderness, and the Warm Springs Reservation in Central Oregon. In some areas, entire logs were consumed by the fire. The intense heat generated by the burning of large woody material causes soil oxidation, conspicuously changing the soil color from black to various shades of red. High-severity fires that consume entire logs or stumps affect relatively small areas but can have profound soil impacts. After a severe fire, soils may be void of almost all biological activity, and the length of time for recovery is unknown. (Photo credit: Doni McKay, USDA Forest Service.) (b) Postburn photo of a high-intensity soil burn treat-

ment applied by the combustion of a parallel stack of large downed logs in spring 2013 at the Pringle Falls Experimental Forest, Deschutes National Forest in Central Oregon (Cowan et al. 2016; Smith et al. 2016). (Photo credit: Ariel D. Cowan, Oregon State University, used with permission.) (c) The photo shows postfire morels (*Morchella* spp.) fruiting among the leaves of tanoak (*Notholithocarpus densiflorus*) in the spring following the 2002 Biscuit Fire, a massive wildfire that burned nearly 202,343 ha (2020 km²) in the Siskiyou National Forest in Southern Oregon and Northern California. (Photo credit: Jane E. Smith, USDA Forest Service)

loss of habitat for soil macrofauna and mesofauna, leading to reductions in abundance and diversity (Battigelli et al. 2004). Disturbance is often unavoidable, yet it can be moderated by using best management practices, such as operating on designating skid trails or on the surface of frozen or dry soil (Craigg et al. 2015). Long-standing federal guidelines recommend limiting compaction to less than 15% of the land on an aerial basis (Page-Dumroese et al. 2000). However, many regions have scrapped this well-intentioned yet ineffective guideline because it does not account for slope, microtopography, or variable patterns of surface and subsurface water flow.

Compaction and soil disruption present similar challenges in rangeland systems. In the intermountain West, cattle grazing has impacted riparian soils, leading to bank

erosion, bank compaction, and breaking up of soil crusts (Box 5.2). The Taylor Grazing Act of 1934 put in place guidelines for grazing on public lands, but the balance between grazing and ecosystem protection remains a contentious issue. One long-term study observed that heavy grazing leads to a loss of soil C and to changes in the microbial community (Ingram et al. 2008). Globally, grazing by large herbivores, such as cattle (*Bos taurus*), moose (*Alces alces*), or yak (*Bos grunniens*), decreases soil fauna and microbes, but the magnitude of the response varies by biome (Andriuzzi and Wall 2017). In a long-term experiment, trampling resulted in a decrease in bacterial biomass and a shift particularly in cyanobacterial communities, with a decrease in those that fix N (Steven et al. 2015).

Postfire Biodiversity

Wildfires are a common large-scale disturbance that differentially affects soil communities and productivity, largely depending on soil burn severity (Certini 2005). Most fires are of mixed severity and create a mosaic of burned and unburned patches on the landscape (Keeley 2009; NWCG 2003) (Fig. 5.5a, b). Many soil inhabitants have evolved to live with and survive mixed severity fire. For example, amphibians will burrow into the ground, hide in moist logs, or take cover under rocks. Springtails are able to migrate to deeper soil horizons (Malmström 2012).

Soil microbes tend to congregate in the upper soil profile where nutrient concentrations are the highest (Oliver et al. 2015), yet they respond quickly to ecosystem disturbances caused by fire (Barker et al. 2013; Hebel et al. 2009; Smith et al. 2004, 2005, 2017). When large downed wood burns, the soil beneath is exposed to prolonged, intense heat, which can reach temperatures that exceed the lethal threshold for fine roots and most soil organisms (Busse et al. 2013; Smith et al. 2016). High-severity fires may burn at soil surface temperatures exceeding 300 °C (Smith et al. 2016) and can cause partial to total vegetation mortality aboveground and complete or near-complete loss of belowground soil microbes in the top 10 cm of the soil profile (Hebel et al. 2009; Smith et al. 2017). Soil microbial community responses to high-intensity fires tend to be longer lasting and more extreme. On the other hand, in forest ecosystems that are managed under frequent fire regimes (e.g., Southeastern United States), fire alters soil microbial composition but does not affect total diversity (Brown et al. 2013; Oliver et al. 2015). The fires in such ecosystems are generally lower in severity, due to lower fuel loads. Low-severity fires typically produce temperatures below 100 °C at the surface. These fires mostly remove smaller shrubs and small-diameter trees and leave larger trees. The soil microbes below a soil depth of 5 cm remain intact (Cowan et al. 2016; Reazin et al. 2016). Intact EM fungal communities may contribute to the regeneration success and resilience of forests (Cowan et al. 2016). Fire stimulates the appearance of phoenicoid fungi (Carpenter and Trappe 1985), particularly members of the order Pezizales, which commonly start fruiting as soon as a few weeks after fires and continue to fruit for up to 2 years (Adamczyk et al. 2012; Fujimura et al. 2005; Petersen 1970; Warcup 1990; Wicklow 1975). Morels (*Morchella* spp.) are well-known members of the Pezizales (Fig. 5.5c). Springtails may recover from low-intensity fires after a few years, but high-intensity fires may lead to a long-term shift from fungal to bacterial feeding species (Malmström 2012). The patchiness of fire has also been shown to increase the diversity of forest macrofauna that include members of spiders (order Araneae), isopods, centipedes (class Chilopoda), milli-pedes (class Diplopoda), and click beetle (family Elateridae) larvae (Gorbunova et al. 2017).

Environmental change and fire suppression throughout the twentieth century in the Western United States have created conditions that facilitate high-intensity forest fires (Hessburg et al. 2015). The presence and amount of large down wood influence the extent of extreme soil burning (Busse et al. 2013; Smith et al. 2016). A greater emphasis on restoring and maintaining healthy landscapes and the ecological benefits of fire has increased interest in prescribed fire and manual removal of woody materials from forested areas to decrease fire severity. Studies are ongoing to determine whether soils that experience severe fires remain on a unique soil community trajectory that distinguishes them from areas that burn simultaneously but at a lower intensity or whether both fire intensities converge to a system state similar to that preceding the fire disturbance (Reazin et al. 2016).

Invasive Organisms

Invasive organisms are the second most common cause of native species loss, after habitat destruction (Wilcove et al. 1998). In the case of plant invasion, there are numerous plant-soil feedbacks that affect both the ability of the plant to invade and the composition of the soil community, especially in the rhizosphere. For example, a plant's relationship with mycorrhizal fungi can affect its ability to invade a new ecosystem. If the plant has an obligate symbiosis, it is limited by the distribution of the symbiont (Pringle et al. 2009). Invading plants that lack symbiotic relationships can disperse to a new environment, and once established, they can also affect symbiotic relationships of native plants. For example, in grassland soils invaded by cheatgrass, the arbuscular mycorrhizal communities changed in nearby native species (Hawkes et al. 2006). Disruption of the mycorrhizal networks can lead to differences in plant characteristics and ecosystem functions.

The impact that invasive plants have on the soil community, particularly the rhizosphere community, is likely complex and variable between individuals and their location (Coats et al. 2014). Japanese barberry (*Berberis thunbergii*) is an invasive shrub common throughout northeastern United States forests. The ability of this shrub to invade appears to be most influenced by land use legacy, including reforestation following agriculture (DeGasperis and Motzkin 2007). The growth of Japanese barberry has been shown to alter microbial community composition and function, in part by altering N cycling and also by increasing soil pH (Kourtev et al. 2002). The shrub has additionally been observed to increase the density of earthworms, with the species of worms matching nonnative European species (Kourtev et al. 1999).

Invasive European earthworms (suborder Lumbricina) arrived in North America with European settlers, ship ballast, and ornamental plants (Bohlen et al. 2004; Hendrix et al. 2008). Deforestation in the eastern United States led to habitat loss, which greatly reduced native earthworm populations and facilitated the spread of a dozen European species. Prior to colonization by European species, the hardwood forests of Canada, the Great Lakes region, and New England had been earthworm-free since the end of the last glaciation thousands of years ago (Hendrix 1995). However, since the 1990s, invasive European earthworms have invaded these earthworm-free forests, causing dramatic changes in the forest floor and soil habitats (Bohlen et al. 2004). These impacts have been exacerbated by the recent spreading of three co-invading Asian earthworms (*Metaphire hilgendorfi*, *Amynthas agrestis*, and *Amynthas tokioensis*) (Chang et al. 2018) whose range expansion in the last 15–20 years may have been facilitated by recreational fishing, off-road vehicles, and the transportation of compost, horticultural waste, and mulch (Bellitürk et al. 2015; Chang et al. 2017). Environmental factors including soil properties and litter inputs play an important role in determining the abundance of invasive earthworms (Crumsey et al. 2014; Szlávecz and Csuzdi 2007). Recent studies have also shown that white-tailed deer (*Odocoileus virginianus*) abundance increases invasive earthworm abundance (Dávalos et al. 2015).

In North American forests, the presence of invasive earthworms has led to major changes in vegetation, physical and chemical soil properties, and biogeochemical cycles. They are associated with declines in plant diversity and are known to change plant community composition in favor of graminoids and nonnative plants (including Japanese barberry) over native species (Craven et al. 2017). Through feeding, burrowing, and casting, invasive earthworms reduce the understory vegetation and leaf litter layer (Dempsey et al. 2011; Dobson and Blossey 2015; Hale et al. 2005, 2006; Nuzzo et al. 2009) and cause soil mixing and translocation of forest floor C from the O horizon into the soil, resulting in increased litter decomposition rates, a thicker A horizon, and increased aggregate formation (Bohlen et al. 2004; Greiner et al. 2012; Snyder et al. 2011; Szlavecz et al. 2011). Nonnative earthworms change the size of various C and N pools in the soil (Fahey et al. 2013; Ma et al. 2013; Yavitt et al. 2015) and increase CO_2 and nitrous oxide (N_2O) efflux (Eisenhauer et al. 2007). Their activity also affects soil nutrient (calcium, magnesium, potassium, and phosphorus) concentrations (Resner et al. 2015) and increases soil pH.

Altogether, these changes in the soil habitat profoundly affect soil microfauna and mesofauna. By eliminating the thick leaf litter layer (O horizon) on the soil surface, invasive earthworms reduce fungal biomass and increase the ratio of bacteria to fungi (Dempsey et al. 2011, 2013). However, reported results from case studies describing impacts on microbial biomass in the overall soil profile have ranged from positive (Groffman et al. 2004, 2015), to neutral (Snyder et al. 2011), to negative (Eisenhauer et al. 2007). Some of the observed differences can likely be attributed to earthworm species identities and interspecific interactions (Chang et al. 2016). By changing the soil structure, soil organic matter properties, and soil microbial community, invasive earthworms may change C use efficiency of soil microbes, increase soil microbial biomass carrying capacity, and promote C metabolism (Dempsey et al. 2013; Groffman et al. 2015). In general, they also have negative impacts on soil microarthropods, such as springtails and mites (Eisenhauer et al. 2007; Gao et al. 2017).

Climate Change and Belowground Biodiversity

Climate change effects include a number of specific conditions that directly and indirectly alter the soil biota. Three conditions that have been measured across multiple ecosystems are CO_2 enrichment, increased temperature, and altered precipitation patterns. A meta-analysis comparing 75 soil biology studies found that altered precipitation results in the most dramatic and consistent effects across ecosystems (Blankinship et al. 2011). Increased precipitation resulted in an increase in biomass for microbiota and fauna, but the size of the effect was larger in drought conditions. Especially in forested systems, drought can result in lower biomass (Blankinship et al. 2011) and a decrease in the rate of litter decomposition (Lensing and Wise 2007). One reason for the more dramatic effect in forest ecosystems may be related to a greater degree of desiccation of the litter layer (Keith et al. 2010). When precipitation was increased in a rangeland experiment, visible soil crusts decreased, but cyanobacteria increased in DNA analysis, suggesting that altered precipitation in dry ecosystems may dramatically change the basic soil community structure (Steven et al. 2015).

In the case of CO_2 enrichment, shifts in microbial and microfauna populations have been observed in some cases, but not others (Blankinship et al. 2011; García-Palacios et al. 2015). In the decade-long Free Air CO_2 Enrichment (FACE) experiments, differences in responses were also noted between deciduous and coniferous forests. During the early years of the studies, net primary production increased, but later the coniferous forests became increasingly N limited, and net production decreased. This was not true of the deciduous forest, which continued to experience enhanced net primary production (Walker et al. 2015). The CO_2 enrichment also did not lead to increased mycorrhizal colonization in the conifers at the FACE site (Pritchard et al. 2014), although increased mycorrhizal biomass has been observed in other studies (Treseder 2004). The complexity of responses associated with CO_2 enrichment in part stems from varying responses across different taxa and functional groups (García-Palacios et al. 2015); for example, increased abundance of detritivores but not herbivores has been noted across multiple studies (Blankinship et al. 2011).

Higher soil temperatures increase the rate at which soil organisms decompose and respire organic matter (Zhang et al. 2005), including C that has persisted in the soil organic matter pool (Briones et al. 2010; Treseder et al. 2016). These functional changes are accompanied by shifts in the soil biota. García-Palacios and others (2015) observed an increase in fungal biomass when a grassland was warmed, and others have reported a similar shift in favor of fungi over bacteria (Zhang et al. 2005). More detailed phylogenetic analysis has not yielded consistent results across ecosystems, with numerous microbial functional and taxonomic groups changing in abundance in different studies (Pold and DeAngelis 2013). Differences in plant response likely contribute to these varying results, but these cross-ecosystem comparisons are also complicated by the use of different analytical methods across studies.

Forest Management

Harvesting

Recent results from a long-term soil study in North American offer keen insight on the effects of forest harvesting on soil microbial diversity. The Long-Term Soil Productivity (LTSP) study was installed in the early 1990s in coniferous and hardwood forests across the United States and Canada, with the goal of understanding the effects of clear-cut harvesting and site preparation disturbance on soil and vegetation productivity (Powers 2006). Common treatments across sites included soil compaction (none, moderate, severe) and biomass removal during harvesting (bolewood only, whole tree, whole tree plus forest floor). Exhaustive genetic analysis of bacterial and fungal communities identified numerous changes in community composition, including the loss of EM fungal diversity and an increase in heat- and desiccation-tolerant organisms in harvested plots across 18 North American sites (Wilhelm et al. 2017b). The majority of taxa, however, were unaffected by harvesting, leading the authors to state, "Changes resulting from harvesting were relatively minor in comparison to the variability between soil layers and among geographic regions." Results specific to subboreal LTSP sites also discovered that harvesting produced a patchy distribution of fungal populations (Hartmann et al.

2012), stronger effects of compaction on fungal than bacterial communities (Hartmann et al. 2012), modest effects of harvesting on hemicellulolytic populations (Leung et al. 2016), and distinct changes in microbial communities responsible for organic matter decomposition (Cardenas et al. 2015).

It is interesting to compare these results with the 10-year vegetation responses across the long-term soil productivity study network and ask whether there is a common or parallel response between soil and plant communities. Unlike the varied responses shown by soil communities, few changes in vegetation growth have been found across the network of sites and treatments (Ponder et al. 2012). Cardenas and others (2015) suggest that "soil microbial communities are more sensitive than above-ground biomass to harvesting and might be responsive indicators of disturbance." This also reaffirms that the link between soil biodiversity and forest function or productivity has not been conclusively established (Grigal 2000) and that it remains an important area for continued research.

Fuel Reduction Practices (Burning, Thinning, Mastication)

Wildfire mitigation is a priority in United States forests and rangelands, with thinning and burning the most common best management practices in use. These treatments are designed to (1) reduce hazardous fuel loads, (2) lessen wildfire severity and spread, (3) improve forest health and resilience to natural disturbances, and (4) improve forest productivity. Any ecological consequences affecting soil biodiversity are unintentional yet require scrutiny. For example, in a meta-analysis of 139 published studies of forest disturbance, Holden and Treseder (2013) identified a nearly 30% reduction in soil microbial biomass due to both natural and management-caused disturbances, including thinning and prescribed fire.

Most evidence from United States forests points to benign or transient effects of single-application thinning or burning on soil microbial diversity and function (Grayston and Rennenberg 2006; Overby and Gottfried 2017; Overby and Hart 2016; Overby et al. 2015). In the case of fungal diversity, Cairney and Bastias (2007) note that treatment effects are site specific and, in particular, fire-intensity specific, with community responses greatest in the upper soil horizons. Repeated burning, by comparison, has been shown in separate studies to produce positive changes that favor fire-selected fungal communities (Oliver et al. 2015) and reduced activity of cellulolytic fungi that help drive soil C turnover (Bastias et al. 2009).

A recent study of ponderosa pine (*Pinus ponderosa*) forests in Central Oregon provides insight on the importance of fire intensity and time since burning on soil fungal community responses. Treatments included high-intensity burning of log piles, low-intensity broadcast burning, and an unburned control (Cowan et al. 2016). Within 1 week of burning, a strong response in fungal community turnover favoring ascomycetes and fire-responsive populations was found on high-intensity plots, whereas low-intensity broadcast burning showed only moderate changes relative to the control (Reazin et al. 2016). Within 1 year of burning, however, little difference in EM fungal richness, diversity, or composition was found among treatments (Cowan et al. 2016), suggesting that there is value in mixed severity burning to ensure the survival of fungal refugia and meet fuel reduction objectives. Other studies have noted a variety of postfire responses by EM communities, ranging from essentially no change to substantial changes in community composition, with no consistent trend relative to fire intensity (Glassman et al. 2016; Smith et al. 2004; Southworth et al. 2011; Trappe et al. 2009).

New Approaches to Understand Soil Biodiversity

Methods used to study soil biodiversity have changed dramatically in the last several decades. The most significant of these changes has been the adoption of molecular methods that target the biomolecules that compose the soil and the increasingly sophisticated techniques to analyze the biomolecules so as to better describe the soil community. By the early 2000s, most soil microbiology laboratories had begun to include measurements of lipids or DNA fingerprinting, but technologies were such that even a few hundred sequences were considered a large number, even though it was suspected this was far from representative of the community. With today's increased accessibility to next-generation sequencing (Metzker 2010), most microbial DNA studies now include several thousand sequences per sample, and more studies of soil fauna include sequencing to assess biodiversity (Creer et al. 2016). This shift toward sequencing has resulted in new challenges associated with data storage and interpretation, and a new generation of soil biologists is spending more time coding than peering through a microscope. These new tools provide great opportunities and allow the community to ask new and exciting questions: Which organisms are active? What are the functions of all the organisms that we have sequenced but have not cultured or observed extensively? How many cryptic species are there? When does soil biology composition influence soil function? How do we determine the appropriate scale for study?

Over the next several years, we will undoubtedly continue to see studies that take a molecular approach, but those studies will go beyond characterizing communities based solely on ribosomal DNA. We will see increased emphasis on connections of structure and function. One avenue for exploring function is through the use of shotgun metagenomics. This method has been applied to examine the abundance of broad classes of functional genes (Myrold et al. 2014). For example, Uroz and others (2013) characterized the organic horizon soil of Norway spruce (*Picea abies*) as being enriched in genes that degrade carbohydrates compared to the mineral horizon, and soil samples representing a range of different pH were analyzed to determine the physiological adaptations to acid and basic conditions (Malik et al. 2017). Recently, mitochondrial metagenomic analysis has been used to study soil animals and, among other findings, has revealed Collembola dispersal patterns in forested islands (Cicconardi et al. 2017).

Sequencing that targets DNA is only one of many methods collectively referred to as "omics." Targeting messenger RNA, the community's transcriptome, can be used to determine short-term responses, such as the genetic interplay between tree roots and newly colonizing mycorrhizae (Kurth et al. 2015). The first report of a soil metatranscriptome was from a method development study designed to elucidate the functional diversity of eukaryotic microorganisms in forest soil (Bailly et al. 2007). Among the complementary DNA (cDNA) sequences recovered, phosphate transporters, and glutamine synthetases, researchers gained a new understanding of nutrient acquisition. The key feature of metatranscriptomic analysis involves random sequencing of microbial community RNA in the absence of predefined primer or probe specificity; metatranscriptomics thus has a great potential for the discovery of novel genes.

Enzyme activity within soil has been studied by measuring degradation of specific organic molecules since the nineteenth century, but new technologies can target the protein molecules themselves (Bastida et al. 2009). Metaproteomics was used to identify the fungal enzymes important in litter decomposition and uncovered the seasonal shift from ascomycetes to basidiomycetes (Schneider et al. 2012). Both soil metaproteomics and metabolomics (Koek et al. 2006; Kind et al. 2009) are still under development, but both are promising tools that will be applied in the near future to better link microbial communities to function.

Aside from using molecular methods alone, several studies have linked molecular methods with stable isotopes to target specific functional communities. By using labeled substrates (e.g., ^{18}O, ^{13}C, or ^{15}N), stable isotope probing (SIP) distinguishes metabolically active members of a soil community from the inactive ones. Sufficient incorporation of the label into the biomolecule of interest (e.g., lipids, nucleic acids, proteins) allows the "heavy" fraction to be separated from the "lighter" one. Stable isotope probing identifies active consumers in environmental samples (Dumont and Murrell 2005), linking metabolic capacity to phylogenetics

(Hungate et al. 2015) and genomics (Chen and Murrell 2010). The SIP method has already been applied to examine some food web structures. For example, the microbes associated with methanotrophic communities were tracked using DNA-SIP (Maxfield et al. 2012), and Collembola eating habits were investigated using labeled lipids (Menzel et al. 2017). A rapidly developing area of research combines molecular gut content analysis, stable isotope technique, and fatty acid analysis. This approach was used to investigate feeding group, trophic level, or food resource preference of earthworms (Ferlian et al. 2014; Chang et al. 2016), Collembola (Ferlian et al. 2015), and mites (Pollierer et al. 2009). With high-throughput sequencing, molecular gut content analysis will be a much more powerful tool for understanding basic predator-prey connections.

The technologies described here will undoubtedly continue to shape the study of soil biodiversity. Future progress in sequencing technologies and analysis may lead to powerful alternative strategies, such as combining mRNA-SIP with metatranscriptomic analysis (Dumont et al. 2011; Jansson et al. 2012). Higher-resolution sequencing in space and time will allow researchers to apply network analysis (Faust and Raes 2012) and structural equation modeling to community analysis so as to uncover novel interactions between community members and their environment. These new methods should not be adapted at the expense of methods used to better describe species, however. A large number of soil microbes still exist that have not been cultured or characterized, and there is dwindling expertise in soil invertebrates, including microfauna. As new molecular technologies open many exciting avenues, it should also be a priority to retain taxonomic knowledge and pass it to a new generation of scientists.

Conclusions

The forest and rangeland soils of the United States are home to a myriad of biological diversity. Diverse soil habitats, encompassing a spectrum of nutrient availability, moisture content, gas diffusion, pore sizes, and inputs from the surface, provide ample spaces for biological specialization and interaction. These habitats are dynamic through time as well, responding to natural and human-generated disturbances, such as compaction, wildland fire, invasive species, and climate change. People shape these biological interactions through how we choose to manage forests and rangelands. Although definitive evidence linking biodiversity responses to management practices is far from complete, we know that changes in microbial species richness, diversity, and community composition are common consequences of management practices. Intensive harvesting leads to greater changes in biodiversity compared to less-intense practices such as

thinning, and the strongest biodiversity effects are seen in the O horizon and the upper mineral soil. Short-term changes in biodiversity (0–5 years postdisturbance) can be expected, but long-term responses are less common, suggesting community resilience (Box 5.1).

Recent molecular methodological advances have increased knowledge of soil biodiversity, especially microorganisms. Yet challenges remain to identify active soil communities, interactions among organisms and trophic levels, and functional importance to the whole ecosystem. Improved understanding of biological interactions within soil is key to sustaining and protecting our forests and rangelands now and for the future. Increased knowledge of the soil biodiversity has the potential to improve functional predictions. A recent meta-analysis of 82 environmental datasets revealed that 44% of variations in process rates could be explained by environmental variables such as temperature, moisture, and pH, but in most cases, models for C and N cycling were improved when microbial community data were included (Graham et al. 2014). The challenge for the next several years will be determining the best biological parameters to measure and exactly how to integrate those data into existing and new models.

In a rapidly changing world, there has been a great deal of conversation about biodiversity loss. This conversation has focused on plants and animals, and it remains an open question if we are also losing soil biodiversity. Overall, high levels of belowground biodiversity do not necessarily correspond with hot spots of aboveground biodiversity, such as the Tropics. This means prioritizing conservation efforts to areas with high aboveground biodiversity may not protect soil organisms and the benefits we derive from their interactions and resulting ecosystem functions and services. A recent meta-analysis of soil biodiversity studies concluded that because of functional redundancy, a loss of biodiversity is not likely to lead to changes in the C cycle unless key organisms are lost (Nielsen et al. 2011). The problem with this conclusion is we lack a clear understanding of what those key organisms might be, and gaining this knowledge will likely take many more years of research. In the meantime, should we protect soil biodiversity? Likely the functional redundancy in soil does increase its resistance and resilience, and if we continue neglect protections for soil biodiversity, we run the risk of losing key species before they are recognized (Turbé et al. 2010).

Key Findings

- Soil is home to a myriad of biological diversity, accounting for about 25–30% of all species on Earth. Community members range vastly in size, mobility, ecological function, and response to disturbance, and collectively they flourish in a multitude of physical and chemical soil habitats.

- Changes in species richness, diversity, and community composition are common consequences of management practices. Although short-term changes (0–5 years post-disturbance) can be expected, long-term responses are still unknown.
- Intense harvesting and severe burning lead to greater changes in species diversity and community composition compared to less-intense practices such as thinning or low-severity prescribed burning. Best management practices that limit soil compaction, severe soil heating, and exposure of bare mineral soil help to sustain community diversity and resilience.
- Major disturbances such as wildfire, invasive plants and animals, and climate change are likely to modify the health and function of soil organisms. However, the extent of such changes is difficult to generalize (beyond site-specific responses), as studies have yielded inconsistent results across ecosystems and disturbance severities.
- Links between plant diversity and soil biodiversity are ambiguous in US forests and rangelands. Thus, prioritizing conservation efforts to areas with high plant diversity may not protect soil organism diversity or function.
- Improved understanding of soil biodiversity, composition, function, and resilience is a pressing need to assist efforts to sustain and protect our forests and rangelands. Until recently, the use of traditional research techniques offered limited insight toward filling this knowledge gap. New and expanding molecular technologies now provide an unprecedented capability to address current and future ecological questions for the benefit of land stewardship.

Key Information Needs

Science

- **Increased knowledge of soil biodiversity**—currently, soil diversity is largely underdescribed. The groups with the most undescribed taxa are bacteria and archaea, but many groups, including fungi, nematodes, and insects, have numerous undescribed taxa.
- **Distribution of soil organisms across ecosystems in North America**—such information could help identify areas at risk from pathogens and categorize communities that perform specific ecosystem functions well (e.g., C storage, plant production, water infiltration).
- **Continuation of taxonomic expertise, particularly for invertebrate taxa**—fewer young scientists are being trained in these areas of expertise, and the field is at risk of becoming stagnant or even losing knowledge.
- **More detailed information on how diverse soil communities contribute key ecosystem functions, including** water filtration and storage, nutrient and C cycling, and wildlife habitat—combining emerging techniques like high-throughput sequencing and stable isotope probing (SIP) will deepen our understanding of these areas.
- **Determine climate change impacts on soil communities**—this is an ongoing area of research, and work needs to continue.

Management Questions

- **How do soil organisms respond to management regimes such as thinning, prescribed fire, and grazing?** This document provides an initial synthesis of some responses reported in the literature, but comprehensive synthesis reports and distilled fact sheets could further inform management decision-making processes.
- **What invasive soil organisms are present and how do they spread?** Some monitoring efforts are currently underway for invasive earthworms and flatworms, but there is a need for sustained, systematic monitoring.
- **How can forest and rangeland management anticipate climate change and protect or enhance soil biodiversity to promote ecosystem resistance and resilience?**

Literature Cited

Adamczyk JJ, Kruk A, Penczak T, Minter D (2012) Factors shaping communities of pyrophilous macrofungi in microhabitats destroyed by illegal campfires. Fungal Biol 116(9):995–1002

Allison SD, Martiny JB (2008) Resistance, resilience, and redundancy in microbial communities. Proc Natl Acad Sci 105(Suppl 1):11512–11519

Andriuzzi WS, Wall DH (2017) Responses of belowground communities to large aboveground herbivores: meta-analysis reveals biome-dependent patterns and critical research gaps. Glob Chang Biol 23(9):3857–3868

Arredondo JT, Johnson DA (1999) Root architecture and biomass allocation of three range grasses in response to nonuniform supply of nutrients and shoot defoliation. New Phytol 143(2):373–385

Bach EM, Williams RJ, Hargreaves SK et al (2018) Greatest soil microbial diversity found in micro-habitats. Soil Biol Biochem 118:217–226

Bailly J, Fraissinet-Tachet L, Verner M-C et al (2007) Soil eukaryotic functional diversity, a metatranscriptomic approach. ISME J 1(7):632–642

Bais HP, Fall R, Vivanco JM (2004) Biocontrol of *Bacillus subtilis* against infection of *Arabidopsis* roots by *Pseudomonas syringae* is facilitated by biofilm formation and surfactin production. Plant Physiol 134(1):307–319

Bardgett RD, van der Putten WH (2014) Belowground biodiversity and ecosystem functioning. Nature 515(7528):505–511

Barker JS, Simard SW, Jones MD, Durall DM (2013) Ectomycorrhizal fungal community assembly on regenerating Douglas-fir after wildfire and clearcut harvesting. Oecologia 172(4):1179–1189

Barrios E (2007) Soil biota, ecosystem services and land productivity. Ecol Econ 64(2):269–285

Bastias BA, Anderson IC, Rangel-Castro JI et al (2009) Influence of repeated prescribed burning on incorporation of ^{13}C from cellulose by forest soil fungi as determined by RNA stable isotope probing. Soil Biol Biochem 41(3):467–472

Bastida F, Moreno JL, Nicolás C et al (2009) Soil metaproteomics: a review of an emerging environmental science—significance, methodology and perspectives. Eur J Soil Sci 60(6): 845–859

Bastow J (2012) Chapter 3.1: Succession, resource processing, and diversity in detrital food webs. In: Wall DH, Bardgett RD, Behan-Pelletier V et al (eds) Soil ecology and ecosystem services. Oxford: Oxford University Press, pp 117–135

Bates ST, Berg-Lyons D, Caporaso JG et al (2011) Examining the global distribution of dominant archaeal populations in soil. ISME J 5(5): 908–917

Battigelli JP, Spence JR, Langor DW, Berch SM (2004) Short-term impact of forest soil compaction and organic matter removal on soil mesofauna density and oribatid mite diversity. Can J For Res 34(5):1136–1149

Bellitürk K, Görres JH, Kunkle J, Melnichuk RDS (2015) Can commercial mulches be reservoirs of invasive earthworms? Promotion of ligninolytic enzyme activity and survival of *Amynthas agrestis* (Goto and Hatai, 1899). Appl Soil Ecol 87:27–31

Belnap J, Lange OL (2001) Structure and functioning of biological soil crusts: a synthesis. In: Belnap J, Lange OL (eds) Biological soil crusts: structure, function, and management, Ecological Studies 150. Springer, Berlin/Heidelberg, pp 471–479

Belnap J, Ludwig JA, Wilcox BP et al (2012) Introduced and invasive species in novel rangeland ecosystems: friends or foes? Rangel Ecol Manag 65(6):569–578

Blanc C, Sy M, Djigal D, Brauman A et al (2006) Nutrition on bacteria by bacterial-feeding nematodes and consequences on the structure of soil bacterial community. Eur J Soil Biol. 42(Suppl 1): S70–S78

Blankinship JC, Niklaus PA, Hungate BA (2011) A meta-analysis of responses of soil biota to global change. Oecologia 165(3):553–565

Bohlen PJ, Scheu S, Hale CM et al (2004) Nonnative invasive earthworms as agents of change in northern temperate forests. Front Ecol Environ 2(8):427–435

Bonkowski M, Villenave C, Griffiths B (2009) Rhizosphere fauna: the functional and structural diversity of intimate interactions of soil fauna with plant roots. Plant Soil 321(1–2):213–233

Bothe H, Słomka A (2017) Divergent biology of facultative heavy metal plants. J Plant Physiol 219:45–61

Brantley SL, Megonigal JP, Scatena FN et al (2011) Twelve testable hypotheses on the geobiology of weathering: hypotheses on geobiology of weathering. Geobiology 9(2): 140–165

Briones MJI, Garnett MH, Ineson P (2010) Soil biology and warming play a key role in the release of 'old C' from organic soils. Soil Biol Biochem 42(6):960–967

Briske DD (ed) (2017) Rangeland systems, Springer series on environmental management. Springer, Cham

Bronick CJ, Lal R (2005) Soil structure and management: a review. Geoderma 124(1):3–22

Brown SP, Callaham MA, Oliver AK, Jumpponen A (2013) Deep Ion Torrent sequencing identifies soil fungal community shifts after frequent prescribed fires in a southeastern US forest ecosystem. FEMS Microbiol Ecol 86(3):557–566

Burns RG, DeForest JL, Marxsen J et al (2013) Soil enzymes in a changing environment: current knowledge and future directions. Soil Biol Biochem 58:216–234

Busse MD, Shestak CJ, Hubbert KR, Knapp EE (2010) Soil physical properties regulate lethal heating during burning of woody residues. Soil Sci Soc Am J 74(3):947–955

Busse MD, Shestak CJ, Hubbert KR (2013) Soil heating during burning of forest slash piles and wood piles. Int J Wildland Fire 22(6):786

Butenko KO, Gongalsky KB, Korobushkin DI et al (2017) Forest fires alter the trophic structure of soil nematode communities. Soil Biol Biochem. 109(Suppl C):107–117

Cairney JWG, Bastias BA (2007) Influences of fire on forest soil fungal. Can J For Res 37(2):207–215

Cardenas E, Kranabetter JM, Hope G, et al (2015) Forest harvesting reduces the soil metagenomic potential for biomass decomposition. ISME J 9(11):2465–2476

Carini P,; Marsden PJ, Leff JW et al (2016) Relic DNA is abundant in soil and obscures estimates of soil microbial diversity. Nat Microbiol 2: 6242

Carpenter S, Trappe JM (1985) Phoenicoid fungi: a proposed term for fungi that fruit after heat treatment of substrates. Mycotaxon 103:203–206

Carpenter D, Hodson ME, Eggleton P, Kirk C (2007) Earthworm induced mineral weathering: preliminary results. Eur J Soil Biol 43:S176–S183

Certini G (2005) Effects of fire on properties of forest soils: a review. Oecologia 143(1):1–10

Chang C-H, Szlavecz K, Buyer JS (2016) Species-specific effects of earthworms on microbial communities and the fate of litter-derived carbon. Soil Biol Biochem 100:129–139

Chang C-H, Szlavecz K, Buyer JS (2017) *Amynthas agrestis* invasion increases microbial biomass in Mid-Atlantic deciduous forests. Soil Biol Biochem 114:189–199

Chang C-H, Johnston MR, Görres JH et al (2018) Co-invasion of three Asian earthworms, *Metaphire hilgendorfi, Amynthas agrestis* and *Amynthas tokioensis* in the USA, Biol Invasions 20: 843–848

Chen Y, Murrell JC (2010) When metagenomics meets stable-isotope probing: progress and perspectives. Trends Microbiol 18(4):157–163

Cicconardi F, Borges PAV, Strasberg D et al (2017) MtDNA metagenomics reveals large-scale invasion of belowground arthropod communities by introduced species. Mol Ecol 26(12): 3104–3115

Coats VC, Pelletreau KN, Rumpho ME (2014) Amplicon pyrosequencing reveals the soil microbial diversity associated with invasive Japanese barberry (*Berberis thunbergii* DC.). Mol Ecol 23(6):1318–1332

Coleman DC, Callaham MA Jr, Crossley DA Jr (2018) Chapter 7: Soil biodiversity and linkages to soil processes. In: Coleman DC, Callaham MA Jr, Crossley DA Jr (eds) Fundamentals of soil ecology, 3rd edn. Academic, London, pp 233–253. https://www.sciencedirect.com/science/article/pii/B9780128052518000077 (January 22, 2018)

Connell JH (1978) Diversity in tropical rain forests and coral reefs. Science 199(4335):1302

Cowan AD, Smith JE, Fitzgerald SA (2016) Recovering lost ground: effects of soil burn intensity on nutrients and ectomycorrhiza communities of ponderosa pine seedlings. For Ecol Manag 378:160–172

Coyle DR, Nagendra UJ, Taylor MK et al (2017) Soil fauna responses to natural disturbances, invasive species, and global climate change: current state of the science and a call to action. Soil Biol Biochem 110:116–133

Craigg TL, Adams PW, Bennett KA (2015) Soil matters: improving forest landscape planning and management for diverse objectives with soils information and expertise. J For 113(3):343–353

Craven D, Thakur MP, Cameron EK et al (2017) The unseen invaders: introduced earthworms as drivers of change in plant communities in North American forests (a meta-analysis). Glob Chang Biol 23(3):1065–1074

Creer S, Deiner K, Frey S et al (2016) The ecologist's field guide to sequence-based identification of biodiversity. Methods Ecol Evol 7(9):1008–1018

Crotty FV, Adl SM, Blackshaw RP, Murray PJ (2012) Protozoan pulses unveil their pivotal position within the soil food web. Microb Ecol 63(4):905–918

Crumsey JM, Le Moine JM, Vogel CS, Nadelhoffer KJ (2014) Historical patterns of exotic earthworm distributions inform contemporary associations with soil physical and chemical factors across a northern temperate forest. Soil Biol Biochem 68:503–514

Dávalos A, Simpson E, Nuzzo V, Blossey B (2015) Non-consumptive effects of native deer on introduced earthworm abundance. Ecosystems 18(6):1029–1042

De Vries FT, van Groenigen JW, Hoffland E, Bloem J (2011) Nitrogen losses from two grassland soils with different fungal biomass. Soil Biol Biochem 43:997–1005

Decaëns T, Jiménez JJ, Gioia C et al (2006) The values of soil animals for conservation biology. Eur J Soil Biol 42:S23–S38

DeGasperis BG, Motzkin G (2007) Windows of opportunity: historical and ecological controls on Berberis thunbergii invasions. Ecology 88(12):3115–3125

Deharveng L (2004) Recent advances in Collembola systematics. Pedobiologia 48(5–6):415–433

Dempsey MA, Fisk MC, Fahey TJ (2011) Earthworms increase the ratio of bacteria to fungi in northern hardwood forest soils, primarily by eliminating the organic horizon. Soil Biol Biochem 43(10):2135–2141

Dempsey MA, Fisk MC, Yavitt JB et al (2013) Exotic earthworms alter soil microbial community composition and function. Soil Biol Biochem 67:263–270

Dlott G, Maul JE, Buyer J, Yarwood S (2015) Microbial rRNA:rDNA gene ratios may be unexpectedly low due to extracellular DNA preservation in soils. J Microbiol Methods 115:112–120

Dobson A, Blossey B (2015) Earthworm invasion, white-tailed deer and seedling establishment in deciduous forests of north-eastern North America. J Ecol 103(1):153–164

Dumont MG, Murrell JC (2005) Stable isotope probing—linking microbial identity to function. Nat Rev Microbiol 3(6):499

Dumont MG, Pommerenke B, Casper P, Conrad R (2011) DNA-, rRNA- and mRNA-based stable isotope probing of aerobic methanotrophs in lake sediment: stable isotope probing of methanotrophs. Environ Microbiol 13(5):1153–1167

Eisenhauer N, Partsch S, Parkinson D, Scheu S (2007) Invasion of a deciduous forest by earthworms: changes in soil chemistry, microflora, microarthropods and vegetation. Soil Biol Biochem 39(5):1099–1110

Ekelund F, Saj S, Vestergård M et al (2009) The "soil microbial loop" is not always needed to explain protozoan stimulation of plants. Soil Biol Biochem 41(11):2336–2342

Ettema CH, Yeates GW (2003) Nested spatial biodiversity patterns of nematode genera in a New Zealand forest and pasture soil. Soil Biol Biochem 35(2):339–342

Fahey TJ, Yavitt JB, Sherman RE et al (2013) Earthworm effects on the incorporation of litter C and N into soil organic matter in a sugar maple forest. Ecol Appl 23(5):1185–1201

Faust K, Raes J (2012) Microbial interactions: from networks to models. Nat Rev Microbiol 10(8):538–550

Ferlian O, Cesarz S, Marhan S, Scheu S (2014) Carbon food resources of earthworms of different ecological groups as indicated by ^{13}C compound-specific stable isotope analysis. Soil Biol Biochem 77:22–30

Ferlian O, Klarner B, Langeneckert AE, Scheu S (2015) Trophic niche differentiation and utilisation of food resources in Collembolans based on complementary analyses of fatty acids and stable isotopes. Soil Biol Biochem 82:28–35

Fernández-Calviño D, Rousk J, Brookes PC, Bååth E (2011) Bacterial pH-optima for growth track soil pH, but are higher than expected at low pH. Soil Biol Biochem 43(7):1569–1575

Fetzer I, Johst K, Schäwe R et al (2015) The extent of functional redundancy changes as species' roles shift in different environments. Proc Natl Acad Sci 112(48):14888–14893

Ficetola GF, Pansu J, Bonin A et al (2015) Replication levels, false presences and the estimation of the presence/absence from eDNA metabarcoding data. Mol Ecol Resour 15(3):543–556

Fierer N, Jackson RB (2006) The diversity and biogeography of soil bacterial communities. Proc Natl Acad Sci 103(3):626–631

Fierer N, Strickland MS, Liptzin D et al (2009) Global patterns in belowground communities. Ecol Lett 12(11):1238–1249

Fisher MC, Henk DA, Briggs CJ et al (2012) Emerging fungal threats to animal, plant and ecosystem health. Nature 484(7393):186–194

Fujimura KE, Smith JE, Horton TR et al (2005) Pezizalean mycorrhizas and sporocarps in ponderosa pine (Pinus ponderosa) after prescribed fires in eastern Oregon, USA. Mycorrhiza 15(2):79–86

Gall JE, Boyd RS, Rajakaruna N (2015) Transfer of heavy metals through terrestrial food webs: a review. Environ Monit Assess 187:201

Gao M, Taylor MK, Callaham MA (2017) Trophic dynamics in a simple experimental ecosystem: interactions among centipedes, Collembola and introduced earthworms. Soil Biol Biochem 115:66–72

García-Palacios P, Vandegehuchte ML, Shaw EA et al (2015) Are there links between responses of soil microbes and ecosystem functioning to elevated CO_2, N deposition and warming? A global perspective. Glob Chang Biol 21(4):1590–1600

Gebremikael MT, Steel H, Buchan D et al (2016) Nematodes enhance plant growth and nutrient uptake under C and N-rich conditions. Sci Rep 6:32862

Geisen S (2016) The bacterial-fungal energy channel concept challenged by enormous functional versatility of soil protists. Soil Biol Biochem 102:22–25

Giller KE, Witter E, McGrath SP (2009) Heavy metals and soil microbes. Soil Biol Biochem 41(10):2031–2037

Glassman SI, Levine CR, DiRocco AM et al (2016) Ectomycorrhizal fungal spore bank recovery after a severe forest fire: some like it hot. ISME J 10(5):1228–1239

Gorbunova AY, Korobushkin DI, Zaitsev AS, Gongalsky KB (2017) Forest fires increase variability of soil macrofauna communities along a macrogeographic gradient. Eur J Soil Biol 80:49–52

Graham EB, Wieder WR, Leff JW et al (2014) Do we need to understand microbial communities to predict ecosystem function? A comparison of statistical models of nitrogen cycling processes Soil Biol Biochem 68:279–282

Grayston SJ, Rennenberg H (2006) Assessing effects of forest management on microbial community structure in a central European beech forest. Can J For Res 36(10):2595–2604

Greiner HG, Kashian DR, Tiegs SD (2012) Impacts of invasive Asian (Amynthas hilgendorfi) and European (Lumbricus rubellus) earthworms in a North American temperate deciduous forest. Biol Invasions 14(10):2017–2027

Griffiths BS, Philippot L (2013) Insights into the resistance and resilience of the soil microbial community. FEMS Microbiol Rev 37(2):112–129

Grigal DF (2000) Effects of extensive forest management on soil productivity. For Ecol Manag 138(1):167–185

Groffman PM, Bohlen PJ, Fisk MC, Fahey TJ (2004) Exotic earthworm invasion and microbial biomass in temperate forest soils. Ecosystems 7(1):45–54

Groffman PM, Fahey TJ, Fisk MC et al (2015) Earthworms increase soil microbial biomass carrying capacity and nitrogen retention in northern hardwood forests. Soil Biol Biochem 87:51–58

Hale CM, Frelich LE, Reich PB, Pastor J (2005) Effects of european earthworm invasion on soil characteristics in Northern Hardwood forests of Minnesota, USA. Ecosystems 8(8):911–927

Hale CM, Frelich LE, Reich PB (2006) Changes in hardwood forest understory plant communities in response to European earthworm invasions. Ecology 87(7):1637–1649

Hartmann M, Howes CG, VanInsberghe D et al (2012) Significant and persistent impact of timber harvesting on soil microbial communities in northern coniferous forests. ISME J 6: 2199

Hawkes CV, Belnap J, D'Antonio C, Firestone MK (2006) Arbuscular mycorrhizal assemblages in native plant roots change in the presence of invasive exotic grasses. Plant Soil 281(1–2):369–380

Hebel CL, Smith JE, Cromack K (2009) Invasive plant species and soil microbial response to wildfire burn severity in the Cascade Range of Oregon. Appl Soil Ecol 42(2):150–159

Hendrix PF (ed) (1995) Earthworm ecology and biogeography in North America. CRC Press, Boca Raton

Hendrix PF, Callaham MA, Drake JM et al (2008) Pandora's box contained bait: the global problem of introduced earthworms. Annu Rev Ecol Evol Syst 39(1): 593–613

Hessburg PF, Churchill DJ, Larson AJ et al (2015) Restoring fire-prone Inland Pacific landscapes: seven core principles. Landsc Ecol 30(10):1805–1835

Hinsinger P (2001) Bioavailability of soil inorganic P in the rhizosphere as affected by root-induced chemical changes: a review. Plant Soil 237(2):173–195

Högberg P, Johannisson C, Yarwood S et al (2011) Recovery of ectomycorrhiza after 'nitrogen saturation' of a conifer forest. New Phytol 189(2):515–525

Holden SR, Treseder KK (2013) A meta-analysis of soil microbial biomass responses to forest disturbances. Front Microbiol 4:163

Hungate BA, Mau RL, Schwartz E et al (2015) Quantitative microbial ecology through stable isotope probing. Appl Environ Microbiol 81(21):7570–7581

Huston MA (2014) Disturbance, productivity, and species diversity: empiricism vs. logic in ecological theory. Ecology 95(9):2382–2396

Hutchens E (2009) Microbial selectivity on mineral surfaces: possible implications for weathering processes. Fungal Biol Rev 23(4):115–121

Ingram LJ, Stahl PD, Schuman GE et al (2008) Grazing impacts on soil carbon and microbial communities in a mixed-grass ecosystem. Soil Sci Soc Am J 72(4): 939

Jansson JK, Neufeld JD, Moran MA, Gilbert JA (2012) Omics for understanding microbial functional dynamics. Environ Microbiol 14(1):1–3

Johansen JR (2001) Impacts of fire on biological soil crusts. In: Belnap J, Lange OL (eds) Biological soil crusts: structure, function, and management, Ecological Studies 150. Springer, Berlin/Heidelberg, pp 385–397

Jones DL, Nguyen C, Finlay RD (2009) Carbon flow in the rhizosphere: carbon trading at the soil–root interface. Plant Soil 321(1–2):5–33

Jury WA, Gardner WR, Gardner WH (1991) Soil physics. Wiley, New York. 328 p

Kardol P, Wardle DA (2010) How understanding aboveground–belowground linkages can assist restoration ecology. Trends Ecol Evol 25(11):670–679

Keeley JE (2009) Fire intensity, fire severity and burn severity: a brief review and suggested usage. Int J Wildland Fire 18(1):116

Keith AM, Brooker RW, Osler GHR et al (2009) Strong impacts of belowground tree inputs on soil nematode trophic composition. Soil Biol Biochem 41(6):1060–1065

Keith DM, Johnson EA, Valeo C (2010) Moisture cycles of the forest floor organic layer (F and H layers) during drying. Water Resour Res 46:7529

Kergunteuil A, Campos-Herrera R, Sánchez-Moreno S et al (2016) The abundance, diversity, and metabolic footprint of soil nematodes is highest in high elevation alpine grasslands. Front Ecol Evol 4:84

Kibblewhite MG, Ritz K, Swift MJ (2008) Soil health in agricultural systems. Philos Trans R Soc B Biol Sci 363:685–701

Kimura M, Jia Z-J, Nakayama N, Asakawa S (2008) Ecology of viruses in soils: past, present and future perspectives. Soil Sci Plant Nutr 54(1):1–32

Kind T, Wohlgemuth G, Lee DY et al (2009) FiehnLib: mass spectral and retention index libraries for metabolomics based on quadrupole and time-of-flight gas chromatography/mass spectrometry. Anal Chem 81(24):10038–10048

King JR, Warren RJ, Bradford MA (2013) Social insects dominate eastern US temperate hardwood forest macroinvertebrate communities in warmer regions. PLoS One 8(10):e75843

Kluber LA, Tinnesand KM, Caldwell BA et al (2010) Ectomycorrhizal mats alter forest soil biogeochemistry. Soil Biol Biochem 42(9):1607–1613

Koek MM, Muilwijk B, van der Werf MJ, Hankemeier T (2006) Microbial metabolomics with gas chromatography/mass spectrometry. Anal Chem 78(4):1272–1281

Kourtev PS, Huang WZ, Ehrenfeld JG (1999) Differences in earthworm densities and nitrogen dynamics in soils under exotic and native plant species. Biol Invasions 1(2–3):237–245

Kourtev PS, Ehrenfeld JG, Häggblom M (2002) Exotic plant species alter the microbial community structure and function in the soil. Ecology 83(11):3152–3166

Kurth F, Feldhahn L, Bönn M et al (2015) Large scale transcriptome analysis reveals interplay between development of forest trees and a beneficial mycorrhiza helper bacterium. BMC Genomics 16:658

Lambers H, Mougel C, Jaillard B, Hinsinger P (2009) Plant-microbe-soil interactions in the rhizosphere: an evolutionary perspective. Plant Soil 321(1–2):83–115

Lennon JT, Jones SE (2011) Microbial seed banks: the ecological and evolutionary implications of dormancy. Nat Rev Microbiol 9(2):119–130

Lennon JT, Placella SA, Muscarella ME (2017) Relic DNA contributes minimally to estimates of microbial diversity. bioRxiv:131284

Lensing JR, Wise DH (2007) Impact of changes in rainfall amounts predicted by climate-change models on decomposition in a deciduous forest. Appl Soil Ecol 35(3):523–534

Leung HTC, Maas KR, Wilhelm RC, Mohn WW (2016) Long-term effects of timber harvesting on hemicellulolytic microbial populations in coniferous forest soils. ISME J 10(2):363–375

Li L, Vogel J, He Z et al (2016) Association of soil aggregation with the distribution and quality of organic carbon in soil along an elevation gradient on Wuyi Mountain in China. PLoS One 11(3):e0150898

Ma JF, Ueno H, Ueno D et al (2003) Characterization of phytosiderophore secretion under Fe deficiency stress in Festuca rubra. Plant Soil 256(1):131–137

Ma Y, Filley TR, Johnston CT et al (2013) The combined controls of land use legacy and earthworm activity on soil organic matter chemistry and particle association during afforestation. Org Geochem 58:56–68

Malik AA, Thomson BC, Whiteley AS et al (2017) Bacterial physiological adaptations to contrasting edaphic conditions identified using landscape scale metagenomics. MBio. 8(4):e00799–e00717

Malmström A (2012) Life-history traits predict recovery patterns in Collembola species after fire: a 10 year study. Appl Soil Ecol 56(Suppl C):35–42

Maxfield PJ, Dildar N, Hornibrook ERC et al (2012) Stable isotope switching (SIS): a new stable isotope probing (SIP) approach to determine carbon flow in the soil food web and dynamics in organic matter pools. Rapid Commun Mass Spectrom 26(8):997–1004

Menzel R, Ngosong C, Ruess L (2017) Isotopologue profiling enables insights into dietary routing and metabolism of trophic biomarker fatty acids. Chemoecology 27(3):101–114

Metzker ML (2010) Sequencing technologies—the next generation. Nat Rev Genet 11(1):31–46

Microbiology by numbers (2011) Nat Rev Microbiol 9(9):628

Miravete V, Roura-Pascual N, Dunn RR, Gómez C (2014) How many and which ant species are being accidentally moved around the world? Biol Lett 10:20140518

Myrold DD, Zeglin LH, Jansson JK (2014) The potential of metagenomic approaches for understanding soil microbial processes. Soil Sci Soc Am J 78(1):3

National Wildlife Coordinating Group [NWCG] (2003) Glossary of wildland fire terminology. https://www.nwcg.gov/glossary-of-wildland-fire-terminology. Accessed 8 Jan 2018

Nemec KT (2014) Tallgrass prairie ants: their species composition, ecological roles, and response to management. J Insect Conserv 18(4):509–521

Ngosong C, Gabriel E, Ruess L (2014) Collembola grazing on arbuscular mycorrhiza fungi modulates nutrient allocation in plants. Pedobiologia 57(3):171–179

Nielsen UN, Ayres E, Wall DH, Bardgett RD (2011) Soil biodiversity and carbon cycling: a review and synthesis of studies examining diversity–function relationships. Eur J Soil Sci 62(1):105–116

Nuzzo VA, Maerz JC, Blossey B (2009) Earthworm invasion as the driving force behind plant invasion and community change in northeastern North American forests. Conserv Biol 23(4):966–974

Oliver AK, Callaham MA, Jumpponen A (2015) Soil fungal communities respond compositionally to recurring frequent prescribed burning in a managed southeastern US forest ecosystem. For Ecol Manag 345:1–9

Orgiazzi A, Bardgett RD, Barrios E et al (eds) (2016) Global soil diversity atlas. Luxembourg: European Commission. https://doi.org/10.2788/2613

Overby ST, Gottfried GJ (2017) Microbial and nitrogen pool response to fuel treatments in pinyon-juniper woodlands of the southwestern USA. For Ecol Manag 406(Suppl C):138–146

Overby S, Hart S (2016) Short-term belowground responses to thinning and burning treatments in southwestern ponderosa pine forests of the USA. Forests 7(2):45

Overby ST, Owen SM, Hart SC et al (2015) Soil microbial community resilience with tree thinning in a 40-year-old experimental ponderosa pine forest. Appl Soil Ecol 93:1–10

Page-Dumroese D, Jurgensen M, Elliot W et al (2000) Soil quality standards and guidelines for forest sustainability in northwestern North America. For Ecol Manag 138(1):445–462

Park O, Kim J, Ryu C-M, Park C-S (2004) Colonization and population changes of a biocontrol agent, Paenibacillus polymyxa E681, in seeds and roots. Plant Pathol J 20(2):97–102

Petersen PM (1970) Danish fireplace fungi: an ecological investigation on fungi on burns. Dansk Botanisk Arkiv 27(3):1–97

Philippot L, Raaijmakers JM, Lemanceau P, van der Putten WH (2013a) Going back to the roots: the microbial ecology of the rhizosphere. Nat Rev Microbiol 11(11):789–799

Philippot L, Spor A, Hénault C et al (2013b) Loss in microbial diversity affects nitrogen cycling in soil. ISME J 7(8):1609–1619

Pimentel D (ed) (2014) Biological invasions: economic and environmental costs of alien plant, animal, and microbe species. CRC Press, Boca Raton. 463 p

Pold G, DeAngelis K (2013) Up against the wall: the effects of climate warming on soil microbial diversity and the potential for feedbacks to the carbon cycle. Diversity 5(2):409–425

Pollierer MM, Langel R, Scheu S, Maraun M (2009) Compartmentalization of the soil animal food web as indicated by dual analysis of stable isotope ratios ($^{15}N/^{14}N$ and $^{13}C/^{12}C$). Soil Biol Biochem 41(6):1221–1226

Ponder F, Fleming RL, Berch S et al (2012) Effects of organic matter removal, soil compaction and vegetation control on 10th year biomass and foliar nutrition: LTSP continent-wide comparisons. For Ecol Manag. 278(Suppl C):35–54

Powers RF (2006) Long-Term Soil Productivity: genesis of the concept and principles behind the program. Can J For Res 36(3):519–528

Powers RF, Scott DA, Sanchez FG et al (2005) The North American Long-Term Soil Productivity experiment: findings from the first decade of research. For Ecol Manag 220(1):31–50

Pringle A, Bever JD, Gardes M et al (2009) Mycorrhizal symbioses and plant invasions. Annu Rev Ecol Evol Syst 40(1):699–715

Pritchard SG, Taylor BN, Cooper ER et al (2014) Long-term dynamics of mycorrhizal root tips in a loblolly pine forest grown with free-air CO_2 enrichment and soil N fertilization for 6 years. Glob Chang Biol 20(4):1313–1326

Quirk J, Beerling DJ, Banwart SA et al (2012) Evolution of trees and mycorrhizal fungi intensifies silicate mineral weathering. Biol Lett 8(6):1006–1011.

Raynaud X, Nunan N (2014) Spatial ecology of bacteria at the microscale in soil. PLoS One 9(1):e87217

Reazin C, Morris S, Smith JE et al (2016) Fires of differing intensities rapidly select distinct soil fungal communities in a northwest US ponderosa pine forest ecosystem. For Ecol Manag 377:118–127

Resner K, Yoo K, Hale C et al (2011) Elemental and mineralogical changes in soils due to bioturbation along an earthworm invasion chronosequence in northern Minnesota. Appl Geochem 26:S127–S131

Resner K, Yoo K, Sebestyen SD et al (2015) Invasive earthworms deplete key soil inorganic nutrients (Ca, Mg, K, and P) in a northern hardwood forest. Ecosystems 18(1):89–102

Robin A, Vansuyt G, Hinsinger P et al (2008) Iron dynamics in the rhizosphere: consequences for plant health and nutrition. Adv Agron 99:183–225

Ruess L, Schütz K, Migge-Kleian S (2007) Lipid composition of Collembola and their food resources in deciduous forest stands: implications for feeding strategies. Soil Biol Biochem 39(8):1990–2000

Schneider K, Maraun M (2009) Top-down control of soil microarthropods: evidence from a laboratory experiment. Soil Biol Biochem 41(1):170–175

Schneider T, Keiblinger KM, Schmid E et al (2012) Who is who in litter decomposition? Metaproteomics reveals major microbial players and their biogeochemical functions. ISME J 6:1749

Shade A, Peter H, Allison SD et al (2012) Fundamentals of microbial community resistance and resilience. Front Microbiol 3:417

Shestak CJ, Busse MD (2005) Compaction alters physical but not biological indices of soil health. Soil Sci Soc Am J 69(1):236–246

Sigler WV, Zeyer J (2004) Colony-forming analysis of bacterial community succession in deglaciated soils indicates pioneer stress-tolerant opportunists. Microb Ecol 48(3):316–323

Singh BK, Bardgett RD, Smith P, Reay DS (2010) Microorganisms and climate change: terrestrial feedbacks and mitigation options. Nat Rev Microbiol 8(11):779–790

Six J, Elliott ET, Paustian K (2000) Soil macroaggregate turnover and microaggregate formation: a mechanism for C sequestration under no-tillage agriculture. Soil Biol Biochem 32(14):2099–2103

Sławska M, Bruckner A, Sławski M (2017) Edaphic Collembola assemblages of European temperate primeval forests gradually change along a forest-type gradient. Eur J Soil Biol 80:92–101

Smith DP, Peay KG (2014) Sequence depth, not PCR replication, improves ecological inference from next generation DNA sequencing. PLoS One 9(2):e90234

Smith JE, McKay D, Niwa CG et al (2004) Short-term effects of seasonal prescribed burning on the ectomycorrhizal fungal community and fine root biomass in ponderosa pine stands in the Blue Mountains of Oregon. Can J For Res 34(12):2477–2491

Smith JE, McKay D, Brenner G et al (2005) Early impacts of forest restoration treatments on the ectomycorrhizal fungal community and fine root biomass in a mixed conifer forest: prescribed fire and EMF species richness. J Appl Ecol 42(3):526–535

Smith JE, Cowan AD, Fitzgerald SA (2016) Soil heating during the complete combustion of mega-logs and broadcast burning in central Oregon USA pumice soils. Int J Wildland Fire 25(11):1202

Smith JE, Kluber LA, Jennings TN (2017) Does the presence of large down wood at the time of a forest fire impact soil recovery? For Ecol Manag 391:52–62

Snyder BA, Callaham MA, Hendrix PF (2011) Spatial variability of an invasive earthworm (*Amynthas agrestis*) population and potential impacts on soil characteristics and millipedes in the Great Smoky Mountains National Park, USA. Biol Invasions 13(2):349–358

Southworth D, Donohue J, Frank JL, Gibson J (2011) Mechanical mastication and prescribed fire in conifer–hardwood chaparral: differing responses of ectomycorrhizae and truffles. Int J Wildland Fire 20(7):888

Stefanowicz AM, Stanek M, Woch MW (2016) High concentrations of heavy metals in beech forest understory plants growing on waste heaps left by Zn-Pb ore mining. J Geochem Explor 169:157–162

Stephenson SL (2011) From morphological to molecular: studies of myxomycetes since the publication of the Martin and Alexopoulos (1969) monograph. Fungal Divers 50(1):21–34

Steven B, Kuske CR, Gallegos-Graves LV et al (2015) Climate change and physical disturbance manipulations result in distinct biological soil crust communities. Appl Environ Microbiol 81(21):7448–7459

Szlávecz K, Csuzdi C (2007) Land use change affects earthworm communities in eastern Maryland, USA. Eur J Soil Biol 43:S79–S85

Szlavecz K, McCormick M, Xia L et al (2011) Ecosystem effects of nonnative earthworms in Mid-Atlantic deciduous forests. Biol Invasions 13(5):1165–1182

Taylor Grazing Act of 1934; Act of June 28, 1934; as amended June 26, 1936; 43 U.S.C. 315

Tedersoo L, Nara K (2010) General latitudinal gradient of biodiversity is reversed in ectomycorrhizal fungi. New Phytol 185(2):351–354

Trappe MJ, Cromack K Jr, Trappe JM et al (2009) Interactions among prescribed fire, soil attributes, and mycorrhizal community structure at Crater Lake National Park, Oregon, USA. Fire Ecol 5(2):30–50

Treseder KK (2004) A meta-analysis of mycorrhizal responses to nitrogen, phosphorus, and atmospheric CO_2 in field studies. New Phytol 164(2):347–355

Treseder KK, Marusenko Y, Romero-Olivares AL, Maltz MR (2016) Experimental warming alters potential function of the fungal community in boreal forest. Glob Chang Biol 22(10):3395–3404

Turbé A, De Toni A, Benito P et al (2010) Soil biodiversity: functions, threats and tools for policy makers. Bioemco 00560420

Tyler G, Påhlsson A-MB, Bengtsson GE et al (1989) Heavy-metal ecology of terrestrial plants, microorganisms and invertebrates. Water Air Soil Pollut 47(3):189–215

Ulyshen MD, Wagner TL, Mulrooney JE (2014) Contrasting effects of insect exclusion on wood loss in a temperate forest. Ecosphere 5(4):1–15

Urbanová M, Šnajdr J, Baldrian P (2015) Composition of fungal and bacterial communities in forest litter and soil is largely determined by dominant trees. Soil Biol Biochem 84:53–64

Uroz S, Calvaruso C, Turpault M-P, Frey-Klett P (2009) Mineral weathering by bacteria: ecology, actors and mechanisms. Trends Microbiol 17(8):378–387

Uroz S, Ioannidis P, Lengelle J et al (2013) Functional assays and metagenomic analyses reveals differences between the microbial communities inhabiting the soil horizons of a Norway spruce plantation. PLoS One 8(2):e55929

Uroz S, Buée M, Deveau A et al (2016) Ecology of the forest microbiome: highlights of temperate and boreal ecosystems. Soil Biol Biochem 103:471–488

Walker AP, Zaehle S, Medlyn BE et al (2015) Predicting long-term carbon sequestration in response to CO_2 enrichment: how and why do current ecosystem models differ? Glob Biogeochem Cycles. 29(4):2014GB004995

Wall DH, Bradford MA, St. John MG et al (2008) Global decomposition experiment shows soil animal impacts on decomposition are climate-dependent. Glob Chang Biol 14(11):2661–2677

Wall DH, Bardgett RD, Behan-Pelletier V et al (eds) (2012) Soil ecology and ecosystem services. Oxford: Oxford University Press. 464 p

Wall DH, Nielsen UN, Six J (2015) Soil biodiversity and human health. Nature 528(7580):69–76

Wang K-H, McSorley R, Bohlen P, Gathumbi SM (2006) Cattle grazing increases microbial biomass and alters soil nematode communities in subtropical pastures. Soil Biol Biochem 38(7):1956–1965

Warcup JH (1990) Occurrence of ectomycorrhizal and saprophytic discomycetes after a wild fire in a eucalypt forest. Mycol Res 94(8):1065–1069

Warren S (1995) Chapter 11: Ecological role of microphytic soil crusts in arid environments. In: Allsopp A, Colwell RR, Hawksworth DL (eds) Microbial diversity and ecosystem function. CAB International, Wallingford, pp 199–209

Warren SD, Eldridge DJ (2001) Chapter 10: Biological soil crusts and livestock in arid ecosystems: are they compatible? In: Belnap J, Lange OL (eds) Biological soil crusts: structure, function, and management. Springer, Berlin/Heidelberg, pp 401–415

Warren SD, St. Clair LL, Leavitt SD (2017) Aerobiology and passive restoration of biological soil crusts. Biogeosci Discuss. https://doi.org/10.5194/bg-2017-430

Weber B, Burkhard Büdel B, Belnap J (eds) (2016) Biological soil crusts: an organizing principle in drylands. Springer, Cham

Wicklow DT (1975) Fire as an environmental cue initiating ascomycete development in a tallgrass prairie. Mycologia 67(4):852–862

Wilcove DS, Rothstein D, Dubow J et al (1998) Quantifying threats to imperiled species in the United States. Bioscience 48(8):607–615

Wilhelm RC, Cardenas E, Leung H et al (2017a) A metagenomic survey of forest soil microbial communities more than a decade after timber harvesting. Sci Data 4:170092

Wilhelm RC, Cardenas E, Maas KR et al (2017b) Biogeography and organic matter removal shape long-term effects of timber harvesting on forest soil microbial communities. ISME J 11(11):2552

Williamson KE, Fuhrmann JJ, Wommack KE, Radosevich M (2017) Viruses in soil ecosystems: an unknown quantity within an unexplored territory. Ann Rev Virol 4(1):201–219

Wilshire HG (1983) The impact of vehicles on desert soil stabilizers. In: Webb RH, Wilshire HG (eds) Environmental effects of off-road vehicles: impacts and management in arid regions. Springer, New York, pp 31–50

Yarwood SA, Högberg MN (2017) Soil bacteria and archaea change rapidly in the first century of Fennoscandian boreal forest development. Soil Biol Biochem 114:160–167

Yavitt JB, Fahey TJ, Sherman RE, Groffman PM (2015) Lumbricid earthworm effects on incorporation of root and leaf litter into aggregates in a forest soil, New York State. Biogeochemistry 125(2):261–273

Young IM, Crawford JW (2004) Interactions and self-organization in the soil-microbe complex. Science 304(5677):1634–1637

Youssef NH, Couger MB, McCully AL et al (2015) Assessing the global phylum level diversity within the bacterial domain: a review. J Adv Res 6(3):269–282

Zhang W, Parker KM, Luo Y et al (2005) Soil microbial responses to experimental warming and clipping in a tallgrass prairie. Glob Chang Biol 11(2):266–277

Zhang X, Song C, Mao R et al (2015) Comparing differences in early-stage decay of macrophyte shoots between in the air and on the sediment surface in a temperate freshwater marsh. Ecol Eng 81:14–18

Wetland and Hydric Soils

6

Carl C. Trettin, Randall K. Kolka, Anne S. Marsh,
Sheel Bansal, Erik A. Lilleskov, Patrick Megonigal,
Marla J. Stelk, Graeme Lockaby, David V. D'Amore,
Richard A. MacKenzie, Brian Tangen, Rodney Chimner,
and James Gries

Introduction

Soil and the inherent biogeochemical processes in wetlands contrast starkly with those in upland forests and rangelands. The differences stem from extended periods of anoxia, or the lack of oxygen in the soil, that characterize wetland soils; in contrast, upland soils are nearly always oxic. As a result, wetland soil biogeochemistry is characterized by anaerobic processes, and wetland vegetation exhibits specific adaptations to grow under these conditions. However, many wetlands may also have periods during the year where the soils are unsaturated and aerated. This fluctuation between aerated and nonaerated soil conditions, along with the specialized vegetation, gives rise to a wide variety of highly valued ecosystem services.

Wetlands were once considered unproductive lands that were a barrier to agricultural, transportation, and urban development. As a result, approximately 53% of the wetlands in the conterminous United States have been converted to other land uses over the past 150 years (Dahl 1990). Most of those losses are due to draining and conversion to agriculture. States in the Midwest such as Iowa, Illinois, Missouri, Ohio, and Indiana have lost more than 85% of their original wetlands, and California has lost 96% of its wetlands. In the mid-1900s, the importance of wetlands in the landscape started to be understood, and wetlands are now recognized for their inherent value that is realized through the myriad of ecosystem services, including storage of water to mitigate flooding, filtering water of pollutants and sediment, storing and sequestering carbon (C), providing critical habitat for wildlife, and recreation.

Historically, drainage of wetlands for agricultural development has been the largest threat to wetlands. While conversion to agriculture still persists, currently the primary threat is from urbanization (US EPA 2016). The current rate of wetland loss is approximately 23 times less than historic

C. C. Trettin (✉)
U.S. Department of Agriculture, Forest Service, Southern Research Station, Cordesville, SC, USA
e-mail: Carl.trettin@usda.gov

R. K. Kolka
U.S. Department of Agriculture, Forest Service, Northern Research Station, Center for Research on Ecosystem Change, Grand Rapids, MN, USA

A. S. Marsh
U.S. Department of Agriculture, Forest Service, Research and Development, Bioclimatology and Climate Change Research, Washington, DC, USA

S. Bansal
U.S. Geological Survey, Jamestown, ND, USA

E. A. Lilleskov
U.S. Department of Agriculture, Forest Service, Northern Research Station, Climate, Fire and Carbon Cycle Sciences, Houghton, MI, USA

P. Megonigal
Smithsonian Institute, Smithsonian Environmental Research Center, Edgewater, MD, USA

M. J. Stelk
Association of State Wetland Managers, Windham, ME, USA

G. Lockaby
School of Forestry & Wildlife, Auburn University, Auburn, AL, USA

D. V. D'Amore
U.S. Department of Agriculture, Forest Service, Pacific Northwest Research Station, Juneau, AK, USA

R. A. MacKenzie
U.S. Department of Agriculture, Forest Service, Pacific Southwest Research Station, Institute of Pacific Islands Forestry, Hilo, HI, USA

B. Tangen
U.S. Geological Survey, Northern Prairie Wildlife Research Center, Lincoln, NE, USA

R. Chimner
School of Forest Resources and Environmental Science, Michigan Technological University, Houghton, MI, USA

J. Gries
U.S. Department of Agriculture, Forest Service, Eastern Region, Milwaukee, WI, USA

© The Author(s) 2020
R. V. Pouyat et al. (eds.), *Forest and Rangeland Soils of the United States Under Changing Conditions*,
https://doi.org/10.1007/978-3-030-45216-2_6

99

rates, with wetland restoration and creation nearly offsetting current losses (US FWS 2011). Changes in precipitation patterns and extreme events may affect wetlands, especially those with perched water tables and those along the coast (Amatya et al. 2016). The combination of warming temperatures and variable precipitation, most notably lower precipitation, may lead to the drying of wetlands, which could dramatically change vegetation communities and soil processes and reduce the ability of wetlands to sustain valued ecosystem services.

Goods and Services Derived from Wetlands

In addition to basic ecosystem functions, wetlands also provide valued goods and services (Sarukhán et al. 2005). Those benefits depend on factors such as hydrology, vegetation, soils, and the condition of the wetland, as well as the position of the wetland within the landscape (Brander et al. 2006; Woodward and Wui 2001). De Groot and others (2012) estimated the mean global value of coastal and inland wetlands at \$193,845 ha^{-1} year^{-1} and \$25,682 ha^{-1} year^{-1}, respectively. Higher inland wetland values were found in areas with high gross domestic product and population density, indicating a particular demand for services in these areas. Many values are external to markets and are considered a public good that cannot be traded; however, C storage is a service that has been commoditized. Soils are integral to the provision of ecosystem services, and hence, the derived value. Water supply, water quality, habitat, and provision of goods are major ecosystem services that are sensitive to soils and soil processes.

Water Storage and Supply

The geographic setting of the wetland and soil type affect water storage, flood control, and mediation of water supply. Riverine wetlands can store floodwaters temporarily, which lessens streamflow to downstream areas and reduces flood events (Hey et al. 2002). Wetland vegetation intercepts and slows water flowing through the wetland and takes up water through transpiration (Vepraskas and Craft 2016). In coastal areas, wetland vegetation can dissipate wave energy from major storm events (e.g., hurricanes, tsunamis) and, in the case of mangroves (*Avicennia* spp.), protect against wind damage during storms (Das and Crepin 2013; Gedan et al. 2011; Koch et al. 2009; Spaulding et al. 2014). It is estimated that because of wetlands, as much as \$625 million in damages were avoided during Hurricane Sandy (Narayan et al. 2017). In addition, wetlands connected to groundwater may recharge or discharge critical water resources, a particularly important function in wetlands in dry regions with limited surface water connectivity (Van der Kamp and Hayashi 1998).

Water Quality

Eutrophication is the over-enrichment of water by nutrients and other pollutants that can result in excessive algal blooms. Because wetlands have both anaerobic and aerobic biogeochemical soil processes, they can be effective at ameliorating nutrient runoff, thereby reducing the risk of eutrophication (Hemond and Beniot 1988; Johnston et al. 1990). Microbial-mediated processes are the principal mechanism affecting nutrients and pollutants in runoff. Plant structures such as stems, roots, or trunks provide structural support for microbes, and direct uptake by vegetation is also a mechanism affecting nutrient and pollutant removal (Furukawa et al. 1997; Gosselink and Turner 1978). Removal of excessive nitrogen (N) is particularly important since it is a common constituent in runoff from agriculture and urban lands. As nitrate, N can be removed through denitrification; the microbial transformation to N gas (Reddy et al. 1989). In contrast, phosphate may be complexed to iron and aluminum minerals by chemical reactions (Smil 2000). Suspended sediment particles in water flowing through wetlands settle out of the water column as water flow slows, and these sediments are then trapped by plant stems, roots, and trunks (Furukawa et al. 1997; Gosselink and Turner 1978). The water purification processes of wetlands have been engineered into constructed wetlands that are used to treat municipal wastewater (Kadlec and Wallace 2009; Vymazal 2011).

Carbon

Wetland soils are a natural sink for C, and per unit area they store much more C than upland soils. This is especially true for organic soils that globally store approximately 30% of soil C on approximately 3% of the land area (Page et al. 2011). Similarly, mangroves can store three to five times more C than temperate or tropical forest soils (Alongi 2014; Donato et al. 2012). Accordingly, the value of wetlands for C storage, and thus climate change mitigation, is widely recognized. The ability of wetlands to store large amounts of soil C is due in part to high plant productivity that removes carbon dioxide (CO_2) from the atmosphere (Kolka et al. 2018; Nahlik and Fennessy 2016; Schuur et al. 2015; Windham-Myers et al. 2018) and the decomposition of plant matter that is hindered by the lack of oxygen, which results in the accumulation of organic C (Mcleod et al. 2011). In some freshwater wetlands, the production of methane (CH_4) and nitrous oxide (N_2O), two gases with a greater greenhouse gas warming potential than CO_2, can partially or fully offset this climate mitigation capacity (Bridgham et al. 2013; Mitsch et al. 2013; Neubauer 2014; Smith et al. 2003).

Wildlife Habitat

Wetlands support almost half of the threatened and endangered species in the United States, as well as commercially valuable species of fish and shellfish, and mammals trapped for pelts (Flynn 1996; Niering 1988). Wetland soils provide important habitat for benthic algae and invertebrates that are an important food resource for many resident and migratory birds and fish (Currin et al. 1995; MacKenzie 2005; MacKenzie and Dionne 2008; MacKenzie et al. 2015). Other wetland species that feed on wetland soil resources are valued for recreational purposes (Bergstrom et al. 1990; Creel and Loomis 1992; Jenkins et al. 2010). For example, the Prairie Pothole Region, which extends across the northern Great Plains into Canada, provides breeding habitat for 50–80% of waterfowl in North America (Batt et al. 1989). Wetlands are particularly popular locations for fishing, birdwatching, and hunting. Because of their open spaces and natural vegetation, wetlands are also often valued for their aesthetics, which can increase nearby property values (Doss and Taff 1996; Frey et al. 2013; Mahan et al. 2000).

Commodities

Wetlands provide marketable goods. Wetland soils can be highly productive for agriculture and silviculture, but the intensity of management may affect other ecosystem services, and excessive drainage can result in wetland loss (Dahl 2011). In the United States, approximately 6500 km² of organic soil wetlands are being used for crop production (ICF International 2013). Intensive silviculture in wetlands is most common in the southeastern United States, where species such as loblolly pine (*Pinus taeda*), cypress (*Taxodium* spp.), and black gum (*Nyssa sylvatica*) are harvested for lumber, mulch, pulpwood, and other purposes (Beauchamp 1996; Wear and Greis 2002). Forested wetlands are also managed in the Great Lakes region, and wetlands support commercial and recreational fisheries, including both fish and shellfish (Feierabend and Zelazny 1987; Lellis-Dibble et al. 2008).

Objective and Scope

In this chapter, we consider organic and mineral soil wetlands in the United States and its Island Territories that are forested or nonforested in tidal and nontidal settings. This chapter does not consider constructed wetlands (e.g., wastewater treatment systems or stormwater detention basins) or lakes and rivers. We synthesize the state of wetland soil science relative to ecosystem functions that regulate valued ecosystem services. We build on volumes dedicated to wetlands (e.g., Mitsch and Gosselink 2015) and wetland soils (e.g., Richardson and Vepraskas 2001) to assess the role of soils in wetland processes and to identify information and tools to address threats posed to the sustainability of ecosystem services provided by wetlands. Because wetlands are so different from upland forests and rangelands, a background on wetland ecology and biology is needed to provide context

for understanding the role of wetland soils in ecosystem services and their socioeconomic values.

Wetland Soil Types

In the United States, wetlands are defined on the basis of hydrology, soil properties, and composition of the vegetation. Specifically wetlands are lands that have a hydrologic regime where the site is either flooded permanently, periodically during the year, or the water table is at the soil surface during the growing season, the soils are hydric, and the vegetation is dominated by plants adapted for living in saturated soil conditions (US EPA 2015). Although the regulatory definition of wetlands by the US Army Corps of Engineers has regional variants,[1] the requirement for hydric soils is consistent across the country. Hydric soils are defined as those "formed under conditions of saturation, flooding, or ponding long enough during the growing season to develop anaerobic conditions in the upper part" (Federal Register 13 July 1994). The USDA Natural Resources Conservation Service (hereafter, NRCS) maintains the list of hydric soils in the United States[2], along with characteristics of their physical and chemical properties, and soil surveys show the spatial distribution of hydric soils.

For consideration here, hydric soils may be categorized as either mineral or organic (Fig. 6.1). This distinction reflects conditions associated with the hydrogeomorphic setting and vegetation, which influences soil properties and processes. Accordingly, we consider mineral and organic soils separately. Hydric mineral soils occur in each of the soil taxonomic orders; in contrast, organic soils are represented by the Histosol order, which represents soils with a thick (>40 cm) accumulation of organic matter on top of mineral sediments or rock (Vasilas et al. 2016). Most of these soils within the Histosol order are formed under anoxic conditions (e.g., peat soils), but some soils form under aerated conditions, and hence they are not wetland soils and are distinctly recognized (e.g., suborder Folists) and not considered here. Histosols are also termed peatlands in the international literature, but the required thickness of the organic layer varies among countries. Mineral soils may also have a thin (<40 cm) accumulation of organic materials on the surface; when the organic layer is 15–40 cm thick, the soils are histic-mineral soils (Trettin and Jurgensen 2003).

---

[1]For regional supplements of the wetland definition, see https://www.usace.army.mil/Missions/Civil-Works/Regulatory-Program-and-Permits/reg_supp/.

[2]For a list and description of hydric soils in the United States, see https://www.nrcs.usda.gov/wps/portal/nrcs/main/soils/use/hydric/.

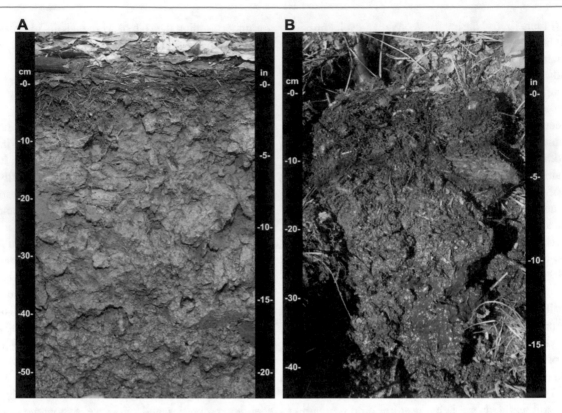

Fig. 6.1 Examples of hydric soils: (**a**) Mineral soils have a characteristic mottling; (**b**) organic soils typically lack mottling but have a thick accumulation of organic matter that is also indicative of anaerobic conditions. (Source: Vasilas et al. (2016))

Tidal and Nontidal Wetlands

The hydrologic regime of wetlands controls the anaerobic conditions in the soil, which are requisite for wetland biogeochemical functions. Nontidal wetlands are regulated by freshwater hydrology, with the source of the water being either precipitation or groundwater. These wetlands occur across a broad range of geomorphic surfaces from the tropics to the arctic. Correspondingly, the hydrologic regime is influenced by landscape position and climate, with the period of saturation varying from a few weeks to continuous. The National Wetlands Inventory classifies these nontidal freshwater wetlands into the Palustrine category, which does not convey information about the soil.

Tidal wetlands are those that have a hydrologic setting that is mediated by a combination of tidal waters, groundwater, and precipitation. Tidally influenced wetlands may occur in either marine or freshwater settings. The freshwater tidal zone occurs in low-gradient landscapes that have a large tidal amplitude, whereby the freshwater drainage is impeded by the oscillating tide. The salinity of marine-influenced wetlands varies from less than 0.2 to over 35 parts per thousand (ppt), depending on the position within the estuary or coastal zone. Wetland communities along the salinity gradient include a variety of plant communities (Fig. 6.2). Mangroves are the only forested wetland that occurs in saltwater tidal landscapes.

Distribution of Wetlands

There are approximately 420,462 km^2 of freshwater wetlands and 23,347 km^2 of tidal wetlands in the conterminous United States (Dahl 2011), with another 585,346 km^2 of freshwater and tidal wetlands in Alaska (Clewley et al. 2015). Wetlands are widely distributed throughout the United States, but they are concentrated in Alaska and also in the Atlantic coastal plain, Mississippi Valley, the upper Great Lakes, and the Prairie Pothole Regions (Fig. 6.3). While a breakdown of the soil type has not been assessed for all wetlands, approximately 80% of the freshwater nontidal wetlands in the conterminous United States are mineral soil wetlands, with the balance being peatlands; 54% of freshwater wetlands are forested (Kolka et al. 2018). The mineral soil wetlands are widely distributed across the country, while organic soil wetlands or peatlands occur primarily in the upper Great Lakes region, the southeastern Atlantic coastal plain, and Alaska.

Role of Soils in Wetland Ecosystem Functions

An important consideration of terrestrial wetlands is that they receive inputs from uplands, and they discharge outputs to groundwater and adjacent waterways and uplands. Those inputs and outputs are predominantly conveyed through the soil by hydrologic forces. The soil water moves material

Cooper River Tidal Reach

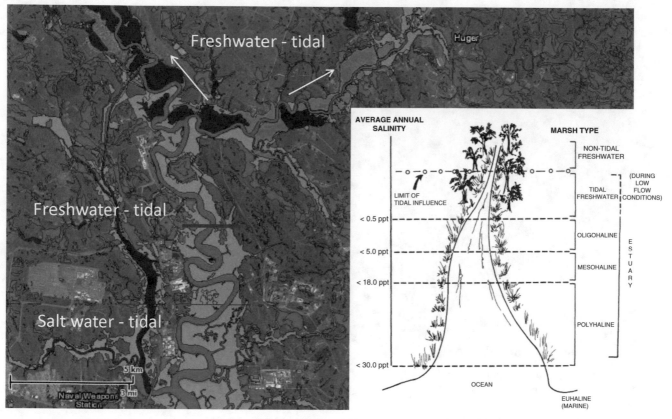

Fig. 6.2 Gradient in wetland communities from the tidal coast to the nontidal terrestrial zone along the East Branch Cooper River in South Carolina. (Source: Inset from Cowardin et al. (1979))

through the soil and also regulates soil aeration (Fig. 6.4). The continually shifting anoxic zone changes biogeochemical processes and rates, reflecting the highly dynamic and sensitive environment. Accordingly, hydrologic processes are inextricably linked to the soil biogeochemical processes. Resources are available that describe the details of wetland hydrology (Richardson et al. 2001; Verry 1997; Winter 1988; Winter and Woo 1990), hydric soil biogeochemistry (Vepraskas and Faulkner 2001), hydric soil biology (Craft 2001), hydric soil properties (Vepraskas 1996; Tiner 1999), wetland C cycle (Trettin and Jurgensen 2003), and wetland nutrient cycling (Lockaby and Walbridge 1998). Here we build on those fundamentals to synthesize soil processes across the range of basic soil materials (mineral-organic), water salinity (freshwater-saltwater), and hydrologic forcing (nontidal-tidal).

Nontidal Wetlands

Mineral Soils

Mineral soil wetlands (MSW) are found throughout the United States in various geomorphic settings, including river floodplains and deltas, glacially formed or aeolian-formed

environments, sedimentary plains, and mountain ranges. Generally, MSW are characterized by restrictive drainage soils overlain by ponded water during a portion of the year. Substrates of MSW are characterized by intermittently to perpetually saturated or anoxic conditions, or both (Vepraskas and Craft 2016). Soils of MSW often have a mottled appearance (see Fig. 6.1) that is caused by the reduction, translocation, and oxidation of iron and manganese oxides. Compared to organic soil wetlands, the organic matter concentration, water holding capacity, porosity, and cation exchange capacity are generally lower in MSW. Conversely, MSW typically have greater soil bulk density, soil pH, and nutrient availability (Mitsch and Gosselink 2015). Hydraulic conductivity can range from low to high, depending on soil properties. Primary water sources for MSW include precipitation and groundwater. Water losses are principally attributed to evapotranspiration, surface flow, and seepage or groundwater recharge (Hayashi et al. 2016; Winter 1989). Hence, the annual period of inundation, soil conditions, biotic communities (e.g., vegetation), and abiotic characteristics (e.g., water chemistry) of MSW can vary widely depending on climate, land use, hydrology, geomorphic setting, and vegetation (Euliss et al. 2004, 2014; Mushet et al. 2015).

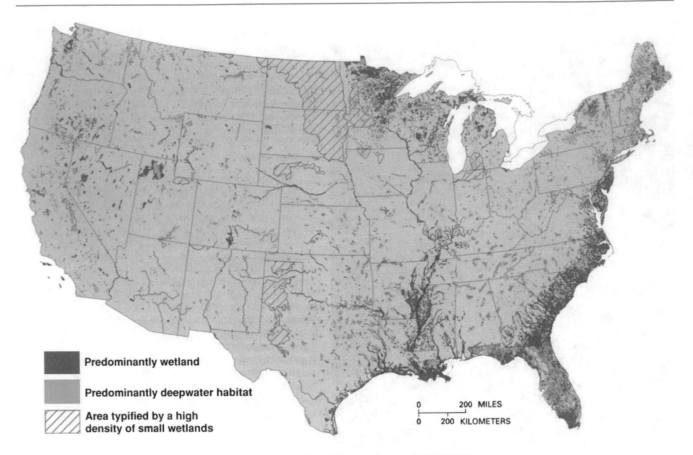

Fig. 6.3 Current distribution of wetlands in the conterminous United States. (Source: Dahl (2011))

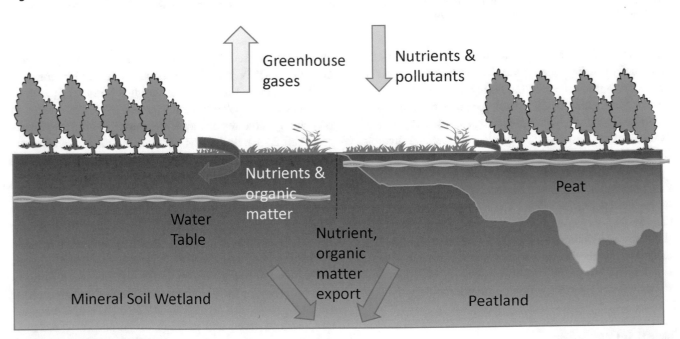

Fig. 6.4 Landscape cross section of uplands, wetlands, and water showing the water inputs, the water table, and the shifting anoxic zone in mineral and organic peat soils

Mineral soil wetlands often are distinguished by high primary and secondary productivity, which can be augmented through inputs of organic and mineral matter and nutrients from anthropogenic activities. The characteristic anoxic soil conditions of MSW result in relatively slow rates of decomposition of organic matter; thus, soils of

MSW function as a long-term C sequestration or storage site (Euliss et al. 2006; Wickland et al. 2014). These anoxic soil conditions, when combined with abundant C stores, also provide conditions for microbial CH_4 production (Bridgham et al. 2013; Mitsch et al. 2013). However, since saturated conditions can be short-lived for seasonally saturated wetlands, CH_4 production can be extremely variable both temporally and spatially (Finocchiaro et al. 2014; Tangen et al. 2015). Moreover, some MSW, such as prairie potholes, can be characterized by high sulfate concentrations that can inhibit methanogenesis, even during times of extended saturated and anoxic conditions (Pennock et al. 2010). Many MSW, especially those embedded within agricultural landscapes, produce N_2O through denitrification and nitrification processes that occur across the naturally occurring soil-moisture gradients (Bedard-Haughn et al. 2006; Tangen et al. 2015).

Prairie Potholes

The Prairie Pothole Region contains the largest concentration of MSW ecosystems in North America (see Box 6.1). It covers nearly 800,000 km^2 of the north-central United States (North Dakota, South Dakota, Minnesota, Iowa, and Montana) and south-central Canada (Manitoba, Alberta, and Saskatchewan) (Dahl 2014). The retreat of the Pleistocene glaciers (approximately 12,000 years ago) from the northern Great Plains left an undulating, hummocky landscape, draped by sediment and dotted with melting, stagnant ice blocks that eventually formed small depressions. The deposited glacial till had extremely low hydraulic conductivity. This allowed these depressions to remain saturated for extended periods, undergo oxidation-reduction processes (e.g., gleying), accumulate organic C, and develop into functional wetlands (referred to as "potholes"). In total, there are an estimated 58 million potholes (Beeri and Phillips 2007) covering approximately 70,000 km^2 of surface area (Dahl 2014; Euliss et al. 2006) within a matrix of croplands and grasslands. The majority of these wetlands are less than 0.01 km^2 in area and have ponded water that is more than 2 m deep (Dahl 2014). Periods of inundation range from ephemeral to permanent, and salinities span a gradient from fresh to hypersaline (Euliss et al. 2004; Goldhaber et al. 2014; Niemuth et al. 2010; Winter and Rosenberry 1998) (Box Fig. 6.1).

Box 6.1

Box Fig. 6.1 A seasonal wetland in a native prairie grassland setting in the Prairie Pothole Region. The underlying glacial till in the region has extremely low hydraulic conductivity, allowing water to pond in small, shallow depressions referred to as "potholes." Pothole wetlands can function as groundwater discharge, recharge, or flow-through sites that vary in size, water chemistry, water permanence, and biotic communities. These wetlands provide a range of ecosystem services, including breeding habitat for migratory waterfowl, pollinator and wildlife habitat, flood mitigation, filtration of pollutants, groundwater recharge, C sequestration, and nutrient retention. (Photo credit: Lawrence D. Igl, U.S. Geological Survey)

Wetlands provide a range of ecosystem services, including breeding habitat for migratory waterfowl, pollinator and wildlife habitat, flood mitigation, filtration of pollutants, groundwater recharge, C sequestration, and nutrient retention. The gleyed soils of Prairie Pothole Region wetlands make them hotspots of biogeochemical activity due to the geologic history and critical zone processes in the region (Badiou et al. 2011; Bansal et al. 2016; Dalcin Martins et al. 2017; Goldhaber et al. 2014, 2016; Tangen et al. 2015).

Approximately 70 million years ago, an inland sea formed organic-rich marine shales that provided a source of pyrite in the glacial till underlying the Prairie Pothole Region. As groundwater slowly flows from topographically higher to lower wetlands (i.e., groundwater recharge to discharge, respectively), oxidation of pyrite and gypsum and subsequent dissolution of carbonates in the till result in an accumulation of ions such as sulfate (SO_4^{2-}), bicarbonate (HCO_3^-), calcium (Ca^{2+}), and magnesium (Mg^{2+}) in wetland soils and waters. Wetland chemistry is further controlled by evapotranspiration, mineral precipitation and dissolution, and biogeochemical processes. The concentrations of the various chemical constituents of Prairie Pothole Region wetlands have a strong bottom-up influence on the wetland plant and invertebrate communities (Euliss et al. 1999; Stewart and Kantrud 1972), which in turn influence fish, bird, and other wildlife populations (McLean et al. 2016a, b).

During the growing season, extensive stands of emergent macrophytes, combined with algal and submerged aquatic vegetation, contribute to high concentrations of dissolved organic carbon (DOC) and particulate organic C in standing water and sediment porewater of Prairie Pothole Region wetlands. These C inputs fuel some of the highest rates of CH_4 flux (as high as 0.75 g CH_4 m^{-2} day^{-1}) and SO_4^2 reduction reported from freshwater wetlands (Dalcin Martins et al. 2017). Research suggests that wetlands with elevated SO_4^2 concentrations tend to have reduced CH_4 emissions (Pennock et al. 2010), similar to marine tidal wetlands (Pennock et al. 2010; Poffenbarger et al. 2011). Denitrification and nitrification are also important processes in Prairie Pothole Region wetland soils. Wetlands that receive runoff from surrounding agricultural lands can produce large pulses of N_2O emissions (Tangen et al. 2015). As C storage and sequestration sites and as sources of greenhouse gases (GHGs), Prairie Pothole Region wetlands have the potential to impact net GHG budgets of the United States (Badiou et al. 2011; Bridgham et al. 2006; Creed et al. 2013; Euliss et al. 2006; Tangen et al. 2015).

Forested Mineral Soil Wetlands

Forested mineral soil wetlands are found from the tropical coastal margin to the boreal zone. Lugo and others (1990) categorized these into four major forested wetland types within the United States.

Deep Water Swamps Deep water swamps are forested wetlands with standing water throughout all or most of the year (Conner and Buford 1998). Their distribution is primarily the coastal plain of the Southern United States, extending up the Mississippi River Valley into Southern Illinois. Vegetation is mainly baldcypress (*Taxodium distichum*), water tupelo (*Nyssa aquatic*), pond cypress (*Taxodium ascendens*), and swamp tupelo (*Nyssa biflora*) (Harms et al. 1998). While swamp soils are primarily histic, inclusions of Alfisols, Entisols, and Inceptisols occur on microsites where flooding is intermittent (Conner and Buford 1998). As an example, small mesic islands occur in the Great Dismal Swamp of Virginia, and forest communities on these sites resemble those of nearby uplands (Carter 1990). On a smaller scale, red maple (*Acer rubrum*) and black ash (*Fraxinus nigra*) swamps in the Northeastern United States may exhibit mineral soils with near-neutral pH and high levels of Ca and Mg (Ehrenfeld 2012).

Major Alluvial Floodplains Alluvial landforms are most often associated with morphologically younger soils (Entisols, Inceptisols). However, older soils (Ultisols, Alfisols, Mollisols) have been mapped in floodplains along the Atlantic Coast. Mineral soils occur in these floodplains on level and convex microrelief, while mineral soils approaching histic-mineral and organic soils dominate in concave positions (Anderson and Lockaby 2007).

Two types of forested floodplains, black-water and red-water, are recognized in the Eastern United States and reflect the source of the sediments carried by the river. Black-water rivers originate in the coastal plain and have a dark color due to the dissolved and suspended organic matter. Red-water rivers originate in the piedmont or mountains, with the color reflecting the suspended mineral sediments; sometimes these are called brown or even white water, depending on the color of the suspended clay and silt. Red-water rivers tend to exhibit relatively steep gradients that provide more energy to carry suspended sediment loads in comparison with the lower gradient black-water streams, which carry less sediment. Consequently, red-water systems often exhibit higher fertility than their black-water counterparts (Furch 1997; Schilling and Lockaby 2006). Although floodplains are generally more fertile than corresponding uplands, deficiencies of N, phosphorus, and base cations are common and may restrict forest growth (Schilling and Lockaby 2006). These nutrient inadequacies are particularly common on black-water sites. In spite of differences in nutrient content, Schilling and Lockaby (2006) were not able to discern major differences in net primary productivity (NPP) between two such floodplains in Georgia. The Mississippi Delta, the foremost alluvial floodplain in the United States, can be very productive and has exhibited some of the highest estimates of annual forest NPP in North America (Nelson et al. 1987).

Floodplains that occur in the arid Southwestern United States are distinct from those of humid regions. Streams associated with these systems may carry heavy sediment loads, and soil surfaces can be hydrophobic with sparse vegetation present (Stromberg et al. 2012). Soils vary from sandy loams to clay loams and may show little horizon development (Stromberg et al. 2012).

Minor Alluvial Floodplains Soils in the small alluvial floodplains are usually of local origin, in contrast to the soils of major floodplains (Hodges 1998). Soil texture varies with microrelief and distance from the stream. Typically, a natural levee or front occurs nearest the stream and is composed of sandy material. Behind that front, flats and lessor convex surfaces occur and may be clayey. Concave microrelief in the form of narrow sloughs may exhibit higher deposits of organic matter (Hodges 1998). In piedmont areas and near fall lines, minor floodplains are often buried in clay sediments deposited as a result of poor agricultural practices on uplands in the eighteenth, nineteenth, or early twentieth century (Trimble 1974). These sediment deposits can be very deep (>2 m) and usually consist of clays.

Wet Flatwoods In the Atlantic coastal plain, flatwoods sites are predominantly occupied by loblolly (*Pinus taeda*) and slash pine (*Pinus elliottii*), with inclusions of pond pine (*Pinus serotina*) on wetter areas. However, on slightly higher elevations and if fire has been excluded, hardwoods may dominate with swamp laurel (*Quercus laurifolia*), water oak (*Quercus nigra*), and willow oak (*Quercus phellos*), along with red maple, green ash (*Fraxinus pennsylvanica*), and American elm (*Ulmus americana*) (Harms et al. 1998). Soil diversity is generally high, and Aquods are often present with wet Alfisols and Ultisols (Harms et al. 1998). Many of these sites exhibit pronounced phosphorus deficiency, and productivity is often increased dramatically by the addition of phosphorus fertilizer (Harms et al. 1998).

Organic Soils

Organic soils are the distinguishing characteristic of peatlands as a wetland class. Peatlands accumulate significant amounts of soil organic matter, or peat, when plant production exceeds decomposition over time (Kolka et al. 2011a; Sahrawat 2004). In the United States, nonfrozen peatlands occupy approximately 163,000 km², with Alaska contributing about 48% of that area. Although nonfrozen peatlands only represent about 23% of the total nontidal wetland area in the United States, they store about 46% (14.2×10^{12} kg, or 14.2 petagrams [Pg]) of wetland C, with approximately 41% in Alaska (Kolka et al.

2018). Peatlands in the United States are also important sources of CH_4 to the atmosphere, contributing 3.2×10^9 kg or 3.2 teragrams (Tg) C year^{-1} (Kolka et al. 2018). Peatland accumulation rates range from 0.18 to 7.8 mm year^{-1}, with lower rates associated with higher latitudes (Kolka et al. 2011a, b). In both forested and nonforested peatlands, C is mainly sequestered in the organic soils, with much less sequestered aboveground in trees, shrubs, and sedges.

Generally, there are two types of peatlands: fens, which are those that are connected to the regional groundwater, and bogs, which have a perched water table and are not connected to the regional groundwater. In both hydrologic conditions, forested and nonforested peatlands exist. Those that are connected to the regional groundwater tend to be more nutrient rich, have a higher pH, and support more diverse plant communities.

Nonforested Peatlands

Nonforested peatlands include wetland types such as marshes and some fens and open bogs. Nonforested peatlands are usually a result of high water tables that are consistently near the surface. These high water tables are generally a function of being connected to the regional groundwater system leading to a fen ecosystem (Winter and Woo 1990), but they can also have high perched water tables, as in the case of open bogs. Because of the connection to regional groundwater that tends to be higher in nutrients, cations, and anions, these plant communities tend to be more productive and diverse than those that are only fed by precipitation (see discussion in Forested Peatlands section). Because many nonforested peatlands are connected to the regional water table, the large supply of water makes them less susceptible than forested peatlands to changes in precipitation regimes. This connection to groundwater can also cause pH, exchangeable cations, N, and P to vary widely. Open bogs are generally at the lower pH and nutrient end of the gradient, with pH, cations, and other nutrients ranging from intermediate to high in fens. Under normal conditions, fens and bogs are net sinks for CO_2, but they can be sources of CH_4 (Turetsky et al. 2014). Ecosystem storage of C ranges from approximately 400 to 1500 Mg C ha^{-1}, while CH_4 fluxes range from 0.02 to 0.07 Mg C ha^{-1} year^{-1} (Kolka et al. 2018).

Forested Peatlands

Forested peatlands include wetland types such as forested bogs, forested fens, bottomland hardwood swamps, pocosins, and Carolina bays. Forested peatlands are usually a result of variable water tables that typically draw down at times, especially during the growing season. Peatlands that have the most variable water tables are generally those that are perched and fed mainly by precipitation. Water table drawdown allows an aerobic soil/root system that enhances growth of the plant community and allows shrubs and trees to grow. As in nonforested peatlands, forested peatlands that are only fed by precipitation (e.g., forested bogs) tend to be have lower concentrations of nutrients, cations, anions, and pH than those fed primarily by groundwater (forested fens) or surface water (swamps). Forested peatlands associated with a variable water table fed only by precipitation are more susceptible to long-term changes in precipitation regimes, especially droughts (Zoltai et al. 1998). Ecosystem storage of C ranges from approximately 410 to 1820 Mg C ha^{-1}, while CH$_4$ fluxes range from 0.02 to 0.08 Mg C ha^{-1} year^{-1} (Kolka et al. 2018).

Tidal Wetlands

Tidal wetlands occupy the intertidal margins of shorelines, extending from shallow coastal bays and estuaries into reaches of coastal rivers. They share tidal hydrology arising from ocean tides that propagate inland into estuaries, bays, and rivers and therefore vary widely in salinity from full-strength seawater to brackish to freshwater. Tidal marshes are dominated by herbaceous plant species, while tidal forests are dominated by mangroves in the marine setting and bottomland hardwood swamp species in the freshwater tidal areas. Shrub-dominated tidal wetlands are known as shrub-scrub ecosystems. Tidal wetlands occur on both mineral and organic soils; soil type is a characteristic that affects the physical stability of the system against accelerated sea level rise and biogeochemical processes such as suppression of CH$_4$ emissions by iron (Fe (III)) respiration (Neubauer et al. 2005).

Tidal wetlands are zones of intense biogeochemical activity within the coastal plain landscape due to regular tidal flooding and direct connectivity to rivers and estuaries. They are dynamic environments where high rates of plant production, microbial activity, and hydrologic exchange influence adjacent aquatic ecosystems, groundwater, and the atmosphere (Megonigal and Neubauer 2009). Tidally driven hydrology generally produces open element cycles compared to nontidal wetlands, a concept embodied in the Outwelling hypothesis, which links tidal wetland and aquatic biogeochemical cycles (Odum et al. 1995). Tidal wetlands are an important source of C to estuaries (Nixon 1980), which supports microbial metabolism and nutrient cycling and regulates ultraviolet light penetration of the water column.

Element cycles in freshwater, brackish, and saline tidal wetlands are constrained by many of the same factors. Flooding frequency is the most important factor regulating primary production in tidal marshes; peak production of herbaceous plants at an intermediate flooding frequency varies by species. Production is limited by N, P, or both N and P,

depending on the geomorphic setting and anthropogenic nutrient loading. Coastal eutrophication tends to increase plant growth, decrease root-shoot ratio, and trigger changes in microbial metabolism and decomposition. The net effect of these responses on C sequestration, elevation gain, and ecosystem stability can be positive or negative for reasons that remain unclear (Megonigal and Neubauer 2009).

Tidal wetlands are referred to as "blue carbon ecosystems" because they hold large stocks of soil and plant C and support rapid rates of C sequestration (Chmura et al. 2003; McLeod et al. 2011). Marshes, mangroves, and seagrass meadows account for about half of the total marine soil C budget (Duarte et al. 2005) and bury an amount of C equivalent to that stored by terrestrial forests, despite occupying just 2.5% of all land area (McLeod et al. 2011). These high rates of C sequestration are attributed to interactions among three primary factors: rapid plant production, slow decomposition, and sea level rise (Kirwan and Megonigal 2013). As rates of sea level rise accelerate, coastal wetlands have the potential to sequester soil C at increasingly rapid rates as long as plants survive flooding and contribute to soil building. Compared to upland soils, the sequestration potential of tidal wetland soils is extremely high because rising sea level increases the potential soil volume over time. Coastal wetlands have only recently been recognized as important C sinks, and therefore, their response to global change is largely unexplored. The future stability of these systems is uncertain because global change drivers such as temperature and elevated CO$_2$ perturb the complex biotic and abiotic feedbacks that drive high rates of soil C sequestration (Kirwan and Mudd 2012).

Tidal wetlands have the capacity to gain elevation by a combination of sediment trapping and in situ addition of organic C, the major processes driving soil development and C stocks. Accordingly, the difference in elevation between sea level rise and sediment accretion reflects the net change in submergence depth. The relative importance of mineral versus organic material to elevation gain depends on a variety of hydrogeomorphic variables such as tidal range, soil surface elevation, and sediment supply and ecological variables such as plant community composition, net primary production, and the relative contributions of aerobic versus anaerobic microbial respiration. Organic matter accounts for up to 80% of soil mass in tidal wetlands, occupying about twice the volume of mineral matter on a mass basis and contributing at least twice as much to accretion as an equal mass of minerals (Neubauer 2008; Turner et al. 2000). Tidal wetlands with organic soils are typically perched high in the tidal frame, which decreases their vulnerability to rapid sea level rise; however, they are also more vulnerable to elevation loss caused by plant stress due to increased flooding or salinity.

Tidal Freshwater Wetlands

Tidal forcing by oceans can extend upstream into coastal rivers, creating a unique combination of freshwater with flood-ebb tidal cycles. Tidal freshwater wetlands (TFWs) are common on all US coasts where nontidal rivers and tidal estuaries meet. Elevation transitions mark changes in plant community composition, with low and intermediate elevations dominated by herbaceous species (e.g., marshes) and high elevations dominated by trees (e.g., tidal freshwater swamps). The primary production of trees in TFW is limited by frequent flooding, while herbaceous production is comparable to brackish and salt marshes.

TFWs have several unique features. TFW hydrology lacks the strong seasonal variability of nontidal freshwater due to tidal forcing, which means that hydrologic exchanges and greenhouse gas emissions are relatively stable. The lack of salinity leads to plant communities that are far more diverse than saline tidal wetlands. One consequence of low salinity is a limited supply of sulfate, a terminal electron acceptor that effectively suppresses production of CH_4 emissions. As a result, TFWs support higher CH_4 emissions than saline wetlands (Bartlett et al. 1987; Poffenbarger et al. 2011). Dissolved inorganic C and alkalinity exported from TFWs to rivers can potentially have a relatively large effect on river carbonate chemistry because freshwaters are poorly buffered compared to saline estuarine water.

Tidal Marine Wetlands

The combination of tidal hydrology and high salinity gives rise to wetlands characterized by low species diversity and microbial respiration dominated by sulfate reduction. At salinity levels more than 0.5 practical salinity units (psu), all tree species and most herbaceous species in TFWs are replaced by a limited number of plant species with specialized adaptations to salinity stress. Temperature is an important factor in the distribution and function of tidal marine ecosystems because it determines whether mangrove species are present. Mangroves are restricted to tropical and subtropical climates but can occupy intertidal areas regardless of salinity. Tidal freshwater swamps and tidal freshwater mangroves are distinctions of plant species composition and latitude and are not necessarily meaningful for understanding differences in element cycling.

Salinity affects microbial processes that regulate C cycling and greenhouse gas emissions. Sulfate contributes about 8% of the salts in seawater, and it is a dominant substrate for microbial respiration in marine tidal wetland soils. Sulfate-reducing bacteria outcompete microorganisms that produce CH_4, effectively suppressing CH_4 emissions from tidal wetlands in proportion to sulfate availability. The relationship between salinity and CH_4 emissions is used to predict the point at which the radiative cooling effect of CO_2 capture through soil C sequestration is balanced by the heat-ing effect of CH_4 emissions in tidal wetlands of varying salinity. Above 18 psu, tidal wetlands are net sinks of CO_2 by one conservative estimate (Poffenbarger et al. 2011), but this remains an area of significant uncertainty.

Sustainability of Wetland Functions and Ecosystem Services with Changing Conditions

As previously summarized, wetlands and associated soil processes are sensitive to their inherent setting and associated hydrologic, biotic, and abiotic conditions. Here we summarize the sensitivities of wetlands to changing conditions resulting from both natural and anthropogenic causes.

Long-Term Climate Variability

Projected increases in temperature and changes in the frequency and intensity of precipitation are likely to impact wetland soil function, primarily as a result of changes in the water regime. Over the coming decades, average annual temperatures are expected to warm by 1.4 °C, and even more in northern latitudes (Vose et al. 2017). Higher temperatures can alter hydrology by increasing evapotranspiration, melting permafrost, and changing the amount and timing of runoff and groundwater recharge from snowmelt and glaciers. Wetland hydrology may be further impacted by precipitation that contributes to hydrologic inputs, including a projected decrease in winter and spring precipitation in the Southwest and an increase in precipitation in the northern United States and Alaska (Easterling et al. 2017). Heavy precipitation is likely to increase by as much as 10–20% by the end of the century over the entire United States, with the potential to contribute to extreme flooding in some wetlands (Easterling et al. 2017). Warming is expected to increase drought conditions and alter the frequency and severity of wildfires, particularly in the western United States and Alaska, where peatlands are vulnerable to new fire regimes (Young et al. 2016).

Long-Term Shifts in Temperature and Precipitation

Vegetation Response

Shifts in temperature, precipitation, and the combination of both will alter species composition and diversity and, in some instances, ecosystem function of vegetated wetlands. In northern alpine wetlands, an increase in air temperature is expected to result in faster-growing species from lower latitudes outcompeting and replacing slower-growing species (Burkett and Kusler 2000). In salt marshes, warming has

---

---

been shown to change the native plant assemblage (Gedan and Bertness 2009; McKee et al. 2012). Warming will also allow tropical mangrove species to expand their northern range in the United States. A decrease in the number of freezing events has decreased and allowed the black mangrove (*Avicennia germinans*) to expand into salt marsh ecosystems (Saintilan et al. 2014; Simpson et al. 2017), thus increasing mangrove area by more than 70% along the Texas coastline (Armitage et al. 2015). A decrease in precipitation will shift freshwater wetland plant communities to communities that include plant species that are more tolerant of drier conditions, or even replace them entirely with upland species (Burkett and Kusler 2000; Johnson and Poiani 2016; Mortsch 1998).

Changes in temperature and precipitation will also affect wetland biomass production. Marsh productivity increased by 10–40% as temperatures naturally increased by 4 °C along a latitudinal gradient (Kirwan et al. 2009). Experimental warming increased aboveground growth of two salt marsh plants, saltmeadow cordgrass (*Spartina patens*) and smooth cordgrass (*S. alterniflora*), by 14–45% (Gedan et al. 2011). This increase in productivity will result in increased litter production, thereby affecting organic matter turnover, greenhouse gas emissions, and soil organic matter (Ellison 2000). However, once temperatures go beyond plant thresholds (e.g., –40 °C for mangroves), production will decrease (Clough et al. 1982; Mortsch 1998). Decreased precipitation can alleviate stress from waterlogged conditions in some marshes (Charles and Dukes 2009), but when combined with warmer temperatures, it will also lower water tables, increase soil salinity, and expose sediments to aerobic conditions that increase organic matter decomposition (Burkett and Kusler 2000; Clair et al. 1995; Morrissey et al. 2014; Mortsch 1998). These conditions may decrease plant productivity (Field 1995; Day et al. 2008) as well as increase aerobic decomposition of wetland soils, and thus, CO_2 fluxes to the atmosphere (Field 1995; Moor et al. 2015). In mangroves and salt marsh systems that need to maintain their elevation relative to sea level, increased decomposition of peat has impacted their ability to keep up with sea level (Bridgham et al. 1995; Burkett and Kusler 2000; Rogers et al. 2014; Woo 1992). Alternatively, increased precipitation increases plant growth, seedling survival, and seed germination by reducing salinity and osmotic stress of plants and trees (Clough et al. 1982; Ellison 2000; Field 1995; Noe and Zedler 2001). Increased precipitation will also increase peat and sediment accumulation through increased delivery of sediments from flooding, peat production through root growth, and organic matter accumulation from litter inputs (Rogers et al. 2014; Sanders et al. 2016; Scavia et al. 2002). This increased peat and sediment accumulation has allowed mangroves and salt marshes to keep up with sea level (Eslami-Andargoli et al. 2009; Rogers et al. 2014; Snedaker 1995).

Changes in community composition and production may affect functionality of wetland soils. Positive impacts include increased C stocks and C accumulation rates that can potentially offset climate change impacts (Donato et al. 2012; MacKenzie et al. 2016). For example, mangrove migration into salt marshes has significantly increased aboveground and belowground C stocks by 2.3 Mg C ha^{-1} year^{-1} (Kelleway et al. 2016; Simpson et al. 2017). Shifts in plant communities in fens would result in greater root depths and higher C storage. Alternatively, this could result in lower C storage if roots increase peat/soil oxygenation that fuels organic matter decomposition (Moor et al. 2015). Shifts in community composition can also impact accretion and sedimentation rates of wetlands. Increased sediment loads and belowground C accumulation are two mechanisms that help wetlands maintain their position relative to sea level (Krauss et al. 2014). Thus, decreased productivity and decreased litter and root inputs will negatively impact the ability of wetlands to maintain their position relative to sea level. Shifts in wetland plant communities can also decrease nutrient soil storage or increase peat decomposition.

Altered Hydrology

Changes in long-term precipitation and temperature regimes are expected to alter wetland hydrology, impacting soil function and ecosystem feedbacks. Differences in wetland landscape position and regional variability in temperature and precipitation suggest that responses are likely to vary widely. In some cases, the hydrologic conditions that define wetlands may cease to exist, and wetlands will transition to uplands or open water systems.

Increased evaporative losses, particularly when coupled with decreases in precipitation, may lead to declines in the water table and less frequent flooding in riparian wetlands. Wetlands in arid to mesic environments may be particularly vulnerable (Johnson et al. 2005; Springer et al. 2015). In prairie pothole wetlands, climate simulations indicate that a 3 °C increase in temperature combined with a 20% decrease in precipitation would increase the amount of emergent vegetation, shifting the community to a dry marsh with significantly less open water available for waterfowl (Johnson et al. 2005).

In peatlands, droughts have been found to increase soil microbial activity and C loss (Griffis et al. 2000; Laiho 2006). Drying can increase the oxidation of peat, which enhances microbial growth and promotes enzymatic activity that reduces phenolic compounds, further contributing to microbial decomposition and growth (Fenner and Freeman 2011). Warmer temperatures increase enzymatic activation rates, but responses may vary with the complexity of the substrate (Davidson and Janssens 2006). Warmer, drier peatland soils may also promote increased recruitment and growth of shrubs (Waddington et al. 2015), which in turn may increase transpiration interception losses.

Greater amounts of precipitation may increase the period and depth of wetland flooding, if not offset by evaporative losses. Depressional wetlands may be particularly susceptible to increased precipitation because of their topography and drainage (Johnson et al. 2004; Mitsch et al. 2013). Long-term changes in extreme flooding in riverine wetlands and direct links to climate change are less clear, as changes in streamflow can be affected by land use, dams, and other factors (Walsh et al. 2014).

Increased flooding may also result in decreased CO_2 and increased CH_4 and N_2O emissions, as microbial respiration shifts from aerobic to anaerobic soil conditions (Yu et al. 2008). Increases in CH_4 and N_2O emissions may be of particular concern because these trace gases have greater global warming potentials than CO_2. However, much is still unknown about the release of CH_4 from wetland soils, including the capacity for CH_4 oxidation as it diffuses through the soil column (Segarra et al. 2015).

Thawing of Permafrost Wetlands

The primary threat to continuous permafrost, discontinuous permafrost, and sporadic and isolated zones in wetland soils in northern latitudes is long-term climatic warming that leads to soil thawing. Permafrost has a seasonal cycle where the near-surface ice in the soil melts during warmer seasons. The active layer defines the extent of this thaw zone and varies depending on aspect, topography, and soil conditions. The depth of this thaw also governs the physical stability of the ground. Extensive thawing of frozen soils can destabilize the landscape, resulting in changes to the landscape structure that affect the fate of the soil system when it refreezes and the active layer is once again frozen. Extensive thawing can result in features such as patterned ground and ice-wedge polygons that can change so dramatically that structures and vegetation communities are altered.

A major concern of the increased soil thaw is the exposure of C pools to decomposition. Overall, the northern latitude soils of the arctic and boreal regions contain 2050 Pg C (to a depth of 3 m) (Schuur et al. 2015). Thawing and exposure of previously frozen soil horizons can release C either as a gas through respiration or secondary products as dissolved organic carbon (DOC). The enhanced losses of these C pools from the soil have both global implications for the balance of greenhouse gases in the atmosphere and terrestrial realms, as well as for the associated aquatic communities that face increased amounts of dissolved organic matter mobilized from the soil pool.

In thawed permafrost soils, the presence of melt water can maintain anaerobic conditions that deter the rapid oxidation of organic matter (Shurr et al. 2018). However, the potential for CH_4 flux increases, leading to enhanced emissions of both CH_4 and CO_2 from the active layer of permafrost soils. Saturated soils can also enhance the movement of DOC from the soil. Subsidence of soil surface horizons is common upon melt. The lowering of the surface coincides with a rise in the water table, so anaerobic conditions are maintained and gaseous losses are mitigated. However, the potential drainage of soil water due to collapse features can expand these features, which is especially threatening to physical structures, such as roads and buildings.

Sea Level Rise

Many coastal wetlands will face increased inundation, saltwater intrusion, and salinity stress as sea level rise approaches 1–4 m by 2100 (Mendelssohn and Morris 2002). Many of these wetlands will be subject to warmer ocean temperatures and increased exposure to hurricanes and associated storm surges. These impacts may be compounded by shifts in freshwater, sediment, and nutrient inputs to coastal wetlands; northward migration of species such as mangroves; and changes in the productivity of wetland plants (Megonigal et al. 2016).

Flooding frequency is the single most important factor that governs tidal wetland ecology and biogeochemistry, and it is currently changing at an accelerated pace due to a combination of sea level rise and land subsidence. Sea level rose more than 2 m over the last 4000 years, and it is forecast to rise by the same amount over the next 200 years. Simultaneously, land is subsiding at variable rates that are increasing in some locations due to factors such as groundwater withdrawal. Considerable uncertainty remains as to the capacity of tidal wetlands to keep pace with relative sea level rise through a combination of elevation gain and horizontal migration into uplands. However, it is well established that the most vulnerable systems are those with small tidal ranges and low sediment supplies, conditions that are typical of large areas of tidal marine wetlands, such as those in Chesapeake Bay and the Mississippi River delta. Tidal freshwater wetlands may be less vulnerable than some tidal marine wetlands because they typically occur near upstream sediment sources.

Sea level rise is changing the boundary between tidal freshwater wetlands and tidal marine wetlands, with uncertain consequences for C stocks and element cycling. Organic accretion rates tend to be higher in TFWs than tidal marine wetlands (Craft 2007), suggesting that increasing salinity intrusion may cause soil C sequestration to decline. However, studies on the direct effects of sulfate on soil organic matter decomposition rates have reported both accelerated rates and no change in rates (D'Angelo and Reddy 1999; Weston et al. 2006, 2011) but provide little mechanistic insights to explain these differences (Sutton-Grier and Megonigal 2011). There are also perplexing observations that are not adequately explained by our present understanding of anaerobic decomposition, such as reports that increased inundation caused faster rates of soil organic matter decomposition (Kirwan et al. 2013; Mueller et al. 2016) and that increased salinity stimulated CH_4 emissions (Weston et al. 2011).

Salinity changes can affect a number of nutrient cycling processes, including P adsorption, denitrification, and nitrification (e.g., Caraco et al. 1989; Howarth 1988). The intrusion of saltwater and associated SO_4^{2-} can lead to the breakdown of the "iron curtain" described by Chambers and Odum (1990), which releases sediment-bound P (Caraco et al. 1989). Elevated salinity can also lead to decreased sorption of NH_4^+ to soil particles, and increased NH_4 concentrations may suppress N_2 fixation (e.g., Howarth 1988). The physiological effects of salinity on nitrifying and denitrifying microbes reduce the activity of these organisms (Furumai et al. 1988; MacFarlane and Hebert 1984; Stehr et al. 1995). Much of the research on the effects of rising salinity on soil and sediment biogeochemistry has focused on transient effects. Over longer time periods, salt-sensitive plants, animals, and microbes will likely be replaced by salt-tolerant species (e.g., Magalhães et al. 2005). Relatively little is known from direct manipulations of salinity about the direction of these longer-term effects.

Sea level rise will increase the landward progression of tidal freshwater wetlands. The resulting conversion of nontidal to tidal wetlands will result in significantly increased periods of soil anoxia, which would be expected to cause corresponding changes in the C and nutrient cycles.

Extreme Events

Fire

Increased temperatures and evapotranspiration, when not offset by precipitation, are expected to lead to drier fuels and lower water tables in many northern peatlands (Flannigan et al. 2009), which in turn are likely to increase wildfire frequency, particularly in summers with extreme heat and low moisture availability (Young et al. 2016). This response is evident in Alaska, where fire frequency, duration, and intensity have increased over the past several decades (Flannigan et al. 2009; Kasischke and Turetsky 2006; Partain et al. 2016; Sanford et al. 2015).

Increased wildfire activity in peatlands will reduce soil C stocks, releasing CO_2 and other greenhouse gases into the atmosphere (McGuire et al. 2009). Peatland fires in western Canada alone are estimated to emit up to 6 Tg C year^{-1} (Turetsky et al. 2004). In black spruce (*Picea mariana*) stands of Alaska, decadal fire emissions exceeded ecosystem uptake by as much as 86 ± 16 Tg C (Turetsky et al. 2011). The increase in severity of fires has been shown to increase the depth of organic soil burning, which can result in particularly high C emissions (Turetsky et al. 2011), and their smoldering may contribute to atmospheric pollution (Flannigan

et al. 2009). Smoke may also increase albedo, which may help offset radiative forcing from the fire's greenhouse gases (Lyon et al. 2008; Randerson et al. 2006).

Floods Resulting from Increased Incidence of Severe Storms

Extreme precipitation events are becoming more severe and frequent (Min et al. 2011) in many parts of the United States (Groisman et al. 2012; Kunkel et al. 1999) as well as globally (Groisman et al. 2005), with consequences to structures and infrastructure, including houses, dams, roads, sewer and stormwater drainage, and drinking water systems (Gariano and Guzzetti 2016; Mamo 2015). While the role of wetlands within the watershed on storm runoff is recognized (Johnston et al. 1990), extreme events suggest the need to consider the role of wetlands especially in urbanizing landscapes. The interactions of road and stormwater conveyance infrastructure and other water management systems with extreme precipitation events are relatively unknown, presenting risks to downslope resources. While flooding is the obvious risk (Amatya et al. 2016), storms may significantly increase the sediment transported by the waterways as well (Shelby et al. 2005).

Atmospheric Effects

Humans have altered the chemistry of the atmosphere via a number of processes, including fossil fuel combustion, land use change, volatilization from agricultural activities, incineration, and other industrial processes (Monks et al. 2009). These atmospheric changes can affect wetland soils either via direct positive and negative effects on plants (e.g., CO_2, nitrogen and sulfur oxides, ozone) or via wet and dry deposition onto soils (e.g., reduced and oxidized N, sulfur oxides, and Hg). Wetland-specific soil impacts often derive from the interaction of these pollutants with anoxic sediments.

Elevated CO_2

In addition to its effect on climate, increasing CO_2 can have a direct effect on photosynthesis because CO_2 is one of the key substrates for this process. Numerous upland studies have shown that elevated CO_2 can increase plant photosynthetic rates and subsequently increase productivity, if no other conditions, such as N, other nutrients, or light, are limiting (Johnson 2006). Without limitations, elevated CO_2 leads to higher plant production, longer growing seasons, and higher leaf area (Hyvönen et al. 2007), as well as lower stomatal conductance and transpiration and higher water use and light use efficiency (Drake et al. 1996). Those principals are presumed for wetlands, which have not received the degree of study compared to uplands, and they are supported by several studies indicating similar responses for wetland plant communities. For example, in sedge wetlands, photosynthetic rates and biomass yields increase, N in plant tissues is reduced, and overall ecosystem C accumulation increases under increased CO_2 (Rasse et al. 2005).

Increased salinity by rising sea levels in coastal wetlands tends to lessen the effect of elevated CO_2 on plant communities and soils by stressing the plant communities, which counteracts the gains from elevated CO_2 (Erickson et al. 2007). However, others have found that productivity gains and soil additions related to responses to elevated CO_2 have led to an increase in marsh elevation that have counterbalanced sea level rise (Langley et al. 2009).

A secondary response to elevated CO_2 in wetlands is greater CH_4 emissions (Vann and Megonigal 2003), although the mechanism of the response is not clear. Vann and Megonigal (2003) speculated that recent elevated carbohydrates (due to higher photosynthetic rates) become available for fermentation through root exudates that drive the methanogen community. Another secondary response of wetlands, especially peatlands, to elevated CO_2 is higher production and transport of DOC (Freeman et al. 2004). Similar to the proposed mechanism for elevated CH_4 production, higher DOC fluxes result from greater plant production, especially belowground, leading to higher available C in the soil pore water.

Pollutants and Nutrients in Deposition

Mercury

Mercury (Hg) is a toxin that can affect neurological development in animals and humans (Zillioux et al. 1993). The major anthropogenic sources of atmospheric Hg are coal- and oil-fired power plants, gold mining, metal manufacturing, cement production, waste disposal, and caustic soda production (Pirrone et al. 2010). Emissions in North America are declining (Weiss-Penzias et al. 2016) and are currently estimated at 9% of the global total, compared with 64% in Asia (Pirrone et al. 2010). Deposition trends within the United States are mixed, with some regions experiencing declines and others experiencing increases, with the latter attributed, at least in part, to long-distance transport from Asia (Weiss-Penzias et al. 2016).

Mercury that enters wetlands is predominantly in an unreactive form, but it can be transformed via microbial action under anoxic conditions to a readily bioavailable form, methyl mercury, when in wetland soils or sediments. As a result, wetlands can be sinks for total Hg and major sources of methyl mercury export to freshwater ecosystems (Driscoll

et al. 2007; Rudd 1995; St. Louis et al. 1994, 1996). Methyl mercury bioaccumulates in food webs and has a broad range of impacts on aquatic biota (Zillioux et al. 1993).

Sulfur

The major anthropogenic source of S emissions is fossil fuel combustion and other industrial activities (Klimont et al. 2013). Europe and North America have been experiencing declines over the last two decades as a result of control measures, whereas increases have occurred in the developing world, especially India and China (Klimont et al. 2013; Vet et al. 2014).

Sulfur oxide deposition has multiple effects on wetlands, most notably interactions of the S with other biogeochemical cycles. In particular, deposition of S to wetlands can result in an increase in Hg bioavailability (e.g., Jeremiason et al. 2006), decreases in CH_4 production (Dise and Verry 2001; Gauci et al. 2002, 2004; Vile et al. 2003), and decreases in DOC export (de Wit et al. 2007; Montieth et al. 2007; Oulehle et al. 2013). Mercury methylation increases in response to SO_4^{2-} additions to wetlands, presumably by the activity of sulfate-reducing bacteria stimulated by SO_4^{2-} additions (Kolka et al. 2011b). Sulfate-mediated decreases in CH_4 production could be reducing the wetland contribution to global climate change (Gauci et al. 2004), and hence declines in SO_4^{2-} deposition could have the unintended consequence of increasing global warming derived from wetland CH_4 sources. Similarly, declines in SO_4^{2-} deposition to watersheds could be contributing to observed increasing trends in DOC in streams and lakes of northern North America and Europe (Erlandsson et al. 2008; Montieth et al. 2007).

Nitrogen

Biologically available N is deposited primarily as ammonium or nitrate, which may lead to eutrophication and acidification of ecosystems (Galloway et al. 2008). Nitrogen deposition in the United States was historically dominated by nitrate, but recent reductions in nitrate and increases in ammonium deposition have reversed this pattern (Li et al. 2016). Wetlands act as major sinks for N, either storing or denitrifying a significant fraction of atmospheric and hydrologic inputs of N (Baron et al. 2013). As a result of the denitrification process, some of the transformed N may be released in the form of N_2O, a potent greenhouse gas (Baron et al. 2013). This N storage and removal capacity is sensitive to hydrologic conditions. Nitrogen deposition can also lead to increased CH_4 emissions from peatlands (Aerts and Caluwe 1999), perhaps by shifting plant communities to sedges that transport CH_4 from deeper peat (Eriksson et al. 2010; Nykänen et al. 2002).

In nutrient-poor sphagnum (*Sphagnum* spp.) peatlands, N deposition has been found to affect both productivity and decomposition, which together largely define the net C balance of peatlands. In these peatlands, the effect of elevated N deposition on C balance appears to depend on deposition rate, with increased rates of net C accumulation in areas of moderate deposition such as eastern Canada (Turunen et al. 2004) but decreased net C accumulation in areas of higher deposition as a result of enhanced decomposition (Bragazza et al. 2006) and reduced sphagnum moss litter production (Bragazza et al. 2012). The negative effects of N deposition on sphagnum can be at least partially alleviated by higher P availability (Limpens et al. 2004).

Land Use and Land Management

Many wetlands have been altered by a variety of land use and land management practices that have affected inherent soil functions and provision of essential ecosystem goods and services. Some of the primary land uses that are associated with wetland alteration include urban and infrastructure development, agriculture, and forest management.

Urban and Infrastructure Development

Wetlands in urban landscapes are typically confined to detention basins and riparian zones. The surrounding mosaic of impervious surfaces is the principal factor affecting the hydrology and water quality of urban wetlands. However, like other wetlands, urban wetlands act as sponges, retaining excess rainwater and reducing flood risk to nearby homes and businesses. Although wetlands benefit water quality by absorbing many of the pollutants that would otherwise flow directly into rivers, streams, and other water bodies (e.g., pet wastes, pesticides, fertilizers, heavy metals, hydrocarbons, and road salts), they can also be negatively impacted by the volume of pollutants that enter them (Mitsch and Gosselink 2015). Coastal wetlands are particularly at risk from urban development, which can cause excessive nutrient overload from stormwater runoff as well as wastewater treatment effluent that can cause algal blooms in estuarine wetlands (Dahl and Stedman 2013). Excessive loads of heavy metals can also accumulate in wetland soils, causing increased mortality of aquatic animals and increasing risks to public health.

Stormwater runoff that passes over warmed impervious surfaces also increases the water temperature and decreases the amount of dissolved oxygen content. Warmer water with low dissolved oxygen can negatively impact aquatic species and the organisms that depend on them for food, shelter, or both. Additionally, rising water temperatures can increase the release of P from sediment, further increasing the risk for eutrophication of lakes, ponds, and estuaries (O'Driscoll et al. 2010).

Road and bridge construction across wetlands can cause water impoundment to occur even when culverts are installed, fundamentally changing the wetland hydrology. Construction within a wetland usually leads to an increase in sediment load. This may result in the fragmentation of critical wildlife corridor connectivity and can increase the spread of opportunistic invasive species that thrive in degraded wetlands, leading to further degradation of the soil quality (Czech et al. 2000). Road and bridge maintenance activities and practices can also increase the amount of road salt, herbicides, heavy metal, and many other toxic chemicals that end up in wetlands (Braun and Osterholz 2007).

Similarly, dams have had widespread impacts on associated floodplains and riparian areas by stimulating channel incision, reducing peak flows, elevating low flows, and increasing deposition (King et al. 2012). Levees and dike reduce overbank flooding, thereby changing the hydroperiod for much of a floodplain and, correspondingly, the nature of ecosystem functions. Also, increased flood depths and velocities occur in riparian areas nearest the streams that are not enclosed by the levees, and this can result in scouring and changes in forest species composition. Berms for roads that cross floodplains in a perpendicular angle to the main direction of streamflow may increase flooding (and decrease velocity) on the upstream side while reducing flooding and sheet flow on the downstream side. Consequently, disruption of natural sheet flow on the upper side induces a ponding effect and may cause decreases in forest NPP in response to the damming effect of the berm (Young et al. 1995).

Agriculture

Many wetlands are associated with agricultural land uses, either directly for cropping and grazing or indirectly when adjacent to cultivated lands. In some instances, wetland hydrology is disrupted through drainage and tillage activities that effectively result in the loss of wetland properties. In other instances, wetlands remain hydrologically intact within, adjacent to, or downstream from agricultural fields and thus receive nutrient-enriched runoff.

Sediment, Nutrient, and Chemical Runoff

Wetlands nested within, adjacent to, or downstream from agricultural fields tend to receive increased deposition of sediment, nutrients, and agrichemicals through wind and water erosion. These types of impacts have the potential to affect provisioning of various wetland ecosystem services. For example, sedimentation has been shown to affect wetland biotic communities through associated effects (e.g., burial) on plant seed and invertebrate egg banks (Gleason and Euliss 1998; Gleason et al. 2003). Wetlands in cropland settings receive greater amounts of precipitation runoff than those in a grassland setting (Euliss and Mushet 1996; van der Kamp et al. 1999, 2003), and upper soil horizons (e.g., O and A horizons) of wetlands with a cropland history have been identified that were overlain by considerable amounts of sediments (exceeding 1 m) deposited from adjacent uplands (Tangen and Gleason 2008). A study in the Prairie Pothole Region showed that wetlands surrounded by agricultural lands had five times higher clay and two times higher P transport rates than wetlands surrounded by grasslands (Martin and Hartman 1987). Elevated soil nitrate concentrations and N_2O emissions have been identified in cropland associated with wetlands, ostensibly due to inputs of N-based fertilizers (Bedard-Haughn et al. 2006; Tangen et al. 2015). Moreover, various agrichemicals (e.g., herbicides and fungicides) have been identified in sediments and tissues of wetland biota (e.g., Belden et al. 2012; McMurry et al. 2016; Venne et al. 2008). McMurry and others (2016) found that glyphosate concentrations in soils were more than four times greater in cropland wetlands than grassland wetlands. Other major wetland systems of the United States, such as the Florida Everglades (largest subtropical wetland in the United States) and wetlands associated with the Chesapeake Bay (largest estuary in the United States) and the Great Lakes (largest freshwater lakes in the world with over 1200 km of coast wetlands), receive excessive amounts of N, P, and S through fertilizer runoff, causing water quality issues by promoting harmful algal blooms and hypoxia, which can be detrimental to native flora and fauna as well as human health (e.g., contaminated drinking water).

Livestock Grazing

The impacts of livestock grazing on wetland soils have been shown to be both beneficial and detrimental, depending on the site and grazing management practices. Livestock grazing can directly impact wetland soils through biomass removal, trampling, soil compaction, erosion, altered microtopography, and altered soil nutrient status. Livestock grazing has been found to alter soil C and N storage in grasslands. Whether grazing increases or decreases soil C and N storage is a matter of debate and in most cases depends on the location of the study and the grazing management system (Li et al. 2011). Generally, studies have found that grazing reduces soil nutrient levels (Roberson 1996).

Soil compaction is arguably the most severe impact of livestock grazing, restricting root growth by reducing oxygen availability and space for roots to grow. Soil compaction also reduces the soil infiltration rate, leading to increased rates of precipitation runoff into nearby surface waters. Increased runoff coupled with increased stream bank erosion can cause serious water quality issues. Erosion can lead to channel downcutting and lowered water tables, which affect soil chemistry. However, in cases where keystone herbivores have been removed, livestock soil disturbances have been shown to benefit ephemeral wetlands by controlling invasive species and maintaining a more open canopy (Marty 2005).

Cropland Drainage

Wetlands have been used for crop production as a result of their naturally high organic C and nutrient content. However, alteration of the soil aeration is needed to support cropping, and that is accomplished through drainage. This practice of draining wetlands for crop production is the principal cause of wetland loss in the United States. Effectively, the drainage systems remove the hydrologic conditions necessary to support wetland soil processes. One of the most direct impacts of aeration is accelerated respiration that leads to the loss of stored C, as CO_2, to the atmosphere. For example, soil C stores were reduced by up to 50% when prairie pothole wetlands were converted from native grasslands to croplands (Bedard-Haughn et al. 2006; Gleason et al. 2008, 2009; Tangen et al. 2015), and the effect of drainage on soil loss extends for decades, especially on peatlands (Bridgham et al. 2006). Drainage and tillage also affect soil structure and erosional processes and alter the distribution of nutrients and metals (Skaggs et al. 1994).

Forest Management

Forest management practices may affect wetlands through disturbance regimes (e.g., harvesting), changes in vegetation community, and alterations of hydrology, and while minor drainage is allowed for silvicultural purposes, it must not alter the wetland status. These disturbance regimes may be imposed singularly or in combination, depending on the forest management objectives. Two types of silvicultural system are typically used in forested wetlands: natural regeneration and plantation management. Stands intended for natural regeneration are either clear-cut or selectively harvested. The effects of harvesting and natural regeneration activities on floodplain functions and soils have been shown to be positive (e.g., in terms of increasing productivity) or neutral (Lockaby et al. 1997). In a comparison of logging practices in bottomland hardwoods, there was no deterioration of hydric soil processes in a red-river bottom (Aust et al. 2006; McKee et al. 2012). If best management practices are followed, studies indicate that there is no change in nutrient source, sink, or transformation relationships or soil productivity. Habitat may be altered for some species of wildlife until canopy closure is regained (Clawson et al. 1997).

In wet pine flats, management may be more intensive and involves site preparation including treatments that reduce wetness in the upper soil, such as bedding. Bedding involves creating an elevated planting bed that is formed by disking soil from shallow trenches (15–50 cm) on either side of the bed. As a result, the aerated soil volume is greater in a bedded soil, resulting in enhanced organic matter decomposition and nutrient cycling (Grigal and Vance 2000), and productivity also tends to be higher on beds (Neaves et al. 2017a). While there has been concern about this site preparation method, especially with respect to hydrology, C storage, and nutrient cycling, it has been considered inconsequential for many wetland functions (Harms et al. 1998). Studies in Michigan (Trettin et al. 2011) and South Carolina (Neaves et al. 2017b) on soil property responses to bedding and plantation establishment have shown disturbance effects to be short term (e.g., <20 years).

Minor drainage may also be incorporated into the silvicultural system to reduce the surface water level in advance of harvesting, so soil disturbance is minimized during the logging; the lowered water table may also increase seedling survival and enhance early stand growth (Fox et al. 2015; Skaggs et al. 2016). In the southeastern United States, minor drainage has been incorporated into water management systems, whereby the degree of drainage can be controlled. This enables the imposition of the normal wetland hydrology following stand establishment. While minor silvicultural drainage is allowed under the Clean Water Act, the drainage should not alter the jurisdictional status of the wetland. Most of the minor drainage applications are historical, with current activities primarily limited to ditch cleaning and maintenance. In general, the effects of silvicultural drainage on hydric soils are typically to reduce soil C stocks, particularly on peatlands, and alter greenhouse gas emissions, but productivity may also be enhanced, which can increase soil C (Minkkinen et al. 1999). As a result of the lowered water table, methane emissions are typically lowered and CO_2 emissions increase (Moore and Knowles 1989; Nykänen et al. 1998).

Restoration and Mitigation

Wetland restoration, both freshwater and tidal, is an evolving practice that can be employed to replace wetlands and wetland functions that have been lost on the landscape. While there is no "cookbook" approach for achieving desired performance outcomes due to wetland and regional diversity, there are features common to all wetlands that should be considered when attempting to restore them.

While the number of potential functions and services provided by wetlands is very broad, they can be combined to fall under a small number of categories: hydrologic, soil biogeochemical, habitat, and landscape. The functions that any particular type of wetland can provide can be determined and used to set the restoration project's ecological performance goal(s). The most effective performance goals and criteria are to be SMART (specific, measurable, achievable, results-oriented, and time-fixed); address hydrology, soils, and vegetation; and reflect incremental change.[3] For voluntary projects that require a permit, the goals as defined in the permit application help to define the benefits of the project, and also its limits. For mitigation and mitigation banks, the goal is to replace what was lost, based on desired functional replacement. Incorporating adaptive management into every step of the process, from planning to design, through construction, completion, and long-term management, can help to optimize achievement of the restoration goals (Stelk et al. 2017).

Reliance on biotic criteria over a short timeframe for measurement of success may create problems establishing achievable performance criteria. Short-term monitoring data can describe initial conditions and suggest a site's potential for sustaining itself, but a long-term management plan is needed to ensure that the site remains on its trajectory (Kolka et al. 2000). In recent years, there has been increasing recognition that longer monitoring timeframes and measurement of abiotic and biotic performance are likely to yield more reliable indicators of progress toward meeting project goals (Environmental Law Institute 2004; Kusler 2006). A longer process is critical to building healthy wetland soils that, in turn, improve the potential for native wetland vegetation to persist over time and reduce vulnerability to invasive species. For example, in restoring longleaf pine (*Pinus palustris*) wet savannas, managers opted to leave slash pine plantations in place to build up litter to carry ground fires that are needed to manage the target longleaf pine (a tree that lives upwards of 300 years) (Kirkman et al. 2001).

Consistently, the single most often cited reason for wetland restoration project failure is the inability to correctly assess the restoration site and plan a wetland restoration that could be achieved on that site. Hydrologic sources and constraints, onsite and offsite stressors, soil compaction, and other factors should be examined before determining what kinds of wetland or stream restoration are achievable at a specific location. If the site is not properly assessed, then the design, plan, and onsite construction actions will not achieve the desired outcomes. Thoroughly researching soil condition is imperative (Stelk et al. 2017). In 1998–1999, an estimated 676 birds died in the North Shore Restoration Area of Lake Apopka in Florida after the surrounding farmland was restored. The restoration was focused on reflooding the farm

---

[3] SMART does not have a universal definition, and the words within the acronym have changed over time to fit the specific situation. It was first coined in 1981 by George T. Doran, a consultant and former Director of Corporate Planning for Washington Water Power Company, in a published paper titled "There's a S.M.A.R.T. Way to Write Management's Goals and Objectives." Here we use the terms and definitions recommended by Peter Skidmore in a 2017 white paper on wetland restoration (Stelk et al. 2017).

fields and the elimination or breaching of the levees that separated the fields from the main body of the lake. Organochlorine pesticide residues remained in the soils and ended up accumulating in the fish that the birds were eating, which caused the mortality event (Industrial Economics 2004).

Another important consideration regarding wetland soils and restoration is C storage. In general, the more the water table is lowered, the faster soil C is lost (Couwenberg et al. 2009). Dry wetland soils (e.g., peatlands) are also susceptible to frequent fires that can further cause large soil C losses (Turetsky et al. 2011).

Restoration strongly modifies hydrology and vegetation of restored wetlands, which directly influences soil properties. Restoring wetlands has been found to reverse the loss of soil C from draining (Järveoja et al. 2016). For instance, restored mineral soil wetlands in the Prairie Pothole Region accumulated 0.74 Mg C ha^{-1} $year^{-1}$ over a 55-year period (Ballantine and Schneider 2009). Guidelines from the Intergovernmental Panel on Climate Change for mineral soil wetlands state that cultivation leads to losses of up to 71% of the soil organic C in the top 30 cm of soil over 20 years, and restoration increases depleted soil C stocks by 80% over 20 years and by 100% after 40 years (Wickland et al. 2014). An increase in soil C is also regularly measured after restoring organic soil wetlands (Lucchese et al. 2010). Projects that restore or preserve soil C pools in tidal wetlands can be issued C credits through voluntary and regulatory programs (Emmett-Mattox et al. 2011).

Wetland restoration typically lowers DOC export from wetlands (Strack and Zuback 2013); however, there may be an initial flush after restoration activities. Rewetting or creating freshwater wetlands may increase CH_4 emissions (Badiou et al. 2011; Strack and Zuback 2013), although some studies have found that restoration did not increase CH_4 emissions (Richards and Craft 2015). Methane emissions appear to be especially high in restored wetlands located in agricultural settings or in deep water areas with emergent vegetation (Schrier-Uijl et al. 2014; Strack and Zuback 2013). In the long term, the climate benefits of increasing soil C sequestration through restoring degraded wetlands appear to be a positive for greenhouse gas mitigation (Strack and Zuback 2013), especially in saline tidal wetlands where the presence of sulfate in floodwater suppresses CH_4 production (Poffenbarger et al. 2011).

In addition to improving greenhouse gas mitigation, wetland restoration also has the potential to modify soil physical properties (Price et al. 2003). For instance, degraded wetlands are often compacted, with high bulk density and low hydraulic conductivity (Wüst-Galley et al. 2016), but restoration has been found to reverse that trend (Ballantine and Schneider 2009). However, the timeframe for soil restoration is much longer and may take many decades compared to other ecosystem components such as hydrology and vegeta-

tion that can recover in months to years (Lucchese et al. 2010; Schimelpfenig et al. 2014).

Tools

Interest in modeling wetland soil processes developed as researchers considered ecological processes (Mitsch 1988). Subsequently, models to consider nutrient removal in riparian zones (e.g., Lowrance et al. 2000), C cycling (e.g., Zhang et al. 2002), and peat accumulation (Frolking et al. 2010) have been developed, and recently USDA published a framework for C accounting in managed wetlands (Ogle et al. 2014). A major consideration for models applicable to hydric soils is the ability to consider anaerobic conditions explicitly, as it is mediated by water table, soil moisture, inherent soil physical and chemical properties, and vegetation. However, many wetland soil models take a simpler or nonmechanistic approach (Trettin et al. 2001). Correspondingly, the extent to which wetland hydrology is characterized in models varies widely. For most of the models, estimates of wetland hydrology are only based on precipitation and potential evapotranspiration. Few models consider groundwater input or lateral flow. Accordingly, the functionality to estimate DOC and inorganic C, as well as dissolved nutrient transport, is lacking.

To consider the role of wetlands on the nutrient and C cycling, water quality, and hydrology of a watershed or other large area, the wetland must be functionally integrated with adjoining ecosystems and water bodies. This implies then that a watershed-scale model is needed to simulate the hydrologic conditions throughout the watershed so that the biogeochemical processes within the uplands and wetlands reflect the hydrologic conditions within the component ecosystems. The challenge is that modeling watershed hydrology is a difficult endeavor, hence warranting dedicated hydrologic models; accordingly, incorporating watershed hydrology is beyond the scope of most existing biogeochemical models. Dai and others (2011) used a linked modeling framework to address this issue at the watershed scale, running the hydrologic simulation using the MIKE SHE modeling system and taking that response to drive the soil biogeochemistry simulation in the forest denitrification-decomposition (Forest-DNDC) model. That study demonstrated the importance of considering the wetland biogeochemistry explicitly to characterize the C balance, even though wetlands make up only a portion of the watershed.

Conversely, watershed-scale models are available to assess nutrient and C cycling and are typically applied to manage watersheds or basins. The Soil Water Assessment Tool (SWAT) is a continuation of models developed by the USDA Agricultural Research Service (USDA ARS) (Arnold et al. 2012) designed to simulate hydrology and nutrient loads within a basin. However, it was developed primarily for uplands and typically operates at large scales that obscure the presence of wetlands. Currently, modifications to SWAT that explicitly consider anaerobic soil conditions and shallow water tables are being developed. In contrast, the Riparian Ecosystem Management Model (REMM), also developed by USDA ARS, facilitates assessments of uplands and riparian zones, including wetlands and streams, on a small scale. The Soil and Water Integrated Model (SWIM) is another semi-spatially distributed model capable of assessing the contributions of wetlands to regional-scale hydrology and nutrient cycling (Hatterman et al. 2008).

Key Findings

- Wetland soil processes are different from those of upland soils and inextricably linked to the hydrologic regime of the site and the vegetation; hence they are sensitive to changes in environmental conditions and management regimes. Organic and mineral wetland soils exhibit distinct physical and chemical properties.
- While wetlands account for a relatively small portion of the terrestrial landscape, they provide highly valued ecosystem services that are largely mediated by soil processes, including water purification and C sequestration. However, these services are dependent on stability of the hydrologic regime, so changes in hydrology driven by climate, drainage, or other disturbances can eliminate key functions and services (e.g., turning peatlands from a net C sink to a source). Similarly, changes in the chemical environment (e.g., from air or water pollution) can overwhelm or eliminate key services like water purification.
- Wetlands are sensitive to changes in temperature, moisture regime, and external inputs, whereby altered soil processes and vegetative communities may be expected to alter functions and hence ecosystem services.
- Wetland restoration can offset loss of natural wetlands. However, functional wetland restoration requires a comprehensive assessment of site and environmental conditions as the foundation for an adaptive management plan to ensure long-term viability of the system.

Key Information Needs

- Soil processes of tidal freshwater wetlands, especially forests, are not well understood as a result of the complex hydrologic setting mediated by tidal and terrestrial

processes. This information is important because these wetlands are at the interface of the changing sea level and commonly occur within rapidly urbanizing regions (e.g., southeastern United States). Watershed-scale studies are needed that focus on the dynamics and transfer of organic matter, nutrients, and chemicals through tidal freshwater wetlands and the linkages to adjoining uplands and tidal waters.
- There is considerable uncertainty whether hydric soil processes in restored wetlands are comparable to natural systems after the establishment phase, thereby providing the capacity to provision ecosystem services. This gap exists because of the paucity of long-term monitoring data. Accordingly, long-term monitoring of restored sites across the gradient of common wetland types is warranted.
- Silvicultural drainage systems must sustain functional properties of the wetland; however, there are uncertainties about the interactions of management regimes, climate, and stand development on the functionality of pine plantations in wetlands. Assessment of long-term response of soil processes in forested wetlands where silvicultural drainage has been used is needed. Opportunities should focus on established (>10 years) sites that have some prior information from studies or monitoring that could provide a foundation for further assessment, with the goal being to better understand long-term responses to ensure sustainability.
- There is a need for manipulation experiments to assess the interactions of management regimes and extreme events on wetland soil processes that affect surface water quality. Particularly important are systems sensitive to mercury methylation, nutrient loading, and burning.
- Although pools of C are important to measure, we need more CO_2 and CH_4 measurements to understand both temporal and spatial variation in greenhouse gas fluxes and how change (e.g., both land use and climate) is influencing those fluxes.
- The interactions of key wetland ecosystem and soil stressors such as climate change, air pollution, disturbances like fire, and land use change may interact in ways that are hard to predict and, in combination, could have important unforeseen impacts on wetland functions and services. This knowledge gap speaks to the need for robust, process-based models to better understand uncertainties and risks so that management and protective measures may be developed. Robust mechanistic models that predict soil biogeochemical processes in response to changing environmental conditions and management regimes are needed to provide tools for assessing and managing ecosystem services.

Literature Cited

Aerts R, de Caluwe H (1999) Nitrogen deposition effects on carbon dioxide and methane emissions from temperate peatland soils. Oikos 84(1):44–54

Alongi DM (2014) Carbon cycling and storage in mangrove forests. Annu Rev Mar Sci 6:195–219

Amatya DM, Harrison CA, Trettin CC (2016) Hydro-meteorologic assessment of October 2015 extreme precipitation event on Santee Experimental Forest watersheds. J S C Water Resour 3(1):19–30

Anderson C, Lockaby BG (2007) Soils and biogeochemistry of tidal freshwater forested wetlands. In: Conner WH, Doyle TW, Krauss KW (eds) Ecology of tidal freshwater forested wetlands of the southeastern United States. Springer, Dordrecht, pp 65–88

Armitage AR, Highfield WE, Brody SD, Louchouarn P (2015) The contribution of mangrove expansion to salt marsh loss on the Texas Gulf Coast. PLoS One 10:e0125404

Arnold JG, Moriasi DN, Gassman PM et al (2012) SWAT: model use, calibration and validation. Trans Am Soc Agric Biol Eng 55:1491–1508

Aust WM, Fristoe TC, Gellerstedt LA et al (2006) Long-term effects of helicopter and ground-based skidding on soil properties and stand growth in a tupelo-cypress wetland. For Ecol Manag 226:72–79

Badiou P, McDougal R, Pennock D, Clark B (2011) Greenhouse gas emissions and carbon sequestration potential in restored wetlands of the Canadian prairie pothole region. Wetl Ecol Manag 19:237–256

Ballantine K, Schneider R (2009) Fifty-five years of soil development in restored freshwater depressional wetlands. Ecol Appl 19:1467–1480

Bansal S, Tangen B, Finocchiaro R (2016) Temperature and hydrology affect methane emissions from prairie pothole wetlands. Wetlands 36(supplement 2):371–381

Baron JS, Hall EK, Nolan BT et al (2013) The interactive effects of excess reactive nitrogen and climate change on aquatic ecosystems and water resources of the United States. Biogeochemistry 114(1–3):71–92

Bartlett KB, Bartlett DS, Harriss RC, Sebacher DI (1987) Methane emissions along a salt marsh salinity gradient. Biogeochemistry 4:183–202

Batt BDJ, Anderson MG, Anderson CD, Caswell FD (1989) The use of prairie potholes by North American Ducks. In: van der Valk AG (ed) Northern prairie wetlands. Iowa State University Press, Ames, pp 204–227

Beauchamp SK (1996) Cypress: from wetlands and wildlife habitat to flowerbeds and front yards. University of Florida, Institute of Food and Agricultural Sciences, Gainesville

Bedard-Haughn A, Matson AL, Pennock DJ (2006) Land use effects on gross nitrogen mineralization, nitrification, and N_2O emissions in ephemeral wetlands. Soil Biol Biochem 38:3398–3406

Beeri O, Phillips RL (2007) Tracking palustrine water seasonal and annual variability in agricultural wetland landscapes using Landsat from 1997 to 2005. Glob Chang Biol 13:897–912

Belden JB, Hanson B, McMurry ST et al (2012) Assessment of the effects of farming and conservation programs on pesticide deposition in High Plains wetlands. Environ Sci Technol 46:3424–3432

Bergstrom JC, Stoll JR, Titre JP, Wright VL (1990) Economic value of wetlands-based recreation. Ecol Econ 2:129–147

Bragazza L, Freeman C, Jones T et al (2006) Atmospheric nitrogen deposition promotes carbon loss from peat bogs. Proc Natl Acad Sci 103(51):19386–19389

Bragazza L, Buttler A, Habermacher J et al (2012) High nitrogen deposition alters the decomposition of bog plant litter and reduces carbon accumulation. Glob Chang Biol 18(3):1163–1172

Brander LM, Florax RJGM, Vermaat JE (2006) The empirics of wetland valuation: a comprehensive summary and a meta-analysis of the literature. Environ Resour Econ 33:223–250

Braun RJ, Osterholz LC (eds) (2007) Indiana storm water quality manual. Indiana Department of Environmental Management, Indianapolis, 15 p

Bridgham SD, Johnston CA, Pastor J, Updegraff K (1995) Potential feedbacks of northern wetlands on climate change. Bioscience 45:262–274

Bridgham SD, Megonigal JP, Keller JK et al (2006) The carbon balance of North American wetlands. Wetlands 26:889–916

Bridgham SD, Cadillo-Quiroz H, Keller JK, Zhuang Q (2013) Methane emissions from wetlands: biogeochemical, microbial, and modeling perspectives from local to global scales. Glob Chang Biol 19:1325–1346

Burkett V, Kusler J (2000) Climate change: potential impacts and interactions in wetlands of the United States. J Am Water Resour Assoc 36:313–320

Caraco NF, Cole JJ, Likens GE (1989) Evidence for sulfate-controlled phosphorus release from sediments of aquatic systems. Nature 341:316–318

Carter V (1990) The Great Dismal Swamp: an illustrated case study. In: Lugo AE, Brinson M, Brown S (eds) Ecosystems of the world 15: forested wetlands. Elsevier, Amsterdam, pp 201–211

Chambers RM, Odum WE (1990) Porewater oxidation, dissolved phosphate and the iron curtain. Biogeochemistry 10:37–52

Charles H, Dukes JS (2009) Effects of warming and altered precipitation on plant and nutrient dynamics of a New England salt marsh. Ecol Appl 19:1758–1773

Chmura GL, Anisfeld SC, Cahoon DR, Lynch JC (2003) Global carbon sequestration in tidal, saline wetland soils. Glob Biogeochem Cycles 17(4):1111

Clair TA, Warner BG, Roberts R et al (1995) Canadian inland wetlands and climate change. In: Environment Canada (ed) Canadian country study: climate impacts and adaptations. Environment Canada, Ottawa, pp 189–218

Clawson RG, Lockaby BG, Jones R (1997) Amphibian responses to helicopter harvesting in forested floodplains of low order, blackwater streams. For Ecol Manag 90(2–3):225–236

Clewley D, Whitcomb J, Moghaddam M et al (2015) Evaluation of ALOS PALSAR data for high-resolution mapping of vegetated wetlands in Alaska. Remote Sens 7:7272–7297

Clough BF, Andrews TJ, Cowan IR (1982) Physiological processes in mangroves. In: Clough BF (ed) Mangrove ecosystems in Australia: structure, function and management. Australian National University Press, Canberra, pp 193–210

Conner WH, Buford MA (1998) Southern deepwater swamps. In: Messina MG, Conner WH (eds) Southern forested wetlands: ecology and management. CRC Press, Boca Raton, pp 261–287

Couwenberg J, Dommain R, Joosten H (2009) Greenhouse gas fluxes from tropical peatlands in south-east Asia. Glob Chang Biol 16:1715–1732

Cowardin LM, Carter V, Golet FC, LaRoe ET (1979) Classification of wetlands and deepwater habitats of the United States, Report FWS/OBS-79/31. U.S. Department of the Interior, Fish and Wildlife Service, Washington, DC. http://www.fws.gov/wetlands/Documents/classwet/index.html. Accessed 5 Apr 2019

Craft CB (2001) Biology of wetland soils. In: Richardson JL, Vepraskas MJ (eds) Wetland soils. Lewis Publishers, Boca Raton, pp 107–136

Craft CB (2007) Freshwater input structures soil properties, vertical accretion and nutrient accumulation of Georgia and United States (U.S.) tidal marshes. Limnol Oceanogr 52:1220e1230

Creed IF, Miller J, Aldred D et al (2013) Hydrologic profiling for greenhouse gas effluxes from natural grasslands in the Prairie Pothole Region of Canada. J Geophys Res Biogeosci 118:680–697

Creel M, Loomis J (1992) Recreation value of water to wetlands in the San Joaquin Valley: linked multinomial logit and count data trip frequency models. Water Resour Res 28(10):2597–2606

Currin CA, Newell SY, Paerl HW (1995) The role of standing dead *Spartina alterniflora* and benthic microalgae in salt marsh food webs: considerations based on multiple stable isotope analysis. Mar Ecol Prog Ser 121:99–116

Czech B, Krausman PR, Devers PK (2000) Economic associations among causes of species endangerment in the United States. Bioscience 50(7):593–601

D'Angelo EM, Reddy KR (1999) Regulators of heterotrophic microbial potentials in wetland soils. Soil Biol Biochem 31:815–830

Dahl TE (1990) Wetlands losses in the United States: 1780's to 1980's. U.S. Department of the Interior, Fish and Wildlife Service, Washington, DC, 13 p. Available at: http://www.fws.gov/wetlands/Status-And-Trends/index.html. Accessed 5 Apr 2019

Dahl TE (2011) Status and trends of wetlands in the conterminous United States 2004 to 2009. U.S. Department of the Interior, Fish and Wildlife Service, Washington, DC, 108 p. Available at: http://www.fws.gov/wetlands/Status-And-Trends/index.html. Accessed 5 Apr 2019

Dahl TE (2014) Status and trends of prairie wetlands in the United States 1997 to 2009. U.S. Fish and Wildlife Service, Washington, DC. Available at http://www.fws.gov/wetlands/Status-And-Trends/index.html. Accessed 5 Apr 2019

Dahl TE, Stedman SM (2013) Status and trends of wetlands in the coastal watersheds of the conterminous United States 2004 to 2009. U.S. Department of the Interior, Fish and Wildlife Service and National Oceanic and Atmospheric Administration, National Marine Fisheries Service. 46 p. Available at: http://www.fws.gov/wetlands/Status-And-Trends/index.html. Accessed 5 Apr 2019

Dai Z, Trettin CC, Li C et al (2011) Effect of assessment scale on spatial and temporal variations in CH_4, CO_2, and N_2O fluxes in a forested wetland. Water Air Soil Pollut 223(1):253–265

Dalcin Martins P, Hoyt DW, Bansal S et al (2017) Abundant carbon substrates drive extremely high sulfate reduction rates and methane fluxes in prairie pothole wetlands. Glob Chang Biol 23(8):3107–3120

Das S, Crepin AS (2013) Mangroves can protection against wind during storms. Estuar Coast Shelf Sci 134:98–107

Davidson EA, Janssen IA (2006) Temperature sensitivity of soil decomposition and feedbacks to climate change. Nature 440:165–173

Day JW, Christian RR, Boesch DM et al (2008) Consequences of climate change on the ecogeomorphology of coastal wetlands. Estuar Coasts 31:477–491

de Groot R, Brander L, van der Ploeg S et al (2012) Global estimates of the value of ecosystems and their services in monetary units. Ecosyst Serv 1(1):50–61

De Wit HA, Mulder J, Hindar A, Hole L (2007) Long-term increase in dissolved organic carbon in streamwaters in Norway is response to reduced acid deposition. Environ Sci Technol 41(22):7706–7713

Dise NB, Verry ES (2001) Suppression of peatland methane emission by cumulative sulfate deposition in simulated acid rain. Biogeochemistry 53(2):143–160

Donato DC, Kauffman JB, MacKenzie A et al (2012) Whole-island carbon stocks in the tropical Pacific: implications for mangrove conservation and upland restoration. J Environ Manag 97:89–96

Doss CR, Taff SJ (1996) The influence of wetland type and wetland proximity on residential property values. J Agric Resour Econ 21:120–129

Drake BG, Muehe MS, Peresta G et al (1996) Acclimation of photosynthesis, respiration and ecosystem carbon flux of a wetland on Chesapeake Bay, Maryland to elevated atmospheric CO_2 concentration. Plant Soil 187(2):111–118

Driscoll CT, Han YJ, Chen CY et al (2007) Mercury contamination in forest and freshwater ecosystems in the northeastern United States. Bioscience 57(1):17–28

O'Driscoll M, Clinton S, Jefferson A, Manda A, McMillan S (2010) Urbanization effects on watershed hydrology and in-stream processes in the Southern United States. Water 2(3):605–648

Duarte CM, Middelburg JJ, Caraco N (2005) Major role of marine vegetation on the oceanic carbon cycle. Biogeosciences 2:1–8

Easterling DR, Kunkel KE, Arnold JR et al (2017) Precipitation change in the United States. In: Wuebbles DJ, Fahey DW, Hibbard KA et al (eds) Climate science special report: fourth national climate assessment, vol I. U.S. Global Change Research Program, Washington, DC, pp 207–230

Ehrenfeld J (2012) Northern red maple and black ash swamps. In: Batzer DP, Baldwin AH (eds) Wetland habitats of North America: ecology and conservation concerns. University of California Press, Berkeley

Ellison AM (2000) Mangrove restoration: do we know enough? Restor Ecol 8:219–229

Emmett-Mattox S, Crooks S, Findsen J (2011) Gases and grasses: the restoration, conservation, or avoided loss of tidal wetlands carbon pools may help to mitigate climate change. Environ Forum 28:30–35

Environmental Law Institute (2004) Measuring mitigation: a review of the science for compensatory mitigation performance standards. Environmental Law Institute®, Washington, DC. Report prepared for U.S. Environmental Protection Agency. https://www.eli.org/sites/default/files/eli-pubs/revision.pdf. Accessed 5 Apr 2019

Erickson JE, Megonigal JP, Peresta G, Drake BG (2007) Salinity and sea level mediate elevated CO_2 effects on C3–C4 plant interactions and tissue nitrogen in a Chesapeake Bay tidal wetland. Glob Chang Biol 13:202–215

Eriksson T, Öquist MG, Nilsson MB (2010) Effects of decadal deposition of nitrogen and sulfur, and increased temperature, on methane emissions from a boreal peatland. J Geophys Res 115:G04036. https://doi.org/10.1029/2010JG001285

Erlandsson M, Buffam I, Fölster J et al (2008) Thirty-five years of synchrony in the organic matter concentrations of Swedish rivers explained by variation in flow and sulphate. Glob Chang Biol 14(5):1191–1198

Eslami-Andargoli L, Dale P, Sipe N, Chaseling J (2009) Mangrove expansion and rainfall patterns in Moreton Bay, Southeast Queensland, Australia. Estuar Coast Shelf Sci 85:292–298

Euliss NH Jr, Mushet DM (1996) Water-level fluctuation in wetlands as a function of landscape condition in the prairie pothole region. Wetlands 16:587–593

Euliss NH Jr, Wrubleski DA, Mushet DM (1999) Wetlands of the prairie pothole region: invertebrate species composition, ecology, and management. In: Batzer DP, Rader RB, Wissinger SA (eds) Invertebrates in freshwater wetlands of North America: ecology and management. Wiley, New York, pp 471–514

Euliss NH Jr, LaBaugh JW, Fredrickson LH et al (2004) The wetland continuum: a conceptual framework for interpreting biological studies. Wetlands 24:448–458

Euliss NH Jr, Gleason RA, Olness A et al (2006) North American prairie wetlands are important nonforested land-based carbon storage sites. Sci Total Environ 361:179–188

Euliss NH Jr, Mushet DM, Newton WE et al (2014) Placing prairie pothole wetlands along spatial and temporal continua to improve integration of wetland function in ecological investigations. J Hydrol 513:490–503

Feierabend SJ, Zelazny JM (1987) Status report on our Nation's wetlands. National Wildlife Federation, Washington, DC, 50 p

Fenner N, Freeman C (2011) Drought-induced carbon loss in peatlands. Nat Geosci 4(12):895–900

Field CD (1995) Impact of expected climate change on mangroves. Hydrobiologia 295:75–81

Finocchiaro R, Tangen B, Gleason R (2014) Greenhouse gas fluxes of grazed and hayed wetland catchments in the U.S. Prairie Pothole Ecoregion. Wetl Ecol Manag 22:305–324

Flannigan M, Stocks B, Turetsky M, Wotton M (2009) Impacts of climate change on fire activity and fire management in the circumboreal forest. Glob Chang Biol 15:549–560

Flynn K (1996) Understanding wetlands and endangered species: definitions and relationships, Extension Publication ANR–979. Alabama Cooperative Extension System, Florence, 6 p

Fox TR, Jokela EJ, Allen HL (2015) Pine plantation silviculture in the southern United States. In: Rauscher HM, Johnsen K (eds) Southern forest science: past, present, and future, General Techniocal Report SRS-GTR-075. U.S. Department of Agriculture, Forest Service, Southern Research Station, Asheville, pp 63–82

Freeman C, Fenner N, Ostle NJ et al (2004) Export of dissolved organic carbon from peatlands under elevated carbon dioxide levels. Nature 430:195–198

Frey EF, Palin MB, Walsh PJ, Whitcraft CR (2013) Spatial Hedonic valuation of a multiuse urban wetland in Southern California. Agric Resour Econ Rev 42(2):387–402

Frolking S, Roulet NT, Tuittila E et al (2010) A new model of Holocene peatland net primary production, decomposition, water balance, and peat accumulation. Earth Syst Dynam 1:1–21

Furch K (1997) Chemistry of Várzea and Igapó soils and nutrient inventory of their floodplain forests. In: Junk WJ (ed) The Central Amazon floodplain: ecology of a pulsing system, Ecological Studies 126. Springer, Berlin, pp 47–67

Furukawa K, Wolanski E, Mueller H (1997) Currents and sediment transport in mangrove forests. Estuar Coast Shelf Sci 44(3):301–310

Furumai H, Kawasaki T, Futuwatari T, Kusuda T (1988) Effect of salinity on nitrification in a tidal river. Water Sci Technol 20:165–174

Galloway JN, Townsend AR, Erisman JW et al (2008) Transformation of the nitrogen cycle: recent trends, questions, and potential solutions. Science 320(5878):889–892

Gariano SL, Guzzetti F (2016) Landslides in a changing climate. Earth Sci Rev 162:227–252

Gauci V, Dise N, Fowler D (2002) Controls on suppression of methane flux from a peat bog subjected to simulated acid rain sulfate deposition. Glob Biogeochem Cycles 16(1):4-1–4-12

Gauci V, Matthews E, Dise N et al (2004) Sulfur pollution suppression of the wetland methane source in the 20th and 21st centuries. Proc Natl Acad Sci 101(34):12583–12587

Gedan KB, Bertness MD (2009) Experimental warming causes rapid loss of plant diversity in New England salt marshes. Ecol Lett 12:842–848

Gedan KB, Kirwan ML, Wolanski E et al (2011) The present and future role of coastal wetland vegetation in protecting shorelines: answering recent challenges to the paradigm. Clim Chang 106(1):7–29

Gleason RA, Euliss NH Jr (1998) Sedimentation of prairie wetlands. Great Plains Res 8:97–112

Gleason RA, Euliss NH Jr, Hubbard DE, Duffy WG (2003) Effects of sediment load on emergence of aquatic invertebrates and plants from wetland soil egg and seed banks. Wetlands 23:26–34

Gleason RA, Laubhan MK, Euliss NH Jr (eds) (2008) Ecosystem services derived from wetland conservation practices in the United States Prairie Pothole Region with an emphasis on the U.S. Department of Agriculture Conservation Reserve and Wetlands Reserve Programs, U.S. Geological Professional Paper 1745. U.S. Department of the Interior, Geological Survey, Reston, 58 p

Gleason RA, Tangen BA, Browne BA, Euliss NH Jr (2009) Greenhouse gas flux from cropland and restored wetlands in the Prairie Pothole Region. Soil Biol Biochem 41:2501–2507

Goldhaber MB, Mills CT, Morrison JM et al (2014) Hydrogeochemistry of prairie pothole region wetlands: role of long-term critical zone processes. Chem Geol 387:170–183

Goldhaber MB, Mills CT, Mushet DM et al (2016) Controls on the geochemical evolution of Prairie Pothole Region lakes and wetlands over decadal time scales. Wetlands 36(supplement 2):S255–S272

Gosselink JG, Turner RE (1978) The role of hydrology in freshwater wetland ecosystems. In: Good RE, Whigham DF, Simpson RL (eds) Freshwater wetlands. Academic, London, pp 63–78

Griffis TJ, Rouse WR, Waddington JM (2000) Interannual variability of net CO_2 exchange at a subarctic fen. Glob Biogeochem Cycles 14:1109–1121

Grigal DF, Vance ED (2000) Influence of soil organic matter on forest productivity. N Z J For Sci 30:169–205

Groisman PY, Knight RW, Easterling DR et al (2005) Trends in intense precipitation in the climate record. J Clim 18(9):1326–1350

Groisman PY, Knight RW, Karl TR (2012) Changes in intense precipitation over the central United States. J Hydrometeorol 13(1):47–66

Harms WR, Aust WM, Burger JA (1998) Wet flatwoods. In: Messina MG, Conner WH (eds) Southern forested wetlands: ecology and management. CRC Press, Boca Raton

Hatterman F, Krysanova V, Hesse C (2008) Modeling wetland processes in regional applications. Hydrol Sci J 53(5):1001–1012

Hayashi M, van der Kamp G, Rosenberry DO (2016) Hydrology of prairie wetlands: understanding the integrated surface-water and groundwater processes. Wetlands 36(supplement 2):S237–S254

Hemond HF, Beniot J (1988) Cumulative impacts on water quality functions of wetlands. Environ Manag 12(5):639–653

Hey DL, McGuiness D, Beorkem MN et al (2002) Flood damage reduction in the upper Mississippi River basin: an ecological means. McKnight Foundation, Minneapolis, 37 p

Hodges JD (1998) Minor alluvial floodplains. In: Messina MG, Conner WH (eds) Southern forested wetlands: ecology and management. CRC Press, Boca Raton

Howarth RW (1988) Nutrient limitation of net primary production in marine ecosystems. Annu Rev Ecol Syst 19:89–110

Hyvonen R, Agren GI, Linder S et al (2007) The likely impact of elevated [CO_2], nitrogen deposition, increased temperature and management on carbon sequestration in temperate and boreal forest ecosystems: a literature review. New Phytol 173(3):463–480

ICF International (2013) Greenhouse gas mitigation options and costs for agricultural land and animal production within the United States, Report prepared for U.S. Department of Agriculture, Climate Change Program Office under Contract No. AG-3142-P-10-0214. ICF International, Washington, DC

Industrial Economics, Inc (2004) Final Lake Apopka natural resource damage assessment and restoration plan, Final Report prepared for U.S. Department of the Interior, Fish and Wildlife Service. Industrial Economics, Inc./The St. Johns River Water Management District, Cambridge, MA/Palatka

Järveoja J, Peichl M, Maddison M et al (2016) Impact of water table level on annual carbon and greenhouse gas balances of a restored peat extraction area. Biogeosciences 13:2637–2651

Jenkins WA, Murray BC, Kramer RA, Faulkner SP (2010) Valuing ecosystem services from wetlands in the Mississippi alluvial valley. Ecol Econ 9:1051–1061

Jeremiason JD, Engstrom DR, Swain EB et al (2006) Sulfate addition increases methylmercury production in an experimental wetland. Environ Sci Technol 40(12):3800–3806

Johnson DW (2006) Progressive N limitation in forests: review and implications for long-term responses to elevated CO_2. Ecology 87(1):64–75

Johnson WC, Poiani KA (2016) Climate change effects on prairie pothole wetlands: findings from a twenty-five year numerical modeling project. Wetlands 36:273–285

Johnson WC, Boettcher SE, Poiani KA, Guntenspergen G (2004) Influence of weather extremes on the water levels of glaciated prairie wetlands. Wetlands 24(2):385–398

Johnson WC, Millett BV, Gilmanov T (2005) Vulnerability of Northern Prairie wetlands to climate change. Bioscience 55(10):863–872

Johnston CA, Detenbeck NE, Neimi GJ (1990) The cumulative effects of wetlands on stream water quality and quantity: a landscape approach. Biogeochemistry 10:105–141

Kadlec RH, Wallace SD (2009) Treatment wetlands, 2nd edn. CRC Press, Boca Raton, 366 p

Kasischke ES, Turetsky MR (2006) Recent changes in the fire regime across the North American boreal region—spatial and temporal patterns of burning across Canada and Alaska. Geophys Res Lett 33:L09703

Kelleway JJ, Saintilan N, Macreadie PI et al (2016) Seventy years of continuous encroachment substantially increases 'blue carbon' capacity as mangroves replace intertidal salt marshes. Glob Chang Biol 22:1097–1109

King SL, Battaglia LL, Hupp R et al (2012) Floodplain wetlands of the Southeastern coastal plain. In: Batzer DP, Baldwin AH (eds) Wetland habitats of North America: ecology and conservation concerns. University of California Press, Berkeley

Kirkman LK, Mitchell RJ, Helton RC, Drew MB (2001) Productivity and species richness across an environmental gradient in a fire-dependent ecosystem. Am J Bot 88:2119–2128

Kirwan ML, Megonigal JP (2013) Tidal wetland stability in the face of human impacts and sea-level rise. Nature 504:53–60

Kirwan ML, Mudd SM (2012) Response of salt-marsh carbon accumulation to climate change. Nature 489:550–553

Kirwan ML, Guntenspergen GR, Morris JT (2009) Latitudinal trends in *Spartina alterniflora* productivity and the response of coastal marshes to global change. Glob Chang Biol 15:1982–1989

Kirwan ML, Langley JA, Guntenspergen GR, Megonigal JP (2013) The impact of sea-level rise on organic matter decay rates in Chesapeake Bay brackish tidal marshes. Biogeosciences 10:1869–1876

Klimont Z, Smith SJ, Cofala J (2013) The last decade of global anthropogenic sulfur dioxide: 2000–2011 emissions. Environ Res Lett 8(1):014003

Koch EW, Barbier EB, Silliman BR et al (2009) Non-linearity in ecosystem services: temporal and spatial variability in coastal protection. Front Ecol Environ 7(1):29–37

Kolka RK, Nelson EA, Trettin CC (2000) Conceptual assessment framework for forested wetland restoration: the Pen Branch experience. Ecol Eng 15:17–21

Kolka RK, Rabenhorst MC, Swanson D (2011a) Chapter 33.2: Histosols. In: Huang PM, Li Y, Sumner ME (eds) Handbook of soil sciences properties and processes, 2nd edn. CRC Press, Boca Raton, pp 33.8–33.29

Kolka RK, Mitchell CP, Jeremiason JD et al (2011b) Mercury cycling in peatland watersheds. CRC Press, Boca Raton, pp 349–370

Kolka R, Trettin C, Tang W et al (2018) Terrestrial wetlands. In: Cavallaro N, Shrestha G, Birdsey R et al (eds) Second state of the carbon cycle report (SOCCR2): a sustained assessment report. U.S. Global Change Research Program, Washington, DC, pp 507–567

Krauss KW, McKee KL, Lovelock CE et al (2014) How mangrove forests adjust to rising sea level. New Phytol 202:19–34

Kunkel KE, Andsager K, Easterling DR (1999) Long-term trends in extreme precipitation events over the conterminous United States and Canada. J Clim 12(8):2515–2527

Kusler J (2006) Developing performance standards for the mitigation and restoration of Northern forested wetlands. The Association of State Wetland Managers, Windham. https://www.aswm.org/pdf_lib/forested_wetlands_080106.pdf

Laiho R (2006) Decomposition in peatlands: reconciling seemingly contrasting results on the impacts of lowered water levels. Soil Biol Biochem 38(8):2011–2024

Langley JA, McKee KL, Cahoon DR et al (2009) Elevated CO_2 stimulates marsh elevation gain, counterbalancing sea-level rise. Proc Natl Acad Sci 106(15):6182–6186

Lellis-Dibble KA, McGlynn KE, Bigford TE (2008) Estuarine fish and shellfish species in U.S. commercial and recreational fisheries: economic value as an incentive to protect and restore estuarine habitat, NOAA Tech. Memo. NMFS–F/SPO–90. U.S. Department of Commerce, Washington, DC, 94 p

Li W, Huang H-Z, Zhang Z-N, Wu G-L (2011) Effects of grazing on the soil properties and C and N storage in relation to biomass allocation in an alpine meadow. J Soil Sci Plant Nutr 11:27–39

Li Y, Schichtel BA, Walker JT et al (2016) Increasing importance of deposition of reduced nitrogen in the United States. Proc Natl Acad Sci 113(21):5874–5879

Limpens J, Berendse F, Klees H (2004) How phosphorus availability affects the impact of nitrogen deposition on Sphagnum and vascular plants in bogs. Ecosystems 7(8):793–804

Lockaby BG, Walbridge MR (1998) Biogeochemistry. In: Messina MG, Conner WH (eds) Southern forested wetlands: ecology and management. CRC Press, Boca Raton, pp 149–172

Lockaby BG, Stanturf JA, Messina M (1997) Harvesting impacts on functions of forested floodplains: a review of existing reports. For Ecol Manag 90(2–3):93–100

Lowrance R, Altier LS, William RG et al (2000) REMM: the riparian ecosystem management model. J Soil Water Conserv 55:27–34

Lucchese M, Waddington JM, Poulin M et al (2010) Organic matter accumulation in a restored peatland: evaluating restoration success. Ecol Eng 36:482–488

Lugo AE, Brinson M, Brown S (1990) In: Lugo AE, Brinson M, Brown S (eds) Ecosystems of the world, 15: forested wetlands. Elsevier Science Publishers, Amsterdam, 527 p

Lyon EA, Jin Y, Randerson T (2008) Changes in surface albedo after fire inboreal forest ecosystems of interior Alaska assessed using MODIS satellite observations. J Geophys Res 113:G02012

MacFarlane GT, Hebert RA (1984) Effect of oxygen tension, salinity, temperature, and organic matter concentration on the growth and nitrifying activity of an estuarine strain of Nitrosomonas. FEMS Microbiol Lett 23:107–111

MacKenzie RA (2005) Spatial and temporal patterns in insect emergence from a southern Maine salt marsh. Am Midl Nat 153:63–75

MacKenzie RA, Dionne M (2008) Habitat heterogeneity: the importance of salt marsh pools and high marsh surfaces to fish production in two Gulf of Maine salt marshes. Mar Ecol Prog Ser 368:217–230

MacKenzie RA, Dionne M, Miller J et al (2015) Community structure and abundance of benthic infaunal invertebrates in Maine fringing marsh ecosystems. Estuar Coasts 38:1317–1334

MacKenzie RA, Foulk PB, Klump JV et al (2016) Sedimentation and belowground carbon accumulation rates in mangrove forests that differ in diversity and land use: a tale of two mangroves. Wetl Ecol Manag 24:245–261

Magalhães CM, Joye SB, Moreira RM et al (2005) Effect of salinity and inorganic nitrogen concentrations on nitrification and denitrification rates in intertidal sediments and rocky biofilms of the Douro River estuary, Portugal. Water Res 39:1783–1794

Mahan BL, Polasky S, Adams RM (2000) Valuing urban wetlands: a property price approach. Land Econ 76(1):100–113

Mamo TG (2015) Evaluation of the potential impact of rainfall intensity variation due to climate change on existing drainage infrastructure. J Irrig Drain Eng 141(10):05015002

Martin DB, Hartman WA (1987) The effect of cultivation on sediment composition and deposition in prairie pothole wetlands. Water Air Soil Pollut 34:45–53

Marty JT (2005) Effects of cattle grazing on diversity in ephemeral wetlands. Conserv Biol 19:1626–1632

McGuire AD, Anderson LG, Christensen TR et al (2009) Sensitivity of the carbon cycle in the Arctic to climate change. Ecol Monogr 79:523–555

McKee SE, Aust WM, Seiler JR et al (2012) Long-term site productivity of a tupelo-cypress swamp 24 years after harvesting disturbances. For Ecol Manag 265:172–180

McLean KI, Mushet DM, Renton DA, Stockwell CA (2016a) Aquatic-macroinvertebrate communities of prairie-pothole wetlands and lakes under a changed climate. Wetlands 36(supplement 2):423–435

McLean KI, Mushet DM, Stockwell CA (2016b) From "duck factory" to "fish factory": climate induced changes in vertebrate communities of prairie pothole wetlands and small lakes. Wetlands 36(supplement 2):407–421

Mcleod E, Chmura GL, Bouillon S et al (2011) A blueprint for blue carbon: toward an improved understanding of the role of vegetated coastal habitats in sequestering CO_2. Front Ecol Environ 9:552–560

McMurry ST, Belden JB, Smith LM et al (2016) Land use effects on pesticides in sediments of prairie pothole wetlands in North and South Dakota. Sci Total Environ 565:682–689

Megonigal JP, Neubauer SC (2009) Biogeochemistry of tidal freshwater wetlands. In: Perillo GME, Wolanski E, Cahoon DR, Brinson M (eds) Coastal wetlands: an integrated ecosystem approach. Elsevier Science, Amsterdam, pp 535–562

Megonigal JP, Chapman S, Crooks S et al (2016) Impacts and effects of ocean warming on tidal marsh and tidal freshwater forest ecosystems. In: Laffoley D, Baxter JM (eds) Explaining ocean warming: causes, scale, effects and consequences. IUCN, Gland, pp 105–122

Mendelssohn IA, Morris JT (2002) Eco-physiological controls on the productivity of Spartina alterniflora Loisel. In: Weinstein MP, Kreeger DA (eds) Concepts and controversies in tidal marsh ecology. Springer, Dordrecht

Min SK, Zhang XB, Zwiers FW, Hegerl GC (2011) Human contribution to more-intense precipitation extremes. Nature 470(7334):378–381

Minkkinen K, Vasander H, Jauhiainen S et al (1999) Post-drainage changes in vegetation composition and carbon balance in Lakkasuo mire, Central Finland. Plant Soil 207:107–120

Mitsch WJ (1988) Productivity-hydrology-nutrient models of forested wetlands. In: Mitsch WJ, Straškraba M, Jorgensen SE (eds) Wetland modelling. Elsevier, Amsterdam, pp 115–132

Mitsch W, Gosselink J (2015) Wetlands, 5th edn. Wiley, New York

Mitsch WJ, Bernal B, Nahlik AM et al (2013) Wetlands, carbon and climate change. Landsc Ecol 28:583–597

Monks PS, Granier C, Fuzzi S et al (2009) Atmospheric composition change—global and regional air quality. Atmos Environ 43:5268–5350

Montieth DT, Stoddard L, Evans CD et al (2007) Dissolved organic carbon trends resulting from atmospheric deposition chemistry. Nature 450(7169):537–540

Moor H, Hylander K, Norberg J (2015) Predicting climate change effects on wetland ecosystem services using species distribution modeling and plant functional traits. Ambio 44:113–126

Moore TR, Knowles R (1989) The influence of water table levels on methane and carbon dioxide emissions from peatland soils. Can J Soil Sci 69:33–38

Morrissey EM, Gillespie JL, Morina JC, Franklin RB (2014) Salinity affects microbial activity and soil organic matter content in tidal wetlands. Glob Chang Biol 20:1351–1362

Mortsch LD (1998) Assessing the impact of climate change on the Great Lakes shoreline wetlands. Clim Chang 40:391–416

Mueller P, Jensen K, Megonigal P (2016) Plants mediate soil organic matter decomposition in response to sea level rise. Glob Chang Biol 22(1):404–414

Mushet DM, Goldhaber MB, Mills CT et al (2015) Chemical and biotic characteristics of prairie lakes and large wetlands in south-central North Dakota—Effects of a changing climate, U.S. Geological Survey Scientific Investigations Report 2015–5126, 55 p. https://doi.org/10.3133/sir20155126

Nahlik AM, Fennessy MS (2016) Carbon storage in US wetlands. Nat Commun 7(1):13835. https://doi.org/10.1038/ncomms13835

Narayan S, Beck MW, Wilson P et al (2017) The value of coastal wetlands for flood damage reduction in the Northeastern USA. Sci Rep 7:9463. https://doi.org/10.1038/s41598-017-09269-z

Neaves CM, Aust M, Bolding MC et al (2017a) Loblolly pine (Pinus taeda L.) productivity 23 years after wet site harvesting and site preparation in the lower Atlantic coastal plain. For Ecol Manag 401:207–214

Neaves CM, Aust M, Bolding MC et al (2017b) Soil properties in site prepared loblolly pine (Pinus taeda L.) stands 25 years after wet weather harvesting in the lower Atlantic coastal plain. For Ecol Manag 404:344–353

Nelson LE, Switzer GL, Lockaby BG (1987) Nutrition of Populus deltoides plantations during maximum production. For Ecol Manag 20:25–41

Neubauer SC (2008) Contributions of mineral and organic components to tidal freshwater marsh accretion. Estuar Coast Shelf Sci 78:78–88

Neubauer SC (2014) On the challenges of modeling the net radiative forcing of wetlands: reconsidering Mitsch et al. 2013. Landsc Ecol 29:571–577

Neubauer SC, Givler K, Valentine S, Megonigal JP (2005) Seasonal patterns and plant-mediated controls of subsurface wetland biogeochemistry. Ecology 86:3334–3344

Niemuth ND, Wangler B, Reynolds RE (2010) Spatial and temporal variation in wet area of wetlands in the Prairie Pothole Region of North Dakota and South Dakota. Wetlands 30:1053–1064

Niering WA (1988) Endangered, threatened, and rare wetland plants and animals of the continental United States. In: Hook DD Jr, Smith WHM, Gregory HK et al (eds) The ecology and management of wetlands. Volume 1: The ecology of wetlands. Timber Press, Portlands, pp 227–238

Nixon SW (1980) Between coastal marshes and coastal waters—a review of twenty years of speculation and research on the role of salt marshes in estuarine productivity and water chemistry. In: Hamilton P, MacDonald K (eds) Estuarine and wetland processes. Plenum Press, New York, pp 438–525

Noe GB, Zedler JB (2001) Spatio-temporal variation of salt marsh seedling establishment in relation to the abiotic and biotic environment. J Veg Sci 12:61–74

Nykänen H, Alm J, Silvola J et al (1998) Methane fluxes on boreal peatlands of different fertility and the effect of long-term experimental lowering of the water table on flux rates. Glob Biogeochem Cycles 12:53069

Nykänen H, Vasander H, Huttunen JT, Martikainen PJ (2002) Effect of experimental nitrogen load on methane and nitrous oxide fluxes on ombrotrophic boreal peatland. Plant Soil 242(1):147–155

Odum WE, Odum EP, Odum HT (1995) Nature's pulsing paradigm. Estuaries 18:547–555

Ogle SM, Hunt P, Trettin CC (2014) Chapter 4: Quantifying greenhouse gas sources and sinks in managed wetlands systems. In: Eve M, Pape D, Flugge M et al (eds) Quantifying greenhouse gas fluxes in agriculture and forestry: methods for entity-scale inventory, Tech. Bull. 139. U.S. Department of Agriculture, Office of the Chief Economist, Washington, DC

Oulehle F, Jones TG, Burden A et al (2013) Soil–solution partitioning of DOC in acid organic soils: results from a UK field acidification and alkalization experiment. Eur J Soil Sci 64(6):787–796

Page SE, Rieley J, Banks CJ (2011) Global and regional importance of the tropical peatland carbon pool. Glob Chang Biol 17:798–818

Partain JL Jr, Alden S, Bhatt US et al (2016) An assessment of the role of anthropogenic climate change in the Alaska fire season of 2015. Bull Am Meteorol Soc 97(12):5 p

Pennock D, Yates T, Bedard-Haughn A et al (2010) Landscape controls on N$_2$O and CH$_4$ emissions from freshwater mineral soil wetlands of the Canadian Prairie Pothole region. Geoderma 155:308–319

Pirrone N, Cinnirella S, Feng X et al (2010) Global mercury emissions to the atmosphere from anthropogenic and natural sources. Atmos Chem Phys 10(13):5951–5964

Poffenbarger HJ, Needelman BA, Megonigal JP (2011) Salinity influence on methane emissions from tidal marshes. Wetlands 31:831–842

Price JS, Heathwaite AL, Baird AJ (2003) Hydrological processes in abandoned and restored peatlands: an overview of management approaches. Wetl Ecol Manag 11:65–83

Randerson JT, Liu H, Flanner MG et al (2006) The impact of boreal forest fire on climate warming. Science 314:1130–1132

Rasse DP, Peresta G, Drake BG (2005) Seventeen years of elevated CO$_2$ exposure in a Chesapeake Bay Wetland: sustained but contrasting responses of plant growth and CO$_2$ uptake. Glob Chang Biol 11:369–377

Reddy KR Jr, Patrick WH, Lindau CW (1989) Nitrification-denitrification at the plant root-sediment interface in wetlands. Limnol Oceanogr 34(6):1004–1013

Richards B, Craft CB (2015) Greenhouse gas fluxes from restored agricultural wetlands and natural wetlands, northwestern Indiana. In: Vymazal J (ed) Role of natural and constructed wetlands in nutrient cycling and retention on the landscape. Springer, Cham, pp 17–32

Richardson JL, Vepraskas MJ (eds) (2001) Wetland soils. Lewis Publishers, Boca Raton

Richardson JL, Arndt JL, Montgomery JA (2001) Hydrology of wetland and related soils. In: Richardson JL, Vepraskas MJ (eds) Wetland soils. Lewis Publishers, Boca Raton, pp 35–84

Roberson E (1996) The impacts of livestock grazing on soils and recommendations for management. California Native Plant Society, Sacramento. https://www.cnps.org/cnps/archive/letters/soils.pdf

Rogers K, Saintilan N, Woodroffe CD (2014) Surface elevation change and vegetation distribution dynamics in a subtropical coastal wetland: implications for coastal wetland response to climate change. Estuar Coast Shelf Sci 149:46–56

Rudd JW (1995) Sources of methyl mercury to freshwater ecosystems: a review. Water Air Soil Pollut 80(1):697–713

Sahrawat KL (2004) Organic matter accumulation in submerged soils. Adv Agron 81:169–201

Saintilan N, Wilson NC, Rogers K et al (2014) Mangrove expansion and salt marsh decline at mangrove poleward limits. Glob Chang Biol 20:147–157

Sanders CJ, Maher DT, Tait DR et al (2016) Are global mangrove carbon stocks driven by rainfall? J Geophys Res Biogeosci 121:2600–2609

Sanford T, Wang R, Kenwa A (2015) The age of Alaskan wildfires. Climate Central, Princeton, 32 p

Sarukhán J, Whyte A, MA Board of Review (eds) (2005) Ecosystems and human well-being: synthesis. Island Press, Washington, DC

Scavia D, Field JC, Boesch DF et al (2002) Climate change impacts on US coastal and marine ecosystems. Estuaries 25:149–164

Schilling EB, Lockaby GB (2006) Relationships between productivity and nutrient circulation within two contrasting southeastern U.S. floodplain forests. Wetlands 26:181–192

Schimelpfenig DW, Cooper DJ, Chimner RA (2014) Effectiveness of ditch blockage for restoring hydrologic and soil processes in mountain peatlands. Restor Ecol 22:257–265

Schrier-Uijl AP, Kroon PS, Hendriks DMD et al (2014) Agricultural peatlands: towards a greenhouse gas sink—a synthesis of a Dutch landscape study. Biogeosciences 11:4559–4576

Schuur EA, McGuire AD, Schadel C et al (2015) Climate change and the permafrost carbon feedback. Nature 520(7546):171–179

Segarra KEA, Schubotz F, Samarkin V et al (2015) High rates of anaerobic methane oxidation in freshwater wetlands reduce potential atmospheric methane emissions. Nat Commun 6:7477

Shelby JD, Chescheir GM, Skaggs RW, Amatya DM (2005) Hydrologic and water-quality response of forested and agricultural lands during the 1999 extreme weather conditions in eastern North Carolina. Trans Am Soc Agric Biol Eng 48(6):2179–2188

Shurr EAG, McGuire D, Romanovsky V et al (2018) Chapter 11: Arctic and boreal carbon. In: Cavallaro N, Shrestha G, Birdsey R et al (eds) Second state of the carbon cycle report: a sustained assessment report. U.S. Global Change Research Program, Washington, DC, pp 428–468

Simpson L, Osborne T, Duckett L, Feller I (2017) Carbon storages along a climate induced coastal wetland gradient. Wetlands 37(6):1023–1035

Skaggs RW, Bravé MA, Gilliam JW (1994) Impacts of agricultural drainage on water quality. Crit Rev Environ Sci Technol 24(1):1–32

Skaggs RW, Tian S, Chescheir GM et al (2016) Forest drainage. In: Amatya et al (eds) Forest hydrology: processes, management and assessment. CABI Publishers, Wallingford, pp 124–140

Smil V (2000) Phosphorus in the environment: natural flows and human interferences. Annu Rev Energy Environ 25:53–88

Smith KA, Ball T, Conen F et al (2003) Exchange of greenhouse gases between soil and atmosphere: interactions of soil physical factors and biological processes. Eur J Soil Sci 54:779–791

Snedaker SC (1995) Mangroves and climate change in the Florida and Caribbean region: scenarios and hypotheses. In: Asia-Pacific symposium on mangrove ecosystems. Springer, Dordrecht, pp 43–49

Spaulding MD, Ruffo S, Lacambra C et al (2014) The role of ecosystems in coastal protection: adapting to climate change and coastal hazards. Ocean Coast Manag 90:50–57

Springer KB, Manker CR, Pigati JS (2015) Dynamic response of desert wetlands to abrupt climate change. Quat Int 387:145–146

St. Louis VL, Rudd JW, Kelly CA et al (1994) Importance of wetlands as sources of methylmercury to boreal forest ecosystems. Can J Fish Aquat Sci 51(5):1065–1076

St. Louis VL, Rudd JW, Kelly CA et al (1996) Production and loss of methylmercury and loss of total mercury from boreal forest catchments containing different types of wetlands. Environ Sci Technol 30(9):2719–2729

Stehr G, Böttcher B, Dittberner P et al (1995) The ammonia-oxidizing, nitrifying population of the River Elbe estuary. FEMS Microbiol Ecol 17:177–186

Stelk MJ, Christie J, Weber R et al (2017) Wetland restoration: contemporary issues and lessons learned. Association of State Wetland Managers, Windham

Stewart RE, Kantrud HA (1972) Vegetation of prairie potholes, North Dakota, in relation to quality of water and other environmental factors, Professional Paper 585–D. U.S. Geological Survey, Washington, DC. https://doi.org/10.3133/pp585D

Strack M, Zuback YCA (2013) Annual carbon balance of a peatland 10 yr following restoration. Biogeosciences 10:2885–2896

Stromberg JC, Anderson DC, Scott ML (2012) Riparian floodplain wetlands of the arid and semiarid Southwest. In: Batzer DP, Baldwin AH (eds) Wetland habitats of North America: ecology and conservation concerns. University of California Press, Berkeley

Sutton-Grier AS, Megonigal JP (2011) Plant species traits regulate methane production in freshwater wetland soils. Soil Biol Biochem 43(2):413–420

Tangen BA, Gleason RA (2008) Reduction of sedimentation and nutrient loading. In: Gleason RA, Laubhan MK, Euliss NH Jr (eds) Ecosystem services derived from wetland conservation practices in the United States Prairie Pothole Region with an emphasis on the U.S. Department of Agriculture Conservation Reserve and Wetlands Reserve Programs, Professional Paper 1745. U.S. Department of the Interior, Geological Survey, Reston, 58 p

Tangen BA, Finocchiaro RG, Gleason RA (2015) Effects of land use on greenhouse gas fluxes and soil properties of wetland catchments in the Prairie Pothole Region of North America. Sci Total Environ 533:391–409

Tiner RW (1999) Wetland indicators. CRC Press, Boca Raton

Trettin CC, Jurgensen MF (2003) Carbon cycling in wetland forest soils. In: Kimble JM, Heath LS, Birdsey RA, Lal R (eds) The potential of US forest soils to sequester carbon and mitigate the greenhouse effect. CRC Press, Boca Raton, pp 311–332

Trettin CC, Song B, Jurgensen MF, Li C (2001) Existing soil carbon models do not apply to forested wetlands, Gen. Tech. Rep. SRS-46. U.S. Department of Agriculture, Forest Service, Southern Research Station, Asheville, 10 p

Trettin CC, Jurgensen MF, Gale MR, McLaughlin JW (2011) Carbon and nutrient pools in a northern forested wetland 11 years after harvesting and site preparation. For Ecol Manag 262:1826–1833

Trimble SW (1974) Man-induced soil erosion on the southern Piedmont 1700–1970. Soil and Water Conservation Society, Ankeny

Turetsky MR, Manning SW, Wieder RK (2004) Dating recent peat deposits. Wetlands 24:324–356

Turetsky MR, Donahue WF, Benscoter BW (2011) Experimental drying intensifies burning and carbon losses in a northern peatland. Nat Commun 2:1–5

Turetsky MR, Kotowska A, Bubier J et al (2014) A synthesis of methane emissions from 71 northern, temperate, and subtropical wetlands. Glob Chang Biol 20(7):2183–2197

Turner RE, Swenson EM, Milan CS (2000) Organic and inorganic contributions to vertical accretion in salt marsh sediments. In: Weinstein M, Kreeger DA (eds) Concepts and controversies in tidal marsh ecology. Kluwer Academic Publishing, Dordrecht, pp 583–595

Turunen J, Roulet NT, Moore TR, Richard PJ (2004) Nitrogen deposition and increased carbon accumulation in ombrotrophic peatlands in eastern Canada. Glob Biogeochem Cycles 18(3):GB3002

U.S. Environmental Protection Agency [US EPA] (2015) Section 404 of the Clean Water Act and swampbuster: wetlands on agricultural lands. https://www.epa.gov/cwa-404/section-404-and-swampbuster-wetlands-agricultural-lands. Accessed 18 Jan 2019

U.S. Environmental Protection Agency [US EPA] (2016) National wetland condition assessment 2011: a collaborative survey of the nation's wetlands. EPA Report EPA-843-R-15-005. Washington, DC: U.S. Environmental Protection Agency, Office of Wetlands, Oceans and Watersheds; Office of Research and Development. https://www.epa.gov/national-aquatic-resource-surveys/nwca

U.S. Fish and Wildlife Service # [US FWS] (2011) Status and trends of wetlands in the conterminous United States 2004 to 2009, Report to Congress (https://www.fws.gov/wetlands/status-and-trends/)

van der Kamp G, Hayashi M (1998) The groundwater recharge function of small wetlands in the semi-arid northern prairies. Freshwater functions and values of prairie wetlands. Great Plains Res 8(1):39–56

van der Kamp G, Stolte WJ, Clark RG (1999) Drying out of small prairie wetlands after conversion of their catchments from cultivation to permanent brome grass. Hydrol Sci J 44:387–397

van der Kamp G, Hayashi M, Gallén D (2003) Comparing the hydrology of grassed and cultivated catchments in the semi-arid Canadian prairies. Hydrol Process 17:559–575

Vann CD, Megonigal JP (2003) Elevated CO_2 and water depth regulation of methane emissions: comparison of woody and non-woody wetland plant species. Biogeochemistry 63:117–134

Vasilas LM, Hurt GW, Berkowitz JF (eds) (2016) Field indicators of hydric soils in the United States, Version 8.0. U.S. Department of Agriculture, Natural Resources Conservation Service, Washington, DC, 55 p

Venne LS, Anderson TA, Zhang B et al (2008) Organochlorine pesticide concentrations in sediment and amphibian tissue in playa wetlands in the southern High Plains, USA. Bull Environ Contam Toxicol 80:497–501

Vepraskas MJ (1996) Redoximorphic features for identifying aquic conditions, Tech. Bull. 301. North Carolina State University, Agricultural Research Service, Raleigh

Vepraskas MJ, Craft CB (2016) Wetland soils, genesis hydrology and classification, 2nd edn. CRC Press, Boca Raton

Vepraskas MJ, Faulkner SP (2001) Redox chemistry of hydric soils. In: Richardson JL, Vepraskas MJ (eds) Wetland soils. Lewis Publishers, Boca Raton, pp 85–106

Verry ES (1997) Hydrological processes of natural northern forested wetlands. In: Trettin CC, Jurgensen MF, Grigal DF, Gale MR, Jeglum JK (eds) Northern forested wetlands: ecology and management. CRC Press, Boca Raton, pp 163–188

Vet R, Artz RS, Carou S (2014) A global assessment of precipitation chemistry and deposition of sulfur, nitrogen, sea salt, base cations, organic acids, acidity and pH, and phosphorus. Atmos Environ 93:3–100

Vile MA, Bridgham SD, Wieder RK, Novák M (2003) Atmospheric sulfur deposition alters pathways of gaseous carbon production in peatlands. Glob Biogeochem Cycles 17(2):1058

Vose RS, Easterling DR, Kunkel KE et al (2017) Temperature changes in the United States. In: Wuebbles DJ, Fahey DW, Hibbard KA et al (eds) U.S. Global Change Research Program. Climate science special report: fourth national climate assessment, vol I. U.S. Global Change Research Program, Washington, DC, pp 185–206

Vymazal J (2011) Constructed wetlands for wastewater treatment: five decades of experience. Environ Sci Technol 45(1):61–69

Waddington JM, Morris PJ, Kettridge N et al (2015) Hydrological feedbacks in northern peatlands. Ecohydrolology 8:113–127

Walsh J, Wuebbles D, Hayhoe K et al (2014) Chapter 2: our changing climate. In: Mellilo JM, Richmond TC, Yohe GW (eds) Impacts in the United States: the third national climate assessment. U.S. Global Change Research Program, Washington, DC, pp 19–67

Wear DN, Greis JG (eds) (2002) Southern forest resource assessment, General Technical Report SRS-GTR-53. U.S. Department of Agriculture, Forest Service, Southern Research Station, Asheville, 635 p

Weiss-Penzias PS, Gay DA, Brigham ME et al (2016) Trends in mercury wet deposition and mercury air concentrations across the US and Canada. Sci Total Environ 15(568):546–556

Weston NB, Dixon RE, Joye SB (2006) Ramifications of increased salinity in tidal freshwater sediments: geochemistry and microbial pathways of organic matter mineralization. J Geophys Res 111:G01009

Weston NB, Vile MA, Neubauer SC, Velinsky DJ (2011) Accelerated microbial organic matter mineralization following salt-water intrusion into tidal freshwater marsh soils. Biogeochemistry 102:135–151

Wickland KP, Krusche AV, Kolka RK et al (2014) Chapter 5: Inland wetland mineral soils. In: Hiraishi T, Krug T, Tanabe K et al (eds) Intergovernmental panel on climate change supplement to the 2006 guidelines for national greenhouse gas inventories: wetlands. Intergovernmental Panel on Climate Change, Geneva

Windham-Myers L, Cai WJ, Megonigal P et al (2018) Chapter 15: Tidal wetlands and estuaries. In: Cavallaro N, Shrestha G, Birdsey R et al (eds) Second state of the carbon cycle report (SOCCR2): a sustained assessment report. U.S. Global Change Research Program, Washington, DC

Winter TC (1988) A conceptual framework for assessing cumulative impacts on hydrology of nontidal wetlands. Environ Manag 12:605–620

Winter TC (1989) Hydrologic studies of wetlands in the northern prairie. In: van der Valk AG (ed) Northern prairie wetlands. Iowa State University Press, Ames, pp 193–211

Winter TC, Rosenberry DO (1998) Hydrology of prairie pothole wetlands during drought and deluge: a 17-year study of the cottonwood Lake wetland complex in North Dakota in the perspective of longer term measured and proxy hydrological records. Clim Chang 40:189–209

Winter TC, Woo MK (1990) Hydrology of lakes and wetlands. In: Wolman MG, Riggs HC (eds) The geology of North America,

Vol. 1: Surface water hydrology. Geological Society of America, Boulder, pp 159–187

Woo MK (1992) Impacts of climate variability and change on Canadian wetlands. Can Water Resour J 17:63–69

Woodward RT, Wui Y (2001) The economic value of wetland services: a meta-analysis. Ecol Econ 37(2):257–270

Wüst-Galley C, Mössinger E, Leifeld J (2016) Loss of the soil carbon storage function of drained forested peatlands. Mires Peat 18:1–22

Young PJ, Keeland BD, Sharitz RR (1995) Growth response of bald cypress to an altered hydrologic regime. Am Midl Nat 133:206–212

Young AM, Higuera PE, Duffy PA, Hu FS (2016) Climatic thresholds shape northern high-latitude fire regimes and imply vulnerability to future climate change. Ecography 39:1–12

Yu K, Faulkner SP, Baldwin MJ (2008) Effect of hydrologic conditions on nitrous oxide emissions, methane and carbon dioxide dynamics in a bottomland hardwood forest and its implications for carbon sequestration. Glob Chang Biol 14:798–812

Zhang Y, Li C, Trettin CC et al (2002) An integrated model of soil, hydrology, and vegetation for carbon dynamics in wetland ecosystems. Glob Chang Biol 16(4):1061

Zillioux EJ, Porcella DB, Benoit JM (1993) Mercury cycling and effects in freshwater wetland ecosystems. Environ Toxicol Chem 12(12):2245–2264

Zoltai SC, Morrissey LA, Livingston GP, de Groot WJ (1998) Effects of fires on carbon cycling in North American boreal peatlands. Environ Rev 6:13–24

Urban Soils

7

Richard V. Pouyat, Susan D. Day, Sally Brown,
Kirsten Schwarz, Richard E. Shaw, Katalin Szlavecz,
Tara L. E. Trammell, and Ian D. Yesilonis

Introduction

The global population is expected to exceed 11 billion before the end of the twenty-first century (United Nations 2015). Populations within urban areas are also increasing, with the number of mega-sized (ten million people or more) cities expected to increase from 10 in 1990 to 41 in 2030 (United Nations 2015). In the United States, the human population is not growing as fast as the rest of the world, but the expansion of urban areas has proportionally kept pace, or exceeded, global estimates, with land devoted to urban uses growing by more than 34% between 1980 and 2000 alone (USDA NRCS 2001). Additionally, by 2010 almost 250 million people lived in urban areas, which roughly accounts for 81% of the total population of the United States (U.S. Census 2010).

The relatively rapid expansion of urban areas in the United States, along with the juxtaposition of more than 80% of the population to the soils located in these areas, suggests the increasing national importance of "urban" soils. In this chapter we define and describe the role soil plays in urban landscapes and discuss the importance of these soils within the context of densely populated areas. Additionally, we provide an overview of what is known about the characteristics of urban soils and their role in the provisioning of ecosystem services. Finally, we assess the current state of knowledge of urban soils and provide a list of future informational needs.

What Is an Urban Soil?

The term "urban soil" was first used by Zemlyanitskiy (1963) to describe the characteristics of highly disturbed soils in urban areas. Urban soil was later defined by Craul (1992) as "a soil material having a nonagricultural, man-made surface layer more than 50 cm thick that has been produced by mixing, filling, or by contamination of land surface in urban and suburban areas." This definition was derived from, and is thus similar to, earlier definitions by Bockheim (1974) and Craul and Klein (1980). Since these earlier characterizations, Evans and others (2000) and later Capra and others (2015) use the term "anthropogenic soil," which places urban soils in a broader context of human-altered soils rather than limiting the definition to densely populated urban and suburban areas alone. To recognize a broader set of observations, Effland and Pouyat (1997), Lehmann and Stahr (2007), and more recently Morel and others (2017) more broadly defined urban soils to include soils that are relatively undisturbed yet altered by urban environmental changes, such as the deposition of atmospheric pollutants.

R. V. Pouyat (✉)
Northern Research Station, USDA Forest Service, Newark, DE, USA

S. D. Day
Faculty of Forestry, University of British Columbia, Vancouver, BC, Canada

S. Brown
School of Forest Resources, University of Washington, Seattle, WA, USA

K. Schwarz
Departments of Urban Planning and Environmental Health Sciences, University of California, Los Angeles, Los Angeles, CA, USA

R. E. Shaw
U.S. Department of Agriculture, Natural Resources Conservation Service, New Jersey State Office, Somerset, NJ, USA

K. Szlavecz
Morton K. Blaustein Department of Earth and Planetary Sciences, Johns Hopkins University, Baltimore, MD, USA

T. L. E. Trammell
Department of Plant and Soil Sciences, University of Delaware, Newark, DE, USA

I. D. Yesilonis
U.S. Department of Agriculture, Forest Service, Northern Research Station, Baltimore Field Station, Baltimore, MD, USA

© The Author(s) 2020
R. V. Pouyat et al. (eds.), *Forest and Rangeland Soils of the United States Under Changing Conditions*,
https://doi.org/10.1007/978-3-030-45216-2_7

127

A Range of Soil Conditions

Soil conditions in urban areas generally correspond to a range of anthropogenic effects from relatively low influence (e.g., native forest or grassland soil) to those impacted by urban environmental effects such as patches of urban forest to soil types that are derived from human created materials, sealed by impervious surfaces, or altered by physical disturbances and management (Pouyat et al. 2009; Morel et al. 2017). The latter include massive or highly disturbed soils without structure (Short et al. 1986), human-transported materials (Shaw and Isleib 2017), sealed soils (Scalenghe and Marsan 2009), engineered soils such as green roof media and street tree pit soils (e.g., Grabosky et al. 2002), and soils that were once disturbed, but are now managed, such as public or residential lawns (Trammell et al. 2016). Therefore, comparisons among these soil uses should reflect the relative impact of urban effects such as site disturbances (site grading, use of sealed surfaces), subsequent management activities (fertilization, irrigation), intensity of use (trampling), plant cover, urban environmental changes (air pollution, habitat isolation) that are often novel, and site history (Pouyat et al. 2017; Burghardt 2017).

Habitat for Soil Organisms

Contrary to the generally held belief, urban soils are alive and may harbor a rich diversity of microorganisms and invertebrates. Urban soil communities are a unique combination of both native species that survive or thrive in the urban landscape and species that have been introduced from other regions or continents. Management practices also contribute to the uniqueness of urban soil communities. For instance, irrigation can overcome the lack of soil moisture as a serious site limitation for soil biota, and pesticides can eliminate nontarget species (Szlavecz et al. 2018). Soil sealing limits many soil organisms, although some taxonomic groups (e.g., earthworms [phylum Annelida] and ants [family Formicidae]) can survive under impervious surfaces or pavement (Youngsteadt et al. 2015). Landscaping practices, such as the removal of woody debris and leaf litter, deprive many species of shelter and food resources, while composting and mulching create new ones. In engineered soil environments such as green roofs and tree pit soils, entirely novel communities may assemble over time. The success of soil organisms in such circumstances depends on the constructed substrate, the connectivity among existing green roof habitats, and the age of the habitat (Burrow 2017; Madre et al. 2013).

What Is the Role of Soil in Urban Ecosystems?

Urban soils play multiple, and sometimes conflicting, roles within urban ecosystems (Setälä et al. 2014). Despite the high levels of disturbance typical of most urban soils, they, like their rural counterparts, have the potential to support plant, animal, and microbial organisms and to mediate hydrological and biogeochemical cycles (Pouyat et al. 2010). Each of these functions, however, must be evaluated specifically for urban conditions. Urban soil functions are often significantly altered from those of their rural counterparts, due not only to their modified characteristics but also because of their context within the urban landscape. As an example, consider a compacted soil that has been degraded by building construction or demolition. The permeability of this soil is impaired by compaction (a change in function), and its location in a mostly sealed environment results in the soil receiving much higher volumes of water via runoff (due to landscape position).

Soils play other critical functions that are unique to urban landscapes. For example, they provide a stable base for built structures such as buildings and roads. Additionally, urban soils provide physical support and a convenient and accessible location for underground utilities. They may serve roles in processing waste, whether from septic systems or from food and yard waste recycling programs. All soils have the capacity to accumulate various nutrients such as phosphorus (P) and nitrogen (N) that if transported to surface waters can cause environmentally damaging algal blooms. Soils also store significant levels of toxicants associated with urban environments (e.g., lead [Pb] and arsenic [As]), along with a host of macro and microartifacts (Rossiter 2007). These soils may endanger public health if humans subsequently come in contact with the soils when they are used for other purposes (e.g., urban agriculture or recreation).

Importance of Soil in an Urban Context

Soil provides vital and life-sustaining ecosystem services but is often underappreciated as a natural resource. This is particularly true for urban soils, which are assumed to be highly altered and thus not capable of providing the same ecosystem services as native, unaltered soils. However, as previously mentioned, urban soils can provide many of the same ecosystem services as nonurban soils (Morel et al. 2015; Pavao-Zuckerman 2012). In fact, in many cases the importance of these ecosystem services may actually be enhanced since they are, by definition, closely associated with high densities of people living in urban areas (e.g., Herrmann et al. 2017). In the following sections, we provide examples and present a conceptual framework for the provisioning of ecosystem services by urban soils.

Juxtaposition of People and Soil: An Educational Opportunity

Urban food production has provided a means to connect urban communities with urban soil systems. As an example, for over a decade, Growing Power in Milwaukee, WI, successfully used urban agriculture to build community involvement around food security, healthy food systems, and food justice. Programs, including youth engagement and training, showed how community food systems connect people and soil (Royte 2009). Another example is the Edible Schoolyard Project in Berkeley, CA, which has leveraged the process of growing food to teach school students not only about food production and preparation but also about soil ecosystems, ecological processes, and nutrition (https://edibleschoolyard. org/).

Although Growing Power and the Edible Schoolyard Project were pioneering projects, countless other urban agriculture endeavors have followed, and many have connected communities to valuable soil resources and educational opportunities (e.g., City Slicker Farms, Oakland, CA, and Detroit Black Community Food Network, Detroit, MI). These endeavors, however, are sometimes met with the challenges of legacy pollutants that can reduce or prevent realization of potential ecosystem services. An unfortunate example is the potential for Pb to be found in relatively high concentrations in urban agricultural and garden soils (Schwarz et al. 2012). Contamination of soil by Pb, however, can be ameliorated through soil amendments that reduce the Pb bioavailability or, in more severe cases, through soil removal (Kumpiene et al. 2008). Community-level stewardship and investment by communities in the protection and improvement of soils to produce food in urban areas can be facilitated by network building and information exchange (Schwarz et al. 2016).

Community participation in urban soil stewardship relies on educational resources that allow people to gain a better understanding of the benefits of soils. Professional societies can play a role in providing these resources. For example, the Soil Science Society of America (SSSA) has developed a curriculum that addresses how humans have shaped soil systems and how people can work to protect soil (https://www. soils4teachers.org/). In addition to creating and overseeing K-12 curricula and hosting online seminars, the SSSA's year-long observance of the 2015 International Year of Soils was recognized for its innovation in communication. In New York City, NY, the Urban Soils Institute was created to provide soils information, testing, and education for green infrastructure, community gardening, urban agriculture, and

restoration efforts (Fig. 7.1). Art has also been a creative means for communicating the value of soils. For example, the Hundred Dollar Bill Project, initiated by the nonprofit organization Fundred, is a collective art project with the goal of bringing awareness to the dangers of soil Pb and the importance of investing in a solution to improve both human health and soil (https://fundred.org/).

An Ecosystem Services Framework for Urban Soils

In the provision of ecosystem services, urban soil plays a unique role as the "brown infrastructure" of urban ecological systems, much in the same way urban vegetation is thought of as green infrastructure (Heidt and Neef 2008; Pouyat et al. 2007). While green infrastructure provides services attributed to vegetation, such as the moderation of energy fluxes by tree canopies (Akbari 2002; Heidt and Neef 2008), brown infrastructure provides ecosystem services attributed to soil, such as those previously mentioned.

Traditional engineering approaches, especially those used in urban areas, typically address ecosystem service deficits with built or gray infrastructure or, alternatively, may actually move the ecosystem service function off-site. An example is the collection of wastes from urban areas through a sewer system of pipes to a sanitary treatment plant (i.e., gray infrastructure). Subsequently, processed sanitary wastes may be transported to rural areas and applied as a soil amendment, effectively recycling wastes using soil processes; but transporting wastes off-site can result in the emissions of greenhouse gases from the vehicles transferring the waste. More innovative engineering approaches, however, utilize the recycling function of urban soils within the urban ecosystem rather than creating emissions by transporting the waste to a rural area. Stormwater management is another soil function traditionally addressed by gray infrastructure. Urban soil, however, can be utilized as a permeable media to reduce runoff (e.g., rain gardens) and as such works with green and gray infrastructure to reduce stormwater overflow (Kaushal and Belt 2012).

These newer and more innovative roles for urban soils create new opportunities for realizing ecosystem services and for conceptually framing how we consider these services and the disciplinary and management approaches we implement to enhance them (Fig. 7.2). *Provisioning services* include traditional services such as food production but also include novel services such as those that support structures and roads. *Regulating services* include storage of soil C and mitigation of greenhouse gas emissions and novel services such as stormwater retention and waste management. *Supporting services* include soil formation and nutrient cycling but also may include the sequestration of contami-

Working Ecosystems

• Managed provisioning services

>Food
>Fiber
>Fuel

• Profit, subsistence motive

• Agriculture, plantation, short rotation, urban agriculture

Eco-engineered Ecosystems

• Managed regulating services

>Climate
>Flood
>Water purification
>Disease regulation

• Regulatory & service motive

• Restoration, storm water retention, bioremediation…

Amenity Ecosystems

• Managed cultural services

>Recreational
>Aesthetic
>Spiritual
>Educational

• Consumptive, leisure motive

• Public lands (parks, wildlife areas, ornamental gardens, golf courses)

Supporting ecosystem services: nutrient cycling, soil formation, etc.

Fig. 7.1 As with nonurban soils, urban soils provide ecosystem services. Because of the close proximity of urban soils with dense human populations, the importance of ecosystem services is especially magnified for managed regulating services and managed cultural services

Fig. 7.2 Soils training session for environmental stewardship conducted by the USDA Natural Resources Conservation Service and the New York City Urban Soils Institute, Bronx, NY. (Photo credit: Richard Shaw, USDA Natural Resources Conservation Service)

nants to reduce human exposure and improve public health. Finally, *cultural services*, which include the support of public greenspaces used for recreation, aesthetic, or spiritual values, will be enhanced in urban areas because of the high level of human access. Each of the ecosystem services provided by urban soils can be modified by soil management practices and may also be integrated with engineered approaches (Morel et al. 2015).

Anthropogenic Influences on Urban Soils and Their Assessment

As land is converted to urban uses, soil scientists consider both direct and indirect factors that can affect soil characteristics. Direct effects include those typically associated with urban soils, such as physical disturbances, incorporation of human-created materials, and burial or coverage of soil by fill material and impervious surfaces. Indirect effects are less noticeable and involve changes in the abiotic and biotic environment, which can even affect undisturbed soils within urban and periurban or suburban areas (Pouyat et al. 2010). The resultant conditions of both direct and indirect effects embody what ecologists and soil scientists refer to as "novel" ecosystems (Hobbs et al. 2006). These novel ecosystems represent the suite of conditions that have created and formed urban soils over time since humans began congregating and living in densely populated settlements or cities more than 5000 years ago.

Direct Effects

Urbanization is characterized by the built environment—the buildings, roads, and other structures that form the communities where humans live and work—and subsequent human activity in that environment. The conversion of forested and agricultural lands to urban uses often results in a host of rapid changes related to land development and grading. As humans live and work in this environment, they manipulate soils for a wide variety of purposes, which range from stormwater management, lawn maintenance, and landscaping to urban agriculture and recreation. Furthermore, human activity generates waste products, including everything from industrial waste to feces of pets.

Land Use Change and Urbanization

Land development associated with land use change and urbanization occurs not only at the perimeters of urban areas but also in periurban or suburban communities and in the interior of cities through infill development (Setälä et al. 2014). Usually the first phase of development is to create level topography through grading. During this process, the landscape is altered by the removal of surface O- and

A-horizon soils, and the remaining soils are compacted from the use of heavy equipment. Even if surface soils are replaced after construction, soil structure is degraded, considerable soil carbon (C) stocks are lost (e.g., Chen et al. 2013), and hydraulic conductivity is greatly reduced (e.g., Schwartz and Smith 2016). Additionally, lower soil horizons may be severely compacted, especially if grade changes are significant or soils are located in construction staging areas.

Waste Disposal

Dense populations and concentrated activity in urban areas generate a significant amount of waste products. Domestic and industrial waste materials, construction debris, ash from heat and power production, and dredge spoils commonly end up as components of urban soils. In many areas, these materials have been used to fill wetlands or extend the shoreline, but most cities have now created domestic landfills to address part of the waste problem. Many of these human-made or processed artifacts, which can include black C, trace metals, and organic contaminants, have properties that are unlike natural soils, and these materials can have profound effects on soil-forming processes and soil properties (Huot et al. 2015).

Grading and Stormwater Management

Soil and its management can be a useful tool in retaining stormwater during and after the land development process (Shuster et al. 2014; Shuster and Dadio 2017). As an example, grading is used to alter water flow paths to direct overland flow into soil-plant reservoirs such as with bioretention cells and rain gardens. These reservoirs are often filled with engineered "bioretention" soil mixes that allow for rapid infiltration and are highly penetrable by roots (Kaushal and Belt 2012). Additionally, various techniques such as deep ripping followed by compost amendment can significantly increase permeability in soils degraded by compaction (Chen et al. 2014; Schwartz and Smith 2016).

Sealing and Paving

A high proportion of urban land is covered by impervious surfaces, with much of this surface being soil sealed by asphalt or concrete pavement (Scalenghe and Marsan 2009). Few studies of soils beneath pavement have been conducted. Paving or sealing the soil surface interrupts the flow of energy and materials (including detritus) and, therefore, is disadvantageous to the provision of most ecosystem services. As a consequence, the water and nutrient cycles are disrupted, the heat balance is altered, more anoxic conditions prevail, and thus habitat for root growth and many soil organisms is lost. Existing research indicates that C and N contents are lower in sealed areas than in adjacent unsealed soils (Table 7.1) (Piotrowska-Dlugosz and Charzynski 2015; Raciti et al. 2012).

Table 7.1 Soil organic carbon stocks to 1 m for soil types in two different cities

Soil type	C (kg m^{-2})	Land area (%)	Hectares	Total C (Mg)
New York, NY				
Natural[a]	22.2	6.8	5256	1,362,080
HAHT[b]	18.3	25.4	19,626	3,428,416
Sealed soils	6.0	62.6	48,328	2,885,155
Baltimore, MD[c]				
Natural	10.2	38.2	7995	815,490
HAHT	NA	19.2	4021	NA
Sealed soils	3.3	42.6	8898	293,634

[a]Natural soils include brown till, red till, tidal marsh, and outwash
[b]HAHT are human-altered or human-transported soils that include spolic, artifactual, dredgic, and combustic soils
[c]Data from Yesilonis and Pouyat (2012); natural soils do not include wetlands

Due to the high proportion of sealed surfaces in urban areas, more burden is exerted on the fragmented unsealed areas to provide provisioning, regulating, and supporting services (Setälä et al. 2014). Because cities are often situated in areas with few site limitations (e.g., level topography and well-drained soils), many of the most productive agricultural soils end up sealed and disturbed as urban areas expand (Wessolek 2008). The few advantages of soil sealing include rapid removal of stormwater runoff, containment of pollutants, and the preservation of cultural heritage (e.g., the ruins of Pompeii) (Scalenghe and Marsan 2009; Wessolek 2008).

An indirect effect of concrete surfaces is the introduction of secondary or pedogenic carbonates, which are commonly found in many urban soils as a result of the weathering of calcium silicate and hydroxide minerals from concrete (Washbourne et al. 2015). Dust additions and natural parent materials can also supply calcium carbonate, which can lead to a relatively high soil pH that provides buffering against soil acidification (Pouyat et al. 2007).

Soil Replacement and Recycling

Infill development may occur where soils are absent (e.g., displaced by underground structures), severely degraded (e.g., were previously beneath structures or roadways and are compacted and contain a variety of artifacts), or capped because of the accumulation of contaminants. In these cases, soil is often imported from soil blending or recycling facilities. At such facilities, soil is typically screened and blended with sand and compost. The resulting blended soils are sometimes more susceptible to compaction, are poorly structured, and have a lack of aggregate formation; therefore they may have lower water holding capacities than the soils they replace (e.g., Spoor et al. 2003).

Lawn Management

The estimated amount of lawn cover for the conterminous United States is 163,800 km^2 ± 35,850 km^2, which accounts for almost three-quarters of all irrigated cultivated lands (Milesi et al. 2005). To manage turf grass cover, almost half of all residences apply fertilizers (Law et al. 2004; Osmond and Hardy 2004), with some applying fertilizer rates similar to or exceeding those of cropland systems (e.g., >200 kg ha^{-1} year^{-1}) (e.g., Morton et al. 1988). While the potential for losses of C and N in residential areas can be high (Byrne et al. 2008), turf grass systems have shown the capacity to retain a surprising amount of C and N when compared to agricultural soils (Scharenbroch et al. 2018; Trammell et al. 2016). This apparent retention may be due to turf grass management efforts (e.g., irrigation) that maintain high plant productivity, which can lead to organic matter accumulation in soil (Groffman et al. 2009; Pouyat et al. 2002).

Indirect Effects

Humans indirectly influence urban soil conditions through activities that alter urban climatic conditions (e.g., temperature and moisture), chemical inputs (e.g., pollutant and nutrient concentrations), and spread of native and nonnative invasive species (Pouyat et al. 2010). Cities are warmer due to increased heat-generating activities and reduced heat losses because of less evaporative cooling surface (Taha 1997; Zhao et al. 2014), a phenomenon called the "urban heat island" (Oke 1990). Elevated atmospheric deposition of nutrients (Lovett et al. 2000; Rao et al. 2014) and other pollutant inputs, such as ozone (O_3), can indirectly alter the chemical composition of organic matter (OM) inputs to soil as well as the microbial activity. The addition of nonnative species in urban landscapes (McKinney 2006) also affects soil conditions and biogeochemical cycles (Ehrenfeld 2010; Liao et al. 2008). Introduced invasive plants, soil organisms, and plant insect pests alter plant composition and subsequent plant chemical inputs to soil, in turn altering soil microbial community structure and activity. These indirect human influences on soils alter both soil structure (e.g., aggregate formation) and function (e.g., N cycling).

Urban Climate

The urban heat island is linked with changes in plant phenology, including timing and duration of canopy leaf out, leaf budburst, and flowering (Chen et al. 2016; Jochner and Menzel 2015). The urban heat island effect varies by species (Xu et al. 2016) and alters the timing and quantity of OM inputs to the soil environment. Results from experimental soil warming studies suggest that soil microbial activity and soil ecosystem processes in urban landscapes may be accelerated due to the urban heat island effect (Butler et al. 2012; Craine et al. 2010). However, the net effect of soil warming on microbial processes and biogeochemical cycles (e.g., net soil CO_2 flux) will depend on potential compensation from other anthropogenic factors that may restrict microbial activity in urban ecosystems. While the urban heat island effect in cities has been well studied, the alteration of precipitation in cities is complex and is now a focus for atmospheric scientific research (Shepherd and Burian 2003; Song et al. 2016). Numerous studies have identified an "urban rainfall effect" (Shem and Shepherd 2009; Shepherd and Burian 2003), where urban areas experience increased rainfall, snowfall, and convection storm events compared to nearby rural areas (Niyogi et al. 2011; Shem and Shepherd 2009; Taha 1997). The combined effect of altered temperature and precipitation regimes in urban environments can affect plant productivity, and thus, the quantity and quality of OM inputs to the soil, and microbial activity, all of which strongly influence urban soil structure and function.

Urban Atmospheric Chemistry

Human activities in urban areas alter the chemical environment of soils by elevating nutrients (e.g., N deposited from the atmosphere) and increasing atmospheric pollutant concentrations (e.g., O_3), which can strongly influence the chemical composition and the quantity of organic matter inputs to the soil. Urban precipitation contains greater concentrations of inorganic N, calcium (Ca), and magnesium (Mg) that may actually have beneficial effects on plants and microbes via increased nutrient availability; however, the increase in hydrogen ions (acidity) in precipitation may be detrimental for many soil-swelling organisms (Carreiro et al. 2009; Lovett et al. 2000). Urban environments also have elevated CO_2, nitrogen oxides (NO_x), O_3, and other air pollutants in the atmosphere. How altered chemical inputs affect plant growth, and in turn soil processes, in urban areas is not fully understood (Calfapietra et al. 2015). For example, plant productivity may be stimulated by N deposition and elevated CO_2 yet dampened by greater O_3 concentrations (Gregg et al. 2003). Soil N cycling rates in urban environments may increase due to increased N availability (Pardo et al. 2011) or decrease due to the production of more complex phenolic compounds by stressed plants or depressed enzyme activity by fungi, which are important in the decay of organic matter (Carreiro et al. 2000; Findlay et al. 1996). Therefore, the cumulative effects of altered atmospheric deposition as well as ozone have the potential to stimulate or depress soil organic matter and nutrient pools over time.

Nonnative and Invasive Species

Species invasion is a global phenomenon that causes ecological and economic damage to soil ecosystems including those situated in urban areas (Szlavecz et al. 2018). Urban areas are often the epicenter for nonnative species introductions because cities are transportation hubs supporting global commerce and international trade (McKinney 2006). Nonnative invasive species commonly found in urban environments, such as nonnative invasive plants, nonnative earthworms (suborder Lumbricina), and nonnative tree pests, can have significant impacts on soil conditions (e.g., pH) and processes (e.g., N cycling) (Ehrenfeld 2010).

Species introduction of soil invertebrates is usually accidental and happens through transportation of plants or soil. Historically, soil was used as ballast material in ships bound for North America, and many species from the Atlantic and Mediterranean regions crossed the oceanic barrier in this way (Lindroth 1957). Species that generally do well in human-dominated environments were successful. Many common earthworms, terrestrial isopods (order Isopoda), millipedes (family Diplopoda), snails (class Gastropoda), and beetles (order Coleoptera) dominating urban soils in North America are from other biogeographical realms. Greenhouses provide a shelter and "jumping board" for species arriving from different climatic regions. Some species, such as silverfish (*Lepisma saccharina*) and cave crickets (order Orthoptera, family Rhaphidophoridae), remain in close proximity to humans, while others escape and spread through the landscape (Garthwaite et al. 1995). At the same time, remnant patches of natural vegetation can serve as refugia for populations of native invertebrates (Korsós et al. 2002). The dominance of nonnative species over natives is taxon dependent. For instance in the Northeast region, carrion beetles (order Coleoptera, family Silphidae) are native, and their community composition is driven by urban forest patch size and quality, while woodlice (order Isopoda) assemblages are entirely made up of introduced species (Hornung and Szlavecz 2003; Wolf and Gibbs 2004).

Nonnative invasive plants exert pressure on ecosystems by altering plant community composition, plant productivity and phenology, litter decomposition, and soil processes (Jo et al. 2015; Liao et al. 2008; Trammell et al. 2012). Furthermore, invasions by nonnative plants can shift microbial community composition (Arthur et al. 2012; Kourtev et al. 2003) and have cascading effects on belowground nutrient cycles. Similarly, nonnative earthworms alter soil structure and biogeochemical cycles (Szlavecz et al. 2011) and in urban forests were shown to enhance N cycling and leaf litter decay rates (Pouyat and Carreiro 2003; Szlavecz et al. 2006). Hence, greater earthworm abundance in urban soils may be associated with lower soil OM (Sackett et al. 2013; Smetak et al. 2007) and altered microbial composition (Drouin et al. 2016; Scharenbroch and Johnston 2011). While nonnative plants and earthworms directly interact with urban soils, nonnative insect pests can devastate native tree populations in cities, significantly reducing the urban tree canopy, as is the case with emerald ash borer (*Agrilus planipennis*) (Poland and McCullough 2006). This reduction in urban tree foliage can decrease OM inputs to soil; however, how nonnative insect pests influence urban soil conditions or processes is not well understood.

Mapping, Classification, and Interpretation

The past decade has seen an increase in urban soil survey efforts in the United States. Modern soil surveys for New York City, Chicago, IL, Los Angeles, CA, and Detroit now provide detailed information on physical, chemical, and mineralogical properties of human-altered and human-transported soils, enabling more reliable ratings and interpretations for stormwater management, revegetation and restoration efforts, urban agriculture, and better resource inventory. Maps derived from digital elevation models (DEMs) and digitized surface geology along with greater access to engineering logs and reports assist in identifying and understanding anthropogenic alteration and eventually making some predictions about soils and landscapes (Fig. 7.3). In the field, geophysical, nondestructive methods are increasingly being used to characterize spatial variability of soil properties in urban areas. Ground-penetrating radar can identify contrasting materials, discontinuities, and subsurface interfaces (Doolittle et al. 1997). Electromagnetic inductance and magnetic susceptibility can distinguish certain types of artifactual material (Howard and Orlicki 2015). Portable X-ray fluorescence spectrometry can be used to detect and map concentrations of Pb and other trace metals in the soil (Carr et al. 2007).

Cooperation with universities and Federal, state, and municipal agencies helps to address local soil issues and survey needs, provide operational guidance, and offer opportunities for outreach and education. Although there are global similarities in soil characteristics, the urban soil pattern is unique for every city and is affected by history, geology, and geography. For instance, European settlement in New York City began in 1609 in an area with three islands and one peninsula (772 km^2 total) and limited room for expansion. It was accompanied by extensive filling of wetlands and expansion of the shoreline, much of which was done with waste materials such as construction debris and dredge spoils (Fig. 7.4a). Consequently most of the city's soils are human-altered or human-transported (HAHT), many with a considerable artifact content. In contrast, much of the growth and expansion of Los Angeles took place after 1870, and the city had sufficient room to expand (from 73 km^2 to 233 km^2 to 1215 km^2). The predominant form of human disturbance in Los Angeles was land leveling and terracing, much of which involved a relatively small depth (<50 cm) of surface alteration (Fig. 7.4b). As a result, Los Angeles has a lower percentage of sealed surfaces and HAHT soils (Table 7.2).

There have also been recent updates in the classification of urban soils. The World Reference Base for Soil Resources, an international correlation system, added a Technosols reference soil group for soils dominated by technic or artifactual materials (Rossiter 2007). The USDA Soil Taxonomy added definitions of anthropogenic landforms and microfeatures, artifacts, and HAHT materials, along with 12 HAHT family classes (Soil Survey Staff 2014). Both systems are open to revision and likely to change with advances in urban soil research and mapping.

In an alternative approach, Morel and others (2015) proposed a categorization of soils in urban areas according to their capacity to deliver various ecosystem services (provi-

Fig. 7.3 LiDAR (light detection and ranging)-derived digital elevation models (DEMs) use pulsed laser light to measure distances. This high-resolution topographical information can be used for urban soil surveys. (Image by Randy Riddle, USDA Natural Resources Conservation Service)

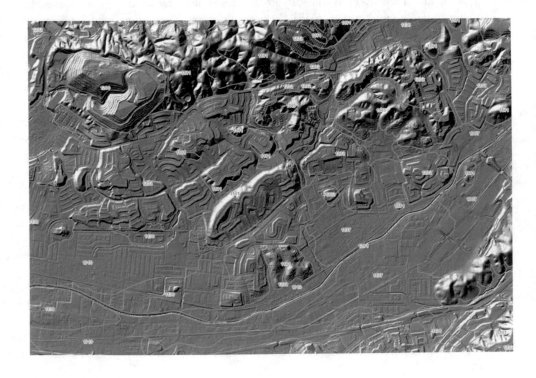

a.

b.

Fig. 7.4 Extensive filling of wetlands and expansion of shorelines often accompanies the development of urban areas. (**a**) Frequently, the fill material is made up of waste materials such as construction debris and dredge spoils. (**b**) In steep topography, the predominant form of

disturbance is often land leveling and terracing, which involves altering a relatively small amount (<50 cm) of the surface. (Photo by Randy Riddle, USDA Natural Resources Conservation Service)

Table 7.2 Soil/land cover type comparison, New York and Los Angeles

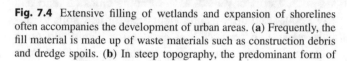

City	Percentage of land area			
	Sealed soil	HAHT[a] soils	Natural soils[b]	Non-soil
New York, NY	63	27	9	1
Los Angeles, CA	43	11	44	2

[a]HAHT are human-altered or human-transported soils that include spolic, artifactual, dredgic, and combustic soils
[b]Natural soils include brown till, red till, tidal marsh, and outwash

sioning, regulating, cultural). The classification was based on a gradient of anthropogenic disturbance and the capacity of the soils to support vegetation. The main objective was to improve the recognition of ecosystem service provision and offer new insight on soil management, design, and engineering in the urban environment.

Ecosystem service delivery can also vary with use-dependent or dynamic soil properties such as soil organic matter, soil structure, bulk density and porosity, pH, electrical conductivity, and nutrient availability. Land use and cover are not regularly differentiated in the soil survey, but for the recent USDA NRCS Soil Survey of New York City, land use-dependent map units were designed in order to assign more precise data on soil physical, chemical, and mineralogical properties. Such data will allow for more reliable interpretations and ratings for green infrastructure, urban agriculture, landscape architecture, and land use classifications, as well as assessment of ecosystem service delivery potential.

Opportunities for Ecosystem Service Enhancements in Cities

As mentioned previously, urban soils can provide many benefits, including C sequestration, reduced stormwater runoff, improved water quality, food production, and recreation. Not all of these outcomes, however, are aligned with one another and in some cases may even conflict. Thus, managing soil in urban areas for ecosystem services requires an interdisciplinary approach. As an example, soil scientists, ecologists, engineers, and horticulturists are actively engaged in assessing and improving the performance of bioinfiltration soil mixes for both pollutant removal and plant growth (for review, see Roy-Poirier et al. 2010). Landscape architects may engage in soil management through specifying custom blended soil mixes for various types of vegetation (Craul 1992). Urban foresters may employ engineered soils to address conflicting soil functions such as supporting pavement while allowing tree root growth (Sloan et al. 2012). As the many benefits of soils are better understood and urban design becomes more multifunctional, the engagement of diverse disciplines in soils research and practice can be expected to expand. There are a number of examples in which scientists, practitioners, and residents have worked together to address the enhancement of ecosystem services provided by urban soils, including the recycling of municipal waste as a soil amendment, engineered soils associated with green roofs and street pits, and the use of plants to maintain or restore the diversity of organisms inhabiting the soil. Here we provide a few examples.

Recycling Municipal Waste to Enhance Urban Soils

Every municipality is charged with handling waste materials that can be used to enhance soils and associated ecosystem functions. Examples of these materials include the organic component of municipal solid waste and the liquid and solid residuals from wastewater treatment. Soils themselves can also be waste materials from construction and dredging projects and road maintenance operations. However, departments within a municipal infrastructure charged with handling wastes and those that could potentially use these same materials are often segregated, with minimal incentive to find common purpose. Even when overlapping interests are found within a single division, there is often no recognition or action based on this overlap. Benefits can be optimized by identifying appropriate end uses for soil-building material and optimizing those uses within the municipal infrastructure, but this requires interdepartmental cooperation and may also involve nontraditional stakeholders. One example of this is the Harvest Pierce County program in Tacoma and Pierce Counties in Washington, a program that manages approximately 80 community gardens and a farm that utilize biosolids and compost as soil amendments (Box 7.1).

Soil Amendments to Reduce Contaminant Bioavailability

Although soils can provide many ecosystem services, soil contaminants can be harmful. Soils can serve as a sink for legacy pollution and are a potential source of exposure to human populations if ingested or inhaled (Schwarz et al. 2012). In some cases, especially in postindustrial urban areas, pollution is widespread, making the removal and disposal of soil impractical (Farfel et al. 2005). As an alternative, researchers have looked at the role that amendments may play in mitigating potential risk to human populations (Kumpiene et al. 2008). In the case of metal(loid)s, amendments can be used to either immobilize or mobilize pollutants by changing the bioavailability of the contaminant (Bolan et al. 2014). For example, chelating agents can be used to mobilize heavy metal(loid)s to encourage plant uptake (Bolan et al. 2014). More commonly, amendments are used to immobilize contaminants to reduce uptake by biological systems. The addition of phosphorus-containing amendments or fertilizers has been of great interest due to their capacity to form pyromorphite, which demonstrates low solubility (Kumpiene et al. 2008; Ryan et al. 2001). The addition of organic matter via compost or biosolids may also tightly bind Pb, making it less bioavailable (Farfel et al. 2005). In addition to changing the bioavailability of pollutants, amendments may also address potential exposure by (1) diluting the concentration of pollutants in the soil and (2) promoting the growth of vegetation, which can serve as a protective barrier between contaminated soil and people. Managing soil contaminants through the use of amendments

has the potential to enhance the ecosystem service of pollutant retention and restore land uses such as recreation and food production (Schwarz et al. 2016). However, unlike soil removal, the process can be slow and dynamic, making sustained maintenance and monitoring a necessity.

Green Roofs: An Opportunity for Ecosystem Service Enhancement

Soils for green roofs are typically engineered substrates designed to be lightweight while still supporting plant life, although natural mineral soils are used in some instances, especially for rooftop farming (e.g., Brooklyn Grange, https://www.brooklyngrangefarm.com). Green roofs provide multiple opportunities for ecosystem service enhancement via soils. Roofs cover 20–25% of some cities (Akbari and Rose 2008) and offer a significant area for new soils to provide ecosystem services and habitat for plants and other biota. Soils and plants on green roofs can mitigate large amounts of stormwater, although green roofs in general cost more than other stormwater controls. They may also be a source for nutrient or pollutant export if green roof soil-plant systems are not carefully designed (Seidl et al. 2013). Furthermore, because green roof soils often require very specific physical properties while maintaining a relatively low density, they have been created from a variety of recycled materials, including roof tiles (Emilsson and Rolf 2005), bricks, glass, and paper pellets (Molineux et al. 2009). Therefore, green roof soils often require nutrient supplements, making them an avenue for utilizing urban waste products. Engineered green roof soils may accumulate a substantial amount of organic matter over time (Schrader and Böning 2006), but soil formation processes in these soils are poorly understood. Green roofs are an example of creation and evolution of novel assemblages of soil organisms. Microbes and invertebrates colonize these spaces and create unique communities over time. Additionally, the variety of green roof designs provides an excellent opportunity to investigate how habitat structure can affect biodiversity (Madre et al. 2013; McGuire et al. 2013). In several instances, arthropod diversity has been shown to be high and include rare species (Kadas 2006; MacIvor and Lundholm 2011).

Diverse Plant Communities Stabilize Ecosystem Service Enhancement

A long-term ecological paradigm is that increased biodiversity enhances ecosystem productivity and stability. Although there is debate about the mechanisms by which diversity influences ecosystem stability, foundational experiments have shown that biodiversity is an important determinant of temporal stability, consumption of limiting resources, and

Box 7.1 Harvest Pierce County Program

The Harvest Pierce County program in Tacoma and Pierce County in Washington manages community gardens and a gleaning program that now serves approximately 80 community gardens and a farm (Box Fig. 7.1). To get started, derelict land owned by the City was inventoried. Most of these properties were then converted to community gardens. Because the area has a history of soil contamination from a former metal smelter, soil testing is provided by the City. Most gardening is done in raised beds, and the materials to construct the beds are provided by the City. Gardeners are also given topsoil amended with biosolids and yard waste compost for use within the beds. Cardboard diverted from the solid waste stream is placed between beds where it acts as a barrier to potentially contaminated soils. Wood waste is diverted from the composting site and used as mulch over the cardboard, which serves as a barrier to weeds and contaminants in the soil.

Box Fig. 7.1 An example of the many ways that municipal waste recycling can enhance urban soils and provide benefits

invasibility by nonnative species (Tilman 1999). Planning and managing for diverse plant communities is considered an important aspect of ecosystem stability in cities. For example, to provide resilience against potential insect pest outbreaks or extreme weather events, current recommendations for urban forest management specify that one genus should not make up more than 5% of the urban forest canopy (Ball 2015). Yet soil disturbance from urbanization can have a homogenizing effect on planted species, especially when combined with cultural, policy, or economic pressures that may favor some species over others (Groffman et al. 2014). Relatively few tree species are well adapted to many of the soils associated with the conversion of agricultural or forested lands to urban uses. Historically, overreliance on a small number of species that are well adapted to such conditions, such as American elm (*Ulmus americana*), has contributed to outbreaks of pests, such as the Dutch elm disease fungus (*Ophiostoma novo-ulmi*), and reduced resilience. Thus, soil protection and management may have the potential to expand habitat for some species. The synergistic effects of altered climate, disturbance regimes, edaphic factors, and species introductions in urban landscapes may lead to unknown outcomes and diverse effects on ecological processes, thus requiring more research on these relationships in urban environments.

The connection between aboveground and belowground biodiversity is poorly understood in general and even more so under urban conditions. In heavily landscaped and managed urban settings, studies indicate that increasing the diversity of plants positively affects soil fauna diversity, leading to enhanced functions such as increased litter decomposition rates and soil organic matter buildup (Byrne 2007; Ossola et al. 2016). Additionally, adding trees to lawn areas, in parks, or in residential areas can lead to significant increases in soil organic matter in only a few decades (e.g., Setälä et al. 2016). Other factors such as the diversity of local habitat types and management inputs like irrigation can also be important drivers of soil biodiversity (Philpott et al. 2014; Smith et al. 2006), suggesting that managers and residents can play an important role in improving ecosystem services of urban soils (Box 7.2).

Key Findings

- Soil conditions in urban areas generally correspond to a range of anthropogenic effects, from relatively low influence (native forest or grassland soil) or indirect urban environmental effects (remnant forest stands) to those derived from human-created materials (landfills), sealed by impervious surfaces (asphalt), or altered by physical disturbances and management (residential yards).
- Despite the high levels of disturbance typically experienced by most urban soils, they, like their rural counterparts, have the potential to support plant, animal, and microbial organisms and mediate hydrological and biogeochemical cycles. The resultant communities of soil organisms are a unique combination of native species surviving or thriving in urban areas and species introduced from other regions or other continents.
- While urban soils can provide many types of ecosystem services (e.g., provisioning, regulating, supporting, and cultural), in many cases they may not be aligned with one

Box 7.2 Plant Species Diversity and Urban Soil Ecosystem Services

How does plant species diversity affect provisioning services provided by urban ecosystems? In forests adjacent to urban interstates in Louisville, KY, the characteristics of stands invaded by *Lonicera maackii*, a nonnative shrub (Box Fig. 7.2), and uninvaded stands (Box Fig. 7.3) varied, with invaded forests having half the plant species richness of uninvaded forests (Box Fig. 7.4). Soil carbon accumulation, an important ecosystem service, will likely be affected by these observed decreases in aboveground litter inputs associated with the lower plant diversity in invaded forests. To determine the impact of the decreased litter on soil C dynamics, soil carbon over a 30-year period was simulated using the CENTURY model. Results showed that soil C accumulation per year in uninvaded forests was almost four times greater than in invaded forests (Box Fig. 7.5).

Box Fig. 7.2 Urban forest in Louisville, KY, that been invaded by the nonnative shrub, *Lonicera maackii*. (Photo credit: Tara Trammell, University of Delaware)

(continued)

Box 7.2 (continued)

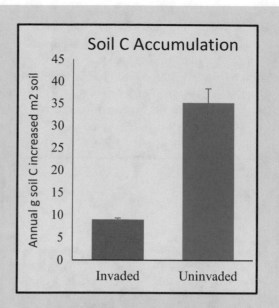

Box Fig. 7.5 Annual soil carbon accumulation (g C m⁻² soil) (mean ± SE) in invaded (9.0 ± 0.35) and uninvaded (35.2 ± 3.20) forests adjacent to urban interstates in Louisville, KY

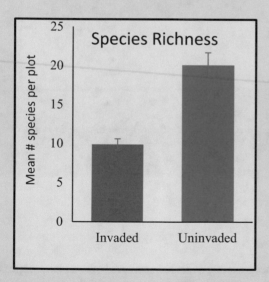

Box Fig. 7.3 Uninvaded urban forest in Louisville, KY. (Photo credit: Tara Trammell, University of Delaware)

Box Fig. 7.4 Mean number of species (±SE) per plot in invaded (9 ± 1) and uninvaded (20 ± 2) forests adjacent to urban interstates in Louisville, KY

another, and in some cases they may even conflict. Therefore, managing soil for ecosystem services in urban areas requires an interdisciplinary approach.

- In the United States, the classification and survey of soils in urban landscapes has advanced tremendously in the last 20 years, with modern soil surveys being conducted in New York City, Chicago, Los Angeles, Detroit, and elsewhere. These surveys have provided detailed information on physical, chemical, and mineralogical properties of human-altered and human-transported soils, enabling more reliable interpretations for stormwater management, revegetation and restoration efforts, urban agriculture, and better resource inventory.

- Many urban environmental effects roughly correspond to changes that are occurring in the overall global climate; therefore, urban areas have been suggested as useful analogues to study the multifactorial effects of climate change on forest and grassland ecosystems.

Key Information Needs

- Novel conditions found in urban landscapes present a challenge for soil scientists who spatially describe urban soil characteristics (e.g., Box 7.3). More data needs to be

Box 7.3 Spatial Heterogeneity of Soil Lead
The amount of lead (Pb) in soil can be highly variable, and proximity to a source is often an important driver that can explain the spatial distribution of soil Pb. Two important sources of lead include vehicular traffic (e.g., Bityukova et al. 2000) because Pb was used as a gasoline additive

until 1986 (Mielke 1999) and interior and exterior lead-based paints, which were used until 1978 (Trippler et al. 1988). Elevated soil Pb levels have been observed next to buildings (Box Fig. 7.6) regardless of building material (wood and brick), and elevated soil lead is often associated with older housing (Schwarz et al. 2012).

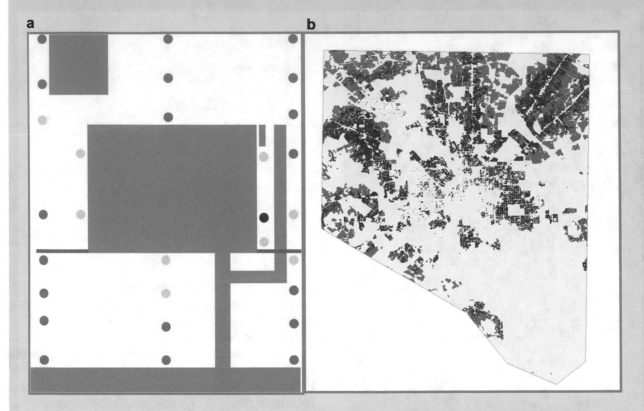

Box Fig. 7.6 Elevated soil lead (Pb) levels are often found near buildings and are especially associated with older housing. (**a**) Gray areas represent the "footprint" for built structures or pavement, green circles represent soil Pb levels below 400 ppm, and yellow and red circles represent soil Pb levels above 400 ppm. (**b**) Parcel-level measurements can be used to create predictive models; areas in red are predicted to exceed 400 ppm of soil Pb, and areas in gray are predicted to fall below 400 ppm of soil Pb. (Source: Schwarz et al. 2013)

obtained at multiple observational scales (site, city, metropolitan area) before this "new heterogeneity" is understood and effectively translated into current classification and mapping systems.

- Urban soils are often degraded with respect to the native soils that they have replaced, and likewise, they very often are covered with sealed surfaces such as asphalt and concrete pavements. More field studies are needed to quantify soil characteristics in urban landscapes as a whole and in particular beneath impervious surfaces.

- The existence of sealed surfaces in urban landscapes often necessitates the need to restore soil functions to a higher level (i.e., on a per unit area basis) than is typically found with native soils. For instance, developing a soil mix for a rain garden will require infiltration rates that are much higher than the rates of the previous native soil. Therefore, research and data are needed for the development of these "hyperfunctioning" soils.

- Studies and field observations are needed to quantify the benefits that are unique to urban soils. These may include, among others, the sustainable use of biosolids and the often unappreciated potential of using urban soils to facilitate STEM (science, technology, engineering, and mathematics) education.

- The role of soil biota in urban soil formation, nutrient cycling, and other functions needs to be addressed more explicitly.

Literature Cited

Akbari H (2002) Shade trees reduce building energy use and CO_2 emissions from power plants. Environ Pollut 116:S119–S126

Akbari H, Rose LS (2008) Urban surfaces and heat island mitigation potentials. J Human–Environ Syst 11:85–101

Arthur MA, Bray SR, Kuchle CR, McEwan RW (2012) The influence of the invasive shrub, *Lonicera maackii*, on leaf decomposition and microbial community dynamics. Plant Ecol 213:1571–1582

Ball J (2015) Using a 5 percent rule for tree selection. American Nurseryman. January 21, online

Bityukova L, Shogenova A, Birke M (2000) Urban geochemistry: a study of element distributions in the soils of Tallinn (Estonia). Environ Geochem Health 22:173–193

Bockheim JG (1974) Nature and properties of highly-disturbed urban soils, Philadelphia, Pennsylvania. Division S-5, Soil Genesis, Morphology and Classification, Annual Meeting of the Soil Science Society of America. Chicago, IL

Bolan N, Kunhikrishnan A, Thangarajan R et al (2014) Remediation of heavy metal(loid)s contaminated soils–to mobilize or to immobilize? J Hazard Mater 266:141–166

Burghardt W (2017) Main characteristics of urban soils. In: Levin MJ, KHJ K, Morel JL et al (eds) Soils within cities—global approaches to their sustainable management—composition, properties, and functions of soils of the urban environment. Schweizerbart Soil Sciences, Stuttgart, pp 19–26

Burrow C (2017) Influence of connectivity and topsoil management practices of a constructed technosol on pedofauna colonization: a field study. Appl Soil Ecol 123:416–419

Butler SM, Melillo JM, Johnson JE et al (2012) Soil warming alters nitrogen cycling in a New England forest: implications for ecosystem function and structure. Oecologia 168:819–828

Byrne LB (2007) Habitat structure: a fundamental concept and framework for urban soil ecology. Urban Ecosyst 10:255–274

Byrne LB, Bruns MA, Kim KC (2008) Ecosystem properties of urban land covers at the aboveground-belowground interface. Ecosystems 11:1065–1077

Calfapietra C, Peñuelas J, Niinemets U (2015) Urban plant physiology: adaptation-mitigation strategies under permanent stress. Trends Plant Sci 20:72–75

Capra GF, Ganga A, Grilli E et al (2015) A review on anthropogenic soils from a worldwide perspective. J Soils Sediments 15:1602–1618

Carr R, Zhang C, Moles N, Harder M (2007) Identification and mapping of heavy metal pollution in soils of a sports ground in Galway City, Ireland, using a portable XRF analyzer and GIS. Environ Geochem Health 30:42–52

Carreiro MM, Sinsabaugh RL, Repert DA, Parkhurst DF (2000) Microbial enzyme shifts explain litter decay responses to simulated nitrogen deposition. Ecology 81:2359–2365

Carreiro MM, Pouyat RV, Tripler C, Zhu W (2009) Carbon and nitrogen cycling in soils of remnant forests along urban-rural gradients: case studies in the New York metropolitan area and Louisville, Kentucky. In: McDonnell MJ, Hahs A, Breuste J (eds) Ecology of cities and towns: a comparative approach. Cambridge University Press, Cambridge, pp 308–328

Chen Y, Day SD, Wick AF (2013) Changes in soil carbon pools and microbial biomass from urban land development and subsequent post-development soil rehabilitation. Soil Biol Biochem 66:38–44

Chen Y, Day SD, Wick AF, McGuire KJ (2014) Influence of urban land development and subsequent soil rehabilitation on soil aggregates, carbon, and hydraulic conductivity. Sci Total Environ 494:329–336

Chen Y, Wang X, Jiang B, Yang N, Li L (2016) Pavement induced soil warming accelerates leaf budburst of ash trees. Urban For Urban Greening 16:36–42

Craine JM, Fierer N, McLauchlan KK (2010) Widespread coupling between the rate and temperature sensitivity of organic matter decay. Nat Geosci 3:854–857

Craul PJ (1992) Urban soil in landscape design. Wiley, New York

Craul PJ, Klein CJ (1980) Characterization of streetside soils of Syracuse, New York. Metria 3:88–101

Doolittle J, Hernandez L, Galbraith J (1997) Using ground penetrating radar to characterize a landfill site. Soil Surv Horizons 38:60–67

Drouin M, Bradley R, Lapointe L (2016) Linkage between exotic earthworms, understory vegetation and soil properties in sugar maple forests. For Ecol Manag 364:113–121

Effland WR, Pouyat RV (1997) The genesis, classification, and mapping of soils in urban areas. Urban Ecosyst 1:217–228

Ehrenfeld JG (2010) Ecosystem consequences of biological invasions. Annu Rev Ecol Evol Syst 41:59–80

Emilsson T, Rolf K (2005) Comparison of establishment methods for extensive green roofs in southern Sweden. Urban For Urban Greening 3:103–111

Evans CV, Fanning DS, Short JR (2000) Human-influenced soils. Agron Monogr 39:33–67

Farfel MR, Orlova AO, Chaney RL et al (2005) Biosolids compost amendment for reducing soil lead hazards: a pilot study of Orgro® amendment and grass seeding in urban yards. Sci Total Environ 340:81–95

Findlay S, Carreiro M, Krischik V, Jones C (1996) Effects of damage to living plants on leaf litter quality. Ecol Appl 6:269–275

Garthwaite RL, Lawson R, Sassaman C (1995) Population genetics of *Armadillidium vulgare* in Europe and North America. Crustacean Issues 9:145–199

Grabosky J, Bassuk N, Marranca BZ (2002) Preliminary findings from measuring street tree shoot growth in two skeletal soil installations compared to tree lawn plantings. J Arboric 28:106–108

Gregg JW, Jones CG, Dawson TE (2003) Urbanization effects on tree growth in the vicinity of New York City. Nature 424:183–187

Groffman PM, Williams CO, Pouyat RV et al (2009) Nitrate leaching and nitrous oxide flux in urban forests and grasslands. J Environ Qual 38:1848–1860

Groffman PM, Cavender-Bares J, Bettez ND et al (2014) Ecological homogenization of urban USA. Front Ecol Environ 12:74–81

Heidt V, Neef M (2008) Benefits of urban green space for improving urban climate. In: Carreiro M, Song Y, Wu J (eds) Ecology, planning, and management of urban forests. Springer, New York

Herrmann DL, Shuster WD, Garmestani AS (2017) Vacant urban lot soils and their potential to support ecosystem services. Plant Soil 413(1–2):45–57

Hobbs RJ, Arico S, Aronson J et al (2006) Novel ecosystems: theoretical and management aspects of the new ecological world order. Glob Ecol Biogeogr 15:1–7

Hornung E, Szlavecz K (2003) Establishment of a Mediterranean isopod (*Chaetophiloscia sicula* Verhoeff, 1908) in a North American temperate forest. Crustaceana Monogr 2:181–189

Howard JL, Orlicki KM (2015) Effects of anthropogenic particles on the chemical and geophysical properties of urban soils, Detroit. Mich Soil Sci 180:154–166

Huot H, Simonnot M-A, Morel JL (2015) Pedogenetic trends in soils formed in technogenic parent materials. Soil Sci 180:182–192

Jo I, Fridley JD, Frank DA (2015) Linking above- and belowground resource use strategies for native and invasive species of temperate deciduous forests. Biol Invasions 17:1545–1554

Jochner S, Menzel A (2015) Urban phenological studies—past, present, future. Environ Pollut 203:250–261

Kadas G (2006) Rare invertebrates colonizing green roofs in London. Urban Habitats 4:66–86

Kaushal SS, Belt KT (2012) The urban watershed continuum: evolving spatial and temporal dimensions. Urban Ecosyst 15(2):409–435

Korsós Z, Hornung E, Kontschán J, Szlavecz K (2002) Isopoda and Diplopoda of urban habitats: new data to the fauna of Budapest. Annales Historico Naturales Musei Nationalis Hungarici 94:193–208

Kourtev PS, Ehrenfeld JG, Häggblom M (2003) Experimental analysis of the effect of exotic and native plant species on the structure and function of soil microbial communities. Soil Biol Biochem 35(7):895–905

Kumpiene J, Lagerkvist A, Maurice C (2008) Stabilization of As, Cr, Cu, Pb and Zn in soil using amendments–a review. Waste Manag 28:215–225

Law NL, Band LE, Grove JM (2004) Nutrient input from residential lawn care practices. J Environ Manag 47:737–755

Lehmann A, Stahr K (2007) Nature and significance of anthropogenic urban soils. J Soils Sediments 7:247–260

Liao C, Peng R, Luo Y et al (2008) Altered ecosystem carbon and nitrogen cycles by plant invasion: a meta-analysis. New Phytol 177:706–714

Lindroth CH (1957) The faunal connections between Europe and North America. Wiley, New York. 344 p

Lovett G, Tranor M, Pouyat RV et al (2000) N deposition along an urban-rural gradient in the New York City metropolitan area. Environ Sci Technol 34:4294–4300

MacIvor JS, Lundholm J (2011) Insect species composition and diversity on intensive green roofs and adjacent level-ground habitats. Urban Ecosyst 14:225–241

Madre F, Vergnes A, Machon N, Clergeau P (2013) A comparison of 3 types of green roof as habitats for arthropods. Ecol Eng 57:109–117

McGuire KL, Payne SG, Palmer MI et al (2013) Digging the New York City skyline: soil fungal communities in green roofs and city parks. PLoS One 8(3):e58020

McKinney ML (2006) Urbanization as a major cause of biotic homogenization. Biol Conserv 127:247–260

Mielke HW (1999) Lead in the inner cities. Am Sci 87:62–73

Milesi C, Running SW, Elvidge CD et al (2005) Mapping and modeling the biogeochemical cycling of turf grasses in the United States. Environ Manag 36:426–438

Molineux CJ, Fentiman CH, Gange AC (2009) Characterizing alternative recycled waste materials for use as green roof growing media in the U.K. Ecol Eng 35:1507–1513

Morel JL, Chenu C, Lorenz K (2015) Ecosystem services provided by soils of urban, industrial, traffic, mining, and military areas (SUITMAs). J Soils Sediments 15:1659–1666

Morel JL, Burghardt W, Kim K-HJ (2017) The challenges for soils in the urban environment. In: Levin MJ, Kim K-HJ, Morel JL et al (eds) Soils within cities—global approaches to their sustainable management—composition, properties, and functions of soils of the urban environment. Schweizerbart Soil Sciences, Stuttgart, pp 1–6

Morton TG, Gold AJ, Sullivan WM (1988) Influence of overwatering and fertilization on nitrogen losses from home lawns. J Environ Qual 17:124–130

Niyogi D, Pyle P, Lei M et al (2011) Urban modification of thunderstorms: an observational storm climatology and model case study for the Indianapolis urban region. J Appl Meteorol Climatol 50:1–13

Oke TR (1990) The micrometeorology of the urban forest. Q J R Meteorol Soc 324:335–349

Osmond DL, Hardy DH (2004) Characterization of turf practices in five North Carolina communities. J Environ Qual 33:565–575

Ossola A, Hahs AK, Nash MA, Livesley SJ (2016) Habitat complexity enhances comminution and decomposition processes in urban ecosystems. Ecosystems 19:927–941

Pardo LH, Fenn ME, Goodale CL et al (2011) Effects of nitrogen deposition and empirical nitrogen critical loads for ecoregions of the United States. Ecol Appl 21:3049–3082

Pavao-Zuckerman MA (2012) Urbanization, soils and ecosystem services. In: Wall DH, Bardgett RD, Behan-Pelletier V, Herrick JE, Jones TH, Ritz K, Six J, Strong DR, van der Putten WH (eds) Soil ecology and ecosystem services. Oxford University Press, Oxford, pp 270–281

Philpott SM, Cotton J, Bichier P et al (2014) Local and landscape drivers of arthropod abundance, richness and trophic composition in urban habitats. Urban Ecosyst 17:513–532

Piotrowska-Dlugosz A, Charzynski P (2015) The impact of the soil sealing degree on microbial biomass, enzymatic activity, and physiochemical properties in the Ekranic Technosols of Torun (Poland). J Soils Sediments 15:47–59

Poland TM, McCullough DG (2006) Emerald ash borer: invasion of the urban forest and the threat to North America's ash resource. J For 104(3):118–124

Pouyat RV, Carreiro MM (2003) Contrasting controls on decomposition of oak leaf litter along an urban-rural land use gradient. Oecologia 135:288–298

Pouyat RV, Groffman PM, Yesilonis I, Hernandez L (2002) Soil carbon pools and fluxes in urban ecosystems. Environ Pollut 116:107–118

Pouyat RV, Yesilonis I, Russell-Anelli J, Neerchal NK (2007) Soil chemical and physical properties that differentiate urban land-use and cover. Soil Sci Soc Am J 71(3):1010–1019

Pouyat RV, Carreiro MM, Groffman PM, Zuckerman M (2009) Investigative approaches to urban biogeochemical cycles: New York metropolitan area and Baltimore as case studies. In: McDonnell MJ, Hahs A, Breuste J (eds) Ecology of cities and towns: a comparative approach. Cambridge University Press, Cambridge/Oxford, pp 329–354

Pouyat RV, Szlavecz K, Yesilonis I et al (2010) Chemical, physical, and biological characteristics of urban soils. In: Aitkenhead-Peterson J (ed) Urban ecosystem ecology. Agronomy Society of America; Crop Science Society of America; Soil Science Society of America, Madison, pp 119–152

Pouyat RV, Setälä H, Szlavecz K et al (2017) Introducing GLUSEEN: a new open access and experimental network in urban soil ecology. J Urban Ecol 3(1):jux002

Raciti SM, Hutyra LR, Finzi AC (2012) Depleted soil carbon and nitrogen pools beneath impervious surfaces. Environ Pollut 164:248–251

Rao P, Hutyra LR, Raciti SM, Templer PH (2014) Atmospheric nitrogen inputs and losses along an urbanization gradient from Boston to Harvard Forest, MA. Biogeochemistry 121:229–245

Rossiter DG (2007) Classification of urban and industrial soils in the World Reference Base for Soil Resources. J Soils Sediments 7:96–100

Roy-Poirier A, Champagne P, Filion Y (2010) Review of bioretention system research and design: past, present, and future. J Environ Eng 136:878–889

Royte E (2009) The street farmer. The New York Times Sunday Magazine. July 1: MM22

Ryan JA, Zhang P, Chou J, Sayers DE (2001) Formation of chloropyromorphite in a lead-contaminated soil amended with hydroxyapatite. Environ Sci Technol 35:3798–3803

Sackett TE, Smith SM, Basiliko N (2013) Indirect and direct effects of exotic earthworms on soil nutrient and carbon pools in North American temperate forests. Soil Biol Biogeochem 57:459–467

Scalenghe R, Marsan FA (2009) The anthropogenic sealing of soils in urban areas. Landsc Urban Plan 90:1–10

Scharenbroch BC, Johnston DP (2011) A microcosm study of the common night crawler earthworm (*Lumbricus terrestris*) and physical, chemical and biological properties of a designed urban soil. Urban Ecosyst 14:119–134

Scharenbroch B, Day S, Trammell T, Pouyat RV (2018) Chapter 6: Urban soil carbon storage. In: Lal R, Stewart BA (eds) Urban soils. Advances in soil sciences, Boca Raton, CRC Press

Schrader S, Böning M (2006) Soil formation on green roofs and its contribution to urban biodiversity with emphasis on collembolans. Pedobiologia 50:347–356

Schwartz SS, Smith B (2016) Restoring hydrologic function in urban landscapes with suburban subsoiling. J Hydrol 543:770–781

Schwarz K, Pickett STA, Lathrop RG et al (2012) The effects of the urban built environment on the spatial distribution of lead in residential soils. Environ Pollut 163:32–39

Schwarz K, Weathers KC, Pickett STA et al (2013) A comparison of three empirically-based, spatially explicit predictive models of residential soil Pb concentrations in Baltimore, Maryland USA: understanding the variability within cities. Environ Geochem Health 35:495–510

Schwarz K, Pouyat RV, Yesilonis ID (2016) Legacies of lead in charm city's soil: lessons from the Baltimore Ecosystem Study. Int J Environ Res Public Health 13:209

Seidl M, Gromaire MC, Saad M, De Gouvello B (2013) Effect of substrate depth and rain-event history on the pollutant abatement of green roofs. Environ Pollut 183:195–203

Setälä H, Birkhofer K, Brady M et al (2014) Urban and agricultural soils: conflicts and trade-offs in the optimization of ecosystem services. Urban Ecosyst 17:239–253

Setälä HK, Francini G, Allen JA et al (2016) Vegetation type and age drive changes in soil properties, nitrogen, and carbon sequestration in urban parks under cold climate. Front Ecol Evol 4:93

Shaw RK, Isleib JT (2017) The case of the New York City Soil Survey Program, United States. In: Levin MJ, K-HJ K, Morel JL et al (eds) Soils within cities: global approaches to their sustainable management. Schweizerbart, Catena Soil Sciences, Stuttgart, pp 107–113

Shem W, Shepherd M (2009) On the impact of urbanization on summertime thunderstorms in Atlanta: two numerical model case studies. Atmos Res 92:172–189

Shepherd JM, Burian SJ (2003) Detection of urban-induced rainfall anomalies in a major coastal city. Earth Interact 7:1–14

Short JR, Fanning DS, Foss JE, Patterson JC (1986) Soils of the mall in Washington, DC: I Statistical summary of properties. Soil Sci Soc Am J 50:699–705

Shuster WD, Dadio S (2017) Chapter 12: An applied hydropedological perspective on the rendering of ecosystem services from urban soils. In: Lal R, Stewart BA (eds) Urban soils. Advances in soil science. CRC Press, Boca Raton

Shuster WD, Dadio S, Drohanc P et al (2014) Residential demolition and its impact on vacant lot hydrology: implications for the management of stormwater and sewer system overflows. Landsc Urban Plan 125:48–56

Sloan JJ, Ampim PAY, Basta NT, Scott R (2012) Addressing the need for soil blends and amendments for the highly modified urban landscape. Soil Sci Soc Am J 76:1133

Smetak KM, Johnson-Maynard JL, Lloyd JE (2007) Earthworm population density and diversity in different-aged urban systems. Appl Soil Ecol 37:161–168

Smith RM, Warren PH, Thompson K, Gaston KJ (2006) Urban domestic gardens (VI): environmental correlates of invertebrate species richness. Biodivers Conserv 15:2415–2438

Soil Survey Staff (2014) Keys to soil taxonomy, 12th edn. U.S. Department of Agriculture, Natural Resources Conservation Service, Washington, DC

Song Y, Liu H, Wang X et al (2016) Numerical simulation of the impact of urban non-uniformity on precipitation. Adv Atmos Sci 33:783–793

Spoor G, Tijink F, Weisskopf P (2003) Subsoil compaction: risk, avoidance, identification and alleviation. Soil Tillage Res 73:175–182

Szlavecz K, Placella SA, Pouyat RV et al (2006) Invasive earthworm species and nitrogen cycling in remnant forest patches. Appl Soil Ecol 32:54–63

Szlavecz K, McCormick M, Xia L et al (2011) Ecosystem effects of nonnative earthworms in mid-Atlantic deciduous forests. Biol Invasions 15:1165–1182

Szlavecz K, Yesilonis I, Pouyat R (2018) Soil as foundation for urban biodiversity. In: Ossola A, Niemelä J (eds) Urban biodiversity: from research to practice. Taylor and Francis, London, pp 18–36

Taha H (1997) Urban climates and heat islands: albedo, evapotranspiration, and anthropogenic heat. Energy Buildings 25:99–103

Tilman D (1999) The ecological consequences of changes in biodiversity: a search for general principles. Ecology 80:1455–1474

Trammell TLE, Ralston HA, Scroggins SA, Carreiro MM (2012) Foliar production and decomposition rates in urban forests invaded by the exotic invasive shrub, *Lonicera maackii*. Biol Invasions 14:529–545

Trammell TLE, Pataki DE, Cavender-Bares J et al (2016) Plant nitrogen concentration and isotopic composition in residential lawns across seven US cities. Oecologia 181:271–285

Trippler DJ, Schmitt MDC, Lund GV (1988) Soil lead in Minnesota. In: Davies BE, Wixson BG (eds) Lead in soil: issues and guidelines. Environmental geochemistry and health. Science Reviews, Northwood

U.S. Census Bureau (2010) Annual estimates of the resident population: April 1, 2010 to July, 2013. http://www.census.gov/population/metro/data/index.html. Accessed 24 Oct 2014

United Nations (2015) World population prospects: the 2015 revision, key findings and advance tables. Working Paper ESA/P/WP.241. United Nations, Department of Economic and Social Affairs, Population Division, New York

USDA Natural Resources Conservation Service (USDA NRCS) (2001) Summary report: 1997 National Resources Inventory (revised December 2000). U.S. Department of Agriculture, Natural Resources Conservation Service/Iowa State University, Statistical Laboratory, Washington, DC/Ames. 89 p

Washbourne CL, Lopez-Capel E, Renforth P, Ascough PL (2015) Rapid removal of atmospheric CO_2 by urban soils. Environ Sci Technol 49:5434–5440

Wessolek G (2008) Sealing of soils. In: Marzluff J, Shulenberger E, Endlicher W et al (eds) Urban ecology, an international perspective on the interaction between humans and nature. Springer, New York, pp 161–179

Wolf JM, Gibbs JP (2004) Silphids in urban forests: diversity and function. Urban Ecosyst 7:371–384

Xu S, Xu W, Chen W et al (2016) Leaf phenological characters of main tree species in urban forest of Shenyang. PLoS One 9:e99277

Yesilonis ID, Pouyat RV (2012) Carbon stocks in urban forest remnants: Atlanta and Baltimore as case studies. In: Lal R, Augustin B (eds) Carbon sequestration in urban ecosystems. Springer, Dordrecht, pp 103–120

Youngsteadt E, Henderson RC, Savage AM et al (2015) Habitat and species identity, not diversity, predict the extent of refuse consumption by urban arthropods. Glob Chang Biol 21:1103–1115

Zemlyanitskiy LT (1963) Characteristics of the soils in the cities. Soviet Soil Sci 5:468–475

Zhao L, Lee X, Smith RB, Oleson K (2014) Strong contributions of local background climate to urban heat islands. Nature 511:216–221

Soil Management and Restoration

8

Mary I. Williams, Cara L. Farr, Deborah S. Page-Dumroese,
Stephanie J. Connolly, and Eunice Padley

Introduction

The destruction of soil is the most fundamental kind of economic loss which the human race can suffer.—The Essential
Aldo Leopold: Quotations and Commentaries

Soils sequester carbon (C), store and regulate water, cycle
nutrients, regulate temperatures, decompose and filter waste,
and support life (Dominati et al. 2010). We depend, and will
continue to depend, on these ecosystem services provided by
soils, services that are products of interactions between and
among abiotic and biotic properties and that are the foundation for self-maintenance in an ecosystem (SER 2004).

But, soil is a limited resource. It takes thousands of years
to develop soil, yet it can lose its productive capacity and
ecological integrity in a fraction of that time as the result of
human activities or natural events (Heneghan et al. 2008;
Hillel 2004). The impacts of management actions and natural events can remain on the landscape for decades and longer, leaving land use and historical legacies (Foster et al.
2003; Morris et al. 2014) that can cause profound ecological
and economic consequences from lost farm, pasture, or forest productivity. Furthermore, climate shifts and environmental stressors affect soil properties and functions, both
directly and indirectly. Rises in temperature affect decomposition and nutrient cycling, biological populations, and soil
hydrologic functions. Flooding is a natural disturbance in
riparian and floodplain ecosystems, but flood sizes and fre-
quencies have been altered by human influences through
damming and channelizing rivers, draining wetlands, and
deforesting floodplains, so that most flooding now often
exceeds the natural range of variation.

As natural resources become limited, the value of managing and restoring aboveground and belowground processes becomes more important. Sustainable soil
management involves the concepts of using, improving,
and restoring the productive capacity and processes of soil
(Lal and Stewart 1992), and we can use ecological restoration, which is intimately linked with soil management, to
ameliorate degraded and disturbed resources, reverse the
trends of soil degradation, and enhance soil properties to
regain ecosystem health. Ecological restoration is one of
several actions that can ameliorate degraded and disturbed
soils, defined as "the process of assisting the recovery of an
ecosystem that has been degraded, damaged, or destroyed"
(SER 2004). The practice of ecological restoration draws
from and integrates many disciplines from agronomy to
wildlife management and from engineering to indigenous
knowledge.

Concerns about ecosystem services (e.g., food, water,
energy, biodiversity conservation) within the context of a
changing climate lead to calls for action, research projects, and eventually the development of new management
and restoration techniques (Adhikari and Hartemink
2016; McBratney et al. 2014). This chapter begins with a
summary of historical forest and rangeland management
with respect to soils and is followed by an overview of the
shifts in policy and planning and advances in management
and restoration. We highlight a few case studies, discuss
monitoring, and end with key findings and information
needs.

M. I. Williams (✉)
Nez Perce Tribe Wildlife Division, Lapwai, ID, USA
e-mail: maryw@nezperce.org

C. L. Farr
U.S. Department of Agriculture, Forest Service, Pacific Northwest
Region, Portland, OR, USA

D. S. Page-Dumroese
Rocky Mountain Research Station, USDA Forest Service,
Moscow, ID, USA

S. J. Connolly
U.S. Department of Agriculture, Forest Service, Northern Research
Station, Newtown Square, PA, USA

E. Padley
U.S. Department of Agriculture, Natural Resources Conservation
Service, Washington, DC, USA

© The Author(s) 2020
R. V. Pouyat et al. (eds.), *Forest and Rangeland Soils of the United States Under Changing Conditions*,
https://doi.org/10.1007/978-3-030-45216-2_8

Context

> Humans are now an order of magnitude more important at moving sediment than the sum of all other natural processes operating on the surface of the planet.—Wilkinson (2005)

Historical Forest Soil Management

The early history of forests in North America does not include a record of soil impacts, but we can infer some aspects of this from land use and population factors. As Europeans began arriving in North America during the mid-seventeenth century, total forest area was estimated at 4.14 million km² (1023 million acres) (Oswalt et al. 2014). Native peoples were soon decimated by disease, and much of their agricultural land reverted to forest naturally (Lewis and Maslin 2015; Mann et al. 1988). The extent of forest cover in what is now the United States, as reconstructed from saw timber inventories, was likely greater from 1650 to 1700 than in any other period (Birdsey et al. 2006). Williams (1989) suggested that forests may have covered "at least four-fifths of the land area east of the Mississippi River." Population increased slowly during the 1700s but jumped from 5.3 million to 76 million people in the 1800s (Fedkiw 1989; MacCleery 2004). To support the increase in population growth, settlers cleared approximately 0.77 million forested km² for farms and pastures between 1850 and 1910, more area than had been cleared in the previous 250 years (Williams 1989). Following the period of intensive land clearing, the remaining 3.05 million km² of forest (Oswalt et al. 2014) was largely saved by technological advances; land used to feed draft animals became available for other crops as animal labor was replaced by motorized equipment, and agricultural production per area boomed as a result of plant breeding, irrigation, and fertilizer. In 2012, forested area has increased only slightly to 3.1 million km² (766 million acres) since 1910 (Oswalt et al. 2014).

Until recently, forest soil management has been characterized primarily by inattention or risk avoidance. Inattention prevailed throughout the 1800s in the absence of regulations, conservation practices, or foresters (Fedkiw 1989; MacCleery 1993); this period of rapid land clearance was also a time of high demand for wood. Land clearing and extraction from remaining forests resulted in an alarming 75% drawdown of sawtimber stocks between 1800 and 1920 (Birdsey et al. 2006), leading to the first forest conservation policies in the 1930s (MacCleery 1993). Wood was enormously important during the nineteenth century as a fuel source for heat and steam power; it was the only material for building fences up until the mid-1800s (MacCleery 1993); and it also served a significant industrial demand for charcoal (Foster and Aber

2004). After 1850, lumber production increased quickly as cities were constructed and farmsteads were built in the Great Plains; railroad expansion also used a large proportion of harvested wood (MacCleery 1993).

Timber extraction prior to the Civil War typically focused on removing high-value trees close to waterways, as logs were heavy and difficult to move (Fedkiw 1989). Logging was mainly accomplished by hand-felling trees and skidding the logs with oxen or horse teams. It was a cumbersome process that left most of the forest unaffected, although damage to streambanks and streams was serious, and effects still exist today (Sedell et al. 1991). When railroads came into widespread use in the latter half of the 1800s, destructive logging practices affected larger areas. The most widespread soil damage of that period was caused by fires ignited by sparks from locomotives. Wildfires ripped through logging slash, destroying the organic horizons of soils and leading to postfire erosion, sedimentation, and nutrient loss (Fedkiw 1989; MacCleery 2004). In the western United States, general policies to pile and burn logging slash were adopted to prevent such wildfires (Lyman 1947). However, slash pile burning comes with its own set of short- and long-term consequences to soil properties that long-term research would later identify (reviewed in Rhodes and Fornwalt 2015). In some regions, farming was attempted on unsuitable logged-over soils that eventually reverted to forest.

Soil fertility, as a component of forest site quality, was gradually recognized during the 1920s and 1930s, and research efforts during the following decades focused on matching forest species to site. The use of site index as a rough measure of productive capacity and the construction of yield tables led to the acceptance of the two soil facts: Soils differ widely in their ability to support tree growth and vegetative growth overall; and forest management as well as other types of resource management, such as wildlife management, could be informed and made more productive by a knowledge of soil (Leopold 1933; Wilde 1958).

Forestry operations moved toward mechanization after World War II (WWII) and progressed in stages as technology changed. Simmons wrote in 1949 that the "tractor, the power saw, and the motortruck are becoming commonplace throughout the country, even on small logging jobs." Truck hauling replaced railroads for moving logs to mills, and ground skidders replaced horse teams for dragging logs to pickup points. Road and skid trail layout became important, utilizing engineering techniques to stabilize roadbeds and manage hydrology. This was primarily to maintain longevity of the roads but also served to limit soil movement and loss. With the advent of heavier tractors and specialized logging equipment, soil compaction became a widespread problem. In the 1960s, researchers began to study soil compaction as a factor responsible for decreased forest growth. Insights as to how soil and ecosystem properties may have been changed by historical logging, fire, and in some locations subsequent tillage or grazing are gradually becoming available. The recognition that soil porosity was a nationwide forestry concern led to including compaction treatments in the North American Long-Term Soil Productivity (LTSP) study initiated in 1989 (Powers et al. 2005), a research effort which has gathered long-term data on the effects and interactions of compaction and organic matter removal relative to forest growth.

Another forest management issue in the post-WWII era was the realization that removing trees from the forest was akin to harvesting crops and that forest soils, as well as agricultural soils, might be susceptible to nutrient loss. As new laboratory techniques developed, researchers were able to quantify amounts of nutrients in various parts of trees, and tree species, and calculate nutrient budgets that estimated removals in timber harvesting (e.g., Perala and Alban 1982). This research led to guidance on limiting certain types of harvests and on the use of forest fertilization in economically viable situations. One of the first reported forest fertilization successes was at the Charles Lathrop Pack Demonstration Forest in Washington, where researchers documented a positive response using potassium fertilization on red pine (*Pinus resinosa*) (Heiberg et al. 1964). Fertility concerns, along with improved laboratory equipment and techniques and the advent of soil surveys in forested areas, led to further examination of forest soils and an increased recognition of the importance of organic material in the forest litter layer and upper mineral soil layers. Nutrient changes have been studied in many locations, notably at Hubbard Brook Experimental Forest in New Hampshire, and studies elucidating soil microbial functions have recently provided insights on historical impacts and considerations for future soil management (Jangid et al. 2011).

Historical Rangeland Soil Management

In the United States, rangelands occupy approximately 35% of the land area (Reeves and Mitchell 2011); major rangelands include the Great Plains, the Desert Southwest, the Great Basin, and the Intermountain Plains and Valleys. Rangeland is "land on which the indigenous vegetation (climax or natural potential) is predominately grasses, grass-like plants, forbs, or shrubs and is managed as a natural ecosystem. If plants are introduced, they are managed similarly. Rangeland includes natural grasslands, savannas, shrublands, many deserts, tundras, alpine communities, marshes and meadows" (SRM 1998). The majority of rangelands are categorized as drylands, which are lands limited by soil water (Hassan et al. 2005) with soils having, in most cases, low organic matter, low fertility, high accumulations of calcium carbonate, and low nutrient resources, such as available nitrogen (N) and phosphorus (P) (Sharma et al. 1992; Stott

and Martin 1989). Despite these limitations, rangelands are expansive and heterogeneous, supporting a diversity of ecosystems that provide ecological, social, and economical services.

Livestock grazing was, and remains to this day, a primary land use of rangelands during the nineteenth and twentieth centuries. Several rangeland research stations, including the Jornada Experimental Range in New Mexico (circa 1912), were developed in the early twentieth century to address and keep pace with the unprecedented and intensive livestock grazing practices in the western United States. Early management was built on assumptions of equilibrium ecology and steady-state management (i.e., the range condition model) (Clements 1916; Dyksterhuis 1949; Sampson 1923), which presumed that livestock grazing controlled plant succession, such that the species composition of plant communities was a linear response to grazing intensity (Briske et al. 2005). For 50 years, the range condition model was the standard protocol and worked moderately well in grasslands dominated by perennial herbaceous forbs and rhizomatous grasses. At the turn of the twentieth century, however, rangeland conditions were considered poor (Gardner 1991), mostly due to improper livestock grazing practices and uninformed management. The inability of the range condition model to account for complex vegetation dynamics such as woody plant encroachment and establishment and spread of nonnative plant species (Westoby et al. 1989), coupled with advances in resilience and state and transition concepts and theories (Briske et al. 2003; Friedel 1991; Holling 1973; Westoby et al. 1989), led to comprehensive reviews of the rangeland profession and management of rangelands. Reviews by the Natural Research Council (1994) and the Society for Range Management (1995) called for and outlined the standardization of monitoring and replacement of the range condition model with a model that could account for multiple states within and across plant communities (i.e., state and transition models) (Briske et al. 2005; Westoby et al. 1989).

Progressive Shifts in Policy and Planning

The starting point must be the soil, or at least the substrate into which plants must establish and root, for although soil can exist without plants, there are few plants that can exist without soil.— Bradshaw (1987)

As we look back on the history of forest and range management, it is apparent that a number of changes in soil management and protection have arisen from new information made possible by advances in research and technology. In recognition of the relationship between aboveground and belowground processes, and with a better understanding of management impacts as well as natural disturbance and recovery processes, several approaches and even shifts in policy have been discussed and developed. The soil ecological knowledge (SEK) approach, for example, acknowledges interactions among principal components of the soil systems as well as feedbacks between aboveground and belowground ecosystem processes (Heneghan et al. 2008).

Currently, forest and rangeland soil management includes approaches that protect soils while seeking to avoid and minimize soil damage. This is still a viable and efficient management approach, as rehabilitation and restoration processes are often costly and impractical, constrained by concerns for tree damage and impeded by accessibility, substrate, and terrain. Statutes, regulations, and guidelines at the federal and state levels have been developed to address many aspects of soil resource protection and management. A more proactive approach to the management and rehabilitation of forest soils has gradually been forming, assembling a variety of approaches that use soil properties to guide land management actions as well as taking direct action toward restoring desirable soil properties.

Forest Service Policy

Limiting the loss of soil productivity has been the focus of soil management on National Forest System (NFS) lands since the 1980s. Under the National Forest Management Act of 1976 (NFMA), all national forests are required to assess the impacts of management actions to ensure that they "will not produce substantial and permanent impairment of the productivity of the land." The NFMA did not define "land productivity," but the USDA Forest Service (hereafter, Forest Service), with guidance from the US Office of General Council, defined it as the capacity of a soil to produce vegetative growth. While land productivity is generally perceived as being broader—including timber, wildlife, watershed, fisheries, and recreation values—soil productivity is essential to the sustained production of all other ecosystem goods and services (Powers et al. 2005). When considering how to monitor land productivity, NFS soil scientists noted the difficulty in detecting a change in productive potential and decided that a change of approximately 15% would be a detectable threshold. Each NFS region developed soil quality standards to detect changes, and some regions developed management guidance that limited soil disturbance to no more than 15%, or 20% on an area basis, reasoning that this would protect land productivity.

Forest Service Manual (FSM) Chapter 2550 Soil Management directs soil resource management on NFS lands (USDA FS 2010). The manual was revised in 2010 to provide a greater focus on ecological functions, with an objective of maintaining or improving soil health on NFS lands "to sustain ecological processes and function so that desired ecosystem services are provided in perpetuity." The FSM defines soil quality as "the capacity of a specific kind of soil to function, within natural or managed ecosystem boundaries, to sustain plant and animal productivity, maintain or enhance water and air quality, and support human health and habitation and ecosystem health" (USDA FS 2010). Soil function is any ecological service, role, or task that soil performs. The FSM identifies six soil functions: soil biology (see also Chapter 5), soil hydrology (see also Chapter 3), nutrient cycling (see also Chapter 4), C storage (see also Chapter 1), soil stability and support, and filtering and buffering. In order to provide multiple uses and ecosystem services in perpetuity, these six soil functions need to be active and effectively working.

With the shift of National policy in the 2010s, several NFS regions and forests have adapted soil quality guidance that reflects a greater focus on managing NFS lands to maintain soil ecological functions as a foundation of planning management actions instead of focusing on soil disturbance as a proxy for maintaining productivity. For example, new regional guidance was created in 2012 for the Eastern Region (Region 9). This guidance directs soil scientists to look at the landscape and determine the site-specific soil properties that are at risk for an ecosystem response from management actions (USDA FS 2012). When high risk is found, actions on those soils are mitigated to protect soil quality and ecosystem function. Monitoring of the soil and ecosystem response on these sites is then fed back into the planning process to better inform the next round. The Colville National Forest also took a similar approach in its recent Forest Land and Resource Management Plan (USDA FS 2018). The forest linked important soil properties to the six soil functions identified in the FSM with the goal of maintaining soil function on the landscape.

Use of Ecological Sites and Associated Information

A more proactive approach to the management and rehabilitation of forest and rangeland soils has gradually been forming, assembling a variety of techniques that use soils and other ecosystem properties to guide management actions, as well as taking direct action toward restoring desirable soil properties. The need for an ecological approach has been recognized for some time. Rowe (1996) stated that, "recognition of land/water ecosystems in a hierarchy of sizes can provide a rational base for the many-scaled problems of protection and careful exploitation in the fields of forestry, agriculture, wildlife and recreation."

The ecological site concept was formed from an integration of previous land classification systems implemented in parts of Canada and Europe during the 1900s (Barnes et al. 1982), along with those previously utilized in the United States, including the Land Systems Inventory (Wertz and Arnold 1972), habitat type classifications pioneered in Finland (Cajander 1926) and widely applied in the

Northwestern United States (Daubenmire 1968), the US Soil Survey, and the Forest Service's ecological classification system (Bailey 1987; Cleland et al. 1997).

An ecological site is "a conceptual division of the landscape that is defined as a distinctive kind of land based on recurring soil, landform, geological, and climate characteristics that differs from other kinds of land in its ability to produce distinctive kinds and amounts of vegetation and in its ability to respond similarly to management actions and natural disturbances" (Caudle et al. 2013). Ecological site descriptions (ESDs) are linked to soil survey polygons and associated soil and site information, such as vegetation type and structure, disturbance processes, successional states, and rates of change, and these attributes inform management decisions for a variety of uses (Herrick et al. 2006b). Within this standardized methodology, ecological sites are the basic land classification units for documenting soil, site, and biological characteristics for current and potential future conditions (USDA 2003). Ecological site descriptions are being developed for rangelands and forests across the United States and provide a standardized communication and planning tool for land managers to assess site function and develop projects.

State and transition models (STMs) are interpretations linked to ecological sites, modeling the plant communities that typically develop in response to ecosystem drivers (Bestelmeyer et al. 2003). The STM concept is an outgrowth of concepts of ecosystem succession developed over the last century and summarized by Christensen (2014), who stated that, "We now understand that there is no single unique or unifying mechanism for successional change, that successional trajectories are highly varied and rarely deterministic, and that succession has no specific endpoint." State and transition models reflect some of the complexity involved in succession by recognizing typical disturbances and feedback loops, continuous and noncontinuous transitions between states, and alternate stable states for each ecological site. As such, they are a valuable tool in providing a framework for the probable outcomes of management decisions. State and transition models and ecological sites are linked to soil map units in a one-to-many relationship, i.e., an ecological site may occur on multiple soil map units, thus providing spatial information needed for land management planning. However, limitations of polygon-based soil maps (Zhu 2000) must be considered when applying ecological site information.

Agencies including the USDA Natural Resources Conservation Service (USDA NRCS), the Department of Interior's Bureau of Land Management (BLM), and Forest Service have jointly adopted ecological sites and STMs to aid in rangeland management (Bestelmeyer et al. 2017; Pellant et al. 2005). The agencies have developed a systematic rangeland assessment, detailed in the Interagency Ecological Site Handbook for Rangelands (Caudle et al. 2013), that incorporates STM concepts of ecosystem function specific to ecological sites.

Advances in Management and Restoration

When exposed to various management practices, soils may lose, retain, or improve their capacity to sustain plant, animal, microbial productivity, health, and vitality while also maintaining balanced hydrologic, C, nutrient cycles and ecosystem functions (Heneghan et al. 2008). Departure from natural ranges of variability due to, for example, loss of organic matter—a key driver of ecosystem function—warrants restoration. Traditional approaches focus on manipulating a single chemical, physical, or biological factor to improve soil function, such as by establishing plant cover on highly degraded sites to prevent erosion. Improved technology since the 1980s has led to rapid and significant advances in understanding soil processes and interactions with other ecosystem components, such that restoration of structure and function requires the integration of multiple factors (physical, biological, and chemical). For example, an understanding of belowground biology and processes precipitated an emphasis on soil health and the capacity of a soil to function as a vital, living ecosystem that sustains water, air, plants, animals, and humans (Doran and Zeiss 2000). Soil health is a term used to emphasize and convey that soil functions are "mediated by a diversity of living organisms that require management and conservation" (Doran and Zeiss 2000). Soil health is similar to the older term "soil quality" but places a greater emphasis on the biological components of soils and their roles in ecological processes, especially the cycling of organic matter and nutrients.

Increased knowledge of the legacies of past land use also led to recognition of degraded steady states and appropriate adjustments of land management planning objectives. The recognition of soil-mediated legacy effects can be useful for developing realistic and practical restoration goals (see Foster et al. 2003; Morris 2011; Morris et al. 2011, 2013, 2014). Another current focus is the recognition of soils as part of an ecosystem, taking on properties from air, water, plants, and minerals and contributing influences to the system. This approach is evident in planning and modeling applications and in developing indicators for sustainability (Jonsson et al. 2016). Although there is much left to discover, the new knowledge contributes in many ways to our ability to manage and restore soil functions. In the following sections, we discuss soils-based management approaches; changes in forest, fire, and mine reclamation practices; invasive species and soils; and innovative techniques in biochar, seed coatings, and soil transplants.

Soils-based Management

Properties observed in soils are a repository of information, referred to as a "soil memory" or "pedomemory" (see references in Nauman et al. 2015a). Biotic and abiotic interactions, such as those between plant communities and soil, over time can develop specific soil properties indicative of site history. Linking soil properties with historical reference plant communities is a foundation for soils-based management frameworks, such as ESDs. Ecological sites and STMs are being used to guide forest and rangeland management and restoration across the United States (Caudle et al. 2013). Mapping soil morphology and ecological sites to estimate historical plant composition has provided guidance for the restoration of historically disturbed red spruce-eastern hemlock (*Picea rubens-Tsuga canadensis*) forests in the central Appalachian Mountains (Nauman et al. 2015a, b). And in the western United States, mapping ecological sites has been useful in managing and restoring shrub-steppe habitats (Williams et al. 2011, 2016a). In the following sections, we briefly discuss other soils-based management approaches, such as resistance and resilience, soil security, and soil sensitivity.

Application of Resistance and Resilience Concepts

A framework for assessing the resistance and resilience of sagebrush (*Artemisia* spp.) and pinyon-juniper (*Pinus - Juniperus*) ecosystems to invasive species and fire threats has been developed for the western United States (Chambers et al. 2014, 2017; Miller et al. 2014). Resistance is "the ability of an area to recover from disturbance, such as wildfire or drought," and resilience is the "ability of an area of land to remain largely unchanged in the face of stress, disturbance, or invasive species" (USDOI 2015) (also see Box 5.1). The concepts are useful to land planning and assessment. Sagebrush communities on cool, moist, more productive sites are more resistant and resilient to drought and species invasions than those at the drier, warmer end of the spectrum (Chambers et al. 2014, 2017; Maestas et al. 2016). While soil differences related to drought resistance are well known in agricultural crops, the concept has not been commonly applied to natural vegetative communities. In the southern Rocky Mountains, particularly in the Four Corners and Upper Rio Grande ecoregions, sagebrush communities are likely to have higher resistance and resilience than those in the Great Basin (Chambers et al. 2014). A substantial portion of the southern Rockies is classified as having moderate to high resistance and resilience due to climate. The monsoonal precipitation pattern also promotes perennial forbs and grasses, leading to higher resistance to invasive species invasion (e.g., Bradford and Laurenroth 2006).

A similar approach is applied to managing pinyon-juniper woodland ecosystems (Miller et al. 2014). Woodlands are long-established, complex ecosystems with a canopy cover of 40% or less and a well-developed understory (Thomas and Packham 2007). Trees in woodlands are often distributed unevenly, with highly variable spatial patterning. Woodland structure is vital to the continued existence of many organisms such as birds and butterflies, and woodlands also provide ecosystem services such as water quality and quantity. Woodlands, like forests, are valued for their wood products, and tree removal could result in changes in soil nutrients, hydrology, and biodiversity.

Soil Security

The concept of soil security has emerged in the past decade as a way to communicate and elevate the urgency of maintaining and improving soil resources to support human needs, biological diversity, and ecosystem structure and function (McBratney et al. 2014). Soil security concerns are placed on par with water, food, and energy security, among other factors, that contribute to sustainability. Addressing soil security requires sustaining the capability or long-term productive capacity of soils, as well as maintaining or restoring their condition. A monetary or capital value placed on soils would ensure their consideration in land allocations based on economics. Stewardship, or connectivity of people to the land, is another aspect of achieving soil security, as is public policy and use regulation. As such, soil security includes considerations of soil health and quality but also adds human elements and synthesizes an overall approach within the context of global sustainability. An example of placing value on soils is the assessment of forest land by Oregon's Department of Revenue (ORS 321.805–855). Large tracts of privately owned forestland are classified into productivity groups using information found in the NRCS Soil Survey. Productive soils receive higher productivity classes and are assessed as more valuable for forest management.

Soil Sensitivity

Several adaptation assessments are using new information derived from soil vulnerability models. Peterman and Ferschweiler (2016) identified soil sensitivity factors that indicate increased vulnerability to drought-related erosion and vegetation loss and predicted changes in vegetation functional groups as a result of climate change. Changing temperature and precipitation regimes directly influence soils, plant productivity and phenology, and ecosystem resilience to invasive species and wildfire. Soil sensitivity is a measure of the ability of a soil to endure or recover from a disturbance. Peterman and Ferschweiler (2016) created vulnerability maps that rank soils according to the presence of vulnerability factors and the potential for change in vegetation type under future climate scenarios. Loss of plant cover and productivity due to drought and warm temperatures can lead to increased erosion and loss of important soil properties, such as water holding capacity, stability, organic matter, and nutrients (Breshears et al. 2003; Munson et al. 2011). Vulnerability maps can help managers identify priority areas for restoration and conservation. For example, Ringo and others (2018) developed a geospatial soil drought model for the Pacific Northwest using soil properties and climate information. These data are being used to examine forest resiliency, wildfire risk, and potential management actions to mitigate undesired ecosystem trajectories.

Forest Management

Forest management has changed in the modern era due to the development of new technologies as well as a better understanding of the ecological impacts of management actions. One major technological change is the mechanization of forest operations, with equipment replacing manpower in harvesting, site preparation, and planting (Silversides 1997). When applied properly, new technology can decrease forestry impacts on soils; for example, the use of low-pressure tires and tracked vehicles produces less compaction, and equipment like the "shovel logger" can reach out as far as 100 m from a road to fell and move trees. Other technology being used in the United States includes tethered logging systems, which can harvest on slopes up to 80% in a ground-based system and can provide access to sites with greater limitations.

Compaction is still the most serious issue today on many forest sites because it alters soil porosity, reduces infiltration, and can increase erosion, thereby leading to reduced movement of water, air, and nutrients through the soil; it also impacts microbial populations and can reduce tree growth (Brussard and van Faasen 1994; Bulmer and Simpson 2005; Page-Dumroese et al. 2009; Stone 2002; Thibodeau et al. 2000; von Wilpert and Schäffer 2006; Wang et al. 2005). Depending on soil texture, the depth of compaction, and site resiliency, soil compaction can also alter plant successional pathways and overall productivity. Soil compaction is greatest in roads, trails, and landings but can occur in general harvest areas, depending on the weight of equipment, number of passes, soil wetness, and other factors. The upper few centimeters of organic soil can recover quickly from light to moderate compaction (Adams 1991; Burger et al. 1985; Hatchell and Ralston 1971; Kozlowski 1999). The weight of logging equipment used in harvesting and site preparation activities increase soil bulk density by compressing soil macropores. Typically, about three passes of heavy equipment are needed to cause a significant increase in soil compaction (Williamson and Neilsen 2000). The change in pore space diminishes root

access to gas exchange, may result in increasingly anaerobic conditions in the soil, limits moisture infiltration and internal drainage, and can lead to increased soil erosion, water run-off, and reduced rooting volume; these changes result in detrimental impacts on seedling establishment and tree growth (Elliot et al. 1998; Greacen and Sands 1980; Williamson and Neilson 2000).

Compaction in mineral soil is not readily ameliorated, and effects can persist for several decades, depending on the severity of compaction and local conditions (Froehlich and McNabb 1984; Greacen and Sands 1980; Landsberg et al. 2003; NCASI 2004; Page-Dumroese et al. 2006). Degree and duration of compaction and effects on tree growth are dependent on climate, moisture regime, soil texture, structure, and organic matter content (Heninger et al. 1997). While new harvest-related equipment and technologies have helped to reduce the effects of compaction and organic matter removal, season of harvest, number of equipment passes, soil texture, and surface organic matter depth all influence the amount of compaction that occurs during harvesting (Page-Dumroese et al. 2009). A number of methods can be used to decrease significant amounts of compaction, including leaving slash in skid trails, increasing equipment operator skill, being aware of soil conditions and properties, or applying biochar and other soil organic amendments (Han et al. 2009; Heninger et al. 2002; Senyk and Craigdallie 1997). Additional actions to mitigate compaction include tilling or ripping and revegetating compacted areas such as haul roads and landings; this is sometimes done on roads that are being abandoned (Luce 1997; NCASI 2004; Sosa-Pérez and MacDonald 2017).

Landscape topography also plays a large role in the amount of disturbance found in forested landscapes. Steep units had less off-trail compaction than flat units because the equipment is usually confined to trails on steeper slopes. Landings, trails, and corridors are usually the locations for soil compaction, and topsoil displacement usually occurs adjacent to many trails (Page-Dumroese et al. 2009). The presence of roads and landings, especially in steep areas, can lead to erosion and other impacts (Neher et al. 2017; Switalski et al. 2004). Occasionally, logging can trigger slope failures (e.g., landslides, mudflows, debris flows) (Guthrie 2002). Indirect disturbances following forestry operations can include changes in microclimate that affect rates of decomposition and nutrient cycling and alter hydrology (Finér et al. 2016; Sun et al. 2017).

The LTSP study found that leaving the forest floor intact after harvesting is critical on many sites to maintain nutrient cycling and C inputs (Powers et al. 2005). Organic matter in woody debris, forest floor detritus, and mineral soil is essential for maintaining ecosystem function by supporting soil C cycling, N availability, gas exchange, water availability, and biological diversity (Jurgensen et al. 1997; Page-Dumroese and Jurgensen 2006). In addition, the buildup of the forest floor may slow the rate of N mineralization, but using fire as a management tool assists in keeping forest floor C/N ratios within ranges more conducive to N mineralization (DeLuca and Sala 2006).

Surface residues often do not increase soil organic matter content (Spears et al. 2003), but using biochar can increase soil C, which can lead to increased soil aggregates and water holding capacity (Page-Dumroese et al. 2017b). Although slash piling is an economical method to dispose of harvest residues and reduce the volume of unmarketable material, pile burning can have short- and long-term impacts that can alter chemical, physical, and biological soil properties and degrade the productive capacity of soils (Rhoades and Fornwalt 2015). These impacts are variable and depend on soil texture, fuel type and loading, weather conditions, and soil moisture (Dyrness and Youngberg 1957; Frandsen and Ryan 1986; Hardy 1996; Rhodes and Fornwalt 2015; Rhodes et al. 2015). In areas where slash piles are plentiful and burned during the fall when conditions are conducive to large heat pulses into the soil, the need for restoration of severely burned soils is key (Page-Dumroese et al. 2017a). Yet, altering the size of slash piles from large (>10 m diameter) to small (<5 m diameter) and adding woodchip mulch have the potential to reduce the need for rehabilitation of burn scars (Rhoades et al. 2015). Guidelines for leaving slash and organic matter in the Pacific Northwest (Forest Guild 2013) and Rocky Mountain forests (Schnepf et al. 2009) are available.

Repeated harvesting without retaining or replacing sufficient amounts of soil nutrients and organic matter leads to continued concerns for loss of fertility and changes in soil biology, particularly on landforms that are weathered from nutrient-poor geological substrates such as granite or quartzite (Bockheim and Crowley 2002; Doran and Zeiss 2000; Federer et al. 1989; Garrison-Johnston et al. 2003; Grigal 2000; Grigal and Vance 2000; Kimsey et al. 2011). Forest fertilization is used in some areas of the United States, typically at the time of planting and during mid-rotation on plantation sites lacking in soil nutrients (Jokela et al. 2010). Fertilizer applications can be cost-effective depending on site conditions, but economic benefits are difficult to predict over the time period involved in growing a stand of timber (Cornejo-Oviedo et al. 2017; Fox et al. 2007; Miller et al. 2016).

WildFire and Prescribed Fire

Wildfires are a keystone process of many forest and rangeland systems, especially in the western United States. Wildfires, whether human-caused or natural, impact the litter layer and associated C and N, alter the environment for soil organisms, and can change the trajectory of forest composition. Fire kills trees and decreases canopy cover, partially or completely burns ground cover, and may form water repellant (hydrophobic) layers in soils, depending on burn severity (DeBano 1981; Madsen et al. 2011). Soil water storage, interception, and evapotranspiration are reduced when vegetation is removed or killed by fire and when organic matter on the soil surface is consumed by fire (Cerda and Robichaud 2009; DeBano et al. 1998; Neary et al. 2005). Fire consumption of ground vegetation and the development of hydrophobic soils increase overland flow erosion and can increase postfire sediment yield (Neary et al. 2005). Some potential secondary effects of severe fires are accelerated erosion and nutrient leaching. However, wildfire can also be beneficial because it reduces hazardous fuel in fire-dominated ecosystems, provides regeneration sites for certain tree and understory species, can increase available water to surviving vegetation, and improves nutrient cycling (Keane and Karau 2010).

As the length of the wildfire season increases due to climate change, the anticipated result is that larger wildfires will occur across the landscape (Abatzoglou and Williams 2016; Jolly et al. 2015). Wildfire concerns have led to an emphasis on accelerated fuel treatments. Busse and others (2014) have summarized the effects of fuel treatments and prescribed fire, noting that in addition to the loss of organic matter, prescribed fire and mechanical thinning operations alter the physical, biological, and chemical properties of the soil. The report encourages land managers to consider the impacts to the soil when planning fuel reduction treatments while acknowledging that these treatments do not make the land completely resistant to wildfire. Similar to the dry western forests and elsewhere, management plans should consider the ecological effects of fuel treatments within a restoration framework to avoid further ecosystem damage.

Many forest stands across the western United States are being thinned to remove fire fuels and reduce the risk of wildfire. Particularly in ponderosa pine (*Pinus ponderosa*) forests, there have been major changes in ecological structure, composition, and processes because of livestock grazing, fire suppression, logging, road construction, and exotic species introductions (Covington and Moore 1994b; Swetnam et al. 1999). These forests are now more susceptible to large, destructive fires that threaten human and ecological communities (Allen et al. 2002). Restoration in these forests requires the need to balance the heterogeneity of ponderosa pine ecosystems and climate fluctuations while also removing large numbers of trees to make the forests more fire resilient and within the natural range of aboveground and belowground C levels (Jurgensen et al. 1997; Rieman and Clayton 1997).

Concerns over large-scale crown fires can be mitigated with hazardous fuel reductions, but these fuel treatments must use ecological principles to limit or prevent further damage (Fulé et al. 2001). Usually, fuel reduction harvesting activities involve cutting and removing small trees with little marketable value (Brown et al. 2004). Residues may be removed and transported to a bioenergy facility (if one is available within a feasible hauling distance), dispersed across a harvest unit by mastication or grinding, or piled and burned (Creech et al. 2012; Jones et al. 2010). Although the impacts of intensive harvests on long-term soil productivity (Powers 2006) are generally known, there is much less known about the impacts of widespread thinning for fire risk reduction. The pattern of disturbance is thought to be different from typical clear-cut and thinning operations (McIver et al. 2003; Miller and Anderson 2002). For example, Landsberg and others (2003) found that the severity of soil compaction on areas thinned for fire risk reduction is dependent on slope.

Wildfire effects and restoration strategies are summarized in Cerda and Robichaud (2009). In the western United States, postfire sediment production may be highly variable, but it can have catastrophic impacts on downstream communities (Moody and Martin 2009). Short-term rehabilitation of burned landscapes tends to be focused on establishing ground cover through mulching or seeding while preventing accelerated erosion. Because of the increased severity and frequency of large wildfires, humans are intervening to assist in postfire ecological recovery efforts (Robichaud et al. 2009). The Forest Service's Burned Area Emergency Response (BAER) program focuses on mitigating unacceptable risks to life, safety, infrastructure, and critical natural and cultural resources on national forests and grasslands. BAER treatments can include erosion control, such as large-scale mulching, to protect municipal watersheds and soil productivity (Beyers 2004; Kruse et al. 2004). The BAER program has been very effective at making timely decisions to protect critical values from short-term damage after fires (e.g., the Hayman Fire in Colorado [Robichaud et al. 2009]). In addition to BAER, long-term postfire restoration focuses on restoring ecosystem function and structure while recovering a level of fire resiliency (Vallejo et al. 2009). Restoration is encouraged in areas where fire is uncommon or fire frequency and severity are outside of the fire regime for the area. Restoration efforts focus on seeding and planting appropriate plant species for the site. In some areas, such as systems managed as wilderness or managed to preserve natural features and ecosystem processes unfettered by humans, it may be desirable to forego restoration efforts after wildland fire. Seeding efforts may inhibit natural regeneration, and soil control measures may lessen the contribution of sediments and associated nutrients in recharging the fertility of streams and lakes located downstream from the burn sites (Christensen et al. 1989).

Restoring fire as an ecosystem process in fire-adapted systems can have beneficial effects (Collins et al. 2009). Over the last century, fire suppression and other management activities have altered the structure and function of forests and rangelands across much of the western United States (Belsky and Blumenthal 1997; Dwire and Kauffman 2003; Hessburg et al. 2005). Forest structure and composition has been most significantly altered due to the lack of fire disturbance. The disruption of the natural fire intervals of the past has resulted in higher stand densities, multilayered stands of mostly one species in some places, and the encroachment of conifers into meadows and grasslands. Dramatically higher stand densities and the development of ladder fuels have increased the risk of uncharacteristically severe wildfire, bark beetle infestations, and in some areas, successional replacement by shade-tolerant competitors. These changes across the landscape increase the probability for disturbances to affect large contiguous areas in uncharacteristic ways. By restoring fire into these ecosystems, generally as part of a management system that includes mechanical thinning and fuel reduction, the forests are restored to lower density stands with higher resiliency to large wildfires and other natural disturbances (Covington and Moore 1994a; Johnstone et al. 2016). Sites that were traditionally savanna ecosystems with widely-spaced trees and grassy understories need more frequent fire to maintain the forest structure (Peterson and Reich 2001). While soil resources are impacted by these management actions, the impacts tend to be less than the results of uncharacteristic large-scale fire disturbances (Hessburg et al. 2005; Johnstone et al. 2016).

Mine Reclamation

After agriculture and infrastructure development, mining is a major driver of deforestation and land degradation in the Americas (FAO 2016; Hosonuma et al. 2012). Millions of hectares of NFS lands are leased for oil, gas, coal, and geothermal operations. In fiscal year (FY) 2015, $1.6 billion worth of products were produced by large mines on NFS lands (USDA FS 2017). As many as 39,000 abandoned mines may be located on NFS lands (USDA FS 2017), and the Forest Service works to minimize or eliminate threats to human health and the environment from these mine sites. Under the BLM, the Abandoned Mine Lands (AML) program aims to restore degraded water quality, clean up mine waste and heavy metal, remediate other environmental issues affecting public lands, and mitigate safety concerns on mine sites abandoned prior to January 1, 1981; currently there are roughly 53,000 abandoned mine sites on BLM lands (BLM 2017). In 2011, the US Government Office of Accountability estimated that the cost of reclaiming 161,000 abandoned mines on public lands was in the range of $10–21 billion. These unproductive abandoned areas are often surrounded by productive forests. They are also usually located in rural areas with rugged terrain and limited access. Eight percent of these abandoned lands contain only physical hazards or limitations and no environmental contamination (American Geosciences Institute 2011).

Several federal and state rules and regulations govern the mining process, from approvals, planning, operations, and finally to reclamation and closure. Reclamation is defined as "the process by which derelict or very degraded lands are returned to productivity and by which some measure of biotic function and productivity is restored" (Brown and Lugo 1994). Mine reclamation rules and regulations vary largely by land ownership and type of extraction and can range from weak to stringent. The Forest Service adopted guidance in 2016 to address short- and long-term postmining maintenance and monitoring after reclamation (USDA FS 2017). Surface coal mining, governed by the Surface Mining Control and Reclamation Act of 1977, has very specific guidelines for topsoil salvage and stabilization. State agencies provide further guidance and regulations for reclamation and bond release. For example, shrub standards for reclaimed coal-mined lands in Wyoming require a minimum of one shrub per m^2 on lands if land use includes wildlife habitat (Wyoming Department of Environmental Quality 1996).

Early research in mine reclamation focused on soil and water protection, thus single-factor approaches (i.e., manipulating one physical, chemical, or biological factor) were employed to protect soil from erosion. Topsoil salvage and reestablishing an adequate plant community to prevent erosion were emphasized and met with varying results (Schuman 2002; Schuman et al. 1998). There are millions of hectares in the United States, including hundreds of thousands of hectares of Forest Service lands that are reclaimed but not restored. Soil quality is still often highly degraded on reclaimed sites. More modern approaches, transpired from years of reclamation practice and research, call for integrated and innovative approaches that target abiotic and biotic processes so that systems are functional (refer to Heneghan et al. 2008; Herrick et al. 2006b; Hild et al. 2009; King and Hobbs 2006; Lamb et al. 2015; McDonald et al. 2016; Stanturf et al. 2014).

Integration of soil stability, hydrology, nutrient cycling, plant functional traits, species turnover and regeneration, and wildlife interactions will not only help unite research with management but can place reclamation within the context of ecosystem function. Commonly, most projects do not have a level of topsoil or subsoil that is reflective of what was on site prior to the disturbance. Overburden generated from open-pit mining can be low in organic matter, soil microorganisms, and plant nutrients such as N and phosphorus and can lack soil structure and texture that are vital to soil fertility

and water-holding capacity (Allen 1989; Feagley 1985). Soil organic matter additions are valuable both for their C and the microbes contained in the material and provide substantial benefits to the affected site. The erosion and sediment control industry has started to address topsoil limitations by providing products that help to compensate for the loss of topsoil and setback from topsoil storage (see Abdul-Kareem and McRae 1984). One way they are doing this is by adding compost and other organic amendments that address several aspects of soil health, including the microbiological aspect, particularly ectomycorrhizae (Harvey et al. 1979). Reclaiming some areas may require building soil over rocks that have been dredged from local streams (Page-Dumroese et al. 2018). There are several options to initiate the soil-building process, but applying a combination of biochar, municipal biosolids, and wood chips offers one way to use local resources to begin to restore site productivity. Operations should weigh the costs and benefits of creating soils or adding soil amendments, as these types of treatments can be expensive.

Soils and Problematic Species

One of the most troubling developments related to the ease and speed of travel is the movement of nonnative, invasive species among continents. Across North America, these "problematic" insects, pathogens, and plants pose serious threats to forest and rangeland ecosystems because, unlike natural, abiotic disturbances like wind and fire, they are efficient at changing species composition by targeting specific species or outcompeting native species. Insects that cause substantial tree mortality, such as the mountain pine beetle (*Dendroctonus ponderosae*) that attacks pines (*Pinus* spp.) (Bentz et al. 2010), the gypsy moth (*Lymantria dispar*) (Potter and Conkling 2017), and emerald ash borer (EAB) (*Agrilus planipennis*) (Knight et al. 2012), are powerful drivers of ecosystem and economic change that causes not only shifts in species composition but also changes soil organic matter production, increases coarse woody debris, alters nutrient and water uptake, and changes understory light and temperature (Lovett et al. 2006). In addition, nonnative plants, such as the invasive shrub European buckthorn (*Rhamnus cathartica*) in the midwestern United States, can alter soil chemical and hydrological properties, leaving legacy effects and management challenges (Heneghan et al. 2006). Similar to postfire landscapes, ecosystems altered by problematic species are subject to flash flooding, soil erosion, and sediment loading.

Management responses to insect and disease impacts are often elusive, but in some cases biocontrols and combinations of treatments can be effective (Havill et al. 2016; Margulies et al. 2017) or sites can be transitioned into other vegetation types (D'Amato et al. 2018). Understanding the feedbacks between plant species and soil may help to combat invasion, as exemplified by current research on soil fungal pathogens and cheatgrass (*Bromus tectorum*) (Meyer et al. 2016). The role of mycorrhizal fungi and other fungal species in combating invasive plant species establishment and spread is also noteworthy (Bellgard et al. 2016; Padamsee et al. 2016). Biochar has been shown to increase the growth of native prairie grasses while decreasing or not affecting invasive perennials (Adams et al. 2013). Some technological advances have had adverse effects on forest soils, and management struggles to adapt. Soil ecological knowledge (SEK) has been applied to prevent or reduce the invasion by exotic species during restoration by adding C to promote microbial immobilization of available and mineralized N in abandoned agricultural land (see Heneghan et al. 2006 and references therein), and this has been shown to reduce nonnative plant species cover and colonization (Baer et al. 2003). Carbon addition is a tool to assist community recovery and assembly (nonnative to native). Because biochar is rich in C, it can improve soil quality and increase vegetation growth. In addition, biochar can limit or reduce the growth of invasive species by limiting N availability (Adams et al. 2013; Page-Dumroese et al. 2017b).

Innovative Approaches

Biochar

Land management stresses such as soil compaction, invasive species, disease or insect outbreaks, and wildfire are being exacerbated by a changing climate (Dale et al. 2001). This, coupled with the need to remove encroaching biomass that has little to no value, is increasing operational expenses (Rummer et al. 2005). However, biochar can be one by-product that brings a high value to traditionally low-value biomass. Biochar is a by-product of pyrolysis of materials such as wood, waste organic materials, and agricultural crop residues at temperatures above 400 °C under complete or partial elimination of oxygen (Beesley et al. 2011; Lehmann 2007). Because of its porous structure, large surface area, and negatively charged surface (Downie et al. 2009; Liang et al. 2006), biochar has the potential to increase water holding capacity and plant-nutrient retention in many soils (Basso et al. 2013; Gaskin et al. 2007; Kammann et al. 2011; Laird et al. 2010) and is used to amend food crop soils (Blackwell et al. 2009). Biochar retains much of the C of the original biomass, which can offset the use of fossil fuels and can reduce greenhouse gas emissions from soil (Jones et al. 2010). When used as a soil amendment, biochar contributes to increase C sequestration, enhances the cation exchange capacity, increases pH, and reduces soil bulk density and resistance to gas and water movement (Mukherjee and Lal

2013). All of these changes have been shown to enhance plant growth (Atkinson et al. 2010).

Biochar may be useful for restoring or revitalizing degraded forest, rangeland, and urban soils. It may also provide a method for increasing soil water holding capacity to improve tree health and reduce the incidence of disease and insect attack (Page-Dumroese et al. 2017b). For example, biochar additions of 25 Mg ha^{-1} resulted in a 10% increase in available water in August on coarse-textured soil in central Montana (Page-Dumroese et al. 2017b). The biggest benefit of biochar, however, may be in facilitating reforestation of degraded or contaminated sites (Page-Dumroese et al. 2017b). Biochar amendments have the potential to reduce leaching and bioavailability of heavy metals such as copper, zinc, lead, and cadmium (Bakshi et al. 2014; Beesley and Marmioli 2011), mainly as a result of changing the soil pH. On mine sites that contain toxic chemicals from decades of activity, establishing vegetation cover to limit erosion and offsite movement of chemicals was successful when biochar was used (Fellet et al. 2011).

Variability in biochar type, application rate, and mode (e.g., top-dressing, tilled, pellets), as well as environmental setting, can play a role in plant response (Barrow 2012; Lehmann 2007; Solaiman et al. 2012; Van Zwieten et al. 2010). Applying biochar to coarse- to medium-textured, unproductive soils at rates less than 100 metric tons ha^{-1} can improve nutrient supply, water holding capacity, and water availability (Chan et al. 2008; Jeffery et al. 2011). When adding biochar to the soil to improve moisture conditions (i.e., water repellency), it is more effective when mixed into the profile rather than surface applied (Page-Dumroese et al. 2015). Biochar's ability to absorb water and adsorb nutrients is also contingent upon its chemical and physical properties, a function of pyrolysis temperature (e.g., pH and surface area increase with temperature to a point) (Downie et al. 2009; Lehmann 2007). In forest soil applications, for example, biochar produced at 550–650 °C was better than other temperatures for absorbing water (Kinney et al. 2012). And in a study of different types, in general biochar enhanced water storage capacity of soils, but it varied with feedstock type and pyrolysis temperature (Novak et al. 2012).

Biochar can be designed with characteristics specific to intended objectives, goals, and environmental settings (Novak and Busscher 2013; Novak et al. 2009). Given enough completed studies and data, decision frameworks could help practitioners decide whether or not to use biochar and to determine what type is appropriate based on initial soil properties and other environmental conditions (Beesley et al. 2011). But production costs may outweigh benefits, and it may not be economically feasible for large-scale production and use. Biochar production can cost $51–$3747 per ton, a wide range that depends on feedstock type, pyrolysis reaction time (slow or fast), temperature, heat source, and transportation (Meyer et al. 2011). Prices for biochar worldwide vary substantially between \$80 and \$13,480 US dollars per ton (Jirka and Tomlinson 2013). Until there are verified benefits to using biochar and investments in less expensive technologies to produce biochar, the market prices for biochar will remain uncertain (Campbell et al. 2018). On-site production of biochar is one approach to alleviate high transportation and material acquisition costs, especially in forest systems where a constant supply of wood material left over from harvesting operations is available (Coleman et al. 2010).

Seed Coating Technologies

Direct seeding in the western United States is a common restoration practice, but germination and seedling emergence can be major barriers to successful revegetation (Chambers 2000; James et al. 2011). Seedbed conditions are highly variable for temperature and moisture (Hardegree et al. 2003), and conditions need to occur that allow seeds to germinate. For some species, the range of temperature and moisture needed for emergence and growth is narrow (Fyfield and Gregory 1989). Seed coatings that facilitate germination and initial growth may be especially useful in situations where nutrients and water are limited (Madsen et al. 2012; Taylor and Harman 1990). Seed coating technologies that use biochar may potentially overcome moisture and temperature limitations that affect native plant germination and growth, especially on arid and semiarid lands, but initial studies show mixed results (Williams et al. 2016b). The cost of seed coatings can be high, which will add to the already expensive price of non-coated native seeds. During 2000–2014, the US Federal Government spent more than \$300 million on native plant seeds used for revegetating land disturbances; in 2013 alone, the cost exceeded \$20.7 million (U.S. Government 2014).

Soil Transplants

Many of the aboveground changes in plant species biomass and diversity are linked to the abundance and composition of microbes within the mineral soil (Smith et al. 2003). Management-induced shifts in soil microbial populations that regulate nutrients or decomposition will result in a concomitant change in aboveground production (Bardgett and McAlister 1999). Early examination of the mechanisms in which soil microbes influence plant succession and competition (e.g., Allen 1989; Allen and Allen 1988, 1990) led to research and development of soil transplants and inoculum to steer restoration efforts. In recent times, studies have transplanted soils or soil inoculant to restore late-successional plant communities (Middleton and Bever 2012; Wubs et al. 2016). In related efforts, researchers are also using biological soil crusts (BSC) to expedite restoration of severely stressed sites (Young et al. 2016). Biological soil crusts are communities of organisms (fungi, lichens, bryophytes, cyanobacteria, and algae) that are intimately associated with the mineral soil surface (Bowker 2007; also see Box 5.2). These communities are most often associated with rangeland sites, but they can be found ephemerally, and sometimes abundantly, within most terrestrial ecosystems. Soil crusts can facilitate succession, and therefore, the assisted recovery of these crusts may help speed succession on degraded lands. Plants and BSCs interact to help restore soil quality through soil stability, runoff to infiltration balances, surface albedo, nutrient capture, and available habitat for microbes (Bowker 2007).

Monitoring Restoration Success

A restored ecosystem should have the following attributes: (1) similar diversity and community structure in comparison to a reference site, (2) presence of indigenous species, (3) presence of functional groups required for long-term stability, (4) capacity of the physical environment to sustain reproduction, (5) normal functioning, (6) integration with the landscape, (7) elimination of potential threats, (8) resilience to natural disturbance, and (9) self-sustainability (SER 2004).

Monitoring is a crucial part of the restoration effort, yet, in practice, only a few attributes, such as plant composition and cover or soil stability, are generally monitored, and usually these attributes are only tracked for a short period of time (<5 years) (Ruiz-Jean and Aide 2005). Studies support that short-term plant community composition monitoring is a necessary but insufficient predictor of long-term success. Examples of long-term monitoring in the western United States show that short-term monitoring alone of plant community composition has detected "false" and "true" failures. In one situation, a project was abandoned after only 4 years and was determined a failure, but decades later the plant community recovered. The lag in plant community response was attributed to soil properties that need more time to recover (i.e., infiltration and nutrient cycling associated with soil organic matter accumulation). The lack of soil organic matter limited the short-term recovery of the system, so it was deemed a reclamation failure (Tongway et al. 2001; Walton 2005; Walton et al. 2001 as cited in Herrick et al. 2006b). In contrast, many restoration projects deemed successful do not persist because one or more processes are absent (Herrick et al. 2006a; Rango et al. 2005). Integration of ecological indicators that reflect soil and site stability, hydrologic function, and biotic integrity (Pellant et al. 2005) has the potential to help avoid identifying false or true failures in restoration (Herrick et al. 2006b). To understand success in ecosystem restoration, we must understand the linkages of aboveground and belowground changes to biotic interactions, plant community effects, aboveground consumers, and the influences of changing species (Bardgett and Wardle 2010).

Qualities of good monitoring programs include being well-designed and standardized, and there must be long-term support that allows for continuous monitoring. Effective monitoring programs address clear questions, use consistent and accepted methods to produce high-quality data, include provisions for management and accessibility of samples and data, and integrate monitoring into research programs that foster continued evaluation and utility of data. There are several steps involved in planning what to monitor: (1) define the goals and objectives of the monitoring; (2) compile and summarize the existing information; (3) develop a conceptual model; (4) prioritize and select indicators; (5) develop the sampling design; (6) develop the monitoring protocols; and (7) establish data management, analysis, and reporting procedures (Fancy et al. 2009; Jain et al. 2012). Monitoring can be expensive in terms of personnel, equipment, and time, but relative to the value of resources that restoration activities protect and the policy it informs, monitoring costs very little. Considerable planning before restoration begins will determine monitoring needs and overall success. Herrick and others (2006b) suggest a ten-step iterative approach to monitoring that begins before restoration is initiated, collects short-term data for use in adjusting restoration efforts, and

lastly, involves long-term monitoring (see Fig. 8.1). This approach allows for short-term monitoring indicators to also be used for long-term efforts.

Monitoring should be scaled both spatially and temporally to the patterns or processes of the response variable, recognizing that patterns often vary with the scale at which a study is conducted (Levin 1992; Wiens 1989). Because processes and populations vary in time and space, monitoring should be designed and conducted at the scale(s) that encompasses the appropriate variation (Bissonette 1997). Given the variety of factors that might influence a response variable, monitoring should be designed to incorporate as much of the variation resulting from those factors as possible. Spatially, this requires sampling at the appropriate scale to detect a biologically meaningful response should one occur. For example, effects may manifest at the landscape level but be obscured at the stand scale, or vice versa (Bestelmeyer et al. 2006). Determining the spatial scale might be particularly relevant when evaluating treatment effects on population trends of selected species (Ritters et al. 1997) because effects discovered within a restoration project area may not extend to the broader population.

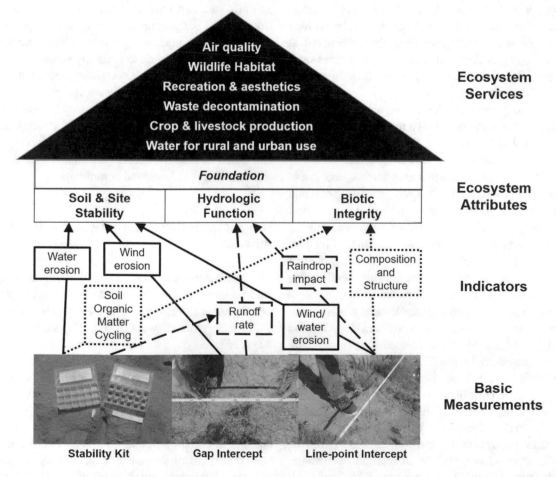

Fig. 8.1 Basic measurements (soil stability, gap intercept, and line-point intercept) used to generate indicators of processes related to ecosystem attributes that serve as the foundation for most ecosystem services and success for restoration projects. (Source: Adapted from Herrick et al. 2006b)

Temporally, the most obvious solution is to conduct studies over a long enough time to detect change if one is occurring (Morrison 1987; Strayer et al. 1986; Wiens 1984). Long-term studies are appropriate when observing slow processes, rare events, subtle processes, and complex phenomena (Strayer et al. 1986). Examples of slow processes include soil organic matter accumulation, plant succession (Bestelmeyer et al. 2006), invasion by exotic species, and long-term population cycles. Rare events include fire, floods, population irruptions of a food item (e.g., insect epidemics resulting in a numerical response by birds), and various environmental crunches (Morrison 1987). Subtle processes are those that may show little change over a short period but whose effects are greater when viewed within a longer timeframe. Complex phenomena are typically the result of multiple interacting factors. This begs the question: How long should monitoring continue? Strayer and others (1986) suggested that "if it continues for as long as the generation time of the dominant organism or long enough to include examples of the important processes that structure the ecosystem under study… the length of study is measured against the dynamic speed of the system being studied." Clearly, "long-term" depends greatly on the response variable and system under study. For some restoration projects, a single scale may be appropriate, whereas others might require monitoring to be done at multiple scales. Even then, relationships observed at one scale may differ from those observed at another (Wiens 1986). Timescale is important, but research projects, especially graduate work, may not be long enough to capture any changes.

Conducting long-term studies is important for understanding management and restoration impacts on ecosystem processes (Bienes et al. 2016). In the United States, many government agencies have committed to long-term monitoring to identify ecological insights that inform ecosystem management. The LTSP study has been generating soil and vegetation responses to management activities for the past 25 years (Powers et al. 2005). Other long-term monitoring efforts (e.g., Long-Term Ecological Research, Fire-Fire Surrogate study) underscore management and scientist commitment to generating long-term datasets. While some of these long-term efforts were not meant to evaluate ecosystem restoration, their datasets can inform research and management. For example, the newly developed Land Treatment Digital Library (LTDL), a spatially explicit database of land treatments by the BLM, serves as a clearinghouse for over 9000 land treatments (e.g., seeding, prescribed fire, weed control, vegetation/soil manipulations) in the western United States spanning more than 75 years (Pilliod et al. 2017). The LTDL can be used to study or query, for example, vegetation and soil response (Knutson et al. 2014), successional patterns in relation to management, trends in treatment types across time (Copeland et al.

2018), or ability of treatments to meet habitat requirements for key species (Arkle et al. 2014).

Case Studies

Mower Tract Ecological Restoration: Monongahela National Forest, West Virginia

The Mower Tract Ecological Restoration project is a landscape-scale restoration effort that combines red spruce restoration, watershed development, and creation of early successional habitat to benefit wildlife (Fig. 8.2) (Barton 2014). In the past, timber removal and coal mining left sites devoid of vegetation with severely compacted soils. These sites were subsequently planted with nonnative plant species such as Norway spruce (Picea abies) and red pine to control erosion and flooding. These activities left the landscape in a state of "arrested succession," where tree growth was stunted and plant recruitment ceased. A suite of restoration activities is now being used to restore habitat and improve water quality; they include soil decompaction, wetland restoration, woody debris loading, and planting of native trees and shrubs. Heavy machinery breaks up compacted soils, tears up grass sod, and knocks down nonnative trees. The dead wood left on the ground creates habitat for plants and animals while new trees are growing. The downed trees also provide perches for birds to naturally spread native seed and encourage natural regeneration. Organic matter from the decaying wood further improves soil conditions and creates a more suitable environment for growth of red spruce and other native plants. Objectives achieved through this project will help conserve and ensure long-term viability of important plants and animal species associated with unique high-elevation forest and wetland communities.

Long-Term Soil Productivity Study: North America

The North American Long-Term Soil Productivity (LTSP) study is novel as it has been collecting data on soil compaction and organic matter removal after clear-cut harvesting for the last 25 years (Powers 2006). The study spans a wide range of forest and climatic regimes. There are several key points: (1) Soils with an already high bulk density are hard to compact more, and (2) coarser-textured soils recover from compaction faster than fine-textured soils (Page-Dumroese et al. 2006). However, soil density recovery was slow, particularly in soils in the frigid temperature regime. These changes in compaction lead to greater short-term (5 years) tree volume growth on coarse-textured soils and less volume growth on fine-textured soil (Gomez et al. 2002). After

Fig. 8.2 The Mower Tract Ecological Restoration Project. (**a**) The landscape was mined and logged in the 1980s. (**b**) Soil compaction caused stunted tree growth and low recruitment. (**c**) Planted nonnative trees were removed to allow growth of native red spruce. (**d**) Restoration involved ripping the soil, removing nonnative trees, and leaving woody debris on site. (**e**) Aerial view of project (c. 2014). (Photo credit: Chris Barton, USDA Forest Service)

20 years and for a range of soil textures (sandy loams to clay loams), soil compaction resulted in a 15% increase in planted tree biomass on the plot-scale basis. This was attributed to increased seedling survival, along with reduced vegetative competition, and was consistent across all the California study sites (Zhang et al. 2017). This same study points to a

near-complete tolerance of forest biomass growth to compaction and soil organic matter removal; similar to results found in Missouri (Ponder et al. 2012).

Many forest sites are resilient because of their inherent high organic matter levels. However, sites with lower soil organic matter, with deficiencies in one or more soil nutrients, or with fine texture may be at risk for productivity declines as these stands reach crown closure (Zhang et al. 2017). For example, a decade after the complete removal of the surface organic matter, reductions in nutrient availability and soil C concentrations were observed to a depth of 20 cm. Soil C storage was not diminished, likely because of changes in bulk density and the decomposition of residual root systems (Powers et al. 2005).

Soil Matters: Deschutes National Forest, Oregon

Craigg and others (2015) highlight how using soil mapping along with inherent and dynamic soil quality information can help guide forest management for multiple uses. They discuss the importance and value of regarding soils as the foundational resource in forest planning processes. By appreciating the differing inherent capabilities of the soil, land managers can match the appropriate land uses to the soils that will support those uses. This allows the soils information to set the stage for land management by defining the landscape potential and project objectives through understanding how the soils are able to support long-term ecosystem outcomes. Management actions are then planned based on the appropriate soil types in order to have the highest success rate of meeting the objectives. Interdisciplinary teams work together to strategize how to integrate the objectives and the actions. Projects are then designed and implemented with dynamic soil properties that require protection measures. Upon implementation, ecosystem responses can be monitored to determine if anticipated results are achieved. This information then feeds back to the beginning of the cycle to refine land management objectives. Soil types differ widely in their inherent capacity to perform various ecological functions as well as in their dynamic response to and recovery from disturbances. Incorporating these concepts into planning processes can greatly enhance the quality of forest management decisions.

The Sisters Area Fuels Reduction Project (SAFRP) on the Deschutes National Forest of eastern Oregon serves as a case study for the application and potential benefits of this soils-based planning tool. Treatments within the planning area had multiple objectives, including improved forest health and resistance to insect epidemics, drought, and serious wildfires in the wildland-urban interface while also providing quality wildlife habitat and other ecosystem services. The Deschutes National Forest soil resource inventory was used to identify three general soil groups within the SAFRP planning area to help assess and match stand-level tree spatial patterns. Each of these soil groups was then paired with the appropriate management objectives in the project design, and treatments were developed to meet desired resource and habitat goals. Post-treatment monitoring has confirmed expectations of desirable stand patterns and vegetation responses where key soil differences were considered. In addition, the project resulted in a successful fuel reduction treatment which aided fire suppression activities during the 2012 Pole Creek Fire.

Key Findings

- Management and restoration approaches have improved in the last few decades, but innovative developments that inform actions in a timely and cost-effective manner remain priorities.
- Integration of physical, biological, and chemical attributes (a multiple-factor approach) will continue to advance our understanding of soil management and restoration.
- Continuous, targeted, and adaptive monitoring is essential to soil management and restoration, and there is a need for both short- and long-term monitoring.
- Nature is always changing—there is no going back. It takes nearly 500 years to build 2.5 cm of soil. Therefore, protecting and restoring our current soil stocks is critical.

Key Information Needs

- What are the boundaries between healthy, at-risk soils and unhealthy soils? It is important to identify thresholds for soil function and structure of soil types and orders.
- How do we determine the best time to take action and what tools are most appropriate?
- How do aboveground and belowground components of the soil interact? A better understanding of how plants, microbes, organic matter, decomposition, and nutrient cycles interact will help improve soil restoration and monitoring efforts.
- How do we design ecological monitoring efforts to detect fluxes and processes at many spatial and temporal scales? Because ecosystems and soils are in a continued state of flux, we must be able to detect these fluxes to ensure ecosystem services are maintained and land is meeting desired ecological conditions.

Acknowledgments We thank Steve Overby and Steve Williams for reviewing this chapter. We also appreciate input from Mike Brown (Bureau of Land Management), Liam Heneghan (DePaul University), Bart Lawrence (Natural Resources Conservation Service), and Cynthia West (Forest Service) while developing the chapter.

Literature Cited

Abatzoglou JT, Williams AP (2016) Impact of anthropogenic climate change on wildfire across western US forests. Proc Natl Acad Sci 113(42):11770–11775

Abdul-Kareem AW, McRae SG (1984) The effects on topsoil of long-term storage in stockpiles. Plant Soil 76(1–3):357–363

Adams PW (1991) Soil compaction on woodland properties. In: The woodland workbook. Extension circular 1109. Revised January 1991. Oregon State University Extension Service, Corvallis

Adams MM, Benjamin TJ, Emery NC et al (2013) The effect of biochar on native and invasive prairie plant species. Invasive Plant Sci Manag 6:197–207

Adhikari K, Hartemink AE (2016) Linking soils to ecosystem services: a global review. Geoderma 262:101–111

Allen MF (1989) Mycorrhizae and rehabilitation of disturbed arid soils: processes and practices. Arid Soil Res Rehabil 3:229–241

Allen EB, Allen MF (1988) Facilitation of succession by the nonmycotrophic colonizer *Salsola kali* (Chenopodiaceae) on a harsh site: effects of mycorrhizal fungi. Am J Bot 75(2):257–266

Allen EB, Allen MF (1990) The mediation of competition by mycorrhizae in successional and patchy environments. In: Grace JB, Tilman D (eds) Perspectives on plant competition. Academic, San Diego, pp 367–389

Allen CD, Savage M, Falk DA et al (2002) Ecological restoration of the southwestern ponderosa pine ecosystems: a broad perspective. Ecol Appl 12:1418–1433

American Geosciences Institute (2011) House Committee on natural resources, subcommittee on energy and mineral resources oversight hearing on mining in America: the administration's use of claim maintenance fees and cleanup of abandoned mine lands. http://www.americangeosciences.org/policy/house-committee-natural-resources-subcommittee-energy-and-mineral-resources-oversight-hearing. Accessed 29 Apr 2018

Arkle RS, Pilliod DS, Hanser SE et al (2014) Quantifying restoration effectiveness using multi-scale habitat models: implications for sage-grouse in the Great Basin. Ecosphere 5(3):1–32

Atkinson CJ, Fitzgerald JD, Hipps NA (2010) Potential mechanisms for achieving agricultural benefits from biochar application to temperate soils: a review. Plant Soil 337(1–2):1–18

Baer SG, Blair JM, Collins SL, Knapp AK (2003) Soil resources regulate productivity and diversity in newly established tallgrass prairie. Ecology 84:724–735

Bailey RG (1987) Suggested hierarchy of criteria for multi-scale ecosystem mapping. Landsc Urban Plan 14:313–319

Bakshi S, He ZL, Harris WG (2014) Biochar amendment affects leaching potential of copper and nutrient release behavior in contaminated sandy soils. J Environ Qual 43:1894–1902

Bardgett RD, McAlister E (1999) The measurement of soil fungal: bacterial biomass ratios as an indicator of ecosystem self-regulation in temperate meadow grasslands. Biol Fertil Soils 29:282–290

Bardgett RD, Wardle DA (2010) Aboveground-belowground linkages: biotic interactions, ecosystem processes, and global change. Oxford University Press, Oxford. 289 p

Barnes BV, Pregitzer KS, Spies TA, Spooner VH (1982) Ecological forest site classification. J For 80:493–498

Barrow CJ (2012) Biochar: potential for countering land degradation and for improving agriculture. Appl Geogr 34:21–28

Barton C (2014) Mower tract ecological restoration: final report 2012–2014. Monongahela National Forest: Lambert Restoration Project. Green Forests Work and Climate Change Response Framework, 22 p

Basso AS, Miguez FE, Laird DA et al (2013) Assessing potential of biochar for increasing water-holding capacity of sandy soils. GCB Bioenergy 5:132–143

Beesley L, Marmiroli M (2011) The immobilization and retention of soluble arsenic, cadmium, and zinc by biochar. Environ Pollut 159:474–480

Beesley L, Moreno-Jiménez E, Gomez-Eyles JL et al (2011) A review of biochars' potential role in the remediation, revegetation and restoration of contaminated soils. Environ Pollut 159(12):3269–3282

Bellgard SE, Padamsee M, Probst CM et al (2016) Visualizing the early infection of *Agathis australis* by *Phytophthora agathidicida*, using microscopy and fluorescent in situ hybridization. For Pathol 46(6):622–631

Belsky AJ, Blumenthal DM (1997) Effects of livestock grazing on stand dynamics and soils in upland forests of the Interior West. Conserv Biol 11(2):315–327

Bentz BJ, Régnière J, Fettig CJ et al (2010) Climate change and bark beetles of the western United States and Canada: direct and indirect effects. Bioscience 60(8):602–613

Bestelmeyer BT, Brown JR, Havstad KM, Alexander R, Chavez G, Herrick JE (2003) Development and use of state-and-transition models for rangelands. J Range Manag 56:114–126

Bestelmeyer BT, Trujillo DA, Tugel AJ, Havstad KM (2006) A multi-scale classification of vegetation dynamics in arid lands: what is the right scale for models, monitoring, and restoration? J Arid Environ 65(2):296–318

Bestelmeyer BT, Ash A, Brown JR et al (2017) Chapter 9: State and transition models: theory, applications, and challenges. In: Briske DD (ed) Rangeland systems, Series on environmental management. Springer, New York, pp 303–345

Beyers JL (2004) Post-fire seeding for erosion control: effectiveness and impacts on native plant communities. Conserv Biol 18(4):947–956

Bienes R, Marques MJ, Sastre B, García-Díaz A, Ruiz-Colmenero M (2016) Eleven years after shrub revegetation in semiarid eroded soils: influence in soil properties. Geoderma 273:106–114

Birdsey R, Pregitzer K, Lucier A (2006) Forest carbon management in the United States. J Environ Qual 35(4):1461–1469

Bissonette JA (1997) Scale-sensitive ecological properties: historical context, current meaning. In: Bissonette JA (ed) Wildlife and landscape ecology: effects of pattern and scale. Springer, New York, pp 3–31

Blackwell P, Riethmuller G, Collins M (2009) Biochar application to soil. In: Lehmann J, Joseph S (eds) Biochar for environmental management: science and technology. Earthscan, London, pp 207–226

Bockheim JG, Crowley SE (2002) Ion cycling in hemlock–northern hardwood forests of the Southern Lake Superior region: a preliminary study. J Environ Qual 31(5):1623–1629

Bowker MA (2007) Biological soil crust rehabilitation in theory and practice: an underexploited opportunity. Restor Ecol 15(1):13–23

Bradford JB, Laurenroth WK (2006) Controls over invasion of *Bromus tectorum*: the importance of climate, soil, disturbance and seed availability. J Veg Sci 17:693–704

Bradshaw AD (1987) The reclamation of derelict land and the ecology of ecosystems. In: Jordan WR, Gilpin ME, Aber JD (eds) Restoration ecology: a synthetic approach to ecological research. Cambridge University Press, Cambridge. 342 p

Breshears DD, Whicker JJ, Johansen MP, Pinder JE III (2003) Wind and water erosion and transport in semi-arid shrubland, grassland and forest ecosystems: quantifying dominance of horizontal and wind driven transport. Earth Surf Process Landf 28:1189–1209

Briske DD, Fuhlendorf SD, Smeins FE (2003) Vegetation dynamics on rangelands: a critique of the current paradigms. J Appl Ecol 40:601–614

Briske DD, Fuhlendorf SD, Smeins FE (2005) State-and-transition models, thresholds, and rangeland health: a synthesis of ecological concepts and perspectives. Rangel Ecol Manag 58:1–10

Brown S, Lugo AE (1994) Rehabilitation of tropical lands: a key to sustaining development. Restor Ecol 2(2):97–111

Brown RT, Agee JK, Franklin JF (2004) Forest restoration and fire: principles in the context of place. Conserv Biol 18(4):903–912

Brussard L, van Faasen HC (1994) Effects of compaction on soil biota and soil biological processes. In: Soane BD, van Ouwerkerk C (eds) Soil compaction and crop production. Elsevier, Amsterdam, pp 215–235

Bulmer CE, Simpson DG (2005) Soil compaction and water content as factors affecting growth of lodgepole pine seedlings on sandy clay soil. Can J Soil Sci 85:667–679

Burger JA, Perumpral JV, Kreh RE, Torbert JL, Minaei S (1985) Impact of tracked and rubber-tired tractors on a forest soil. Trans Am Soc Agric Eng 28(2):369–373

Busse MD, Moghaddas EEY, Hubbert KD (2014) Fuel reduction practices and their effects on soil quality, General Technical Report PSW-GTR-241. U.S. Department of Agriculture, Forest Service, Pacific Southwest Research Station, Albany. 164 p

Cajander AK (1926) The theory of forest types. Acta Forestalia Fenn 29:1–108

Campbell RM, Anderson NM, Daugaard DE, Naughton HT (2018) Financial viability of biofuel and biochar production from forest biomass in the face of market price volatility and uncertainty. Appl Energy 230:330–343

Caudle D, DiBenedetto J, Karl M et al (2013) Interagency ecological site handbook for rangelands. U.S. Department of the Interior, Bureau of Land Management, Washington, DC. 109 p

Cerda A, Robichaud PR (eds) (2009) Fire effects on soils and restoration strategies, vol 5. In: Haigh MJ (series ed) Land reconstruction and management. Science Publishers, CRC Press, Enfield, pp 511–535

Chambers JC (2000) Seed movements and seedling fates in disturbed sagebrush steppe ecosystems: implications for restoration. Ecol Appl 10:1400–1413

Chambers JC, Miller RF, Board DI et al (2014) Resilience and resistance of sagebrush ecosystems: implications for state and transition models and management treatments. Rangel Ecol Manag 67(5):440–454

Chambers JC, Maestas JD, Pyke DA et al (2017) Using resilience and resistance concepts to manage persistent threats to sagebrush ecosystems and greater sage-grouse. Rangel Ecol Manag 70(2):149–164

Chan KY, Van Zwieten L, Meszaros I et al (2008) Using poultry litter biochars as soil amendments. Aust J Soil Res 46:437–444

Christensen NL (2014) An historical perspective on forest succession and its relevance to ecosystem restoration and conservation practice in North America. For Ecol Manag 330:312–322

Christensen NL, Agee JK, Brussard PF et al (1989) Interpreting the Yellowstone fires. Bioscience 39:678–685

Cleland DT, Avers PE, McNab WH et al (1997) National hierarchical framework of ecological units. In: Boyce MS, Haney A (eds) Ecosystem management applications for sustainable forest and wildlife resources. Yale University Press, New Haven, pp 181–200

Clements FE (1916) Plant succession: analysis and development of vegetation. Carnegie Institute, Washington, DC. 512 p

Coleman M, Page-Dumroese D, Archuleta J et al (2010) Can portable pyrolysis units make biomass utilization affordable while using bio-char to enhance soil productivity and sequester carbon? In: Jain TB, Graham RT, Sandquist J (eds) Integrated management of carbon sequestration and biomass utilization opportunities in a changing climate: proceedings of the 2009 national silviculture workshop, General technical report RMRS-P-61. U.S. Department

of Agriculture, Forest Service, Rocky Mountain Research Station, Fort Collins, pp 159–168

Collins BM, Miller JD, Thode AE et al (2009) Interactions among wildland fires in a long-established Sierra Nevada natural fire area. Ecosystems 12(1):114–128

Copeland SM, Munson SM, Pilliod DS et al (2018) Long-term trends in restoration and associated land treatments in the southwestern United States. Restor Ecol 26(2):311–322

Cornejo-Oviedo EH, Voelker SL, Mainwaring DB et al (2017) Basal area growth, carbon isotope discrimination, and intrinsic water use efficiency after fertilization of Douglas-fir in the Oregon coast range. For Ecol Manag 389:285–295

Covington WW, Moore MM (1994a) Postsettlement changes in natural fire regimes and forest structure: ecological restoration of old-growth ponderosa pine forests. J Sustain For 2(1–2):153–181

Covington WW, Moore MM (1994b) Southwestern ponderosa pine forest structure: changes since Euro-American settlement. J For 92:39–47

Craigg TL, Adams PW, Bennett KA (2015) Soil matters: improving forest landscape planning and management for diverse objectives with soils information and expertise. J For 113(3):343–353

Creech MN, Kirkman LK, Morris LA (2012) Alteration and recovery of slash pile burn sites in the restoration of fire-maintained ecosystems. Restor Ecol 20:505–516

D'Amato A, Palik B, Slesak R et al (2018) Evaluating adaptive management options for black ash forests in the face of emerald ash borer invasion. Forests 9(6):348–365

Dale VH, Joyce LA, McNulty S et al (2001) Climate change and forest disturbances: climate change can affect forests by altering the frequency, intensity, duration, and timing of fire, drought, introduced species, insect and pathogen outbreaks, hurricanes, windstorms, ice storms, or landslides. Am Inst Biol Sci Bull 51(9):723–734

Daubenmire RF (1968) Plant communities: a textbook of plant synecology. Harper and Row, New York. 273 p

DeBano LF (1981) Water repellent soils: a state-of-the-art, General Technical Report PSW-GTR-46. U.S. Department of Agriculture, Forest Service, Pacific Southwest Forest and Range Experiment Station, Berkeley. 20 p

DeBano LF, Neary DG, Folliott PF (1998) Fire's effects on ecosystems. Wiley, New York. 335 p

DeLuca TH, Sala A (2006) Frequent fire alters nitrogen transformations in ponderosa pine stands of the inland northwest. Ecology 87(10):2511–2522

Dominati E, Patterson M, Mackay A (2010) A framework for classifying and quantifying the natural capital and ecosystem services of soils. Ecol Econ 69:1858–1868

Doran JW, Zeiss MR (2000) Soil health and sustainability: managing the biotic component of soil quality. Appl Soil Ecol 15:3–11

Downie A, Crosky A, Munroe P (2009) Physical properties of biochar. In: Lehmann J, Joseph S (eds) Biochar for environmental management: science and technology. Earthscan, London, pp 13–32

Dwire KA, Kauffman JB (2003) Fire and riparian ecosystems in landscapes of the western USA. For Ecol Manag 178(1–2):61–74

Dyksterhuis EJ (1949) Condition and management of rangeland based on quantitative ecology. J Range Manag 2:104–105

Dyrness CT, Youngberg CT (1957) The effect of logging and slash burning on soil structure. Soil Sci Soc Am J 21:444–447

Elliot WJ, Page-Dumroese DR, Robichaud PR (1998) The effects of forest management on erosion and soil productivity. In: Lal R (ed) Soil quality and soil erosion. CRC Press, Boca Rattan, pp 195–212

Fancy SG, Gross JE, Carter SL (2009) Monitoring the condition of natural resources in US national parks. Environ Monit Assess 151(1):161–174

Feagley SE (1985) Chemical, physical and mineralogical characteristics of the overburden, Southern Coop series bulletin 294. Arkansas Agricultural Experimental Station, Fayetteville

Federer CA, Hornbeck JW, Tritton LM et al (1989) Long-term depletion of calcium and other nutrients in eastern US forests. Environ Manag 13(5):593–601

Fedkiw J (1989) The evolving use and management of the nation's forests, grasslands, croplands and related resources: a technical document supporting the 1989 RPA assessment, General technical report GTR-RM-175. U.S. Department of Agriculture Forest Service, Rocky Mountain Forest and Range Experiment Station, Washington, DC. 66 p

Fellet G, Marchiol L, Delle Vedove G, Peressotti A (2011) Application of biochar on mine tailings: effects and perspectives for land reclamation. Chemosphere 83:1262–1267

Finér L, Jurgensen M, Palviainen M, Page-Dumroese D (2016) Does clear-cut harvesting accelerate initial wood decomposition? A five-year study with standard wood material. For Ecol Manag 372:10–18

Food and Agriculture Organization of the United Nations [FAO] (2016) State of the world's forests 2016: forests and agriculture—land-use challenges and opportunities. FAO, Rome. 107 p

Forest Guild Pacific Northwest Biomass Working Group [Forest Guild] (2013) Forest biomass retention and harvesting guidelines for the Pacific Northwest. Forest Guild, Santa Fe. 23 p. Available at https://forestguild.org/publications/research/2013/FG_Biomass_Guidelines_PNW.pdf. Accessed 22 April 2019

Foster DR, Aber J (2004) Forests in time: ecosystem structure and function as a consequence of 1000 years of change. Synthesis volume of the Harvard forest long term ecological research program. Yale University Press, New Haven

Foster D, Swanson F, Aber J et al (2003) The importance of land-use legacies to ecology and conservation. Bioscience 53(1):77–88

Fox TR, Allen HL, Albaugh TJ et al (2007) Tree nutrition and forest fertilization of pine plantations in the southern United States. South J Appl For 31(1):5–11

Frandsen WH, Ryan KC (1986) Soil moisture reduces below-ground heat flux and soil temperatures under a burning fuel pile. Can J For Res 16:244–248

Friedel MH (1991) Range condition assessment and concepts of thresholds: a viewpoint. J Range Manag 44:422–426

Froehlich HA, McNabb DS (1984) Minimizing soil compaction in Pacific Northwest forests. In: Stone EL (ed) Forest soils and treatment impacts: proceedings of the 6th North American forest soils conference. University of Tennessee, Knoxville, pp 159–192

Fulé PZ, Waltz AEM, Covington WW, Heinlein TA (2001) Measuring forest restoration effectiveness in reducing hazardous fuels. J For 99(11):24–29

Fyfield TP, Gregory PJ (1989) Effects of temperature and water potential on germination, radical elongation, and emergence of mungbean. J Exp Bot 40:667–674

Gardner BD (1991) Rangeland resources: changing uses and productivity. In: Frederick DD, Sedjo RA (eds) America's renewable resources: historical trends and current challenges. Resources for the Future, Washington, DC, pp 123–166

Garrison-Johnston MT, Moore JA, Cook SP, Niehoff GJ (2003) Douglas-fir beetle infestations are associated with certain rock and stand types in the inland northwestern United States. Environ Entomol 32(6):1354–1363

Gaskin JW, Speir A, Morris LM et al (2007) Potential for pyrolysis char to affect soil moisture and nutrient status of a loamy sand soil. Proceedings of the Georgia water resources conference. University of Georgia, Athens, pp 1–3

Gomez A, Powers RF, Singer MJ, Horwath WR (2002) Soil compaction effects on growth of young ponderosa pine following litter removal in California's Sierra Nevada. Soil Sci Soc Am J 66(4):1334–1343

Greacen EL, Sands R (1980) Compaction of forest soils; a review. Aust J Soil Res 18:163–189

Grigal DA (2000) Effects of extensive forest management on soil productivity. For Ecol Manag 128:167–185

Grigal DF, Vance ED (2000) Influence of soil organic matter on forest productivity. N Z J For Sci 30:169–205

Guthrie RH (2002) The effects of logging on frequency and distribution of landslides in three watersheds on Vancouver Island, British Columbia. Geomorphology 43(3–4):273–292

Han S-K, Han S-H, Page-Dumroese DS, Johnson LR (2009) Soil compaction associated with cut-to-length and whole-tree harvesting of a conifer forest. Can J For Res 39:976–989

Hardegree SP, Flerchinger GN, Van Vactor SS (2003) Hydrothermal germination response and the development of probabilistic germination profiles. Ecol Model 167:305–322

Hardy CC (1996) Guidelines for estimating volume, biomass, and smoke production for piled slash, General Technical Report PNW-GTR-364. U.S. Department of Agriculture Forest Service, Pacific Northwest Station, Portland. 28 p

Harvey AE, Larsen MJ, Jurgensen MF (1979) Comparative distribution of ectomycorrhizae in soils of three western Montana forest habitat types. For Sci 25(2):350–358

Hassan R, Scholes R, Ash N (eds) (2005) Ecosystems and well-being: current state and trends. Island Press, Washington, DC

Hatchell GE, Ralston CW (1971) Natural recovery of surface soils disturbed in logging. Tree Planters Notes 2292:5–9

Havill NP, Shiyake S, Lamb Galloway A et al (2016) Ancient and modern colonization of North America by hemlock woolly adelgid, *Adelges tsugae* (Hemiptera: Adelgidae), an invasive insect from East Asia. Mol Ecol 25(9):2065–2080

Heiberg SO, Madgwick HAI, Leaf AL (1964) Some long-time effects of fertilization on red pine plantations. For Sci 10(1):17–23

Heneghan L, Fatemi F, Umek L et al (2006) The invasive shrub European buckthorn (*Rhamnus cathartica*, L.) alters soil properties in midwestern US woodlands. Appl Soil Ecol 32(1):142–148

Heneghan L, Miller SP, Baer S et al (2008) Integrating soil ecological knowledge into restoration management. Restor Ecol 16(4):608–617

Heninger RL, Terry T, Dobkowski A, Scott W (1997) Managing for sustainable site productivity: Weyerhaeuser's forestry perspective. Biomass Bioenergy 13(4/5):255–267

Heninger R, Scott W, Dobkowski A et al (2002) Soil disturbance and 10-year growth response of coast Douglas-fir on non-tilled and tilled skid roads in the Oregon cascades. Can J For Res 32:233–246

Herrick JE, Havstad KM, Rango A (2006a) Remediation research in the Jornada Basin: past and future. In: Havstad KM, Huenneke LF, Schlesinger WH (eds) Structure and function of a Chihuahuan Desert ecosystem: the Jornada Basin long-term ecological research site. Oxford University Press, New York, pp 278–304

Herrick JE, Schuman GE, Rango A (2006b) Monitoring ecological processes for restoration projects. J Nat Conserv 14:161–171

Hessburg PF, Agee JK, Franklin JF (2005) Dry forests and wildland fires of the inland Northwest USA: contrasting the landscape ecology of the pre-settlement and modern eras. For Ecol Manag 211(1–2):117–139

Hild A, Shaw N, Paige G, Williams M (2009) Integrated reclamation: approaching ecological function. In: Barnhisel RI (ed) Revitalizing the environment: proven solutions and innovative approaches. National Meeting of the American Society of Mining and Reclamation, Billings, pp 578–596

Hillel D (2004) Introduction to environmental soil physics. Elsevier Academic Press, San Diego. 494 p

Holling CS (1973) Resilience and stability of ecological systems. Annu Rev Ecol Syst 4:1–23

Hosonuma N, Herold M, De Sy V et al (2012) An assessment of deforestation and forest degradation drivers in developing countries. Environ Res Lett 7(4):044009

Jain TB, Pilliod DS, Graham RT et al (2012) Index for characterizing post-fire soil environments in temperate coniferous forests. Forests 3(3):455–466

James JJ, Svejcar TJ, Rinella MJ (2011) Demographic processes limiting seedling recruitment in aridland restoration. J Appl Ecol 48(4):961–969

Jangid K, Williams MA, Franzluebbers AJ, Schmidt TM, Coleman DC, Whitman WB (2011) Land-use history has a stronger impact on soil microbial community composition than aboveground vegetation and soil properties. Soil Biol Biochem 43(10):2184–2193

Jeffery S, Verheijen FG, van der Velde M, Bastos AC (2011) A quantitative review of the effects of biochar application to soils on crop productivity using meta-analysis. Agric Ecosyst Environ 144(1):175–187

Jirka S, Tomlinson T (2013) State of the biochar industry: a survey of commercial activity in the biochar field. [Publisher unknown], [Place of publication unknown]. Report of the International Biochar Initiative. 61 p

Johnstone JF, Allen CD, Franklin JF et al (2016) Changing disturbance regimes, ecological memory, and forest resilience. Front Ecol Environ 14(7):369–378

Jokela EJ, Martin TA, Vogel JG (2010) Twenty-five years of intensive forest management with southern pines: important lessons learned. J For 108(7):338–347

Jolly WM, Cochrane MA, Freeborn PH et al (2015) Climate-induced variations in global wildfire danger from 1979 to 2013. Nat Commun 6:7537

Jones G, Loeffler D, Calkin D, Chung W (2010) Forest treatment residues for thermal energy compared with disposal by on-site burning: emissions and energy return. Biomass Bioenergy 34(5):737–746

Jónsson JOG, Davíðsdóttir B, Jónsdóttirb EM et al (2016) Soil indicators for sustainable development: a transdisciplinary approach for indicator development using expert stakeholders. Agric Ecosyst Environ 232:179–189

Jurgensen MF, Harvey AE, Graham RT, Page-Dumroese DS, Tonn JR, Larsen MJ, Jain TB (1997) Impacts of timber harvesting on soil organic matter, nitrogen, productivity, and health of inland northwest forests. For Sci 43(2):234–251

Kammann CI, Linsel S, Gößling JW, Koyro H-W (2011) Influence of biochar on drought tolerance of Chenopodium quinoa Willd and on soil–plant relations. Plant Soil 345(1–2):195–210

Keane RE, Karau E (2010) Evaluating the ecological benefits of wildfire by integrating fire and ecosystem simulation models. Ecol Model 221(8):1162–1172

Kimsey MJ, Garrison-Johnston MT, Johnson L (2011) Characterization of volcanic ash-influenced forest soils across a geoclimatic sequence. Soil Sci Soc Am J 75(1):267–279

King EG, Hobbs RJ (2006) Identifying linkages among conceptual models of ecosystem degradation and restoration: towards an integrative framework. Restor Ecol 14(3):369–378

Kinney TJ, Masiello CA, Dugan B et al (2012) Hydrologic properties of biochars produced at different temperatures. Biomass Bioenergy 41:34–43

Knight KS, Brown JP, Long RP (2012) Factors affecting the survival of ash (Fraxinus spp.) trees infested by emerald ash borer (Agrilus planipennis). Biol Invasions 15(2):371–383

Knutson KC, Pyke DA, Wirth TA et al (2014) Long-term effects of seeding after wildfire on vegetation in Great Basin shrubland ecosystems. J Appl Ecol 51(5):1414–1424

Kozlowski TT (1999) Soil compaction and growth of woody plants. Scand J For Res 14(6):596–619

Kruse R, Bend E, Bierzychudek P (2004) Native plant regeneration and introduction of nonnatives following post-fire rehabilitation with straw mulch and barley seeding. For Ecol Manag 196(2–3):299–310

Laird DA, Fleming P, Davis DD et al (2010) Impact of biochar amendments on the quality of a typical Midwestern agricultural soil. Geoderma 158(3–4):443–449

Lal R, Stewart BA (1992) Need for land restoration. In: Soil restoration. Springer, New York, pp 1–11

Lamb D, Erskine PD, Fletcher A (2015) Widening gap between expectations and practice in Australian minesite rehabilitation. Ecol Manag Restor 16(3):186–195

Landsberg JD, Miller RE, Anderson HW, Tepp JS (2003) Bulk density and soil resistance to penetration as affected by commercial thinning operations in northeastern Washington, Research Paper PNW-RP-551. U.S. Department of Agriculture, Forest Service, Pacific Northwest Research Station, Portland. 35 p

Lehmann J (2007) Bio-energy in the black. Front Ecol Environ 5(7):381–387

Leopold A (1933) Game management. University of Wisconsin Press, 481 p

Levin SA (1992) The problem of pattern and scale in ecology: the Robert H. MacArthur award lecture. Ecology 73(6):1943–1967

Lewis SL, Maslin MA (2015) Defining the anthropocene. Nature 519(7542):171–180

Liang B, Lehmann J, Solomon D et al (2006) Black carbon increases cation exchange capacity in soils. Soil Sci Soc Am J 70(5):1719–1730

Lovett GM, Canham CD, Arthur MA et al (2006) Forest ecosystem responses to exotic pests and pathogens in eastern North America. Bioscience 56(5):395–405

Luce CH (1997) Effectiveness of road ripping in restoring infiltration capacity of forest roads. Restor Ecol 5(3):265–270

Lyman CK (1947) Slash disposal as related to fire control on the national forests of western Montana and northern Idaho. J For 45:259–262

MacCleery DW (1993) American forests: a history of resiliency and recovery, FS-540. U.S. Department of Agriculture, Forest Service; Forest History Society, Durham

MacCleery DW (2004) The historical record for the pathway hypothesis. In: Fedkiw J, MacCleery DW, Sample VA (eds) Pathway to sustainability: defining the bounds on forest management. Forest History Society, Durham, pp 19–42

Madsen MD, Zvirzdin DL, Petersen SL et al (2011) Soil water repellency within a burned piñon–juniper woodland: spatial distribution, severity, and ecohydrologic implications. Soil Sci Soc Am J 75(4):1543

Madsen MD, Kostka SJ, Inouye AL, Zvirzdin DL (2012) Postfire restoration of soil hydrology and wildland vegetation using surfactant seed coating technology. Rangel Ecol Manag 65(3):253–259

Maestas JD, Campbell SB, Chambers JC et al (2016) Tapping soil survey information for rapid assessment of sagebrush ecosystem resilience and resistance. Rangelands 38(3):120–128

Mann LK, Johnson DW, West DC et al (1988) Effects of whole-tree and stem-only clearcutting on postharvest hydrologic losses, nutrient capital, and regrowth. For Sci 34(2):412–428

Margulies E, Bauer L, Ibáñez I (2017) Buying time: preliminary assessment of biocontrol in the recovery of native forest vegetation in the aftermath of the invasive emerald ash borer. Forests 8(10):369–383

McBratney A, Field DJ, Koch A (2014) The dimensions of soil security. Geoderma 213:203–213

McDonald T, Gann GD, Jonson J, Dixon KW (2016) International standards for the practice of ecological restoration—including principles and key concepts. Society for Ecological Restoration, Washington, DC

McIver JD, Adams PW, Doyal JA et al (2003) Environmental effects and economics of mechanized logging for fuel reduction in northeastern Oregon mixed-conifer stands. West J Appl For 18(4):238–249

Meyer S, Glaser B, Quicker P (2011) Technical, economical, and climate-related aspects of biochar production technologies: a literature review. Environ Sci Technol 45:9473–9483

Meyer SE, Beckstead J, Pearce J (2016) Community ecology of fungal pathogens on Bromus tectorum. In: Exotic brome-grasses in arid and semiarid ecosystems of the western US. Springer International Publishing, Cham, pp 193–223

Middleton EL, Bever JD (2012) Inoculation with a native soil community advances succession in a grassland restoration. Restor Ecol 20(2):218–226

Miller D, Anderson H (2002) Soil compaction: concerns, claims, and evidence. In: Baumgartner D, Johnson L, DePuit E, comps. Proceedings on small diameter timber: resource management, manufacturing, and markets. Bulletin MISC0509. Washington State University Cooperative, Spokane, pp 16–18

Miller RF, Chambers JC, Pellant M (2014) A field guide for selecting the most appropriate treatment in sagebrush and piñon-juniper ecosystems in the Great Basin: evaluating resilience to disturbance and resistance to invasive annual grasses, and predicting vegetation response, General Technical Report RMRS-GTR-322-rev. U.S. Department of Agriculture, Forest Service, Rocky Mountain Research Station, Fort Collins. 68 p

Miller RF, Harrington TB, Anderson HW (2016) Stand dynamics of Douglas-fir 20 years after precommercial thinning and nitrogen fertilization on a poor-quality site, Research Paper PNW-RP-606. U.S. Department of Agriculture Forest Service, Pacific Northwest Research Station, Portland. 66 p

Moody JA, Martin DA (2009) Synthesis of sediment yields after wildland fire in different rainfall regimes in the western United States. Int J Wildland Fire 18(1):96–115

Morris LR (2011) Land-use legacies of cultivation in shrublands: ghosts in the ecosystem. Nat Resour Environ Issues 17(1):1–6

Morris LR, Monaco TA, Sheley RL (2011) Land-use legacies and vegetation recovery 90 years after cultivation in Great Basin sagebrush ecosystems. Rangel Ecol Manag 64(5):488–497

Morris LR, Monaco TA, Leger E et al (2013) Cultivation legacies alter soil nutrients and differentially affect plant species performance nearly a century after abandonment. Plant Ecol 214(6):831–844

Morris LR, Monaco TA, Sheley RL (2014) Impact of cultivation legacies on rehabilitation seedings and native species reestablishment in Great Basin shrublands. Rangel Ecol Manag 67(3):285–291

Morrison ML (1987) The design and importance of long-term ecological studies: analysis of vertebrates in the Inyo-White Mountains, California. In: Szaro RC, Severson KE, Patton DR, tech. coord. Management of amphibians, reptiles, and small mammals in North America. General Technical Report RM-GTR-166. U.S. Department of Agriculture, Forest Service, Rocky Mountain Forest and Range Experiment Station, Fort Collins, pp 267–275

Mukherjee A, Lal R (2013) Biochar impacts on soil physical properties and greenhouse gas emissions. Agronomy 3(2):313–339

Munson SM, Belnap J, Okin G (2011) Responses of wind erosion to climate induced vegetation changes on the Colorado plateau. Proc Natl Acad Sci 108:3854–3859

National Council for Air and Stream Improvement, Inc. [NCASI] (2004) Effects of heavy equipment on physical properties of soils and on long-term productivity: a review of literature and current research, Technical bulletin 887. National Council for Air and Stream Improvement, Inc., Research Triangle Park

National Research Council (1994) Rangeland health: new methods to classify, inventory, and monitor rangelands. National Academies Press, Washington, DC. 180 p

Nauman TW, Thompson JA, Teets SJ, Dilliplane TA, Bell JW, Connolly SJ, Liebermann HJ, Yoast KM (2015a) Ghosts of the forest: mapping pedomemory to guide forest restoration. Geoderma 247–248:51–64

Nauman TW, Thompson JA, Teets J et al (2015b) Pedoecological modeling to guide forest restoration using ecological site descriptions. Soil Sci Soc Am J 79(5):1406–1419

Neary DG, Ryan KC, DeBano LF eds (2005) (revised 2008) Wildland fire in ecosystems: effects of fire on soils and water. General technical report RMRS-GTR-42-vol. 4. U.S. Department of Agriculture, Forest Service, Rocky Mountain Research Station, Ogden

Neher DA, Williams KM, Lovell ST (2017) Environmental indicators reflective of road design in a forested landscape. Ecosphere 8(3):01734

Novak JM, Busscher WJ (2013) Selection and use of designer biochars to improve characteristics of southeastern USA coastal plain degraded soils. In: Advanced biofuels and bioproducts. Springer, New York, pp 69–96

Novak JM, Busscher WJ, Laird DL et al (2009) Impact of biochar amendment on fertility of a southeastern coastal plain soil. Soil Sci 174(2):105–112

Novak JM, Busscher WJ, Watts DW et al (2012) Biochars impact on soil moisture storage in an Ultisol and two Aridisols. Soil Sci 177(5):310–320

Oswalt SN, Smith WB, Miles PD, Pugh SA (2014) Forest resources of the United States, 2012: a technical document supporting the forest service 2015 update of the RPA assessment, General Technical Report WO-GTR-91. U.S. Department of Agriculture, Forest Service, Washington Office, Washington, DC. 218 p

Padamsee M, Johansen RB, Stuckey SA et al (2016) The arbuscular mycorrhizal fungi colonising roots and root nodules of New Zealand kauri *Agathis australis*. Fungal Biol 120:807–817

Page-Dumroese DS, Jurgensen MF (2006) Soil carbon and nitrogen pools in mid- to late-successional forest stands of the northwestern USA: potential impact of fire. Can J For Res 36(9):2270–2284

Page-Dumroese DS, Jurgensen MF, Tiarks AE et al (2006) Soil physical property changes at the North American long-term soil productivity study sites: 1 and 5 years after compaction. Can J For Res 36(3):551–564

Page-Dumroese DS, Jurgensen MF, Terry T (2009) Maintaining soil productivity during forest or biomass-to-energy thinning harvests in the western United States. West J Appl For 25(1):5–11

Page-Dumroese DS, Robichaud PR, Brown RE, Tirocke JM (2015) Water repellency of two forest soils after biochar addition. Trans Am Soc Agric Biol Eng 58(2):335–342

Page-Dumroese DS, Busse MD, Archuleta JG et al (2017a) Methods to reduce forest residue volume after timber harvesting and produce black carbon. Scientifica 2017:2745764. 8 p

Page-Dumroese DS, Coleman MD, Thomas SC (2017b) Opportunities and uses of biochar on forest sites in North America. In: Bruckman VJ, Varol EA, Uzun BB, Liu J (eds) Biochar: a regional supply chain approach in view of climate change mitigation. Cambridge University Press, Cambridge, pp 315–334

Page-Dumroese DS, Ott MR, Strawn DG, Tirocke JM (2018) Using organic amendments to restore soil physical and chemical properties of a mine site in northeastern, Oregon, USA. Appl Eng Agric 34:43–55

Pellant M, Shaver P, Pyke DA, Herrick JE (2005) Interpreting indicators of rangeland health, version 4, Technical reference 1734-6. U.S. Department of the Interior, Bureau of Land Management, National Science and Technology Center, Denver. 122 p

Perala DA, Alban DH (1982) Biomass, nutrient distribution and litterfall in *Populus, Pinus* and *Picea* stands on two different soils in Minnesota. Plant Soil 64(2):177–192

Peterman W, Ferschweiler K (2016) A case study for evaluating potential soil sensitivity in arid land systems. Integr Environ Assess Manag 12(2):388–396

Peterson DW, Reich PB (2001) Prescribed fire in oak savanna: fire frequency effects on stand structure and dynamics. Ecol Appl 11(3):914–927

Pilliod DS, Welty JL, Toevs GR (2017) Seventy-five years of vegetation treatments on public rangelands in the great basin of North America. Rangelands 39(1):1–9

Ponder F Jr, Fleming RL, Berch S et al (2012) Effects of organic matter removal, soil compaction, and vegetation control on 10th year biomass and foliar nutrition: LTSP continent-wide comparisons. For Ecol Manag 278:35–54

Potter KM, Conkling BL (eds) (2017) Forest health monitoring: national status, trends, and analysis 2016, General Technical Report SRS-GTR-222. U.S. Department of Agriculture, Forest Service, Southern Research Station, Asheville. 195 p

Powers RF (2006) LTSP: genesis of the concept and principles behind the program. Can J For Res 36:519–529

Powers RF, Scott DA, Sanchez FG et al (2005) The north American long-term soil productivity experiment: findings from the first decade of research. For Ecol Manag 220:31–50

Rango A, Huenneke L, Buonopane M et al (2005) Using historic data to assess effectiveness of shrub removal in southern New Mexico. J Arid Environ 62:75–91

Reeves MC, Mitchell JE (2011) Extent of coterminous U.S. rangelands: quantifying implications of differing agency perspectives. Rangel Ecol Manag 64:585–597

Rhoades CC, Fornwalt PJ (2015) Pile burning creates a fifty-year legacy of openings in regenerating lodgepole pine forests in Colorado. For Ecol Manag 226:203–209

Rhoades CC, Fornwalt PJ, Paschke MW et al (2015) Recovery of small pile burn scars in conifer forests of the Colorado front range. For Ecol Manag 347:180–187

Rieman B, Clayton J (1997) Wildlife and native fish: issues of forest health of sensitive species. Fisheries 22(11):6–15

Ringo C, Bennett K, Noller J et al (2018) Modeling droughty soils at regional scales in Pacific Northwest Forests, USA. For Ecol Manag 424:121–135

Ritters KH, O'Neill RV, Jones KB (1997) Assessing habitat suitability at multiple scales: a landscape-level approach. Biol Conserv 81:191–202

Robichaud PR, Lewis SA, Brown RE, Ashmun LE (2009) Emergency post-fire rehabilitation treatment effects on burned area ecology and long-term restoration. Fire Ecol 5(1):115–128

Rowe JS (1996) Land classification and ecosystem classification. Environ Monit Assess 39(1–3):11–20

Ruiz-Jean MC, Aide TM (2005) Restoration success: how is it being measured? Restor Ecol 13(3):569–577

Rummer B, Prestemon J, May D et al (2005) A strategic assessment of forest biomass and fuel reduction treatments in western states, General Technical Report RMRS-GTR-149. U.S. Department of Agriculture, Forest Service, Rocky Mountain Research Station, Fort Collins. 17 p

Sampson AW (1923) Range and pasture management. Wiley, New York

Schnepf C, Graham RT, Kegley S, Jain TB (2009) Managing organic debris for forest health, Pacific northwest extension publication PNW 609. Univeristy of Idaho, Moscow, p 60

Schuman GE (2002) Mined land reclamation in the northern Great Plains: have we been successful? In: Proceedings 19th annual meeting, American Society of Mining and Reclamation. American Society of Mining and Reclamation, Lexington, pp 9–13

Schuman GE, Booth DT, Cockrell JR (1998) Cultural methods for establishing Wyoming big sagebrush on mined lands. J Range Manag 51:223–230

Sedell JR, Leone FN, Duval WS (1991) Water transportation and storage of logs. In: Meehan WR (ed) Influences of forest and rangeland management on salmonid fishes and their habitats, Special publication 19. American Fisheries Society, Bethesda, pp 325–368

Senyk JP, Craigdallie D (1997) Effects of harvest methods on soil properties and forest productivity in interior British Columbia, Informational report BC-X-367. Natural Resources Canada, Canadian Forest Service, Pacific Forestry Centre, Victoria. 37 p

Sharma BD, Sidhu PS, Nayyar VK (1992) Distribution of micronutrients in arid zone soils of Punjab and their relation with soil properties. Arid Soil Res Rehabil 6:233–242

Silversides C (1997) Broadaxe to flying shear: the mechanization of forest harvesting east of the Rockies. National Museum of Science and Technology, Ottawa. 174 p

Simmons FC (1949) Since the days of Leif Ericson. In: Stefferud A (ed) Trees: the yearbook of agriculture. U.S. Department of Agriculture, Washington, DC, pp 697–694

Smith RS, Shiel RS, Bardgett RD et al (2003) Soil microbial community, fertility, vegetation, and diversity as targets in the restoration management of a meadow grassland. J Appl Ecol 40:51–64

Society for Ecological Restoration [SER] International Science and Policy Working Group [SER] (2004) The SER international primer on ecological restoration. Available at http://www.ser.org/. Accessed 10 Apr 2019

Society for Range Management [SRM] Glossary Update Task Group (1998) Glossary of terms used in range management, 4th edition. Available at https://globalrangelands.org/glossary/

Society for Range Management [SRM] Task Group on Unity in Concepts and Terminology Committee Members (1995) New concepts for assessment of rangeland condition. J Range Manag 48:271–282

Solaiman ZM, Murphy DV, Abbott LK (2012) Biochars influence seed germination and early growth of seedlings. Plant Soil 353(1–2):273–287

Sosa-Pérez G, MacDonald LH (2017) Effects of closed roads, traffic, and road decommissioning on infiltration and sediment production: a comparative study using rainfall simulations. Catena 159:93–105

Spears JD, Holub SM, Harmon ME, Lajtha K (2003) The influence of decomposing logs on soil biology and nutrient cycling in an old-growth mixed coniferous forest in Oregon, USA. Can J For Res 33(11):2193–2201

Stanturf JA, Palik BJ, Dumroese RK (2014) Contemporary forest restoration: a review emphasizing function. For Ecol Manag 331:292–323

Stone DM (2002) Logging options to minimize soil disturbance in the northern Lake States. North J Appl For 19(3):115–121

Stott DE, Martin JP (1989) Organic matter decomposition and retention in arid soils. Arid Soil Res Rehabil 3:115–148

Strayer D, Glitzenstein JS, Jones CG et al (1986) Long-term ecological studies: an illustrated account of their design, operation, and importance to ecology. Institute of Ecosystem Studies, Millbrock, pp 1–38

Sun G, Zhang L, Duan K, Rau B (2017) Impacts of forest biomass removal on water yield across the United States. In: Efroymson RA, Langholtz MH, Johnson KE, Stokes BJ (eds) Billion-ton report: advancing domestic resources for a thriving bioeconomy, volume 2—environmental sustainability effects of select scenarios from volume 1. Oak Ridge National Laboratory, Oak Ridge. 640 p

Surface Mining Control and Reclamation Act; Act of August 3, 1977; 30 U.S.C. 1201 et seq

Swetnam TW, Allen CD, Betancourt JL (1999) Applied historical ecology: using the past to manage for the future. Ecol Appl 9:1189–1206

Switalski T, Bissonette J, DeLuca TH et al (2004) Benefits and impacts of road removal. Front Ecol Environ 2(1):21–28

Taylor AG, Harman GE (1990) Concepts and technologies of selected seed treatments. Annu Rev Phytopathol 28(1):321–339

Thibodeau L, Raymond P, Camiré C, Munson AD (2000) Impact of pre-commercial thinning in balsam fir stands on soil nitrogen dynamics, microbial biomass, decomposition, and foliar nutrition. Can J For Res 30:229–238

Thomas P, Packham J (2007) Ecology of woodlands and forests: description, dynamics, and diversity. Cambridge University Press, Cambridge. 481 p

Tongway DJ, Valentin C, Seghieri J (eds) (2001) Banded vegetation patterning in arid and semiarid environments: ecological processes and consequences for management. Springer-Verlag, New York

U.S. Department of Agriculture, Forest Service [USDA FS] (2012) Forest service manual, FSM 250 watershed and air management, chapter 2550 soil management, R9 supplement. U.S. Department of Agriculture, Northeast Region, Milwaukie

U.S. Department of Agriculture [USDA] (2003) National Range and Pasture Handbook. U.S. Department of Agriculture, Natural Resources Conservation Service, Grazing Lands Technology Institute, Washington, DC. 214 p

U.S. Department of Agriculture, Forest Service [USDA FS] (2010) Forest Service Manual, FSM 2500 watershed and air management, chapter 2550 soil management. WO amendment 2500-2010-1. U.S. Department of Agriculture, National Headquarters, Washington, DC

U.S. Department of Agriculture, Forest Service [USDA FS] (2017) Fiscal year 2018 budget justification. U.S. Department of Agriculture, Washington, DC. 319 p

U.S. Department of Agriculture, Forest Service [USDA FS] (2018) Colville National Forest land management plan (Ferry, Pend Oreille, and Steven Counties, Washington). U.S. Department of Agriculture, Pacific Northwest Region, Portland

U.S. Department of the Interior [US DOI] (2015) Rangeland fire prevention, management, and restoration. Secretary order number 3336. U.S. Department of the Interior, Washington, DC. http://www.forestsandrangelands.gov/rangeland/documents/SecretarialOrder3336.pdf

U.S. Department of the Interior, Bureau of Land Management [BLM] (2017) Budget justifications and performance information, fiscal year 2018. U.S. Department of the Interior, Washington, DC. 446 p

U.S. Government (2014) United States of America spending. Available at http://www.usaspending.gov/

Vallejo R, Serrasolses I, Alloza JA et al (2009) Long-term restoration strategies and techniques. In: Fire effects on soils and restoration strategies. Science Publishers, Enfield, pp 373–398

Van Zwieten L, Kimber S, Morris S et al (2010) Effects of biochar from slow pyrolysis of papermill waste on agronomic performance and soil fertility. Plant Soil 327:235–246

von Wilpert K, Schäffer J (2006) Ecological effects of soil compaction and initial recovery dynamics: a preliminary study. Eur J For Res 125:129–138

Walton M (2005) Spatial patterning of resource accumulation in a 22 year-old water harvesting project in the Chihuahuan Desert. University of Dayton, PhD dissertation, Dayton, OH

Walton M, Herrick JE, Gibbens RP, Remmenga M (2001) Persistence of biosolids in a Chihuahuan Desert rangeland 18 years after application. Arid Land Res Manag 15:223–232

Wang J, LeDoux CB, Edwards P, Jones M (2005) Soil bulk density changes caused by mechanized harvesting: a case study in central Appalachia. For Prod J 55(11):37–40

Wertz WA, Arnold JF (1972) Land systems inventory. U.S. Department of Agriculture, Forest Service, Ogden. 12 p

Westoby M, Walker B, Noy-Meir I (1989) Opportunistic management for rangelands not at equilibrium. J Range Manag 42(4):266–274

Wiens JA (1984) The place for long-term studies in ornithology. Auk 101:202–203

Wiens JA (1986) Spatial scale and temporal variation in studies of shrubsteppe birds. In: Diamond J, Case TJ (eds) Community ecology. Harper and Row, New York, pp 154–172

Wiens JA (1989) Spatial scaling in ecology. Funct Ecol 3(4):385–397

Wilde SA (1958) Forest soils. The Ronald Press Company, New York. 537 p

Wilkinson BH (2005) Humans as geologic agents: a deep-time perspective. Geology 33:161–164

Williams M (1989) Americans and their forests: a historical geography. Cambridge University Press, Cambridge. 624 p

Williams MI, Paige GB, Thurow TL et al (2011) Songbird relationships to shrub-steppe ecological site characteristics. Rangel Ecol Manag 64(2):109–118

Williams CJ, Pierson FB, Spaeth KE et al (2016a) Application of ecological site information to transformative changes on Great Basin sagebrush rangelands. Rangelands 38(6):379–388

Williams MI, Dumroese RK, Page-Dumroese DS, Hardegree SP (2016b) Can biochar be used as a seed coating to improve native plant germination and growth in arid conditions? J Arid Environ 125:8–15

Williamson JR, Neilsen WA (2000) The influence of forest site on rate and extent of soil compaction and profile disturbance of skid trails during ground-based harvesting. Can J For Res 30:1196–1205

Wubs ERJ, van der Putten WH, Bosch M, Bezemer TM (2016) Soil inoculation steers restoration of terrestrial ecosystems. Nat Plants 2(8):16107

Wyoming Department of Environmental Quality (1996) Coal rules and regulations. Wyoming Department of Environmental Quality, Land Quality Division, Cheyenne. Chapter 4

Young KE, Grover HS, Bowker MA (2016) Altering biocrusts for an altered climate. New Phytol 210:18–22

Zhang J, Busse MD, Young DH et al (2017) Aboveground biomass responses to organic matter removal, soil compaction, and competing vegetation control on 20-year mixed conifer plantations in California. For Ecol Manag 401:341–353

Zhu AX (2000) Mapping soil landscape as spatial continua: the neural network approach. Water Resour Res 36(3):663–677

Soil Mapping, Monitoring, and Assessment

Mark J. Kimsey, Larry E. Laing, Sarah M. Anderson,
Jeff Bruggink, Steve Campbell, David Diamond,
Grant M. Domke, James Gries, Scott M. Holub,
Gregory Nowacki, Deborah S. Page-Dumroese,
Charles H. (Hobie) Perry, Lindsey E. Rustad,
Kyle Stephens, and Robert Vaughan

Introduction

Soils are a nonrenewable resource that support a wide array of ecosystem functions. The scope of these functions depends on the nature and properties of the soil at a given location on the Earth. Demand for better soil information has been growing since the development of soil science in the nineteenth century. This recent interest is driven by an increasing recognition of the ecological, economic, and societal benefits of understanding soil properties and the value of that knowledge for realizing management objectives for agriculture, grazing, forestry, and other land uses. Soil surveys are one method for amassing soil data and mapping the extent of various soil types. The Federal Government has singularly been a long-term sponsor of soil surveys in the United States. The history of these surveys is richly documented and illustrated by Helms et al. (2008). Soil surveys describe horizontal (e.g., soil series) and vertical (e.g., horizon depth) properties of soils. Soil mapping enhances assessments of spatial variability in the development and properties of soils as a function of geology, climate, topography, and vegetation. Extensive sampling of soils in concert with other attributes (e.g., forest or rangeland composition) can provide focused estimates and understanding of the linkages between soils and vegetation growth, mortality, and C stocks (O'Neill et al. 2005) (Box 9.1). Thus, soils are not independent of biogeophysical settings and climate, but rather are a result of these variables. Management interpretations of soil functions and processes such as erosion, potential vegetation growth, and

M. J. Kimsey (✉)
College of Natural Resources, University of Idaho,
Moscow, ID, USA
e-mail: kimsey@uidaho.edu

L. E. Laing
U.S. Department of Agriculture, Forest Service, Forest
Management, Range Management, and Vegetation Ecology,
Washington, DC, USA

S. M. Anderson
U.S. Department of Agriculture, Forest Service, Forest
Management, Range Management, and Vegetation Ecology,
Washington, DC, USA

J. Bruggink
U.S. Department of Agriculture, Forest Service, Intermountain
Region, Ogden, UT, USA

S. Campbell
U.S. Department of Agriculture, Natural Resources Conservation
Service, West National Center, Portland, OR, USA

D. Diamond
Greater Yellowstone Coordinating Committee,
Bozeman, MT, USA

G. M. Domke · C. H. (Hobie) Perry
U.S. Department of Agriculture, Forest Service, Northern Research
Station, St. Paul, MN, USA

J. Gries
U.S. Department of Agriculture, Forest Service, Eastern Region,
Milwaukee, WI, USA

S. M. Holub
Weyerhaeuser, Springfield, OR, USA

G. Nowacki
U.S. Department of Agriculture, Forest Service, Eastern Region,
Milwaukee, WI, USA

D. S. Page-Dumroese
Rocky Mountain Research Station, USDA Forest Service,
Moscow, ID, USA

L. E. Rustad
U.S. Department of Agriculture, Forest Service, Northern Research
Station, Center for Research on Ecosystem Change,
Durham, NH, USA

K. Stephens
U.S. Department of Agriculture, Natural Resources Conservation
Service, Portland, OR, USA

R. Vaughan
U.S. Department of Agriculture, Forest Service,
Geospatial Technology Applications Center,
Salt Lake City, UT, USA

Box 9.1 Case Study

Use of Soil Maps for Vegetation Classification and Management in the Southern United States.

The relationship of soils to potential vegetation is a key element in the use of soil maps to develop current vegetation maps. The USDA Natural Resources Conservation Service (USDA NRCS) defines ecological sites as geophysical settings that support similar plant communities under similar management and disturbance regimes, and each ecological site type is unique (see USDA NRCS 2012). In that regard, they are similar to ecological site type and site type phase concepts used by the USDA Forest Service in its hierarchal classification of ecoregions, land types, and site types (ECOMAP 1993).

Statewide digital soil maps served as key input data for mapping ecological systems at relatively fine resolution in Texas and Oklahoma, and similar efforts are underway for Kansas and Nebraska (Diamond and Elliott 2015). Ecological systems represent individual or groups of co-located plant communities (see NatureServe 2019).

In Texas and Oklahoma, digital Soil Survey Geographic (SSURGO) datasets were intersected with land cover from satellite remote sensing to produce relatively fine-resolution (10 m) current vegetation maps (Box Fig. 9.1). Groups of similar ecological site types, informed by site type descriptions (see USDA NRCS n.d.), were used to infer current vegetation (Box Fig. 9.2). The location and extent of ruderal and invasive vegetation types were inferred based on the soil maps. For example, evergreen forest and woodland land cover on prairie soils were classified as "Ruderal Eastern Redcedar Woodland and Forest" (Box Fig. 9.3).

For the national forests and grasslands in Texas, additional information on the plant communities as related to soils and landforms was available, and these data were used to map ecological land types and land type phases (Box Fig. 9.4) (Diamond and Elliott 2010; Van Kley et al. 2007). These maps have been improved and modified and serve as guides to forest management and planning. For example, the location of potential longleaf pine (*Pinus palustris*) restoration areas was mapped (Box Fig. 9.5).

The key element enabling use of soil maps for vegetation mapping and definition of management alternatives is explicitly relating soils to ecological site or ecological land type concepts. Management actions (e.g., timber harvesting, grazing) will have different impacts on the same plant community when it occurs over different soils and ecological sites. Hence, replanting of longleaf pine in southeastern Texas is most likely to be successful on soils and in geophysical settings that once supported

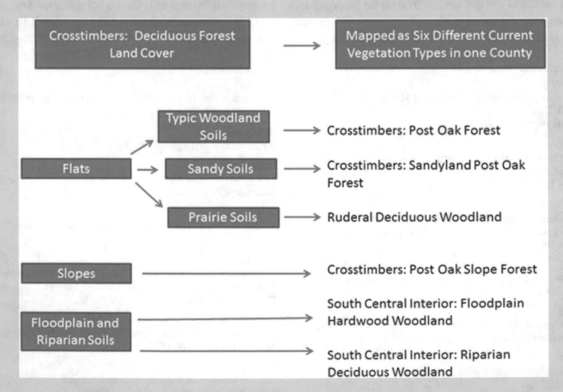

Box Fig. 9.1 Six different deciduous forest ecological system types were mapped for the Crosstimbers in Texas and Oklahoma based on differences in soils and percent slope. (Source: Soil data from NRCS SSURGO maps; land cover from satellite remote sensing)

(continued)

Box 9.1 (continued)

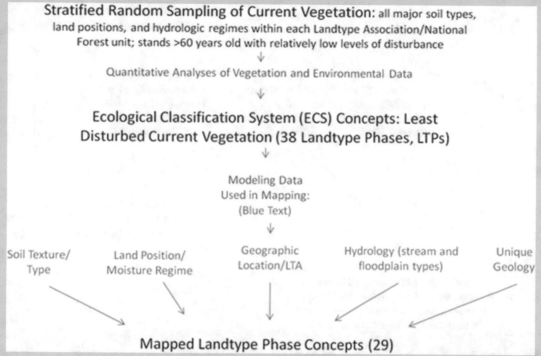

Box Fig. 9.2 Outline of the development of Ecological Classification System (ECS) land type phase (LTP) concepts and modeling data used for mapping these concepts. Soil maps are a key component in this process

Box Fig. 9.3 Ruderal (invasive) deciduous community types (shown in green) and eastern redcedar (*Juniperus virginiana*) community types (shown in red) were mapped in Payne County, Oklahoma, by referencing ecological site descriptions of the historic vegetation as documented in the Soil Survey Geographic (SSURGO) dataset. Urban land cover in Stillwater, OK, (upper left) and Cushing, OK, (lower right) are shown in gray

longleaf pine communities. Harvesting and planting of any given tree species on sands versus clays, or on steep slopes versus flats, may result in different future timber volume, or possibly even in different plant community types, decades later. Management of soils can thus be viewed as inseparable from management of plant communities, and knowledge of the soils (ecological sites) will inform appropriate management options and expectations.

(continued)

Box 9.1 (continued)

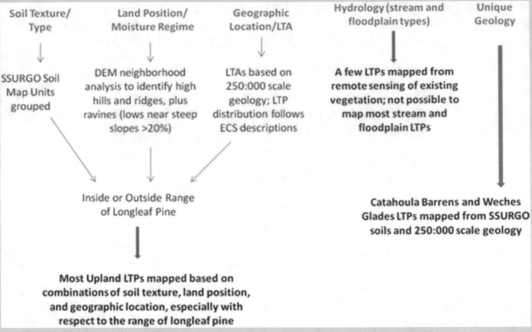

Soil Texture/Type → SSURGO Soil Map Units grouped

Land Position/Moisture Regime → DEM neighborhood analysis to identify high hills and ridges, plus ravines (lows near steep slopes >20%)

Geographic Location/LTA → LTAs based on 250:000 scale geology; LTP distribution follows ECS descriptions

Hydrology (stream and floodplain types) → A few LTPs mapped from remote sensing of existing vegetation; not possible to map most stream and floodplain LTPs

Unique Geology → Catahoula Barrens and Weches Glades LTPs mapped from SSURGO soils and 250:000 scale geology

Inside or Outside Range of Longleaf Pine

Most Upland LTPs mapped based on combinations of soil texture, land position, and geographic location, especially with respect to the range of longleaf pine

Box Fig. 9.4 Primary data layers used for modeling and mapping of Ecological Classification System (ECS) land type phase (LTP) concepts. Twenty-nine of 38 LTP concept types defined from field sampling were mapped. (SSURGO=Soil Survey Geographic database; DEM = digital elevation model)

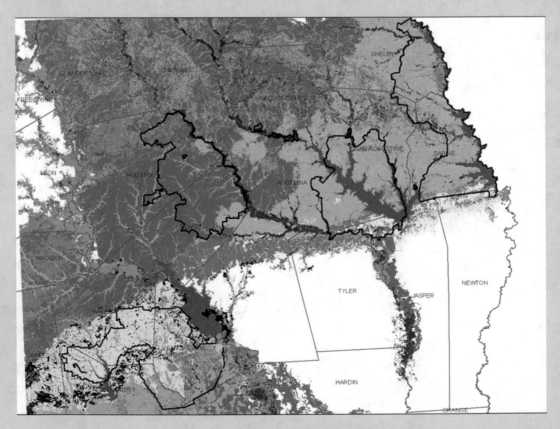

Box Fig. 9.5 Ecological land types for the area surrounding four National Forest Units in southeastern Texas (black outlines). Each color represents a unique ecological site type based on soils and landforms. For example, tan, yellow, and olive colors indicate site types where longleaf pine (*Pinus palustris*) communities historically occurred, and indicate potential locations where this community type could be restored hydrologic function integrate these factors and offer an index for land use limitations and opportunities.

In early soil mapping efforts, flat paper maps were produced by taking information from field surveys and combining it with information from aerial imagery obtained through the use of stereoscopes. Later, these maps have been georeferenced and converted to digital data products. Modern techniques involve more accurate georeferencing, use of digital elevation models from remote sensing or ground measurements, and inclusion of vegetation maps, which help provide more detailed mapping at a variety of scales.

Initially, soil survey efforts in the United States focused on farmlands where the need to understand and manage soils in association with crop production was most keen. Over time, Federal- and state-managed forests and rangelands were mapped. These latter efforts placed greater emphasis on plant communities as they relate to landscape position and soil characteristics, and fine-resolution units were grouped within spatially nested hierarchies (Schoeneberger et al. 2012). Early soil mapping was typically conducted at the scale of 1:24,000 (e.g., 1 map cm = 24,000 ground cm), although some wildlands were mapped at coarser scales (e.g., 1:63,360). Soil surveys are now generally available online and can be accessed through portals such as Web Soil Survey (https://websoilsurvey.sc.egov.usda.gov/), a Web-based soil property mapping program developed by the US Department of Agriculture's Natural Resources Conservation Service (USDA NRCS).

The USDA NRCS and the USDA Forest Service (hereafter, Forest Service) use different multi-scale approaches to capture and provide context to soil information. Each agency has spatially nested, multi-resolution mapping systems, but these are not uniformly complete across the United States. The USDA NRCS products most often focus on soil, whereas the Forest Service explicitly considers soils and vegetation together as an integrated unit for mapping (e.g., Terrestrial Ecological Unit Inventory (TEUI); https://www.fs.fed.us/soils/teui.shtml). Both agencies are moving toward more integrated products that capitalize on an ecological, spatially nested, and multi-resolution context as mapping priorities move toward an all-lands context (see Box 9.2). Variations, where they occur, relate to differing policy and management needs between the two agencies.

In the future, soil mapping will likely be integrated with additional ecological variables and will place a greater emphasis on predicting soil responses to climate change and subsequent implications for agriculture, rangeland and forest management, wetlands, and many associated ecological services. Soils and their associated site characteristics will be used to predict changes in moisture balance and temperature, including the potential responses of vegetation communities to changing climates. Greater use will be made of the various scales of soil-related information. For example, land type associations (see Chap. 8) are used in forest planning to group finer-resolution, integrated soil and vegetation units and provide ecological context to assess ecological integrity and frame desired conditions for management. In turn, the coarser-scale groupings of soil and associated data can be used to support larger-scale planning (e.g., across national forest boundaries).

Because ecosystems are in a state of permanent flux at a variety of spatial and temporal scales, remote sensing plays a critical role in digitally detecting changes in vegetation type and cover amounts over time (Coppin et al. 2004). These data can be used to define elevation and vegetation patterns and to improve the consistency, accuracy, and precision of soil and associated ecological map products. Landform modeling with remotely sensed data can be accomplished at multiple spatial resolutions and can provide new insights at a variety of scales, which is particularly useful when assessing land use and wildfire impacts on factors such as hillslope stability. Computer hardware and software capabilities and efficiencies will continue to advance, ensuring better quality data analyses and increased accessibility to users.

When presented in an ecological context, soils help provide a baseline for future monitoring and assessment following management activities or natural perturbation. According to the Oxford Dictionary, to monitor something means to "observe and check the progress or quality of (something) over a period of time; keep under systematic review," and assessment is defined as "the evaluation or estimation of the nature, quality, or ability of something." Monitoring is the systematic observation and recording of conditions over time, whereas assessments combine monitoring data to inform decision-making and planning.

During monitoring, soil conditions are measured systematically over time to assess changes in soil properties. Sometimes these changes are direct and evident and in other cases they are inferred. For example, one of the goals of the North American Long-Term Soil Productivity (LTSP) study, a program involving the Forest Service and global partners, is to monitor the effects of a pulse change in soil compaction and organic matter removal on tree growth and health, soil recovery, and changes in ecological functions across a range of ecological settings (Ponder et al. 2012). The intent is to validate soil monitoring efforts and evaluate changes in productivity as they relate to soil disturbance, particularly those perceived as detrimental (i.e., slow recovery of soil functions). Furthermore, soil sampling undertaken by the Forest Service's Forest Inventory and Analysis (FIA) program is conducted as a "remeasurement," which facilitates change detection over time and space relative to initial site characterization data. Additional options exist to facilitate intensive

Box 9.2 Soil Inventory Products and Delivery Mechanisms

Product	Description	Delivery system	Agency[a]
Soil Survey Geographic (SSURGO) database	The flagship product of the National Cooperative Soil Survey. Contains field-validated tabular and spatial soil information collected over the last 118 years, compiled to a uniform digital standard. Data was developed at scales ranging from 1:12,000 to 1:63,360 but was predominantly developed at 1:24,000. Available for most areas in the continental United States and the territories, commonwealths, and island nations served by the USDA NRCS. Limited SSURGO data is available for Alaska.	Web Soil Survey Soil data access Geospatial gateway	NRCS
State soil geographic (STATSGO2) database	The National Cooperative Soil Survey (NCSS) general soil map of the United States and its Island Territories. A broad-based inventory mapped at a scale of 1:250,000 in the continental United States, Hawaii, Puerto Rico, and the Virgin Islands and at 1:1,000,000 in Alaska. The only NCSS product that delivers complete coverage of all areas of the United States and its Island Territories. Designed for broad planning and management uses covering state, regional, and multistate areas. The US general soil map is composed of general soil association units and is maintained and distributed as a spatial and tabular dataset.	Web Soil Survey Soil data access Geospatial gateway	NRCS
Gridded Soil Survey Graphic (gSSURGO) database	A rasterized version of SSURGO delivered in the format of an Environmental Systems Research Institute, Inc. (ESRI®) file geodatabase. A file geodatabase has the capacity to store more data and greater spatial extents than the traditional SSURGO product. This makes it possible to offer these data in statewide or even conterminous United States (CONUS) tiles. The gSSURGO database contains all the original soil attribute tables found in SSURGO. All spatial data are stored within the geodatabase instead of externally as separate shapefiles.	Geospatial gateway	NRCS
Raster Component Maps	Component-based soil inventories for a select few areas of the United States delivered as a raster product, in the same format as gSSURGO. Coverage by this relatively new product is currently limited, but is expected to increase in the coming years.	Geospatial gateway	NRCS
National Cooperative Soil Survey (NCSS) soil characterization database	Contains soil characterization data (laboratory data) from the NCSS Kellogg Soil Survey Laboratory and cooperating laboratories. Data can be queried, viewed, and downloaded as comma-delimited text files. Two additional files, a NCSS Microsoft access database containing nearly all results from laboratory analysis and a corresponding ESRI® file geodatabase containing sample locations, are also available for download.	NCSS characterization database	NRCS
Forest Inventory and Analysis (FIA)	FIA reports on the status and trends of the Nation's forest resources across all ownerships. Information collected includes the area of forest land and its location; the species, size, and health of trees; and total tree growth, mortality, and removals by harvest and land use change. The Forest Service has significantly enhanced the FIA program by changing from a periodic survey to an annual survey. The scope of data collection has also expanded to include soil, understory vegetation, tree crown conditions, coarse woody debris, and lichen community indicator of air quality and climate on a subsample of plots. By increasing the capacity to analyze and publish the underlying data and information and knowledge products derived from the data, more information is now readily available.	FIA DataMart	USFS
Soil Resource Inventories (SRI) and Land System Inventories (LSI)	The land systems inventories and soil resource inventories are based on 1970s and 1980s Forest Service direction to provide land base integrated inventories to meet management needs. Land system inventories are a mapping effort focused on an integrated ecological inventory based on geology, geomorphology, soils, and vegetation; most of this mapping has been centered in Idaho in the Intermountain Region. The mapping is designed to fit with the national hierarchy of ecological units. These mapping efforts are typically defined at a broad scale by geology and geomorphology and refined more locally by soils and vegetation. Some SRI completed in the Pacific northwest region describes soil characteristics and classifies the soils of a given area, maps the boundaries and spatial patterns of the soils, and makes predictions about the soil behavior. Data and descriptive information for LSI and SRI inventories are available at Forest Service regional and national forest and ranger district offices.	Forest Service regional and forest offices and districts	USFS

[a]USFS, USDA Forest Service; NRCS, USDA Natural Resources Conservation Service

soil monitoring across the Forest Service's Experimental Forests and Rangelands system (EFR), Long-Term Ecological Research (LTER) network, NRCS Natural Resources Inventory, National Ecological Observatory Network (NEON), and the National Park Service's Inventory and Monitoring networks. In addition, national soil monitoring networks can be linked to various international soil monitoring networks for providing global context to changes observed spatially and temporally (see Box 9.3).

Assessments are designed to characterize and quantify impacts to soil related to disturbances, regardless of cause. Repeated assessments over time provide a monitoring framework to measure recovery rates. This assessment-monitoring continuum provides measures and interpretations relating to severe levels of disturbance that impair soil productivity and site sustainability in areas subjected to forest and range management activities, such as timber sale units, range allotment pastures, or fuels treatment areas. The Forest Service, Bureau of Land Management (BLM), and others have developed soil disturbance sampling guides to aid data collection and best management practices. Many assessments are at a finer scale (e.g., soil rutting by equipment) and tend to focus on specific impacts to soil properties (e.g., decreased water infiltration from soil compaction) and the ramifications that these changes have on ecosystem or agricultural services. Other

assessments, such as the Forest Service's Terrestrial Condition Assessment and the Variable Width Riparian Model, ascertain conditions associated with distinct landscapes and their associated soil/vegetation patterns to aid in evaluating ecosystem health and productivity (Abood et al. 2012).

Soil Mapping

Historical Context

The effort to inventory the soils of the United States has been underway since 1899, when Dr. Milton Whitney, the first Chief of the recently formed Division of Agricultural Soil, initiated surveys on tobacco (*Nicotiana tabacum*) farmlands in Maryland and Connecticut. Initially, most soil surveys focused on cultivated lands because early objectives of soil mapping were to better understand chemical soil properties and their effects on crop production. Over the last 118 years, much has changed, and the objectives of the soil inventory have greatly expanded, and they now include soil and water conservation, timber production, grazing, wildlife habitat, shellfish habitat, recreation, air quality, disaster response, land use planning, and ecosystem management. Furthermore, the soil inventory is no longer the purview of the long defunct

Box 9.3 Examples of Soil Monitoring Networks

Network name	Acronym	Primary agency[a]	Number of sites	Year established	Focus	Web address
Comprehensive research programs						
Experimental Forests and Rangelands	EFR	USFS	80	1909	Forests, rangelands, and watersheds: managing and restoring sites; characterizing plant and animal communities; observing and interpreting long-term environmental change	https://www.fs.fed.us/research/efr/
Long Term Ecological Research	LTER	NSF	26	1876	Natural systems: studying the influence of long-term and large-scale phenomenon on different ecosystems	https://lternet.edu/
National Ecological Observatory Network	NEON	NSF	81	2015	Natural terrestrial and aquatic systems: measuring causes and effects of environmental change	http://www.neonscience.org/
National Critical Zone Observatory Network	CZO	NSF	10	2007	Natural, undisturbed systems: studying how chemical, physical, and biological processes interconnect	http://criticalzone.org/national/
Organization of Biological Field Stations	OBFS	Various	241	1963	Understanding natural processes at every scale	http://www.obfs.org/
Targeted research networks						
Forest Inventory and Analysis	FIA	USFS	1 per 6000 acres	1930	US forests	https://www.fia.fs.fed.us/
Soil Climate and Analysis Network	SCAN	NRCS	219	1991	Agricultural areas	https://www.wcc.nrcs.usda.gov/scan/

[a]*USFS* USDA Forest Service, *NSF* National Science Foundation, *NRCS* USDA Natural Resources Conservation Service

Division of Agricultural Soil, but instead it is the responsibility of the National Cooperative Soil Survey (NCSS), a nationwide partnership of Federal, regional, state, and local agencies, along with private entities and institutions. The Soil Science Division, a subdivision of the USDA NRCS, is the only agency charged with inventorying the soils of all lands, and as a result, it is one of the lead partners of the NCSS. To meet the modern-day soil inventory objectives, the NCSS program encompasses all lands of the United States, including wetlands, forested areas, rangelands, and urban areas (Table 9.1).

Early soil scientists were pioneers in their field and, therefore, encountered many challenges. One of the biggest problems they faced was the lack of base layers, such as topographic maps, aerial images, or other resource inventories. With the advent of commercial aviation in the 1920s, aerial photography became increasingly available and was incorporated into the soil survey process. In 1931, the first stereoscopes were used in Michigan, which was a transformative technological leap in the ability to see and segment landforms. The use of stereoscopes was so valuable that they were not phased out until geographic information systems (GIS) were widely incorporated in the early to mid-2000s. Today, almost all soil scientists are trained in the use of GIS and know how to display, create, and manipulate digital base

Table 9.1 Number of hectares of different land categories in the continental United States (CONUS) that are mapped in the Soil Survey Geographic (SSURGO) database

Category	Hectares of SSURGO mapped in CONUS	Description
Rangeland	262 396 147	Includes all areas identified as grassland/herbaceous or shrub/scrub in the National Land Cover Dataset.
Forest	175 809 364	Includes all areas identified as deciduous, evergreen, or mixed forest in the National Land Cover Dataset.
Urban	44 154 084	Includes all areas identified as developed open space, developed low intensity, developed medium intensity, and developed high intensity in the National Land Cover Dataset. The SSURGO data does not always recognize urban land in the map unit concepts and instead recognizes it as native undisturbed soil.
Wetlands	39 313 672	Includes all areas identified as woody or emergent herbaceous wetlands in the National Land Cover Dataset.
Barren	7 559 042	Includes all areas identified as barren land (rock/sand/clay) in the National Land Cover Dataset.
Perennial ice/snow	87 924	Includes all areas identified as perennial ice/snow in the National Land Cover Dataset.

layers for the purposes of developing pre-maps and final publication polygon layers.

With the rapid advancement in digital soil mapping techniques, the next major phase of soil surveying is underway. Much of the United States has been inventoried in the last 118 years, yet many areas lack high-quality data. It is anticipated that as digital soil mapping (DSM) is incorporated into the NCSS program, soil inventories will be generated for these lands at an accelerated pace. For example, in 2011, the NRCS Soil Survey Geographic (SSURGO) program published their first certified DSM mapping effort in Essex County, Vermont (Soil Survey Area VT009). This was quickly followed by DSM or DSM-assisted soil surveys in Florida, Minnesota, North Dakota, Texas, Washington, Wyoming, and Utah. While the NRCS is moving forward with DSM soil mapping, the use of this technique on Forest Service lands has been limited. The most successful use of DSM on Forest Service lands has been a joint effort between NRCS and the Forest Service encompassing Soil Survey Area MN613 within the Boundary Waters Canoe Area Wilderness on the Superior National Forest in the Northeastern Region. This collaborative effort produced a DSM-generated raster map of soil map units and an associated polygon vector map by bringing together several agency soil scientists and employing multiple modeling approaches and high-resolution topography data derived from LiDAR (light detection and ranging) methods (Fig. 9.1). Other DSM mapping efforts are ongoing, including mapping of volcanic ash mantles in the Forest Service's Northern Region, a soil feature that, due to its high water holding capacity, plays a critical role in alleviating late summer plant drought stress.

Methods for managing, storing, and delivering soil information have also evolved over the last 118 years. Surveys were typically conducted on a county-by-county basis, with a strong emphasis placed on matching and correlating information across political boundaries. Federal land management areas, such as national forests, national parks, or BLM districts, were also used in place of county boundaries, particularly in the western states. Hard copy manuscripts were handwritten and printed for each area surveyed. These reports contained the soil maps, which were typically overlaying orthorectified aerial images, along with tables of data information about the soil properties and interpretations. In 1972, the soil survey program entered the computer age, and tabular information began to be stored in a mainframe national database at Iowa State University. In the 1990s, soil maps were transferred to a digital format, and the first version of the Soil Survey Geographic (SSURGO) database was created (Fig. 9.2). In 2003, the Web Soil Survey program was released, which allowed users to easily interface with and access soil information through the Internet (Soil Survey Staff 2017). While improvements continue to be made and new tools and prod-

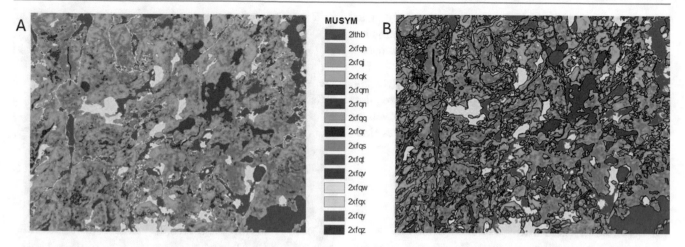

Fig. 9.1 Soil Survey Geographic (SSURGO) soil maps from the Boundary Waters Canoe Area Wilderness, Superior National Forest (MN613; St. Louis County, Minnesota, Crane Lake Part): (**a**) Component-level USDA NRCS SSURGO raster (10 m) soil survey where each pixel represents an individual soil map unit component, and (**b**) SSURGO soil map unit polygons on top of the raster soil map unit component map

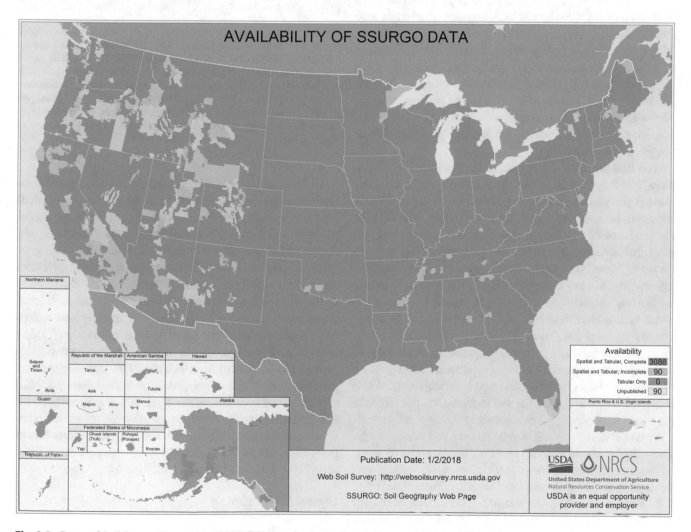

Fig. 9.2 Status of Soil Survey Geographic (SSURGO) mapping in the United States and its Island Territories

ucts are developed, the NCSS soil survey program concurrently seeks to address the form and function of products and services necessary to meet future demands. Fundamentally, this process assesses how NCSS partner organizations collect, develop, manage, and deliver soil information. This transformative process often requires cultural shifts in organizational operations, as well as changes to NCSS standards, a process that requires time, considerable foresight, and agency cooperation.

Methods

Traditional Soil Mapping

Traditional soil maps are typically generated at a 1:24,000 scale (or coarser). This scale can preclude a soil mapper from spatially delineating differing soil characteristics that occur at finer map scales, often a result of changes in topographic features such as slope or aspect. Thus, mappers rely on component mapping. Component mapping creates a single map unit delineation that is composed of more than one soil type, with a specific soil type location described in the attribute data but not represented in the spatial data. An example is a single soil map unit named "Alpha-Beta complex, 3 to 20 percent slopes." This map unit contains two major soils, Alpha and Beta, called components, which have dissimilar properties but cannot be mapped separately at the scale of mapping being used.

The first step in soil mapping is the development of pre-maps. Pre-maps are an inventory of soil map units and their relative position on the landscape based on landform. Next, a substantial amount of time is spent collecting field data for the purposes of validating, updating, and refining the soil landform model. Field work includes describing the soil horizons, identifying the type and cover of plants, identifying any historic pre-European plant communities, denoting landforms, establishing soil sample transects, measuring water tables, recording evidence of flooding and ponding, identifying geologic formations, and traversing landforms. Field work is followed by model adjustment, refinement of soil map unit concepts, final line placement, and generation of tabular soil data from the field and laboratory data. While this is considered "traditional soil mapping," it does include the use of multiple digital data inputs to develop, refine, validate, and finalize predictive models and soil survey maps. These digital inputs include point data layers, aerial imagery, infrared satellite images, topographic layers, geologic layers, vegetation inventories, landform layers, climate data, other soil inventories, digital elevation models (DEMs), and numerous DEM derivatives. The DEM derivatives include products such as hillshade, slope, aspect, wetness index, slope shape, solar radiation models, and customized combination of derivatives.

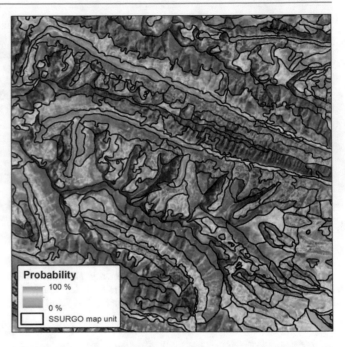

Fig. 9.3 Map showing percent probability of volcanic ash occurrence on the Bitterroot National Forest in Montana overlain by Soil Survey Geographic (SSURGO) map unit polygons (black lines). Red colors indicate a higher probability of ash-mantled soils and blue colors indicate lower probability of occurrence. Mapping individual soil characteristics using digital soil mapping techniques is helping land managers identify timber suitability and reforestation potential in forested landscapes in the western United States. (Source: MT647; Bitterroot National Forest Area, Montana. Raster resolution 30 m)

In the future, the delivery of traditional soil map information will most likely transition from vector (line delineated) products to raster (pixel based) products (Fig. 9.3). Raster products will have finer spatial resolution as modeling and interpretation techniques improve. Soil data for large geographic areas will be available at faster speeds through both desktop and mobile platforms. Data that are now delivered as a map unit will begin to be parsed out into specific soil components for continuous soils classes, or entirely new inventory products will be generated using modern DSM techniques. For example, one of the newer automated mapping products is called SoilGrids1km. This mapping product presents global three-dimensional soil information at a 1 km resolution and contains information on soil properties at six standard depths while also incorporating laboratory analyses of soil carbon (C), pH, sand, silt, clay, depth to bedrock, and more (Hengl et al. 2014). SoilGrids1km can consistently use soil spatial data for input into global models.

Digital Soil Mapping

Digital soil mapping, a relatively new field of soil science initiated in the late twentieth century, continues to evolve as digital technology advances in computational power and memory. Digital soil mapping focuses on developing raster-

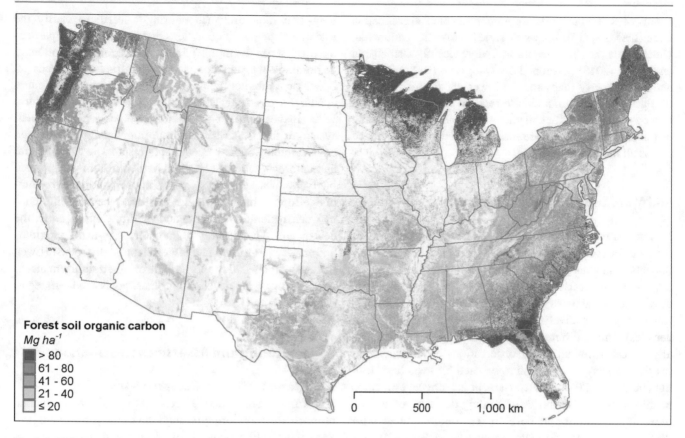

Fig. 9.4 Soil organic C density imputed from USDA Forest Service Forest Inventory and Analysis (FIA) plots in the conterminous United States for 2000–2009. (Source: Wilson et al. 2013, used with permission)

based soil information through computer modeling of field-based soil observations and associated environmental data (Lagacherie and McBratney 2007; Minasny and McBratney 2016; Soil Science Division Staff 2017). A commonly cited basis for DSM is presented by McBratney and others (2003), who propose that soil properties can be spatially modeled by placing a geographic context to Jenny's (1941) soil formation factors; i.e., S = f (cl, o, r, p, t) where natural soil bodies on the landscape (S) are a function of climate (cl), organisms (o), relief or topography (r), soil parent material (p), and time (t). Modern computer software and hardware advances coupled with free remotely sensed imagery (e.g., Landsat TM) and high-resolution DEM models have allowed DSM to move from a purely academic exercise to a fully operational mapping system employed within the NCSS.

A unique feature of DSM products relative to traditional soil maps is the ability to map a suite of soil properties (e.g., map unit component) or a single soil property (e.g., soil organic C) continuously across the landscape at varying pixel sizes (Soil Science Division Staff 2017; Wilson et al. 2013) (Fig. 9.4). Digital soil mapping also provides the capability to rapidly update map products as new information, such as field observations or higher-resolution environmental features, become available. As with all models, some

error will be associated with any prediction. However, DSM provides a distribution of prediction errors for each raster pixel, providing end users an ability to assess map suitability for inclusion in other models or for informing management, regulatory, and policy decisions (Soil Science Division Staff 2017).

A sample of methodologies often employed during the course of a DSM project can include supervised (analyst based) or unsupervised (computer based) spectral classification, linear and logistic regression, kriging, and fuzzy logic. More complex techniques involve classification and regression trees and deep learning algorithms such as neural networks. Most DSM projects do not rely on the use of just a single model or software tool, but use a hybrid approach that requires multiple methodologies based on the training and skills of the soil scientist. Often the model performance and accuracy will dictate the need for the scientist to adjust and try a different approach.

Several recent DSM efforts to model both continuous soil properties and soil classes at continental scales have made major advances in the DSM field by employing cloud computing and high-performance computing (HPC) platforms. Chaney and others (2016) used the parallelized classification regression tree approach called disaggregation and harmonization of soil map units through resampled classification trees (DSMART) to map soil series across the continental United States at a 30 m resolution. More recently, Ramcharan and others (2018) developed both soil class (soil taxonomy) and soil property maps for the lower 48 states at a 100 m resolution, using both a simple random forest classification tree-based approach for soil class maps and model averaging of tree-based models where model predictions are averaged across all model outputs.

Soil Monitoring and Assessment

Monitoring refers to the repeated collection and archiving of data that serves a defined purpose, such as determining impacts from management actions. Monitoring is structured, targets biophysical factors (e.g., soil, water, plant cover, crops), and documents dynamic processes (e.g., hydrologic function) over relatively long periods of time (e.g., years to decades). This is different from research studies, which usually collect a limited set of observations or measurements over a relatively short period to be used for hypothesis testing (Vogt et al. 2011). An assessment is a critical evaluation of information for purposes of guiding decisions on a complex, public issue or to develop restoration strategies for degraded lands. Assessments evaluate the state of a process at a time and place. They are often repeated and can inform policy, but they are usually not prescriptive (Reed et al. 2011). Assessments synthesize complex phenomena such as climate change or land degradation. For this document, monitoring and assessment are treated as a continuum of effort, with one often informing the other.

There are two approaches to monitoring and assessment: (1) operational monitoring and (2) monitoring-to-learn. Operational monitoring (effectiveness monitoring) is designed to inform an ecosystem or species' trajectory and to make timely interventions when necessary (Noon et al. 1999) or to set benchmarks for restoration success (Block et al. 2001). Operational monitoring is a method for keeping an ecosystem within bounds of acceptability and operating within normal ranges. Sometimes this takes the form of compliance or implementation monitoring, where managers monitor and report actions about the landscape or watershed conditions in order to directly inform course corrections. It also provides understanding about the management objectives and implementation efficiency. Operational monitoring also includes monitoring the design and objectives of a management strategy (Herrick et al. 2006).

Monitoring-to-learn is broadly a monitoring approach designed to understand the ecological system; this approach has a fundamentally different purpose than operational monitoring. While operational monitoring connects outcomes to circumstances and actions, dynamic ecosystems require long-term monitoring (monitoring-to-learn) to clarify the role of a sequence of conditions and actions. This is particularly true when the timing of an action becomes important in determining the outcome. Monitoring-to-learn expands on operational monitoring and can include a range of operating conditions or practices so that current actions can inform future management or restoration methods for a given ecosystem or function. Such a monitoring strategy provides an understanding of how ecosystems respond to both natural and anthropogenic disturbances, which in turn can provide intervention when ecosystems are not responding to restoration efforts. This knowledge can often be the difference between restoration success and failure. Consequently, the design and development of monitoring-to-learn programs will make monitoring data more scientifically robust (Ewen and Armstrong 2007) and usable for determining management success or developing other goals and strategies (MacMahon and Holl 2001).

US Monitoring and Assessment Installations

Long-Term Ecological Research Sites

The United States has a rich history of environmental research and monitoring through the Long-Term Ecological Research (LTER) network of place-based field sites supported by a combination of Federal, state, and private organizations that are distributed across the country (Fig. 9.5). The datasets from these sites, some spanning decades and even centuries, have been gathered from a single location with similar methods by overlapping generations of multidisciplinary teams of scientists. LTER sites provide the United States with a valuable network capable of monitoring long-term changes in climate, air quality, soils, water quantity and quality, vegetation distribution and productivity, and the spread of pests, pathogens, and invasive species. These data also provide historical perspectives that are valuable when evaluating ecosystem responses to extreme perturbations, including weather events, fire, pest outbreaks, or pathogen outbreaks. Additionally, they contribute context for scientific field and laboratory experiments and parameterization for ecosystem, regional, and earth system models. Although sometimes criticized as being costly, the LTER program has provided the foundation for important scientific discoveries and critical information for environmental policymaking and decision-making (Lovett et al. 2007).

Forest Inventory and Analysis (FIA) Program

While LTER sites represent intensive place-based monitoring, the Forest Inventory and Analysis (FIA) program provides extensive nationwide environmental monitoring across a variety of spatial scales (Fig. 9.6) and soil attributes (Amacher and Perry 2010; O'Neill et al. 2005; Woodall et al.

Fig. 9.5 Map of the Long-Term Ecological Research (LTER) sites across North America, United States Territories, and Antarctica. (Source: LTER-NCO [CC BY-SA 4.0.], via https://lternet.edu/graphic-resources/ [accessed April 16, 2019])

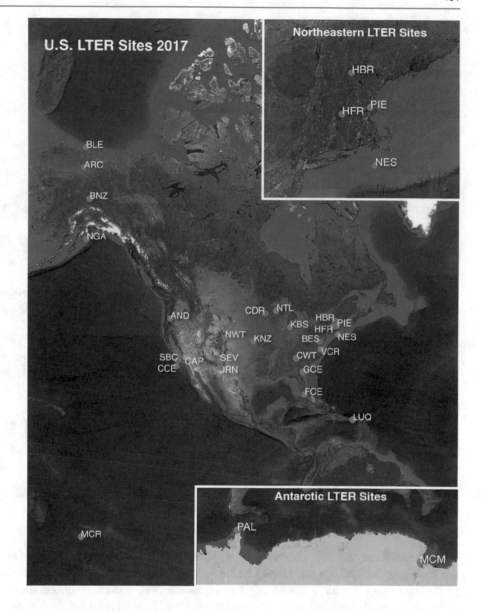

2011). Founded in the 1930s, FIA is a plot-based inventory program. Today, the plot network remains at the heart of the inventory's ability to document the status and trends of the nation's forest resources. The FIA program has a long history of sharing data that support its various assessments. These data have traditionally been provided both in downloadable formats on dedicated websites and through custom tools facilitating a user's quest to generate queries and answers. It's increasingly clear that these data, while publicly available, remain too complex or too cryptic to use by many. To overcome this, FIA is producing authoritative map products with associated accuracies (Riemann et al. 2010; Wilson et al. 2013) that may be downloaded or ingested directly into platforms like Environmental Systems Research Institute, Inc. (ESRI®) ArcGIS Online.

The Forest Inventory and Analysis program also plays a powerful role in providing training data for models leverag-

ing remotely sensed imagery and ancillary data. Domke and others (2016, 2017) leveraged digital representations of soil forming factors (climate, topography and relief, parent material), land cover, and data from the International Soil Carbon Network (http://iscn.fluxdata.org/) to impute observations of soil C on 3636 plots to all plots in the FIA database. Such an approach, in concert with in situ observations, makes the FIA C data more responsive to variation across the landscape and management activities. This is a considerable advancement over past methods, which typically estimated forest floor C as a function of forest type and stand age (Smith and Heath 2002) and forest soil organic C as a fusion of the State Soil and Geographic (STATSGO) database with FIA's forest type groups (Amichev and Galbraith 2004). For example, empirically driven FIA models found that estimated forest litter C stocks were significantly overstated, while soil organic C stocks were significantly understated. This resulted in a

Fig. 9.6 Example of the FIA monitoring framework and the scaling network leading to assessment of forested landscapes. (Source: Amacher and Perry 2010)

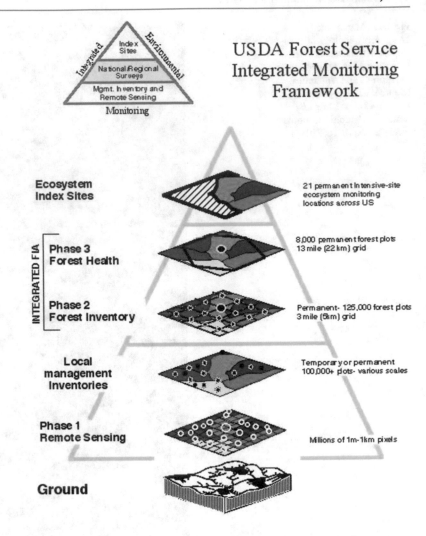

marked decrease in the litter/soil C ratio from 0.30 (18.77 Mg ha^{-1}: 62.87 Mg ha^{-1}) based on the historical models to 0.09 (9.99 Mg ha^{-1}: 109.66 Mg ha^{-1}) (Domke et al. 2016, 2017). Such findings further illustrate the value of FIA as the official source of national forest and soil C stocks at the US Environmental Protection Agency and the United Nations Framework Convention on Climate Change (US EPA 2017).

"Smart" Environmental Sensor Technology

Historically, LTER and FIA monitoring data were obtained across an array of temporal sampling periods by teams of field technicians who relied on mechanical sampling devices. The advent of "smart" environmental monitoring technology with advances in digital sensors, wireless data transmission, and a new generation of data mining and visualization techniques is making high-frequency, high-quality data available on the Internet in near real time, and it is revolutionizing the field of environmental science and monitoring (Fig. 9.7). By providing near real-time data for environmental parameters related to droughts, floods, fires, and other extreme events, these sensor systems offer a new generation of warning devices for known and unknown environmental threats.

These digital devices are routinely deployed at individual research sites across the country and are increasingly being used to connect networks of research sites in real time (e.g., Forest Service's Smart Forest Network). The most ambitious program is the National Science Foundation's NEON program, with 20 domain sites and 40 satellite sites distributed across environmental gradients in the continental United States. Rather than being a bottom-up network of independent sites established historically to address local or regional issues, NEON was envisioned to be an integrated, scientific infrastructure that enables research, discovery, and education about ecological change at a continental scale, equipped with the most advanced digital sensors and monitoring systems.

Guidelines

Effective monitoring and assessment require methods that are adaptable to local and regional ecological settings yet provide standardized protocols that can be uniformly applied regardless of location. To meet this need following the enactment of the National Forest Management Act of 1976, the

Fig. 9.7 Smart sensor array monitoring of soil and atmospheric properties in real-time. (Photo credit: Ian Halm, USDA Forest Service)

Forest Service developed soil quality standards and guidelines to evaluate changes in forest soil productivity and sustainability after land management operations (Page-Dumroese et al. 2000). Additionally, soil productive capacity in the United States is also governed by many policies, but foremost among them are the Multiple Use Sustained Yield Act of 1960, the National Environmental Policy Act of 1969, and the Forest and Rangeland Renewable Resources Planning Act of 1974. Forest industry members of the American Forest and Paper Association must satisfy the requirements of the Sustainable Forestry Initiative (SFI) for soil productivity by using strategies appropriate to soil, topography, and climate. Soil indicators became part of the FIA program's suite of observations (O'Neill et al. 2005) as an outcome of collaboration with the EPA's Environmental Monitoring and Assessment program and the Forest Service's Forest Health Monitoring program. These programs developed the tools needed to monitor and assess the status and trend of national ecological resources at multiple spatial and temporal scales (Pellant et al. 2005).

Like forested regions, guidelines for maintaining rangeland productivity focus on soil and site stability, watershed function, and biotic integrity (Herrick et al. 2005). Rangeland monitoring guidelines generally follow a list of steps to select and interpret soil quality indicators: (1) identify a suite of indicators that are consistently correlated with the func-

tional status of one or more critical ecosystem processes; (2) select indicators based on inherent soil and site characteristics and on site-specific or project-specific resource concerns (e.g., erosion, invasive species); (3) use spatial variability in developing and interpreting indicators to make them more representative of ecological processes; and (4) interpret indicators in the context of an understanding of dynamic, nonlinear ecological processes defined by thresholds (Herrick et al. 2002).

Identifying key soil properties that serve as indicators of soil function is complicated when assessing forest or rangeland soil quality. Soil compaction, erosion, and organic matter losses attributable to a variety of different management actions (Burger 2002; Powers et al. 1990) are the main factors that contribute to declines in ecosystem productivity. These factors alter C allocation and storage, nutrient content and availability, water storage and flux, rhizosphere processes, and insect and disease dynamics (Neary et al. 2010). Practical assessment of soil quality requires consideration of all these functions and their variations in space and time (Amacher et al. 2007; Larson and Pierce 1991).

The FIA soil indicator protocol and other soil monitoring guidelines developed by public and private agencies are designed to specifically measure status of and trends in soil erosion, compaction, internal drainage, organic matter (soil organic C), physiochemical properties, contributions to the global C budget, and accumulations of toxic substances (Amacher and Perry 2010; O'Neill et al. 2005). These data can be used to assess (1) productivity and sustainability of forest or range ecosystems, (2) conservation of soil and water resources, (3) contributions of forest and rangeland soils to the global climate cycle, and (4) accumulation of persistent toxic substances.

Strong policy requirements are increasing the need for effective soil monitoring at many scales. Monitored soil properties represent aggregated soil processes such as decomposition, nutrient cycling, water retention, and microbial populations (Ritz et al. 2004). There are many indicators in the numerous monitoring schemes across the United States. Therefore, the monitoring data are collated into an assessment or framework to facilitate the determination of trends and status at a larger scale (e.g., Perry and Amacher 2009, 2012) in order for the monitoring data to be comprehensible and useful for land managers and policymakers. Monitoring data can be used as an "early warning" system and can steer management away from detrimental impacts to soil quality.

Reliable monitoring protocols have also been identified as critical components of any adaptive management process for forest and rangeland soil conservation programs (Curran et al. 2005; Reeves et al. 2013). These protocols must provide uniform and unambiguous definitions of soil disturbance categories that relate to ecosystem productivity and hydrologic function (Page-Dumroese et al. 2012). For example, the Forest Soil Disturbance Monitoring Protocol (FSDMP) (Page-Dumroese et al. 2009) was designed based on pioneering monitoring efforts in the Pacific Northwest Region (Howes et al. 1983). The FSDMP uses visual soil disturbance classes, a common terminology, and has an accessible database, yet employees can quickly and easily be trained on how to use it (Table 9.2). In an era of declining budgets for monitoring, the database provides critical information for wisely determining where and when to monitor, as well as being able to leverage monitoring efforts across landscapes and ownerships (Page-Dumroese et al. 2012).

In summary, long-term research records, especially when coupled with targeted research networks and remote sensing, are critical tools for monitoring the health of the nation's forests and rangelands while providing the scientific basis for natural resource policy and management decision-making. Long-term data have already served to detect short- and longer-term environmental changes at local, regional, and continental scales and have been a source for scientific discoveries on historical and current environmental issues. A need exists to unify and synthesize data across all these platforms, with the promise to address socio-ecological issues of today, as well as to answer future questions not yet imagined.

Table 9.2 Definitions of visual indicators of detrimental soil disturbances from the Forest Soil Disturbance Monitoring Protocol (Page-Dumroese et al. 2012)

Visual disturbance	Definition
Forest floor impacted	Forest floor material includes all organic horizons above the mineral soil surface.
Topsoil displacement	The surface mineral soil primarily includes the A horizons, but if the A horizon is shallow or undeveloped, it may include other horizons. This disturbance is usually due to machinery but does not include rutting (described below).
Rutting	Ruts vary in depth but are primarily the result of equipment movement. Ruts are defined as machine-generated soil displacement or compression. Often soil puddling is present within the rut.
Burning (light, moderate, or severe)	Burn severity includes only effects on the forest floor and mineral soil, not on aboveground vegetation.
Compaction	Compaction by equipment results in either a compression of the soil profile or increased resistance to penetration (or both).
Platy structure (massive or puddled)	Flat-lying or tabular structure in the mineral soil. Massive indicates no structural units are present and soil material is a coherent mass. Puddled soil is often found after wet weather harvest operations (soil pores are smeared to prevent water infiltration).

Tools and Technology

Web Soil Survey (WSS)

Web Soil Survey provides soil data and information produced by the NCSS. It is operated by the NRCS and provides access to the largest natural resource information system in the world. The Natural Resources Conservation Service has soil maps and data available online for more than 95 percent of the Nation's counties, and they anticipate having data available for 100 percent of the counties in the near future. The site is updated and maintained online as the single authoritative source of soil survey information (https://websoilsurvey.nrcs.usda.gov/).

Gridded Soil Survey Geographic (gSSURGO) Database

The gridded Soil Survey Geographic (gSSURGO) database is similar to the standard NRCS SSURGO database product, but it is in the format of an ESRI® file geodatabase. A file geodatabase has the capacity to store much more data, and thus greater spatial extents, than the traditional SSURGO product, which makes it possible to offer these data in statewide or even conterminous United States (CONUS) tiles. The gSSURGO database contains all the original soil attribute tables in SSURGO. All spatial data are stored within the geodatabase instead of externally as separate shapefiles. Both SSURGO and gSSURGO are considered products of the NCSS partnership. The gSSURGO dataset was created for use in national, regional, and statewide resource planning and analysis of soil data. The raster map layer data can be readily combined with other national, regional, and local raster layers, including the National Land Cover Database (NLCD), the National Agricultural Statistics Service (NASS) Crop Data Layer (CDL), and the National Elevation Dataset (NED) (https://www.nrcs.usda.gov/wps/portal/nrcs/detail/soils/home/?cid=nrcs142p2_053628).

Soil Data Viewer (SDV)

The Soil Data Viewer (SDV) tool, built as an extension to ESRI® ArcMap, allows a user to create soil-based thematic maps. The application can also be run independent of ArcMap, but output is then limited to a tabular report. The soil survey attribute database associated with the spatial soil map is a complicated database with more than 50 tables. Soil Data Viewer provides users access to soil interpretations and soil properties while shielding them from the complexity of the soil database. Each soil map unit, which is typically a set of polygons, may contain multiple soil components that have different uses and management. With SDV, the end user can compute a single soil characteristic value for a map unit and spatially display the results, relieving the burden of querying the soil database, processing the data, and linking it to a spatial map. To ensure appropriate use of the data, SDV also contains processing rules, which provide the user with a tool for quick geospatial analysis of soil data for use in resource assessment and management (https://www.nrcs.usda.gov/wps/portal/nrcs/detailfull/soils/home/?cid=nrcs142p2_053620).

ESRI® ArcGIS Soil Inference Engine (ArcSIE)

ArcSIE is one of the most common software tools used in operational digital soil modeling today. It was developed by Xun Shi at Dartmouth University to utilize a soil scientist's knowledge of soil landscape relationships and fuzzy logic to infer soil classes across the landscape (Shi et al. 2009). ArcSIE is primarily intended for use in areas with existing knowledge of the soil-landscape relationships, but it is being successfully employed in areas undergoing initial mapping, such as the White Mountain National Forest in the Northeast Region (Philippe, J. 2017. Personal communication. 2017. Soil Scientist, USDA NRCS, Soil Science Division, 481 Summer Street, Suite 202, St. Johnsbury, VT 05819.) (http://www.arcsie.com/index.htm).

Terrestrial Ecological Unit Inventory (TEUI) Geospatial Toolkit

The Forest Service's Terrestrial Ecological Unit Inventory (TEUI) program developed the TEUI toolkit system to classify ecosystem types and map ecological units at different spatial scales. The TEUI system distinguishes among land areas that differ in important ecological factors, such as geology, climate, soils, hydrology, and vegetation. Maps and information about ecological units derived through the TEUI process are applied in land use planning to describe land capability and to identify suitability for various uses. The toolkit is used to accelerate the TEUI and soil survey mapping process, but it can also be used for other natural resource mapping efforts. The TEUI system is an ArcGIS extension that assists users in mapping and analyzing landscapes using geospatial data. The toolkit utilizes raster data (e.g., slope, aspect, elevation), polygon data (e.g., map units), and point data (e.g., soil pedon or vegetation plots) to calculate zonal statistics and display the results in tabular or graphical format. This software program was developed and is maintained by the Forest Service Geospatial Technology and Applications Center (GTAC) in Salt Lake City, UT (Fisk et al. 2010; Winthers et al. 2005) (https://www.fs.fed.us/soils/teui.shtml).

Key Findings

- The Natural Resources Conservation Service and the Forest Service provide hierarchically nested, multi-resolution soil mapping systems.
- The Forest Service provides integrated soil and vegetation ecological mapping products.
- Digital modeling continues to evolve, providing refined individual soil property map products.
- Long-term ecosystem monitoring installations provide a wide array of data useful in assessing land use and climate change effects on soil and vegetation patterns at multiple scales.
- Integration of terrestrial monitoring installations with remote sensing is providing near real-time assessments of changes in soil and vegetation properties.
- Guidelines and protocols developed by the Forest Service and other Federal agencies help to assess management impacts on ecosystem function and services and provide guidance in management and policy decision-making.

Key Information Needs

- Uniform, multi-scale soil mapping products for the United States and its Island Territories.

- Data clearinghouses with consistent metadata for collaborative sharing across agencies and organizations.
- Integrated, multi-scale ecological surveys developed from both field and remotely sensed data.
- Standardized and appropriately scaled monitoring data to meet planning and management needs within the context of ecosystem services (providing clean water, healthy forests, and more) and habitat fragmentation.
- Linked monitoring and assessment data to evaluate changes in soil properties and ecosystem function.
- Increased assessments across all land use management projects.
- Acknowledgment at all levels of administration of the importance of assessing land use activity impacts on the soil resource.
- Increased understanding of legacy impacts on current management practices and a differentiation of current impacts from legacy effects.
- Initial and subsequent monitoring, assessment, or both of impacted areas to provide baseline data to guide adaptive management. Impacted areas include drought, wildfire, overgrazing, climate change, and insect/disease outbreaks, all of which have the potential to drastically change western ecosystem dynamics.
- Continued development of sound methods for monitoring, assessing, and managing ecological integrity.

Literature Cited

Abood SA, Maclean AL, Mason LA (2012) Modeling riparian zones utilizing DEMS and flood height data. Photogramm Eng Remote Sens 78(3):259–269

Amacher MC, Perry CH (2010) The soil indicator of forest health in the Forest Inventory and Analysis program. In: Page-Dumroese DS, Neary D, Trettin C (eds) Scientific background for soil monitoring on national forests and rangelands: Proceedings, RMRS-P-59. U.S. Department of Agriculture, Forest Service, Rocky Mountain Research Station, Fort Collins, pp 83–108

Amacher MC, O'Neil KP, Perry CH (2007) Soil vital signs: a new Soil Quality Index (SQI) for assessing forest soil health, Research Paper RMRS-RP-65. U.S. Department of Agriculture, Forest Service, Rocky Mountain Research Station, Fort Collins, 12 p

Amichev BY, Galbraith JY (2004) A revised methodology for estimation of forest soil carbon from spatial soils and forest inventory data sets. Environ Manag 33:S74–S86

Block WM, Franklin AB, Ward JP et al (2001) Design and implementation of monitoring studies to evaluate the success of ecological restoration on wildlife. Restor Ecol 9(3):293–303

Burger JA (2002) Environmental sustainability of forest energy production. 6.2: soil and long-term site productivity values. In: Richardson J, Smith T, Hakkila P (eds) Bioenergy from sustainable forestry: guiding principles and practices. Elsevier, Amsterdam. Chapter 6

Chaney NW, Wood EF, McBratney AB et al (2016) POLARIS: a 30-meter probabilistic soil series map of the contiguous United States. Geoderma 274:54–67

Coppin P, Jonckheere I, Nackaerts K, Muys B (2004) Digital change detection methods in ecosystem monitoring: a review. Int J Remote Sens 25:1565–1596

Curran MP, Miller RE, Howes SW et al (2005) Progress towards a more uniform assessment and reporting of soil disturbance for operations, research, and sustainability protocols. For Ecol Manag 220:17–30

Diamond DD, Elliott LF (2010) Ecological land type association modeling for East Texas, Final Report. U.S. Forest Service, National Forests and Grasslands in Texas, Lufkin

Diamond DD, Elliott LF (2015) Oklahoma ecological systems mapping interpretive booklet: methods, short type descriptions, and summary results. Norman: Oklahoma Department of Wildlife Conservation. Available at https://www.wildlifedepartment.com/lands-andminerals/eco-system-mapping.htm. Accessed 29 May 2020

Domke GM, Perry CH, Walters BF et al (2016) Estimating litter carbon stocks on forest land in the United States. Sci Total Environ 557–558:469–478

Domke GM, Perry CH, Walters BF et al (2017) Toward inventory-based estimates of soil organic carbon in forests of the United States. Ecol Appl 27(4):1223–1235

ECOMAP (1993) National hierarchical framework of ecological units. Unpublished administrative paper. Washington, DC: U.S. Department of Agriculture, Forest Service. Ecoregions of the United States [map, rev. ed.]. Robert G. Bailey, cartog. 1994. Washington, DC: U.S. Department of Agriculture, Forest Service. Scale 1:7,500,000; colored. 20 p

Ewen JG, Armstrong DP (2007) Strategic monitoring of reintroductions in ecological restoration programs. Ecoscience 14(4):401–409

Fisk H, Benton R, Unger C et al (2010) U.S. Department of Agriculture (USDA) TEUI Geospatial Toolkit: an operational ecosystem inventory application. In: Boettinger JL, Howell DW, Moore AC, Hartemink AE, Kienast-Brown S (eds) Digital soil mapping. Progress in Soil Science, vol 2. Springer, Dordrecht, pp 399–410

Forest and Rangeland Renewable Resources Planning Act of 1974; Act of August 17, 1974; 16 U.S.C. 1601

Helms D, Effland ABW, Durana PJ (2008) Profiles in the history of the U.S. soil survey. Wiley, New York, pp 327–331

Hengl T, Mendes de Jesus J, MacMillan RA et al (2014) SoilGrids1km: global soil information based on automated mapping. PLoS One 9(8):e105992

Herrick JE, Brown JR, Tugel AJ et al (2002) Application of soil quality to monitoring and management. Agron J 94(1):3–11

Herrick JE, Van Zee JW, Havstad KM et al (2005) Monitoring manual for grassland, shrubland and savanna ecosystems, Volume I: quick start. volume II: design, supplementary methods and interpretation. U.S. Department of Agriculture, Agricultural Research Service, Jornada Experimental Range, Las Cruces, 206 p

Herrick JE, Schuman GE, Rango A (2006) Monitoring ecological processes for restoration projects. J Nat Conserv 14:161–171

Howes S, Hazard J, Geist MJ (1983) Guidelines for sampling some physical conditions of surface soils, R6-RWM-145. U.S. Department of Agriculture, Forest Service, Pacific Northwest Region, Portland, 34 p

Jenny H (1941) Factors of soil formation. McGraw-Hill, New York

Lagacherie P, McBratney AB (2007) Spatial soil information systems and spatial soil inference systems: perspectives for digital soil mapping. In: Lagacherie P, Mcbratney AB, Voltz M (eds) Digital soil mapping: an introductory perspective. Elsevier, New York, pp 3–24

Larson WE, Pierce FJ (1991) Conservation and enhancement of soil quality. In: Evaluation for sustainable land management in the developing world, Vol. 2. IBSRAM Proc. 12(2). International Board for Research and Management, Bangkok, pp 175–203

Lovett GM, Burns DA, Driscoll CT et al (2007) Who needs environmental monitoring? Front Ecol Environ 5:253–260

MacMahon JA, Holl KD (2001) Ecological restoration: a key to conservation biology's future. In: Soulé ME, Orians GH (eds) Conservation biology: research priorities for the next decade. Island Press, Washington, DC, pp 245–269

McBratney AB, Mendonça Santos ML, Minasny B (2003) On digital soil mapping. Geoderma 117:3–52

Minasny B, McBratney AB (2016) Digital soil mapping: a brief history and some lessons. Geoderma 264:301–311

Multiple-Use Sustained-Yield Act of 1960; Act of June 12, 1960; 16 U.S.C. 528 et seq

National Environmental Policy Act of 1969; Act of January 1, 1970; 42 U.S.C. 4321 et seq

National Forest Management Act of 1976; Act of October 22, 1976; 16 U.S.C. 1600

Natureserve (2019) Terrestrial Ecological Systems of the United States. http://services.natureserve.org. Accessed 6 Mar 2019

Neary DG, Trettin CC, Page-Dumroese DS (2010) Soil quality monitoring: examples of existing protocols. In: Page-Dumroese DS, Neary D, Trettin C (eds) Scientific background for soil monitoring on national forests and rangelands: Proceedings, RMRS-P-59. U.S. Department of Agriculture, Forest Service, Rocky Mountain Research Station, Fort Collins, pp 61–83

Noon BR, Spies TA, Raphael MG (1999) Conceptual basis for designing an effectiveness monitoring program. In: Mulder BS, Noon BR, Spies TA et al (eds) The strategy and design of the effectiveness monitoring program for the Northwest Forest Plan, General Technical Report, PNW-GTR-437. U.S. Department of Agriculture, Forest Service, Pacific Northwest Research Station, Portland, pp 49–68

O'Neill KP, Amacher MC, Perry CH (2005) Soils as an indicator of forest health: a guide to the collection, analysis, and interpretation of soil indicator data in the Forest Inventory and Analysis program, General Technical Report, NC-GTR-258. U.S. Department of Agriculture, Forest Service, North Central Research Station, St. Paul, 53 p

Page-Dumroese D, Jurgensen M, Elliot W et al (2000) Soil quality standards and guidelines for forest sustainability in northwestern. For Ecol Manag 138(1–3):445–462

Page-Dumroese DS, Rice TM, Abbott AM (2009) Forest soil disturbance monitoring protocol: Vol. II—supplementary methods, statistics, and data collection, General Technical Report, WO-GTR-82b. U.S. Department of Agriculture, Forest Service, Washington, DC, 64 p

Page-Dumroese DS, Abbott AM, Curran MP, Jurgensen MF (2012) Validating visual disturbance types and classes used for forest soil monitoring protocols, General Technical Report, RMRS-GTR-267. U.S. Department of Agriculture, Forest Service, Rocky Mountain Research Station, Fort Collins, 17 p

Pellant M, Shaver P, Pyke DA, Herrick JE (2005) Interpreting indicators of rangeland health, ver. 4, Technical reference, 1734-6. U.S. Department of the Interior, Bureau of Land Management, National Science and Technology Center, Denver, 122 p

Perry CH, Amacher MC (2009) Forest soils. In: Smith WB, Miles PD, Perry CH, Pugh SA (Technical coordinators) Forest resources of the United States, 2007. General Technical Report, WO-GTR-78. U.S. Department of Agriculture, Forest Service, Washington Office, Washington, DC, pp 42–44

Perry CH, Amacher MC (2012) Chapter 9: Patterns of soil calcium and aluminum across the conterminous United States. In: Potter KM, Conkling BL (eds) Forest health monitoring: 2008 national technical report, General Technical Report SRS-GTR-158. U.S. Department of Agriculture, Forest Service, Southern Research Station, Asheville, pp 119–130

Ponder F Jr, Fleming RL, Berch S et al (2012) Effects of organic matter removal, soil compaction and vegetation control on tenth year

biomass and foliar nutrition: LTSP continent-wide comparisons. For Ecol Manag 278:35–54

Powers RF, Alban DH, Miller RE et al (1990) Sustaining productivity of north American forests: problems and prospects. In: Gessel SP, Lacate DS, Weetman GF, Powers RF (eds) Sustained productivity of forest soils, proceedings of the seventh North American forest soils conference. University of British Columbia Faculty of Forestry, Vancouver, pp 49–79

Ramcharan A, Hengl T, Nauman T et al (2018) Soil property and class maps of the conterminous US at 100-meter spatial resolution based on a compilation of national soil point observations and machine learning. Soil Sci Soc Am J 82(1):186–201

Reed MS, Buenemann M, Atlhopheng J et al (2011) Cross-scale monitoring and assessment of land degradation and sustainable land management: a methodological framework for knowledge management. Land Degrad Dev 22:261–271

Reeves D, Coleman M, Page-Dumroese D (2013) Evidence supporting the need for a common soil monitoring protocol. J Ecosyst Manage 14(2):1–16

Riemann R, Wilson BT, Lister A, Parks S (2010) An effective assessment protocol for continuous geospatial datasets of forest characteristics using USFS Forest Inventory and Analysis (FIA) data. Remote Sens Environ 114:2337–2352

Ritz K, McNicol JW, Nunan N et al (2004) Spatial structure in soil chemical and microbiological properties in an upland grassland. FEMS Microbiol Ecol 49(2):191–205

Schoeneberger PJ, Wysocki DA, Benham EC, Soil Survey Staff (2012) Field book for describing and sampling soils, version 3.0. Natural Resources Conservation Service, National Soil Survey Center, Lincoln, 300 p

Shi X, Long R, Dekett R, Philippe J (2009) Integrating different types of knowledge for digital soil mapping. Soil Sci Soc Am J 73:1682–1692

Smith JE, Heath LS (2002) A model of forest floor carbon mass for the United States forest types, Research paper NE-722. U.S. Department of Agriculture, Forest Service, Northeastern Research Station, Newtown Square, 37 p

Soil Science Division Staff (2017) Soil survey manual. In: Ditzler C, Scheffe K, Monger HC (eds) USDA handbook 18. Government Printing Office, Washington, DC

Soil Survey Staff (2017) Web soil survey. U.S. Department of Agriculture, Natural Resources Conservation Service Washington, DC. Available at https://websoilsurvey.sc.egov.usda.gov/App/HomePage.htm. Accessed 6 Mar 2019

U.S. Environmental Protection Agency [US EPA] (2017) Land use, land-use change, and forestry. In: Inventory of U.S. greenhouse gas emissions and sinks. EPA 430-P-17-001. Chapter 6

USDA Natural Resources Conservation System [USDA NRCS] (2012) Ecological Site Information System (ESIS): ecological site descriptions. https://esis.sc.egov.usda.gov/About.asp. Accessed 6 Mar 2019

USDA Natural Resources Conservation System [USDA NRCS] (n.d.) Ecological site information system (ESIS). https://esis.sc.egov.usda.gov/. Accessed 6 Mar 2019

Van Kley JE, Turner RL, Smith LS, Evans RE (2007) Ecological classification system for the national forests and adjacent areas of the west gulf coastal plain: second approximation. The Nature Conservancy and Stephen F. Austin State University, Nacogdoches, 379 p

Vogt JV, Safriel U, Von Maltitz G et al (2011) Monitoring and assessment of land degradation and desertification: towards new conceptual and integrated approaches. Land Degrad Dev 22(2):150–165

Wilson B, Tyler R, Woodall CW, Griffith DM (2013) Imputing forest carbon stock estimates from inventory plots to a nationally continuous coverage. Carbon Balance Manag 8(1):1–15

Winthers E, Fallon D, Haglund J et al (2005) Terrestrial ecological unit inventory technical guide. U.S. Department of Agriculture, Forest Service, Washington Office, Ecosystem Management Coordination Staff, Washington, DC, 245 p

Woodall CW, Amacher MC, Bechtold WA et al (2011) Status and future of the forest health indicators program of the USA. Environ Monit Assess 177:419–436

Challenges and Opportunities

10

Linda H. Geiser, Toral Patel-Weynand, Anne S. Marsh,
Korena Mafune, and Daniel J. Vogt

Introduction

There is an inseparable connection between the health of soils and the well-being of people. Disturbances—changing climate and environment, invasive species, increasing wildfire severity and frequency, land management practices, and land use changes—are affecting soil health in complex ways that are only partially understood. This assessment provides a baseline for soil health by synthesizing the most recent research on the properties and processes characterizing forest, rangeland, wetland, and urban soils at regional and national levels. It also summarizes the current state of knowledge about the effects of historical and ongoing disturbances. Finally, it highlights the technologies, networks, and outreach programs central to future research on, and monitoring and management of, the nation's soils. The state of the knowledge indicates that adverse changes to soil properties could, in the long term, undermine the primary ecosystem services provided by soils: abundant water, food, and fiber and moderation of climate warming. Research that can further explain the underlying principles of soil science and the effects of disturbances is needed to inform policies and management practices to optimize soil health on a broad scale and support human health and welfare through the twenty-first century.

L. H. Geiser (✉) · T. Patel-Weynand
Washington Office, U.S. Department of Agriculture, Forest Service, Washington, DC, USA
e-mail: Linda.geiser@usda.gov

A. S. Marsh
U.S. Department of Agriculture, Forest Service, Bioclimatology and Climate Change Research, Washington, DC, USA

K. Mafune
School of Environmental and Forest Science, University of Washington, Seattle, WA, USA

D. J. Vogt
School of Environmental and Forest Science, University of Washington, Seattle, WA, USA

In this chapter, we identify key challenges and opportunities for soils research and management grouped by overarching themes: (1) understanding basic soil properties and processes; (2) understanding disturbance effects; (3) monitoring, modeling, mapping, and data-sharing; (4) training the next generation of scientists; and (5) managing soils. Each theme is introduced with a short recapitulation of its significance and associated issues. Please see the individual chapters and appendixes of this report for a fuller discussion of topical and regional issues, research findings, and information gaps.

A distinct advantage of this particular time in history is that the scientific understanding of soils, the quantity of soils data, and the consistency, organization, and accessibility of these data have never been greater or of higher quality. Numerous federally funded efforts; long-term monitoring networks (Appendix B) and studies; collaborations across multiple agency, state, and private research institutions, together with contributions by nongovernmental organizations and private enterprise; and advances in chemical, physical, and computational techniques are enabling researchers to gather and interpret immense quantities of data. From detailed genomic descriptions of the microbial diversity found in a gram of soil to satellite-mounted instruments allowing the remote assessment of soil moisture across vast swaths of the planet, new tools are yielding information at scales and in quantities that were previously unthinkable. Assessments of these new data along with those from traditional approaches and on-site measurements are informing science-based management recommendations that are setting the stage for site-specific tailoring of soil management practices (Chap. 9). These advances in technology, communication, monitoring, and networking are the prime resources and create the opportunities for addressing the research challenges critical to solving the daunting tasks of growing enough food and sustaining the nation's soil-driven ecosystem services into the future.

It is no coincidence that the words "earth" and "soil" are often used interchangeably in ordinary conversation. It is a crucial concept that a narrow range of conditions for both provides the foundation for life on Earth. Even with modern tools, technologies, and available science and information, there is still so much to learn about the astounding diversity of life in a healthy soil, and there are still so many unanswered questions (Appendix C) about interactive properties and processes in soils. The answers to these questions will be key in sustaining soil health. On the other hand, the terms "dirt" and "soil" are also used interchangeably. The way that resources are verbalized shapes conceptualizations and, down the line, options and actions. If a long-term goal is to optimize the health, sustainability, and productivity of soils, then a simultaneous challenge and opportunity is to reiteratively frame research questions, research and monitoring activities, communications, and management practices in a way that reflects a respect for, and accurate understanding of, the complex interactions among soil physical, chemical, and biological properties, natural and human disturbances, and human health and welfare.

Below are some actions and best management practices that can be undertaken to address challenges and opportunities to further soil science and soil sustainability. They are not intended as recommendations or to be prescriptive in any way, but are intended to encourage thinking and action to enhance soil conditions and their management.

Understanding Basic Soil Properties and Processes

Significance

Information about fundamental soil properties and processes is indispensable to the development of science-based policies and practices. Well-designed policies and practices can sustain soil, water, and air resources; protect soil fertility, water infiltration, and water holding capacity; reduce and prevent erosion and sedimentation; and protect the organic matter, soil moisture, and soil invertebrates and microorganisms essential to soil chemical and physical processes.

Challenges and Opportunities

- **Understanding soil organic matter accumulation and loss** (Chap. 2). Globally, soil C is the largest terrestrial pool. Soils store about 2.3×10^{15} kg or 2300 petagrams (Pg) organic C in the top 2 m and about 500 Pg more below that, representing greater than half the global terrestrial C sink. About 75% of that total is in forest soils, 12% in grassland and shrubland soils, and 10% in tundra soils (Jackson et al. 2017). A critical research frontier is the study of soil organic matter accumulation, its loss from soils, and its complex interactions with soil structure (which can physically protect molecules from decay) and with minerals (many of which are oxygen-sensitive and

whose availability for uptake and oxidation state may change with changing climate). This is particularly important for addressing soil pollution. Additional research challenges are understanding (1) how C dynamics differ between organic and mineral horizons, (2) how to reduce uncertainties in estimations of soil C, and (3) the contribution of deep roots to organic matter inputs below 30 cm.

- **Understanding soil water and hydrologic processes** (Chap. 3). Despite their importance, many soil water processes are still incompletely understood—even in relatively unperturbed ecosystems. Key research challenges include improving hydrologic models to predict soil water dynamics at watershed scales and accurately quantifying water storage capacity and water dynamics at a variety of scales. Because infiltration is important to soil moisture holding capacity, it is necessary to understand the isolated effect of soil water repellency on infiltration and runoff and how those rates can be quantified at the watershed scale given the large temporal and spatial variability of soil water repellency. Related questions include how soil moisture changes with depth and how moisture profiles vary among soils. Understanding soil water can inform soil restoration and basic management practices and mitigation options for drought, flooding, and sea-level rise.

- **Understanding soil fertility and nutrient cycling processes** (Chap. 4). Soil nutrients are essential for plant growth. Because nutrient availability is a key driver of net primary productivity, one of the biggest research challenges of the near future will be to refine conceptual models and techniques for quantifying mineral weathering, a major source of soil nutrients. Understanding local and regional soil differences that affect retention of base cations in soil can help to customize mitigation prescriptions for disturbed soils.

- **Understanding the identity and roles of soil biota** (Chap. 5). Soil organisms account for about 25–30% of all species on Earth (Decaëns et al. 2006) and differ vastly in size, mobility, and ecological function. Collectively they flourish in a multitude of soil physical and chemical habitats—yet only a fraction of soil-dependent species have been identified. Thus, a wide-open opportunity for research is the taxonomy and identity of bacteria and archaea; eukaryotic unicellular and microbial life, including fungi; nematodes, arthropods, and other invertebrate phyla. This knowledge will help advance understanding of how these organisms, and their communities, contribute to ecosystem services such as water filtration, storage, nutrient and C cycling, plant nutrition and drought resistance, plant diversity, and wildlife habitat. Also unresolved are the aboveground and belowground interactions between microbes, organic matter, decomposition, and nutrient cycles. Further, understanding epiphytes and canopy soils in temperate rainforests and their roles in nutrient accumulation and C storage is important (Box 10.1). Such emerging techniques as high-throughput

Box 10.1 Canopy Soils

Old-growth temperate forests, per unit area, represent the longest living and largest stores of carbon on Earth (Urrutia-Jalabert et al. 2015). The old-growth temperate rainforests of the Pacific Northwest region have a large role in carbon emissions and sequestration because high-impact disturbance events are increasing soil respiration losses (Harmon et al. 2011), as well as the release of carbon stored in woody biomass (Luyssaert et al. 2008). These forests are known for their high gross primary productivity, high stores of carbon in soils, large numbers of fungal species, and propensity to host epiphytic and saprophytic species (Franklin et al. 1981; Tejo et al. 2014). Globally, they represent one of the most carbon-dense forest types (Keith et al. 2009), and they hold some of the region's highest biodiversity (Franklin et al. 2002). The resulting high structural variability, rainfall regimes, and associated diversity are conducive to the formation of a wide array of niches for biotic relationships (Frank et al. 2009). Despite this information, the role of abiotic and

biotic interactions in the functioning of carbon and nutrient dynamics of these forests has rarely been described.

It is not widely known that these old-growth forest stands can accumulate significant amounts of canopy soil on tree branches (Van Pelt et al. 2006). Canopy or arboreal soils develop from the accumulation and decomposition of epiphytes and intercepted litter on branches and in bifurcated tree boles in both tropical and temperate rainforests (Haristoy et al. 2014; Nadkami 1981), and they contribute unique structural and functional components in these forests. In some stands, canopy soils were found to contribute as much as 20% and 25% to the nitrogen (N) and carbon (C) storage pools, respectively (Tejo et al. 2014). Riparian species (e.g., big leaf maple *Acer macrophyllum* Pursh.) can have canopy soils reaching depths >35 cm (personal observation). Trees have adapted to the presence of these canopy soils as shown by the extensive growth of adventitious branch roots that are associating with symbiotic fungi (Nadkarni 1981) (e.g., Box Figs. 10.1 and 10.2). It is known that geomor-

Box Fig. 10.1 Adventitious roots in canopy soils of two old-growth big leaf maples in the University of WA study sites located in the Queets (**a**) and Hoh (**b**) Rivers in the Olympia Peninsula. WA. (Photo source: K Mafune)

Box 10.1 (continued)

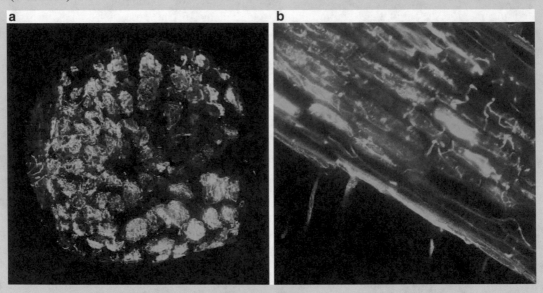

Box Fig. 10.2 (**a**) Confocal scan of an adventitious canopy-soil root cross-section with an extensive amount of intracellular coils, (**b**) lateral scan of intra- and intercellular hyphae within the cortical and epidermal cell layers. The hyphae found in the epidermal cells also extend into the root-hairs. (Photo source: K Mafune)

phology, hydrology, sediment, and riparian-derived large wood shape these riparian forests (Swanson and Lienkaemper 1982; Naiman et al. 2000; Abbe and Montgomery 2003). However, understanding the complexity of biotic and abiotic interactions in canopy soils is important to understanding the function of these forests, as they may strongly affect the characteristics of organic matter and nutrient fluxes into streams, estuaries, and the atmosphere. In particular, studies of the canopy soil habitats are needed to understand how disturbances may impact greenhouse gas emissions and nutrient pools at the temporal and spatial scales of canopy soils in these productive and carbon-dense coastal temperate rainforests.

sequencing and stable isotope probing can provide opportunities to deepen understanding of these effects.

- **Understanding the unique properties and processes of hydric, wetland, and urban soils** (Chaps. 6 and 7). While wetlands and urban areas make up a relatively small portion of the landscape, they provide highly valued ecosystem services that are largely mediated by soil processes, such as water purification and C sequestration (wetlands) and local food, shading, green landscapes, and biodiversity hotspots (urban areas). Key research challenges include understanding the soil processes of tidal freshwater wetlands, particularly C sequestration, and how those processes are mediated by tidal and terrestrial processes. For example, what are the watershed-scale dynamics for the transfer of organic matter nutrients and other chemicals through tidal freshwater wetlands and adjoining uplands and tidal waters? This information is important because tidal wetlands are at the interface of changing sea level and, particularly in the eastern United States, are often within rapidly urbanizing regions. Additional basic research needs for urban soils include understanding processes of soil formation from human-created parent materials (concrete, construction debris, landfill materials) and how to describe, quantify, and map the highly heterogeneous soils characteristic of urban areas.

Understanding Disturbance and Stress Effects

Significance

"Nature is always changing—there is no going back" (Chap. 8). Ecosystems are in a state of flux at many spatial and temporal scales, and soil formation, transformation, and loss are always occurring. Today, however, the pace, spatial scale, and magnitude of disturbances linked to human activities are strikingly different. Disturbances can shift and diminish soil properties and processes such as long-term C storage, nutrient cycling, fertility, and productivity; species composition and diversity; and water infiltration, holding capacity, quality, and abundance and can increase erosion and sedimentation. Repetitive and highly intense disturbances can eliminate key functions, for example, turning peatlands from a net C sink to a net source, and can overwhelm or eliminate key services such as water purification. Understanding how disturbance affects soils is fundamental to the development of management practices that sustain and restore soils.

Disturbance agents identified repeatedly across the chapters, and regional summaries that can adversely affect the properties, functions, and productivity of soils include:

Climate Change
- **Precipitation.** Shifts in global precipitation patterns are producing more frequent and more intense rainfall and increased severity and frequency of both flooding and drought.
- **Temperatures.** Long-term increases in temperature variability, warming air temperatures along with increased solar radiation, and longer growing seasons will raise soil temperatures throughout the United States.
- **Carbon dioxide (CO_2) concentrations.** Atmospheric CO_2 concentrations are increasing globally due to human activities, particularly the combustion of fossil fuels. This increase can indirectly impact soils through the effects of elevated CO_2 on plant productivity. Small changes in soil C can also significantly increase or decrease atmospheric CO_2 concentrations (Chap. 2 Introduction), resulting in mitigation or exacerbation of climate change.

Fire
- Wildfires are becoming more frequent, more intense, and larger in area as more than a century of fire suppression has increased fuel loads and warmer, drier climates have decreased fuel moisture for more extended periods (Abatzoglou and Williams 2016).

Invasive Species, Pests, and Disease
- Pests, pathogens, and invasions of weedy and non-native species (from microbial organisms to macrofauna and macroflora) are all sources of disturbance. Climate change is a significant driver favoring range extensions of some insects and pathogens (Bellard et al. 2018).

Pollution
- Water and air pollution and overuse of fertilizers and pesticides can cause accumulation, leaching, or volatilization of pollutants such as fertilizing N, acidic compounds, mercury, organic pollutants, and toxic metals.

Nonurban Land Uses
- Nonurban soils are affected by land uses such as agriculture, forestry, grazing, road networks, energy development and mining, and recreational infrastructure and activities (Chap. 3).

Urban Land Uses
- Urban soils are disturbed directly by infrastructure development, construction, waste disposal, grading and stormwater management, sealing and paving, soil replacement and recycling, and lawn management and indirectly from urban climate, pollution, and non-native and invasive species (Chap. 7).

Challenges and Opportunities

- **Understanding how to grow C pools and prevent or mitigate soil C loss.** A key research need is understanding how C loss from organic soils can be minimized across organic tundra, boreal, agricultural, forest, and wetland soil types. An analysis of the most important and economically feasible mechanisms for storing C and their importance and potential by region would inform attempts to increase the pace and scale of C storage in soils nationally. Other key research needs include understanding how to minimize soil C loss from wildland fire and the potential value of biochar to sequester C and augment soil fertility and water holding capacity across soil types.
- **Understanding how land use and management affect soil processes and properties.** Many questions remain about how land use influences soils, especially in wetland ecosystems. Particularly important are systems sensitive to mercury methylation, nutrient loading, and burning. Large-scale manipulation experiments can be used to answer questions on the impacts of development and land use patterns on soils.
- **Understanding the effects of climate shifts on soil properties and processes.** Of specific interest are temperature, precipitation, and greenhouse gas effects on the stability of soil C, moisture, nutrients, and biota and the ecological functions provided by them. Research is needed that can help forest and rangeland managers anticipate climate change and protect or enhance soils to promote ecosystem resistance and resilience. Some areas where improved understanding and research are needed are included below.
  - **Understanding how precipitation shifts affect soil water storage, water availability for the growing season, erosion, sedimentation, and water quality.** Reduced snowpack in particular may severely limit water storage and availability during the growing season for much of the western United States. In contrast,

increasing rainfall amounts and intensities are expected to increase runoff and erosion of agricultural soils by about 2.0% and 1.7%, respectively, for each 1% rise in mean annual precipitation (Nearing et al. 2004). What is the prognosis for forest and rangeland soils and what steps can be taken to mitigate projected effects?

– **Understanding how temperature shifts affect soil microbial activity, decomposition rates of soil organic matter, and nutrient loss.** Higher temperatures tend to increase decomposition and soil respiration rates, but how significant is the effect, and what are the quantities of CO_2 released to the atmosphere require further research.

• **Understanding how invasive organisms affect the composition, relationships, and roles of soil biota.** Key research challenges include developing indicators to identify priority conservation areas (given that plant and soil biodiversity is not well correlated in forests or rangelands) and characterizing late-seral soil biology to inform restoration of desired or late-seral plant communities. Systematic monitoring can answer key research questions, such as: What invasive soil organisms are present and how much have they spread? How do plant, animal, insect, and pathogen invasive species impact soil C, and how can managers change practices to alter invasive species effects on ecosystem services?

• **Wetlands—biggest bang for the buck?** Though wetlands make up a relatively small portion of the terrestrial landscape, they provide highly valued ecosystem services that are largely mediated by soil processes, including water purification and C sequestration. Research that provides information to improve wetland restoration practices can have a relatively substantial return for the investment.

• **Using urban soils to study disturbance effects.** Human-created parent materials and effects of urban land use changes can be thought of as a "natural experiment" in soil-forming processes. There is no typical urban soil, and soil horizon development often occurs more rapidly in urban compared to young nonurban soils (Pouyat et al. 2009). Moreover, many urban environmental effects include local climatic shifts. Thus, urban soils may be used to study the multifactorial effects of climate change on forest and grassland remnants situated in urban landscapes.

• **Understanding how soil pollution affects soil properties and biota.** Use and integrate research in soil chemistry modeling, soil mapping, atmospheric deposition and critical loads, and measurements of surface water quality and soil chemistry to better predict soil pollution effects and the ability of soil to recover from these effects. Evaluate spatial extent and risks from soil pollution in forests, grasslands, wetlands, and urban soils.

• **Understanding how increased fire frequency and severity affect soil properties and biota.** Of particular concern are effects on (1) permeability to water, erosion and sedimentation rates, and water quality; (2) amount of soil organic matter lost to combustion and releases of sequestered C; and (3) alterations in mineralogy and chemistry of soils that affect soil biota and availability of plant nutrients. Within these effects, there are questions related to spatial extent of effects, recovery times, and factors that promote resistance and resilience of soils. For example, wildfire increases are projected to double the rates of sedimentation in one-third of 471 large watersheds in the western United States by about 2040 (Neary et al. 2005). Understanding sedimentation effects on stream channel characteristics and water quality can inform management of extreme flows.

Monitoring, Modeling, Mapping, and Data-Sharing: A Key Component of Knowledge Acquisition and Decision-Making for Land Managers

Significance

The right level of investment of resources in monitoring, assessments, and technological development can provide for current and future information needs (Chap. 9). Collection of adequate data to answer pressing and management-centered questions and meet information needs is critical for long-term sustainability in the forest and rangeland sectors.

Challenges and Opportunities

• **Planning strategically.** Strategic planning is needed to review and prioritize existing networks, incorporate new technologies and datasets, and grow and connect new research partnerships.

• **Promoting interdisciplinary teams with access to a range of information.** Data sources range from monitoring networks to field and laboratory experiments, the utilization of genetic and chemical data, light detection and ranging (LiDAR), digitized photographic imagery, and satellite data. An example opportunity for soil, biological, and physical scientists is the study is useful of long-term effects of higher temperatures and CO_2 enrichment on forest and rangeland soils and how to model and manage soil responses. Formation interdisciplinary teams to use long-term forest research sites is useful to develop and test conceptual models and new instrumentation, all of which could improve quantification of weathering and nutrient cycling processes.

- **Sustaining key monitoring networks; collection, storage, and provision of soil network data; and associated support staff.**
  - The United States is a world leader in long-term field studies with large multiagency, multigenerational networks of research sites. Continuous monitoring is a key component of science-based management and restoration. Sustaining and augmenting these networks and providing new opportunities to integrate knowledge across programs will increase benefits from the long-term investment.
  - Examples of key networks include the USDA Forest Service's Forest Inventory and Analysis program's soil indicator, Long-Term Ecological Research sites, Long-Term Soil Productivity network, Critical Zone Observatories, and the National Environmental Observatory Network. See also Appendix B.
  - Supporting improvements in metadata, coverage, standardization, and collation of independent datasets to facilitate data access and interpretation. Support strategic planning to make sure the networks and systems are obtaining the information that will be needed in the future.
  - Strategic integration of existing programs and data-sharing can counter the need for additional funding, although additional sources of stable support for long-term research sites would enhance the pace of research. Explore new public-private partnerships.
- **Conducting long-term experimental studies.** Some questions can be answered with short-term studies, but long-term studies (longer than a rotation) of forest management effects help managers understand trends and anticipate longer-term impacts of decisions on soil properties. Initial and subsequent monitoring provides data to guide adaptive management in areas impacted by drought, wildfire, overgrazing, climate change, and insect and disease outbreaks.
- **Continuing the legacy of twentieth-century paired catchment studies.** Research depends on data from these and other networks to explain underlying soil hydrologic processes that can be used to develop management approaches to reduce erosion and sedimentation, mitigate pollution, and optimize infiltration.
- **Increasing the number of sites with less-intensive monitoring** to improve statistical power to extrapolate across landscapes. It is important to maintain or develop less-intensive monitoring of soil conditions across many sites, because the number of intensive sites does not provide the statistical power needed to extrapolate to actual forest and rangeland soils across landscapes, which is particularly true for measurements lacking in forest and rangeland soils such as soil moisture content. Systematic approaches to soil resampling programs across gradients of natural and human-impacted landscapes can enhance effective environmental monitoring and assessment (Lawrence et al. 2013).

- **Integrating data collected by terrestrial field monitoring installations with digitized aerial and remote sensing data** to provide near real-time assessment of change in soil and vegetation properties. New aerial, satellite, and space technologies are providing ever more detailed, spatially explicit, and temporally explicit information about the planet's soil, water, and air resources. Explore uses of remotely collected soil temperature, soil moisture content, and gas concentration data to answer questions about soil responses and vegetation patterns to ongoing change at multiple scales.
- **Building and supporting data clearinghouses with consistent metadata for collaborative sharing** across agencies and organizations and to assist research and management.
- **Continuing development of sound methods for monitoring, assessing, and managing ecological integrity by improving our understanding of fluxes and processes and developing models to integrate monitoring and process-level studies.** It is critical to make this data and information available for managers and policy makers in an easy-to-use form to promote the maintenance of ecosystem services and inform policies to recover degraded soils and promote healthy soils.
- **Continuing development of more uniform, multi-scale soil mapping products** for the United States to aid research and management efforts. Continue to invest in hierarchically nested, multiresolution soil mapping systems and integrated soil and vegetation ecological mapping products.
- **Using new molecular technologies** to characterize the biological diversity of soils and the effects of climate-induced change on biodiversity. Expanding molecular technologies offers unprecedented possibilities for addressing ecological questions about biological diversity. Improved understanding of soil biodiversity composition, function, and resilience will assist development of best management and adaptive management practices.
- **Using the latest statistical and modeling techniques** to (1) understand shifts in soil-dependent biological communities under natural conditions and under disturbance regimes and (2) identify methods to mitigate soil pollution, control target invasive species, and increase the resilience of soils to climate change.

Training the Next Generation of Scientists

Significance

The skills and abilities of the next generation of soil scientists will be needed to meet the research challenges of the present and future. Yet academic departments; Federal, state, and private-sector jobs; and industries that have focused on soil science are shrinking. Fewer young scientists are being trained in the taxonomy of biological organisms, and the field is at risk of losing knowledge and capacity as older scientists retire.

Challenges and Opportunities

- **Creating broad, inclusive opportunities for educating students and the public at large** about science, technology, engineering, and mathematics (STEM) subjects, including citizen-science education opportunities involving soils. Developing programs with urban, rural, and regional relevance is also useful.
- **Encouraging young scientists to pursue the taxonomy of soil biota.** Describing species—particularly microbes and invertebrates—offers a special opportunity to young scientists looking for an uncrowded field where they can make a difference.
- **Channeling support to successful departments and positions and use advances in technologies to enhance networking opportunities among sites, scientists, and agencies.** The goal is to form a more cohesive national interorganizational task force on understanding forest and rangeland nutrient pools and cycles.

Managing Soils in an Age of Accelerated Disturbance, Land Use, and Environmental Changes

Significance

Sharing of research information with managers has improved forest and rangeland soil management practices significantly over the past 40 years. Limiting loss of soil productivity has been the focus of soil management on national forest lands since the passage of the National Forest Management Act of 1976. Current Forest Service directives (USDA FS 2012) rely on science-based recommendations to guide planning, environmental assessments, and execution of management activities related to national forest and grassland soils. Due to the increasing pace and scale of disturbances, land management decisions on all lands (public and private) rely more

than ever on up-to-date science to sustain soil health and productivity (Chap. 8).

Challenges and Opportunities

- **Planning strategically.** Develop strategies to help overcome past, contemporary, and future stressors. Practice proactive maintenance to avoid more expensive soil remediation later.
- **Linking efforts of researchers, managers, policymakers, communities, nongovernmental organizations, and private enterprise to meet common goals.** Actions might include explore roles that communities, local governments, private enterprise, and nongovernmental organizations can play and how those entities can be integrated into existing networks, data-sharing, and decision-making at management and policymaking levels.
- **Providing opportunities for scientists and managers to develop and test innovative ideas,** and share that information with land managers in a timely manner to support continued advances in management and restoration approaches.
  - **Testing pre-fire treatments to improve resilience of forest soils to wildfire.** Massive efforts are underway to restore watersheds by removing accumulated biomass through mechanical means, slash burning, and prescribed burning. Some questions worth exploring in this arena might be: What happens when fire passes through treated stands relative to untreated stands? How do different restoration prescriptions protect the physical, chemical, and biotic components of the soils? Which soils and localities benefit most from restoration activities and are most easily made resilient to fire? How do legacy disturbances affect response to treatments?
  - **Developing applications of biochar soil amendments in the larger context of hazardous fuel reduction.** Pilot studies to develop and compare technologies for, and to assess the costs and benefits of (1) converting biomass generated by forest restoration and hazardous fuel treatments to biochar, (2) utilizing biochar to sequester C and improve fertility and water-holding capacity of soils, and (3) reducing air pollutants associated with hazardous fuel reduction.
- **Monitoring recovery from disturbance and pollution, and apply understanding of regional and local soil differences to evaluate and mitigate risk.** Some actions might include: standardizing and appropriately scaling monitoring data to meet planning and management needs within the context of ecosystem services and habitat fragmentation. Also useful would be to conduct initial and subsequent postdisturbance monitoring or risk assess-

ments (or both) to guide adaptive management, land management planning, pollution mitigation, silvicultural prescriptions, and other practices that affect soil health and productivity.

- **Developing multifactor, threshold-based best management practices (BMPs).** Some options might be integrating knowledge of the physical, biological, and chemical attributes of managed soils into BMPs. Identifying thresholds for soil function and structure, or "vital" signs of soil health are important as well. Including guidance on how to select appropriate tools and preparing comprehensive synthesis reports and distilled fact sheets to complement BMPs are also useful.
  - **Managing soil organic matter, C, nutrients, and water.** Some ways forward might be to develop and implement BMPs to (1) increase C storage in all land use types, partially mitigate soil C degradation, and promote rebuilding of soil organic carbon (SOC) post-disturbance; (2) promote retention and growth of native vegetation (fuel management, reforestation, and invasive weed control); and (3) leave or add organic residues on-site. Improve assessments of SOC vulnerability to disturbance by considering factors related to quantity and quality of organic matter inputs and soil microbial accessibility and activity.
  - **Managing soils for biotic health and diversity.** To address these parameters developing BMPs to optimize soil biotic communities under management regimes such as thinning, prescribed fire, and grazing could be important. Other considerations might be to promote biodiversity by minimizing management activities that cause the greatest level of disturbance (e.g., severe soil heating or exposure of bare mineral soils), using knowledge of soil resilience to different types of impacts. Conducting cost versus benefit analyses to inform management options would be very useful here.
  - **Managing soils to promote public health and food security in urban areas.** Implementing research and technologies for growing food in urban areas, designing green corridors and parks, and utilizing vegetation for shading and cooling would be useful considerations. These actions can improve food sufficiency and quality and lower the cost of food; improve livability; and reduce costs for cooling, respectively. Preparing management guidelines that conserve water, conserve soil microbiota, prevent soil erosion, prevent or mitigate soil contamination, expand soils and gardens in urban areas, and reduce the C cost related to soil organic matter loss and greenhouse gas emissions of growing food, would be useful as well.
  - **Restoring and managing wetland and hydric soils.** Use comprehensive assessment of site and environmental conditions as the foundation for BMPs to

ensure the long-term viability and functionality of natural wetlands, hydric soils, and restored wetlands.

Conclusions

Soils provide key ecosystem services essential to the health and welfare of humans. Accelerated disturbances, changes in land use and the environment, and continued population growth are increasing pressure on forest, rangeland, wetland, and urban soils to provide sufficient water, food, and fiber and continue to moderate climate change. Sustaining the health, biodiversity, and productivity of these soils into the future is imperative for sustaining the ecosystem services that soils provide. By addressing relevant research questions in relation to soil properties, functions, environmental stress, and disturbance effects, soil science can inform successful policies and practices. This research will be aided by the opportunities provided by new technologies, existing monitoring networks, and advances in mapping and data accessibility and new models to integrate monitoring and process-level studies. Training the next generation of scientists, improving opportunities for interdisciplinary collaboration among scientists and managers, and supporting programs to improve soil health and benefits to both urban and rural communities will help to ensure continued national strength in soils research and applications of its findings.

Literature Cited

Abatzoglou JT, Williams AP (2016) Impact of anthropogenic climate change on wildfire across western US forests. Proc Natl Acad Sci 113(42):11770–11775

Abbe TB, Montgomery DR (2003) Patterns and process of wood debris accumulation in the Queets River basin. Wash Geomorphol 51:81–107

Bellard C, Jeschke JM, Leroy B, Mace GM (2018) Insights from modeling studies on how climate change affects invasive alien species geography. Ecol Evol 8(11):5688–5700

Decaëns T, Jiménez JJ, Gioia C et al (2006) The values of soil animals for conservation biology. Eur J Soil Biol 42:S23–S38

Frank D, Finchkh M, Wirth C (2009) Impacts of land-use on habitat functions of old-growth forests and their biodiversity. In: Wirth C, Gleixner G, Heimann M (eds) Old-growth forests: function, fate and value. Springer, Dordrecht, pp 429–450

Franklin JF, Cromack K Jr, Denison W et al (1981) Ecological characteristics of old-growth Douglas-fir forests, General Technical Report PNW-GTR-118. U.S. Department of Agriculture, Forest Service, Pacific Northwest Forest and Range Experiment Station, Portland, 48 p

Franklin JF, Spies TA, Van Pelt R et al (2002) Disturbances and structural development of natural forest ecosystems with silvicultural implications, using Douglas-fir forests as an example. For Ecol Manag 155:399–423

Haristoy CT, Zabowski D, Nadkarni N (2014) Canopy soils of sitka spruce and bigleaf maple in the Queets River Watershed, Washington. Soil Science Society of America Journal 78(S1):S118-S124

Harmon ME, Bond-Lamberty B, Tang J, Vargas R (2011) Heterotrophic respiration in disturbed forests: a review with examples from North America. J Geophys Res 116:G00K04. https://doi.org/10.1029/2010JG001495

Jackson RB, Lajtha K, Crow SE et al (2017) The ecology of soil carbon: pools, vulnerabilities, and biotic and abiotic controls. Annu Rev Ecol Evol Syst 48(1):419–445

Keith H, Mackey BG, Lindenmayer DB (2009) Re-evaluation of forest biomass carbon stocks and lessons from the world's most carbon-dense forests. Proc Natl Acad Sci 106(28):11635–11640

Lawrence GB, Fernandez IJ, Richter DD et al (2013) Measuring environmental change in forest ecosystems by repeated soil sampling: a North American perspective. J Environ Qual 42(3):623–639

Luyssaert S, Schultze E-D, Borner A et al (2008) Old growth forests as global carbon sinks. Nature 455:213–215

Nadkarni NM (1981) Canopy roots: convergent evolution in rainforest nutrient cycles. Science 214:1023–1024

Naiman RJ, Bilby RE, Bisson PA (2000) Riparian ecology and management in the Pacific Coastal Rain Forest. Bioscience 50:996–1011

National Forest Management Act of 1976; Act of October 22, 1976; 16 U.S.C. 1600

Nearing MA, Pruski FF, O'Neal MR (2004) Expected climate change impacts on soil erosion rates: a review. J Soil Water Conserv 59(1):43–50

Neary DG, Ryan KC, DeBano LF, (eds) 2005 (revised 2008) Fire effects on soil and water. General Technical Report RMRS-GTR-42-vol.4. U.S. Department of Agriculture, Forest Service, Rocky Mountain Research Station, Ogden, 250 p

Pouyat R, Carreiro MM, Groffman PM, Pavao-Zuckerman M (2009) Investigative approaches to urban biogeochemical cycles: New York metropolitan area and Baltimore as case studies. In: McDonnell MJ, Hahs AK, Breuste JH (eds) Ecology of cities and towns: a comparative approach. Cambridge University Press, Cambridge/Oxford, pp 329–354

Swanson FJ, Lienkaemper GW (1982) Interactions among fluvial processes, forest vegetation, and aquatic ecosystems, south fork Hoh River, Olympic National Park. In: Franklin JF, Starkey EE, Matthews JW (eds) Ecological research in National Parks of the Pacific northwest. Oregon State University, Forest Research Laboratory, Corvallis, pp 30–34

Tejo CF, Zabowski D, Nadkami NM (2014) Canopy soils of Sitka spruce and bigleaf maple in the Queets River watershed, Washington. Soil Sci Soc Am J 78(1):118–124

Urrutia-Jalabert R, Malhi Y, Lara A (2015) The oldest, slowest rainforests in the world? Massive biomass and slow carbon dynamics of *Fitzroya cupressoides* temperate forests in southern Chile. PLoS One 10(9):e0137569. https://doi.org/10.1371/journal.pone.0137569

USDA Forest Service [USDA FS] (2012) National Forest System Land Management Planning. 36 CFR Part 219 RIN 0596-AD2. Fed Regis Rules and Regul 77(68):21162–21276

Van Pelt R, O'Keefe TC, Latterell JJ, Naiman RJ (2006) Riparian forest stand development along the Queets River in Olympic National Park, Washington. Ecol Monogr 76:277–298

Appendices

Appendix A: Regional Summaries Northeast

Lindsey E. Rustad

Introduction

Soils are dynamic and, by their very nature, are always under development. This is especially true for soils of the northeastern United States, which have changed dramatically over the past 400 years. The predominant land use has gone from deforestation to reforestation and back to deforestation, and the physical, chemical, and biological environment of soils has been altered accordingly. Alterations in the physical environment have included a rapidly changing climate during the past century and, along the coast, rising sea levels. Alterations in the chemical environment have included decades of elevated atmospheric deposition of sulfur (S)_, nitrogen (N), and mercury, followed by more recent declines in deposition of all three compounds; an increase in the atmospheric concentrations of carbon dioxide (CO_2) and other greenhouse gases; changes in point source inputs of fertilizers, pesticides, and other contaminants associated with agriculture, urbanization, and industrialization; and along the coast, salt water intrusions associated with rising sea levels and storm surges. Alterations in the biological environment include those wreaked by changes in the abundance, virulence, and distribution of pests, pathogens, and invasive species. Managing these soils now and into the future to maintain and optimize their ability to support a vast diversity of life; help sequester and cycle carbon (C), nutrients, and water; and produce food, fiber, and clean water for a growing regional and global population is a challenge. This section describes the environment of the Northeast, highlights the more salient threats to soils of this region, and offers straightforward advice for land managers and forest stewards to be proactive, rather than reactive, in protecting this valuable natural resource. It is critical to act before soils are degraded or altered in new ways, as the costs of remediation in dollars and lost ecosystem services far exceed those of proactive management.

The Environment of the Northeast

With more than 60 million people, the Northeast is the mostly densely populated region of the United States. For this report, this region is defined as the six New England states, New York, New Jersey, Delaware, Pennsylvania, Maryland, and West Virginia (Fig. A1). The region includes the major urban coastal corridor from Washington, DC, to Boston, MA, and north through Portland, ME, as well as suburban and exurban landscapes farther inland and northward. The climate is generally humid continental, with mean annual temperatures ranging from 10 to 16 °C in the southern section to 2 °C in far northern Maine and precipitation ranging from 890 to 1270 mm (Kunkel et al. 2013). Elevation ranges from sea level to 1917 m at the top of Mount Washington in New Hampshire, which has some of the most extreme weather in the country. Over 80% of the land is currently forested (Foster et al. 2010). Forest types include the southern and coastal pines (*Pinus* spp.) and hardwoods of Maryland and New Jersey, the northern hardwoods of New England, the subboreal and boreal spruce-fir (*Picea* spp.-*Abies* spp.) forests of the higher elevations and northern regions, and the Acadian Forest of far northern and coastal Maine. Much of the area from central Pennsylvania northward was glaciated; the retreat of the last glacier occurred about 13,000 years ago. Six soil orders can be found in the region: Alfisols, Entisols, Histosols, Inceptisols, Spodosols, and Ultisols.

Conversion of Forests to Other Land Use

Given the large current population and projections for future growth, one of the most significant threats to forest soils of the Northeast is land use change, particularly conversion of forests to residential, commercial, and industrial uses. Although over 80% of the land is currently forested, this has not always been the case, and the extent of forested land in the Northeast is currently declining (Foster et al. 2010; Olofsson et al. 2016). A quick history underscores the complex changes in land use that have taken place and are likely to happen in the future (Thompson et al. 2013) (Fig. A2). The region was largely forested at the time of the

R. V. Pouyat et al. (eds.), *Forest and Rangeland Soils of the United States Under Changing Conditions*, https://doi.org/10.1007/978-3-030-45216-2

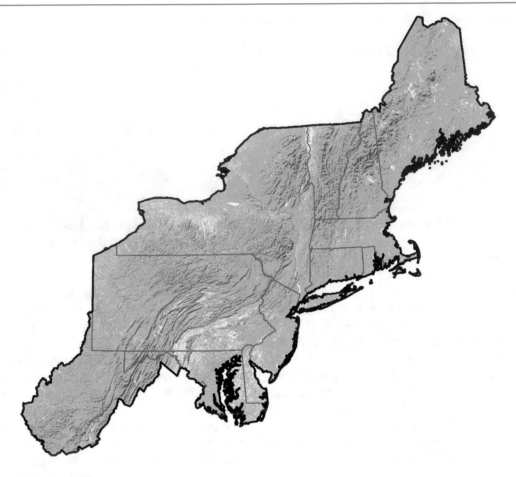

Fig. A1 Map of the Northeast region

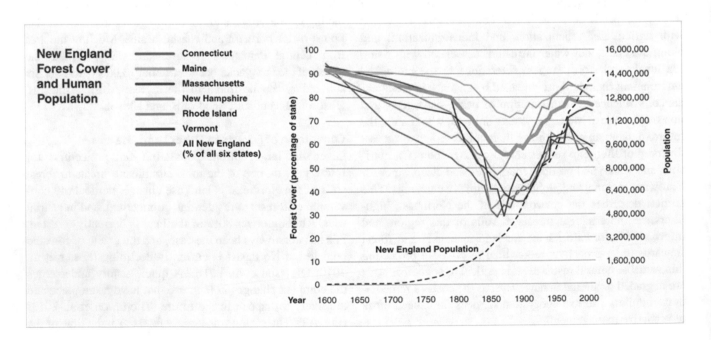

Fig. A2 Forest cover percentage, and human population, in New England, by state, 1600–2000 (Source: Foster et al. 2010)

arrival of the settlers in 1650. During the next two centuries, over 50% of the region was deforested due to intense logging and clearing for agriculture. Forest cover reached a minimum in the mid-nineteenth century. Then industrialization of the Northeast coupled with loss of the agricultural economy to the Midwest led to widespread abandonment of northeastern farms. The resulting reforestation continued until the late twentieth century, after which forest cover began to decline again. The recent decline is largely due to conversion to residential, commercial, and industrial use rather than for agricultural lands. Olofsson et al. (2016), for example, used satellite imagery to estimate land cover change in New England between 1985 and 2011. They found that forest area decreased by about 400 000 ha. More than half of the former forest area was converted to low-density residential development; 15% was converted to high-density residential, commercial, and industrial development; and 31% was converted to other land covers including landfills, agriculture, and recreational parks.

Historically, the large tracts of second-growth forests in the Northeast have provided food and fiber to local and worldwide communities, regulation of hydrologic and nutrient cycles, wildlife habitat, and aesthetic and recreational opportunities. These regrowing forests have also contributed to the sequestration of C in what is called the terrestrial global C sink (Houghton 2002). This sink, variously estimated in the range of 1–3.4 petagrams (Pg) C yr^{-1}, represents the difference between atmospheric concentrations of CO_2 and fossil fuel emissions (Pan et al. 2011). During the latter part of the twentieth century, near the peak of reforestation in the northeastern United States, North American forests and woodlands sequestered 0.6–0.7 Pg C yr^{-1}, with C stored in living biomass, forest products, dead wood, and soil organic matter (SOM) (Goodale et al. 2002). This terrestrial sink has served as a buffer against even greater increases in global atmospheric CO_2 concentrations and has thus moderated the climate change that has occurred to date. The saturation of this terrestrial C sink, or even reversal with the trend toward deforestation in the Northeast, will have implications for the global C cycle. The rate of climate change could be accelerated, with potential negative feedbacks to forest health and productivity and positive feedbacks to a changing climate.

As the previous century of reforestation is reversed due to increasing population and urban, suburban, and industrial development, management strategies that keep forests as forests, maintain soil C, protect soil nutrient capital, reduce soil loss due to erosion, and reduce or manage stressors to forests (e.g., pollution, climate change, invasive species) are warranted. These strategies are needed to protect regional ecosystem services, sequester terrestrial C, and mitigate global climate change.

Climate Change and Extreme Weather Events

Changes in climate, including an increase in the frequency and severity of extreme weather events, have already been observed for the Northeast. These changes are expected to continue (see Box A1). As temperature and moisture drive virtually all physical, chemical, and biological reactions (Rustad and Norby 2002), the projected changes will have profound direct and indirect, as well as short-term and longer-term, impacts on the forest soils of the region (Hayhoe et al. 2007; National Academies of Sciences, Engineering, and Medicine 2016).

Direct impacts include alterations of rates of water and nutrient cycling. Empirical evidence suggests that the rate of biological reactions will approximately double for every 10 °C increase in temperature. Results from syntheses of global change experiments have shown increased rates of soil and ecosystem C and N cycling with increased temperature and increased precipitation (Rustad et al. 2001; Wu et al. 2011) and decreased rates of the same cycling with drought (Wu et al. 2011). A longer growing season will increase the amount of time that plants and microbes are actively cycling soil nutrients. Whether these processes increase or decrease whole ecosystem C storage, however, is unclear. This storage represents the balance of C fixed via photosynthesis and C respired via plant and microbial respiration, both of which will be altered by changes in temperature, moisture, length of growing season, and other confounding factors.

Indirect effects stem from changes in forest species composition (with attendant species-specific and stand-specific changes in water and nutrient cycling); increased fuel loads and occurrence of fire (with impacts on aboveground and belowground C and nutrient stores); and increased abundance, virulence, and range for pests, pathogens, and invasive species (with attendant changes in native plant communities and associated water and nutrient cycling). Declines in snowpack will also have an impact on soil nutrient dynamics. Snow plays an especially important role in much of the northeastern United States, as it serves to insulate soils and protect them from deep frosts. Shrinking snowpacks ironically lead to colder soils in a warmer world (Groffman et al. 2001). Increased soil freezing can result in damage to fine roots, disruption of plant uptake of water and nutrients, and increased leaching of nutrients, including N, from soils to surface waters (Fitzhugh et al. 2001).

Although less is known about the impacts of extreme weather events on forested ecosystems, there is a growing recognition that these types of pulsed events can have a greater impact on natural and managed ecosystems than the more gradual, chronic changes in mean temperature and precipitation that are typically associated with climate change (Arnone et al. 2011; Jentsch et al. 2007). The types of extreme weather events characteristic of the Northeast include wind (hurricanes, nor'easters); intense rain, snow, sleet, hail, and ice storms; and periodic short-, medium-, and long-term droughts. All of these, through temperature and moisture changes, physical mixing, and erosion, have direct effects on the biogeochemistry of forest soils. They also have indirect effects resulting from the aboveground and below-ground forest disturbance and stress associated with these events; changes in the composition of forests toward species more resilient to these events; and introduction of pests, pathogens, and invasive species that can be more competitive following these events.

Of note for the Northeast is the increase in heavy precipitation events separated by longer intervals with no precipitation, resulting in both wetter and drier soils. Given that rates of C and nutrient cycling typically slow down during periods of soil water saturation (if oxygen is limited) and drought (if water is limited) and that periods of rewetting after drought are hotspots for nutrient cycling, the impact of these types of moisture fluctuations on forest soils may be consequential. Ice storms are also characteristic of the region and can cause widespread crown damage with concomitant transfers of C and nutrients from aboveground to belowground and lasting forest damage (Rustad and Campbell 2012). An extreme example was the ice storm of 1998, which affected over 9 million ha of forest land in the northeastern United States and southeastern Canada (Irland 1998). An increase in the intensity of these storms, as has been predicted by Swaminathan et al. (2018), could have a sizable impact on future forests and forest soils of the region.

Managing soils under a changing climate may include strategies to maintain diverse forests with mixtures of species that may be more adapted to expected future climate and more resilient to extreme weather events, minimize erosion and loss of soil C with implementation of best management practices during forest harvest, maintain forests and protected open spaces in developing areas, slow the introduction and movement of invasive species, and minimize other stressors, such as pollutants, that make forests more susceptible to pests and pathogens.

Sea Level Rise and Salt Water Intrusion

With more than 22 531 km of coastline (NOAA 2016), coastal freshwater and marine wetlands are an important landscape feature in the Northeast. Due to complex factors such as ocean currents and land subsidence, sea level rise in the northeastern United States is among the highest in the world. Between 1950 and 2009, the rate of sea level rise along the northeastern seaboard was 3–4 times the global average (Sallenger et al. 2012). Since 1900, sea level has risen by 0.3 m across the northeastern United States and by even more along the coast of New England (Horton et al. 2014a). The rate of sea level rise is expected to accelerate in the future, with sea level rise approaching 0.9–3.7 m by 2100 (Mendelssohn and Morris 2000). Throughout most of the

Northeast, additional sea level rise of 0.6 m would more than triple the frequency of coastal flooding, even without additional storms and associated storm surges (Janowiak et al. 2018).

Given the extensive coastline of the region and particularly rapid rate of sea level rise, higher sea levels and consequent salt water intrusions and salinity stress pose a threat to coastal soils, especially to tidal marine and freshwater wetlands. These wetlands, which receive regular inputs of tidal and land nutrients, are some of the most productive ecosystems in the world and are hotspots for biogeochemical activity (see Chap. 6). Because productivity and element cycling tend to be higher in tidal freshwater wetlands than in marine wetlands, changing boundaries between these two types of wetlands and increased salt water intrusions into freshwater wetlands may decrease C stocks and nutrient cycling along the coasts. These intrusions are compounded by storm surges from hurricanes and nor'easters (Kossin et al. 2007), which are predicted to increase in intensity in the future (Box A1).

Box A1 Changing Climate of the Northeast

The climate of the northeastern United States has been undergoing a period of rapid change, with increases in temperature; longer growing seasons; alterations in the amount, distribution, and intensity of precipitation; decreases in snow and ice cover; and an increase in the frequency and severity of extreme weather events (Dupigny-Giroux et al. 2018). Specifically, Horton et al. (2014b) reported that the climate for the period 1985–2011 was characterized by a 1.1 °C increase in temperature, a 10% increase in precipitation, more frequent extreme precipitation events, longer growing seasons, increases in the percentage of total precipitation falling as rain rather than snow, and declines in depth and duration of winter snowpack. Computer models for the region project more change to come, with mean annual temperatures increasing by another 1.7–5.6 °C (depending on emissions scenario), winter precipitation increasing by another 5–20%, continued increases in the length of the growing season, and declines in or elimination of snowpacks by the end of the twenty-first century (Horton et al. 2014b).

Extreme events characteristic of the region include wind (hurricanes, nor'easters); intense rain, snow, sleet, and hail and ice storms; and periodic short-, medium-, and long-term droughts. Most of these are projected to increase in frequency or severity, or both, in the future (Hayhoe et al. 2007; National Academies

(continued)

Box A1 (continued)

of Sciences, Engineering, and Medicine 2016). A key feature of climate change in the Northeast is that heavy precipitation events, defined as the heaviest 1% of all daily events, have increased by more than 70% between 1958 and 2010 (Melillo et al. 2014). This trend is expected to continue in the future.

Ice storms are another type of extreme event characteristic of the region. Recent climate modeling studies have suggested that while the frequency and distribution of this type of winter weather events are not expected to change, the severity of ice storms may increase (Swaminathan et al. 2018).

Changes in coastal wetland productivity and element cycling are further affected by warming of air and ocean temperatures, increasing fertilization from atmospheric N deposition and rising atmospheric CO_2 concentrations, increased runoff and sediment loads from the large precipitation events characteristic of the region, and coastal development pressures (Kossin et al. 2007; Mueller et al. 2016). Some of these factors, including increased sediment loads and increased productivity due to increased temperature, atmospheric N deposition, and atmospheric CO_2 concentrations, may counterbalance rising seas by increasing coastal wetland elevations (Langley et al. 2009). More research is needed to understand the dynamics between these opposing vectors of change. As sea levels continue to rise, an important part of managing coastal wetland soils will be to protect the remaining wetlands and areas upland from these wetlands from encroaching development.

Air Pollution

The Northeast has received some of the highest pollutant loads in the nation of S, N, mercury (Hg), lead (Pb), and other heavy metals due to both its geographic location downwind from major coal-fired power plants in the Midwest and its dense population, which contributes local sources of pollutants. These pollutants have significantly impacted forest soils of the region, resulting in soil and surface water acidification (Lawrence et al. 2007), base cation depletion (Bailey et al. 2005), N saturation (Aber et al. 1989, 1998), and heavy metal accumulation and toxicity (Friedland and Johnson 1985). Associated declines in forest growth, particularly for iconic species such as red spruce (*Picea rubens*) and sugar maple (*Acer saccharum*), have been well documented (DeHayes et al. 1999; Horsley et al. 2002). Emissions for many of these pollutants peaked by the mid- to late 1980s and have since declined due to effective clean air legislation, the removal of lead-based fuel additives from gasoline, shifts away from coal to other fuel sources, and new technology.

The legacy of these pollutants, however, remains. Soils in some areas lag in recovery for acidity and base cations (due, in part, to release of previously adsorbed sulfate and continued retention of previously deposited Pb, Hg, and other heavy metals) (Driscoll et al. 2007; Johnson et al. 1995). Studies have found, for example, that Spodosols in the northeastern United States can retain Pb for 50–150 years (Kaste et al. 2008). Much of the region is now recovering from these pollutant loads. Forests dominated by red spruce are showing marked rebounds in growth attributed to a combination of reduced atmospheric deposition and warmer temperatures (Kosiba et al. 2018). Continued recovery of the region's soil base cation status will depend on rates of mineral weathering and ongoing reductions in atmospheric inputs of wet and dry deposition.

With trends for increasing population, continued urbanization and industrialization, and the introduction of novel pollutants into the environment such as chemicals from pharmaceuticals and personal care products, several steps are essential to protecting this region's forest soils from further degradation. Continued research on and monitoring of the effects and loads of pollutants, and strategies to minimize input of point and nonpoint source pollutants into the environment, will be needed. The case of "acid rain" is a Northeast success story for this approach. In the United States, the acidity of precipitation due to anthropogenic emissions of S and N (creating sulfuric and nitric acid in the atmosphere) was first identified in the Northeast in the early 1970s (Likens et al. 1972). This discovery was followed by several decades of research and monitoring, the communication of this research to stakeholders and policymakers, and eventually the implementation of air pollution controls with the Clean Air Act Amendments of 1990. Emissions of these pollutants nationwide have since been reduced by 40% for nitric acid and 80% for sulfuric acid (Sullivan et al. 2018), and the research community now studies the "recovery" of ecosystems from these pollutants, rather than further harm.

Urban Soils

Given the already dense population of the Northeast and the projections for future increases, careful attention to urban soils is warranted in this region. Conditions associated with urban soils range from sealed soils, such as under roads, sidewalks, and buildings (e.g., Scalenghe and Marsan 2009), to disturbed soils without structure (Short et al. 1986). They further vary from highly engineered and managed soils with imported materials as are increasingly associated with green roof media, green strips, street tree pits, urban gardens, and public and residential lawns (Edmondson et al. 2012; Shaw and Isleib 2017; Trammell et al. 2016), to relatively unmanaged native soils in protected urban parks and recreation areas. In New York City, the approximate distribution is sealed soils (63%), human-altered or human-transported soils (27%), natural soils (9%), and nonsoil (1%) (Table 7.1). All are impacted to various degrees by local point sources of pollution (sewage effluents, groundwater contaminants, automobile and smokestack emissions), novel pollutants such as those found in pharmaceuticals and healthcare products (Bernhardt et al. 2017), altered hydrology (including irrigation and diversion over pervious and semipervious surfaces to wastewater and gray-water systems), and fertilizer and pesticide applications. Urban soils are further subject to local variations in climate, including increases in temperature and precipitation associated with both the "urban heat island" and "urban rainfall" effects (Butler et al. 2012; Craine et al. 2010; Shem and Shepherd 2009).

Despite their heterogeneity and typically high degree of disturbance, soils play vital beneficial roles in the urban environment. Broadly, they support plant, animal, and microbial populations and biodiversity (including shade trees, parks, community gardens, urban agriculture, green strips and roofs, urban wildlife, and soil microfauna and macrofauna), filter water of chemical pollutants, store water during droughts, and buffer surrounding areas from flooding during extreme rain events (Pouyat et al. 2010). Community participation in urban food production, caring for street trees and green spaces, and providing stewardship for urban soils are also catalysts for renewed connections between urban dwellers and their environment. Strategies that foster these relationships, along with supporting continued research on urban soils and applying this research to current and future development, will be essential for creating sustainable urban communities of the future.

Pests, Pathogens, and Invasive Species

The impact of pests, pathogens, and nonnative invasive species on northeastern forest soils and the ecosystems that they sustain cannot be underestimated (Dukes et al. 2009). Collectively, these organisms have dramatic direct and indirect impacts on forest plant, animal, and microbial species composition, physiology, and function. For example, insect pests such as hemlock woolly adelgid (*Adelgis tsugae*), Asian longhorned beetle (*Anoplophora glabripennnis*), and emerald ash borer (*Agrilus planipennis*) can damage foliage and reduce tree vitality. Pathogens, such as Armillaria root rot (*Armillaria* spp.), chestnut blight (*Cryphonectria parasitica*), and sudden oak death (*Phytophthora ramorum*), can cause widespread mortality of native tree species with resulting alterations in forest species composition and water and nutrient cycling. Invasive plant species, such as Japanese barberry (*Berberis thunbergii*), Oriental bittersweet (*Celastrus orbiculatus*), and glossy buckthorn (*Frangula alnus*), can create dense thickets of shrubs and vines, which effectively eliminate tree regeneration and monopolize light, nutrient, and water resources. Earthworms (phylum *Annelida*) play a particularly important role in soil formation processes and C cycling (Edwards 2004). The introduction of nonnative earthworms has resulted in the elimination of the O horizon and enormous reductions in soil C storage in many invaded forest soils (e.g., Fahey et al. 2013). The cumulative impact of these organisms often leads to extensive ecological and economic damage (Dukes et al. 2009).

The environment of the Northeast is particularly conducive to the spread of pests, pathogens, and invasive species for the following reasons:

- Many of these organisms thrive in warmer, wetter environments, with all phases of their life cycles being accelerated. As the climate of the region warms and gets wetter (Box A1), these species do well (Bale et al. 2002; Harrington et al. 2001; Logan et al. 2003).
- The range of many of these species is limited by winter low-temperature extremes, and as these low temperatures moderate in a warming climate, ranges expand (Shields and Cheah 2005; Trân et al. 2007).
- Many of these species have characteristics that make them more competitive than native species, including traits that allow them to tolerate new climates, undertake long-range dispersal, withstand climatic extremes, and take advantage of elevated atmospheric CO_2 concentrations (Dukes et al. 2009).
- Environmental stressors such as air pollution and climate extremes (drought, floods, and hot temperatures) create forest tree populations more vulnerable to pests, pathogens, and invasive species.
- Population centers along the east coast provide entry points for nonnative species. Shipping and transportation systems help spread the invaders.

More research is needed to understand the biology, physiology, and competitive interactions of pests, pathogens, and invasive species. More monitoring is needed to track the abundance, virulence, and range expansion of these organisms. Better management tools are needed to prevent and control outbreaks of known organisms. New regulations are needed to help prevent introduction of novel organisms.

Northeastern Soils Research, Mapping, and Monitoring in a Changing World

Continued research, mapping, and monitoring are needed for the soils in the complex landscapes of the Northeast. This ongoing work will be critical for developing sound policies that are proactive, rather than reactive, to protect this valuable natural resource. Soils develop over periods of decades to millennia, and costs for protection are typically far less than future costs for remediation. The Northeast has a unique legacy of natural and anthropogenic disturbance. Our ability to understand this legacy and to anticipate the consequences of current and projected known (and unknown) future disturbances, map the current distribution and health of soils and monitor change over time, and communicate this science to land managers and policymakers will require continued investment in basic and applied research, long-term monitoring and mapping, and support for scientific infrastructure and human capital.

Key Findings

- The Northeast has a unique legacy of natural and anthropogenic disturbance, from the glaciers of the past to the cycles of deforestation and afforestation due to logging, agriculture, and development to the inputs of point and nonpoint source pollutants, to recent rapid changes in climate and rising sea levels.
- Land use changed dramatically with the arrival of settlers 400 years ago. As much as 50% of the region was cleared for agriculture and cities by the 1850s. Forest cover currently exceeds 80% due to abandonment of agricultural lands but has been decreasing recently. Forest land is likely to continue decreasing into the future with the pressures of increasing population to convert forests to residential, commercial, industrial, agricultural, and recreational uses.
- The large tracts of second-growth forests in the northeastern United States have provided food and fiber, regulation of hydrologic and nutrient cycles, wildlife habitat, sequestration of C, and aesthetic and recreational opportunities.
- Climate change and extreme weather are altering a myriad of processes including C and N cycling; growing season; microbial activity; species composition; abundance and virulence of pests, pathogens, and invasive species; and insulation of soils by snow.
- Sea level rise and salt water intrusions will be especially problematic for this region due to the extensive coastline (>22 500 km), abundance of wetlands, and the higher rate of rise, which is 3–4 times the global average.
- Reduction of S- and N-containing air pollutants is a regional success story. Emissions have been reduced 40–80% nationwide since the mid-1980s. Recovery of some acidified soils and watersheds will, however, be slow, affecting soil productivity for decades into the future.
- Given the dense human population and projected future population increases in this region, care of urban soils will be especially beneficial to food production, biological diversity, water filtration, green space, and extreme weather buffering. Most urban soils are sealed or disturbed, except in protected parks. Disturbances include pollution, construction, altered hydrology, fertilizers and pesticides, and microclimate intensification.
- The Northeast is particularly conducive to the spread of pests, pathogens, and invasive species and will become more so as the climate warms and gets wetter. Population centers are entry points for nonnative species, aided by shipping and land transportation systems.
- Optimizing management and health of soils into the future will require continued investment in basic and applied science, long-term monitoring and mapping, and support for scientific infrastructure and human capital. Given that soils develop over periods of decades to millennia, future costs for remediation typically far exceed costs for protection.

Literature Cited

Aber JD, Nadelhoffer KJ, Steudler P, Melillo JM (1989) Nitrogen saturation in northern forest ecosystems. Bioscience 39(6):378–386

Aber J, McDowell W, Nadelhoffer K et al (1998) Nitrogen saturation in temperate forest ecosystems. Bioscience 48:921–934

Arnone JA, Jasoni RL, Lucchesi AJ et al (2011) A climatically extreme year has large impacts on C4 species in tallgrass prairie ecosystems but only minor effects on species richness and other plant functional groups. J Ecol 99:678–688

Bailey SW, Horsley SB, Long RP (2005) Thirty years of change in forest soils of the Allegheny Plateau, Pennsylvania. Soil Sci J Am 69(3):681–690

Bale JS, Masters GJ, Hodkinson ID et al (2002) Herbivory in global climate change research: direct effects of rising temperature on insect herbivores. Glob Chang Biol 8:1–16

Bernhardt ES, Rosi EJ, Gessner MO (2017) Synthetic chemicals as agents of global change. Front Ecol Environ 15:84–90

Butler SM, Melillo JM, Johnson JE et al (2012) Soil warming alters nitrogen cycling in a New England forest: implications for ecosystem function and structure. Oecologia 168:819–828

Clean Air Act of 1970, as amended November 1990; 42 U.S.C. s/s 7401 et seq.; Public Law 101–549. November 15, 1990

Craine JM, Fierer N, McLauchlan KK (2010) Widespread coupling between the rate and temperature sensitivity of organic matter decay. Nat Geosci 3:854

DeHayes DH, Schaberg PG, Hawley GJ, Strimbeck GR (1999) Acid rain impacts on calcium nutrition and forest health. Bioscience 49:789–800

Driscoll CT, Han Y-J, Chen CY et al (2007) Mercury contamination in forest and freshwater ecosystems in the northeastern United States. Bioscience 57:17–28

Dukes JS, Pontius J, Orwig D et al (2009) Responses of insect pests, pathogens, and invasive plant species to climate change in the forests of northeastern North America: what can we predict? Can J For Res 39:231–248

Dupigny-Giroux LA, Mecray EL, Lemcke-Stampone MD et al (2018) Northeast. In: Reidmiller DR, Avery CW, Easterling DR et al (eds) Impacts, risks, and adaptation in the United States: Fourth National Climate Assessment, II. U.S. Global Change Research Program, Washington, DC, pp 669–742. Chapter 18

Edmondson JL, Davies ZG, McHugh N et al (2012) Organic carbon hidden in urban ecosystems. Sci Rep 2:963

Edwards CA (2004) Earthworm ecology. CRC Press, Boca Raton

Fahey TJ, Yavitt JB, Sherman RE et al (2013) Earthworms, litter and soil carbon in a northern hardwood forest. Biogeochemistry 114:269–280

Fitzhugh RD, Driscoll CT, Groffman PM et al (2001) Effects of soil freezing disturbance on soil solution nitrogen, phosphorus, and carbon chemistry in a northern hardwood ecosystem. Biogeochemistry 56:215–238

Foster DR, Donahue B, Kittredge DB et al (2010) Wildland and woodlands: a forest vision for New England. Harvard University Press, Cambridge, MA

Friedland AJ, Johnson AH (1985) Lead distribution and fluxes in a high-elevation forest in northern Vermont. J Environ Qual 14(3):332–336

Goodale CL, Apps MJ, Birdsey RA et al (2002) Forest carbon sinks in the Northern Hemisphere. Ecol Appl 12:891–899

Groffman PM, Driscoll CT, Fahey TJ et al (2001) Colder soils in a warmer world: a snow manipulation study in a northern hardwood forest ecosystem. Biogeochemistry 56:135–150

Harrington R, Fleming RA, Woiwod IP (2001) Climate change impacts on insect management and conservation in temperate regions: can they be predicted? Agric For Entomol 3:233–240

Hayhoe K, Wake CP, Huntington TG et al (2007) Past and future changes in climate and hydrological indicators in the U.S. Northeast. Clim Dyn 28:381–407

Horsley SB, Long RP, Bailey SW et al (2002) Health of eastern North American sugar maple forests and factors affecting decline. North J Appl For 19:34–44

Horton BP, Rahmstorf S, Engelhart SE, Kemp AC (2014a) Expert assessment of sea-level rise by AD 2100 and AD 2300. Quat Sci Rev 84:1–6

Horton R, Yohe G, Easterling W et al (2014b) Northeast. In: Melillo JM, Richmond TC, Yohe GW (eds) Climate change impacts in the United States: the Third National Climate Assessment. U.S. Global Change Research Program, Washington, DC, pp 371–395. Chapter 16

Houghton R (2002) Terrestrial carbon sinks—uncertain. Biologist 49(4):155–160

Irland LC (1998) Ice storm 1998 and the forest of the Northeast. J For 96:32–40

Janowiak MK, D'Amato AW, Swanston CW et al (2018) New England and northern New York forest ecosystem vulnerability assessment and synthesis: a report from the New England Climate Change Response Framework project. Gen. Tech. Rep. NRS-173. U.S. Department of Agriculture, Forest Service, Northern Research Station, Newtown Square, 234 p

Jentsch A, Kreyling J, Beierkuhnlein C (2007) A new generation of climate-change experiments: events, not trends. Front Ecol Environ 5:365–374

Johnson CE, Siccama TG, Driscoll CT et al (1995) Changes in lead biogeochemistry in response to decreasing atmospheric inputs. Ecol Appl 5(3):813–822

Kaste Ø, Austnes K, Vestgarden LS, Wright RF (2008) Manipulation of snow in small headwater catchments at Storgama, Norway: effects on leaching of inorganic nitrogen. Ambio 37:29–37

Kosiba AM, Schaberg PG, Rayback SA, Hawley GJ (2018) The surprising recovery of red spruce growth shows links to decreased acid deposition and elevated temperature. Sci Total Environ 637–638:1480–1491

Kossin JP, Knapp KR, Vimont DJ et al (2007) A globally consistent reanalysis of hurricane variability and trends. Geophys Res Lett 34:L04815

Kunkel K, Stevens L, Stevens S et al (2013) Regional climate trends and scenarios for the US National Climate Assessment. Part 1. Climate of the Northeast US. NOAA Tech. Rep. NESDIS 142-1. National Oceanic and Atmospheric Administration, National Environmental Satellite, Data, and Information Service, Washington, DC, 87 p

Langley JA, McKee KL, Cahoon DR et al (2009) Elevated CO_2 stimulates marsh elevation gain, counterbalancing sea-level rise. Proc Natl Acad Sci 106(15):6182–6186

Lawrence GB, Sutherland JW, Boylen CW et al (2007) Acid rain effects on aluminum mobilization clarified by inclusion of strong organic acids. Environ Sci Technol 41(1):93–98

Likens GE, Bormann FG, Johnson NM (1972) Acid rain. Environment 14:33–40

Logan JA, Régnière J, Powell JA (2003) Assessing the impacts of global warming on forest pest dynamics. Front Ecol Environ 1:130–137

Melillo JM, Richmond TC, Yohe GW (eds) (2014) Climate change impacts in the United States: the Third National Climate Assessment. U.S. Global Change Research Program, Washington, DC, 841 p

Mendelssohn IA, Morris JT (2000) Eco-physiological controls on the productivity of *Spartina alterniflora* Loisel. In: Weinstein MP, Kreeger DA (eds) Concepts and controversies in tidal marsh ecology. Springer, Dordrecht, pp 59–80

Mueller P, Jensen K, Megonigal JP (2016) Plants mediate soil organic matter decomposition in response to sea level rise. Glob Chang Biol 22:404–414

National Academies of Sciences, Engineering, and Medicine (2016) Attribution of extreme weather events in the context of climate change. National Academies Press, Washington, DC, 186 p

National Oceanic and Atmospheric Administration [NOAA] (2016) NOAA shoreline website. Homepage. https://shoreline.noaa.gov/faqs.html?faq=2. NOAA Office for Coastal Management. Accessed 27 July 2018

Olofsson P, Holden CE, Bullock EL, Woodcock CE (2016) Time series analysis of satellite data reveals continuous deforestation of New England since the 1980s. Environ Res Lett 11:064002

Pan Y, Birdsey RA, Fang J et al (2011) A large and persistent carbon sink in the world's forests. Science 333(6045):988–993

Pouyat RV, Szlavecz K, Yesilonis ID et al (2010) Chemical, physical, and biological characteristics of urban soils. In: Aitkenhead-Peterson J, Volder A (eds) Urban ecosystem ecology. American Society of Agronomy–Crop Science Society of America–Soil Science Society of America, Madison, pp 119-152

Rustad L, Norby R (2002) Temperature increase: effects on terrestrial ecosystems. In: Mooney HA, Canadell JG (eds) Encyclopedia of global environmental change, Biological and ecological dimensions of global environmental change, vol 2. John Wiley & Sons, Chichester, pp 575–581

Rustad LE, Campbell JL (2012) A novel ice storm manipulation experiment in a northern hardwood forest. Can J For Res 42:1810–1818

Rustad LE, Campbell JL, Marion G et al (2001). A meta-analysis of the response of soil respiration, net nitrogen mineralization, and aboveground plant growth to experimental ecosystem warming. Oecologia 126(4):543–562

Sallenger AH Jr, Doran KS, Howd PA (2012) Hotspot of accelerated sea-level rise on the Atlantic coast of North America. Nat Clim Chang 2:884

Scalenghe R, Marsan FA (2009) The anthropogenic sealing of soils in urban areas. Landsc Urban Plan 90(1–2):1–10

Shaw RK, Isleib JT (2017) The case of the New York City Soil Survey Program, United States. In: Levin MJ, Kim K-HJ, Morel JL et al (eds). Soils within cities: global approaches to their sustainable management. GeoEcology essays. Schweizerbart, Catena Soil Sciences, Stuttgart, pp 107–113

Shem W, Shepherd M (2009) On the impact of urbanization on summertime thunderstorms in Atlanta: two numerical model case studies. Atmos Res 92:172–189

Shields KS, Cheah CAS-J (2005) Winter mortality in *Adelges tsugae* populations in 2003 and 2004. In: Gottschalk KW (ed) Proceedings, 16th U.S. Department of Agriculture interagency research forum on gypsy moth and other invasive species 2005. Gen. Tech. Rep. NE-337. U.S. Department of Agriculture, Forest Service, Northeastern Research Station, Newtown Square, p 73

Short J, Fanning D, McIntosh M et al (1986) Soils of the Mall in Washington, DC: I. Statistical summary of properties. Soil Sci Soc Am J 50(3):699–705

Sullivan TJ, Driscoll CT, Beier CM et al (2018) Air pollution success stories in the United States: the value of long-term observations. Environ Sci Policy 84:69–73

Swaminathan R, Sridharan M, Hayhoe K (2018) A computational framework for modelling and analyzing ice storms. arXiv: 1805.04907. Available at https://arxiv.org/abs/1805.04907. Accessed 15 Apr 15 2019

Thompson JR, Carpenter DN, Cogbill CV, Foster DR (2013) Four centuries of change in northeastern United States forests. PLoS One 8(9):e72540

Trammell T, Pataki DE, Cavender-Bares J et al (2016) Plant nitrogen concentration and isotopic composition in residential lawns across seven US cities. Oecologia 181:271–285

Trân JK, Ylioja T, Billings RF et al (2007) Impact of minimum winter temperatures on the population dynamics of *Dendroctonus frontalis*. Ecol Appl 17:882–899

Wu Z, Dijkstra P, Koch GW et al (2011) Responses of terrestrial ecosystems to temperature and precipitation change: a meta-analysis of experimental manipulation. Glob Chang Biol 17:927–942

Southeast
Mac A. Callaham, Jr. and Melanie K. Taylor

The Setting
Soils in the Southeast region are among the most productive in the United States. The region is often called the wood basket because it provides more timber and forest products than any other region in the country (Wear and Greis 2013). Additionally, southeastern states lead the nation in production of crops such as blueberries (*Vaccinium* spp.), peanuts (*Arachis hypogaea*), pecans (*Carya illinoinensis*), and cotton (*Gossypium hirsutum*) (USDA NASS 2018). This productivity is possible despite the long history of exploitation and degradation of the soil resources.

It is a daunting task to summarize the potential issues for all soils in the Southeast primarily due to the amazing diversity of geology, topography, and climate and the resulting potential vegetation that would be supported in any particular location within the region. This is evident when one considers that the Southeast, spanning 13 states (Alabama, Arkansas, Florida, Georgia, Kentucky, Louisiana,

Mississippi, North Carolina, Oklahoma, South Carolina, Tennessee, Texas, and Virginia; note that Oklahoma and Texas are also discussed in the Great Plains regional summary), includes soils ranging in elevation from sea level up to more than 2030 m and receives annual precipitation ranging from more than 1800 mm at high elevations in the Appalachian Mountains to less than 250 mm in far western Texas. Likewise, soils in the region range in residence time (age) from very short (recently deposited riparian sediments) to ancient (up to 3 million years old for Southern Piedmont soils (Bacon et al. 2012)). These soils are derived from vastly different parent materials, ranging from ocean floor sediment to glacial outwash to sedimentary shales and limestones to igneous rock. Such gradients yield tremendously varying vegetation from high plains grasslands to desert scrub to forests dominated by pines (*Pinus* spp.) to temperate hardwood forest to wet subtropical forests. Given our understanding that soil formation is the result of the interaction among climate, organisms (vegetation and animals), relief (topography), parent material, time, and human influences (Jenny 1980), a region with such wide ranges for each of these factors would be expected to be home to a great variety of soils (Figs. A3 and A4). Indeed, of the 12 recognized soil orders (Soil Survey Staff 2014), 9 are represented in the Southeast (frozen, volcanic, and tropical soils are excluded) (West 2000). We likewise would expect a host of potential threats and challenges to such diverse soils. Here we summarize what we consider to be the most important and pressing challenges facing southeastern soil ecosystems, and we draw from the pertinent literature to provide examples for each issue.

Urbanization

Using the framework, predictions, and forecasts presented in the Southern Forest Futures Project (Wear and Greis 2013), we can overlay many of the challenges to soil resources in the region. Principal among these are changes in land use and land cover and the attendant changes in societal uses and expectations of the resource. In the Futures project, Wear and Greis used four different scenarios (each with different assumptions about changes in population growth and economic conditions) to predict how land use patterns would change between 2010 and 2060, and these predictions were reasonably straightforward to apply to soil-related issues. One of the most important findings—regardless of assumptions—was that a projected area between 11 900 and 17 280 km^2 in the region would be converted to urban use from other uses (forest, pasture, rangeland, cropland). Not surprisingly, most of this urbanization was projected to occur around existing urban centers and along coastlines (e.g., Atlanta, GA; Charlotte, NC; Dallas, TX; Houston, TX; Knoxville, TN; Miami, FL; Nashville, TN). The implication of this land use change would be increased pressure on the remaining, unconverted lands to provide food, fiber, timber, water, and recreational resources adequate to meet the needs of a larger human population with a diminished land base.

Erosion

Certain soils within the Southeast are particularly susceptible to erosion. Chief among these are the Ultisols of the Southern Appalachian Piedmont (Jackson et al. 2005; Trimble 1974) and the loess-derived soils along the edges of the Lower Mississippi Alluvial Valley, along with notable historical erosion of soils in Oklahoma and Texas during the Dust Bowl of the 1930s. Soil conservation-minded agricultural practices, and the planting of millions of hectares of trees on the most highly eroded soils, have reduced erosion rates considerably on much of the land. Nevertheless, soil erosion remains a major issue in many parts of the Southeast, especially in areas with extensive urban development activities such as road building, land clearing, and grading for construction of commercial, industrial, and residential buildings (Fig. A5).

Climate Variability

Climate models suggest that temperatures will become warmer across the Southeast and that precipitation will generally increase, with the exception of the Gulf Coast and the western part of the region, where precipitation is projected to decrease (Kunkel et al. 2013). Precipitation patterns are expected to become increasingly variable with more very wet and very dry conditions (Burt et al. 2017; Li et al. 2013). The effects of these climate variables on soils are dependent on the interplay of climate with soil morphology, vegetation, and human influences. Adding another layer of complexity to these relationships, a changing climate is likely to lead to variation in ranges of tree species (Iverson et al. 2008), with implications for soils where those trees grow. Additionally, hurricanes have become increasingly intense (Emanuel 2005), and hurricane seasons are projected to become longer (Kossin 2008), although there is some uncertainty in these models. In the following discussion, we treat the effects of each of these climate-related variables on soils separately, although many of them are related and will be covered further in the "Interactions" section. Finally, in light of changes in climate, there is intense interest in the global pool of carbon (C) that soils represent, and here we discuss the potential contribution of southeastern soils to these stocks.

Sea Level Rise

It almost goes without saying that sea level rise threatens soils in low-lying areas. A sea level rise of 1.5 m (within the range of projected changes by the year 2100 (Melillo et al. 2014)) would impact an estimated 53 800 km^2 of land lying below 1.5 m elevation in the Southeast. This figure accounts basically for inundation, but does not project the impact on soils

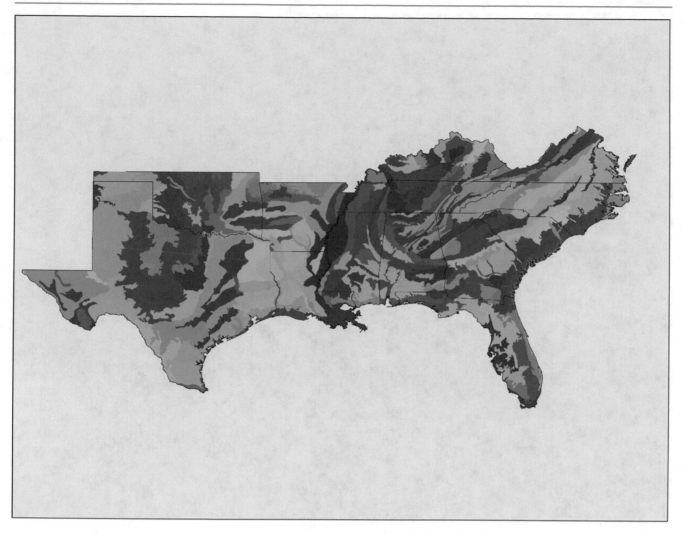

Fig. A3 Map showing diversity of major soils in the Southeast (USDA Forest Service Region 8). There are 259 unique map units depicted here, roughly correlating with the USDA Soil Taxonomy level of "subgroup" (Soil Survey Staff 2014) (Data source: Soil Survey Staff 2014; map compiled by Derek Wallace, USDA Forest Service)

from shifting patterns of marsh vegetation or tidal inundation. Furthermore, intrusion of salt water into inland soils will result in the additional loss of soil resources to sea-level rise phenomena. Again, the primary result of such impacts will be a higher concentration of the human population as residents relocate after being displaced by coastal flooding and increasing demands on the soil resources remaining above sea level.

Hurricanes and Tropical Storms

The Southeast has long been affected by hurricanes, and the intensity of these storms, along with the duration of the Atlantic hurricane season, is expected to continue to increase over the twenty-first century (Emanuel 2005; Kossin 2008). Hurricanes affect soils by increasing flooding and coastal erosion and by inundating soils with salt water caused by the storm surge, thereby impacting soil fertility. Blood et al. (1991) found that Hurricane Hugo in South Carolina inun-

dated forest soils with salt water greater than 150 m from the marsh edge, affecting soil fertility and demonstrating the far-reaching effect of an intense coastal storm. Winds from hurricanes cause changes in plant communities, particularly in forests, and can alter the amount of organic material on the soil surface. Combined, these changes alter the composition of the soil biotic community and can have long-lasting effects on ecosystem stability and productivity (Coyle et al. 2017).

Drought

Increased temperatures and higher variability in precipitation patterns in the Southeast are likely to lead to periods of dryness that are longer or more frequent or both (Burt et al. 2017; Li et al. 2013). In forests of the Southeast, drought has been demonstrated to lead to reduced tree growth, increased mortality, and in some forest types, reduced recruitment (Hu et al. 2017). Together, these effects have the potential to alter

Fig. A4 Fine-resolution map depicting soil diversity using North Carolina as an example. Different colors here represent the USDA Soil Taxonomy level of "family" (Soil Survey Staff 2014). There are 14 great groups, 25 subgroups, and 48 families represented in the soils of North Carolina (Data source: Soil Survey Staff 2014; map compiled by Derek Wallace, USDA Forest Service)

the plant species composition of southeastern forests. Such changes in forest species composition have been linked to changes in soil structure, function, and ability to support fauna (Reich et al. 2005). Decline in plant cover associated with drought exacerbates soil dryness, with generally negative effects on soil fauna, though the magnitude of the effect depends on the taxa in question (Coyle et al. 2017). In addition, drought-stressed plants often become more vulnerable to insect and disease outbreaks (Weed et al. 2013), and drier conditions exacerbate fire potential (see "Interactions" section). Root death associated with drought and fire has a destabilizing effect on soils by making them prone to erosion, which sometimes leads to catastrophic flooding and landslides when rainfall returns (Moody et al. 2013). Parts of the Southeast are heavily dependent on irrigated agriculture, and long periods of drought may subject soils in those areas to increased erosion. In the western part of the Southeast (Texas and Oklahoma), Steiner et al. (2018) suggest that

crops irrigated from the Ogallala aquifer are particularly sensitive to extended periods of drought, and these authors recommend adaptive strategies such as increasing soil organic matter to minimize this sensitivity.

Flooding

Flooding in the Southeast shapes the soil resource by moving and redepositing sediment. Floodplain regions exist because the levels of streams and rivers fluctuate after heavy precipitation events, including tropical storms. Indeed, many plant and animal species exist only on these floodplain soils. In areas that are infrequently flooded, there is some concern that flooding may have negative effects on soil aggregate stability (De-Campos et al. 2009) and soil microbial communities (Unger et al. 2009). Flooding alters soil faunal communities, but outcomes depend largely on where in the soil profile fauna live, their ability to cope with submersion, and the duration and extent of flooding. Severe flooding can

Fig. A5 Examples of erosion of typical Piedmont soils (Ultisols) near Athens, GA, June 2018. (**a**) Erosion associated with roadside utilities installation, and (**b**) erosion associated with new commercial construction. In both cases, workers attempted to slow the erosion process. Sediment was nevertheless making its way into local waterways, resulting in overall soil loss (and diminished surface water quality) (Photo credit: Mac Callaham, USDA Forest Service)

have long-lasting effects on soil invertebrate communities, which can take years to return to preflood levels (Coyle et al. 2017). Although much of the empirical evidence for these relationships comes from examples outside the Southeast, there is no reason to expect different responses of soil organisms to flood-related disturbances. If urbanization in the Southeast increases as projected, special precautions may be needed to mitigate the erosion and potential flooding resulting from more rapid runoff from urban lands.

Carbon Management

Globally, soils are the largest pool of terrestrial C, and the amount of C stored in any given soil is determined by many of the soil-forming factors already discussed, such as climate, organisms (vegetation and fauna), and human influences. In the Southern Forest Futures Project, Wear and Greis (2013) project a 1–9% loss of soil organic C from the Southeast by 2060, due mainly to the conversion of forested land to other land uses, particularly urbanization. The magnitude of this loss is expected to depend on the intensity of urbanization and economic conditions. Though conversion of land from forest to other uses generally means a loss of soil C storage, management options do exist for sustaining

and promoting soil C accumulation. These options are largely focused on agricultural systems and include using agroforestry, manuring, and no-tillage farming practices (Scharlemann et al. 2014). Considering the likely loss of soil C due to conversion of forests to urban landscapes in the Southeast, management options such as reforestation of abandoned agricultural or urban sites may need to be used along with the other management options listed previously for agricultural lands to offset this loss.

Fire

Fire is one of the most important ecological forces in southeastern wildland ecosystems. Fire also plays an important role in many agricultural and rangeland ecosystems within the region. The Southeast experiences more fire in terms of area burned and in terms of total ignitions (prescribed fire and wildfire) than any other region in the United States (Fig. A6) (Mitchell et al. 2014). Fires in the Southeast affect soils in different ways. Most ecosystem types in the region are fire dependent, but others could be classified as fire sensitive. In fire-dependent ecosystems, the predominant soil-related issues result from fire exclusion. For example, when longleaf pine (*Pinus palustris*) ecosystems are allowed to go for long

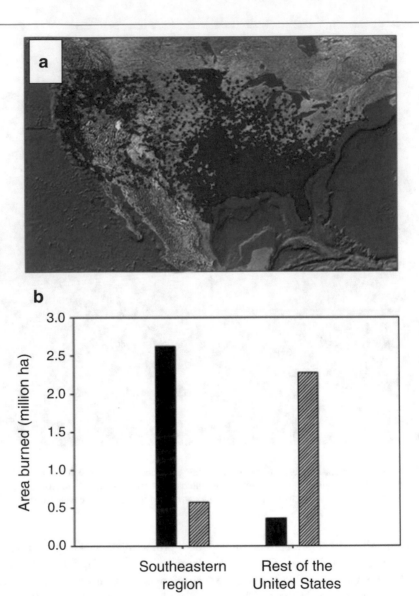

Fig. A6 Fire in the Southeast, compared to the rest of the continental United States. (**a**) Cumulative distribution of Moderate Resolution Imaging Spectroradiometer fire detections in 2010 for the continental United States. Each red point represents a single detection (NASA n.d.). (**b**) Comparison of area burned by prescribed fire and wildfire between the Southeast region and the rest of the United States in 2011. The solid bars indicate prescribed fires (data derived from Melvin 2012); the hatched bars indicate wildfires (NICC n.d.) (Figure redrawn from Mitchell et al. 2014)

periods without fire, the surface soil horizon develops into a thick organic layer composed of decomposing pine leaf litter. This organic layer contains much of the available nutrient pools in these ecosystems; thus, large quantities of fine roots develop in this layer. However, because the material in the O horizon is combustible, this material (along with the associated fine roots) often is consumed when fires do occur, with adverse effects on the overstory trees (O'Brien et al. 2010). There are other ecosystems where fire is uncommon and infrequent (e.g., large areas of wetland soils such as the Okeefenokee swamp in southeastern Georgia). On the rare occasions when these ecosystems do experience fire, it is the slow, smoldering combustion, and consumption of the highly

organic (peat) soils that is the principal long-term ecological impact of fire.

Prescribed fire effects on soils in the eastern United States have been summarized by Callaham et al. (2012). These authors found very little in the literature to suggest that prescribed fires had any long-term effects on soil chemical, physical, or biological components, although there were general patterns for short-term responses. Several of the reviewed studies reported measurable short-term increases in soil pH, which were also associated with increased concentrations of available cations (potassium, calcium, magnesium, and sodium) as components of ash. Likewise, several studies demonstrated that prescribed fires had measurable, but short-term, impacts on forest floor (leaf litter) mass and that these effects were associated with a loss of total nitrogen (N) from the forest floor. On the other hand, mineral forms of N (i.e., plant available forms: nitrate and ammonium) and the microbe-mediated process of N mineralization were all found to increase in soils subjected to prescribed fires.

Another major aspect of wildland fire in the Southeast is its relationship to ecosystem C management (discussed earlier in this summary and in Chap. 2). Unlike some other regions in the United States, the Southeast has many ecosystems that are ecologically adapted to frequent fires. This adaptation makes the management of ecosystem C stocks somewhat counterintuitive relative to other ecosystems.

An often overlooked aspect of fire effects on soils of the Southeast is the effect of fire on soil fauna. Though fauna are, in many cases, drivers of soil processes and fire is a critical component of southeastern ecosystems, there is surprisingly little work on the subject. One exception to the lack of related research is work on the Olustee Experimental Forest in northern Florida. There, Hanula and Wade (2003) found that the effects of prescribed fire on soil arthropods in a longleaf pine forest largely depended on species and fire frequency but that fire overall had a negative effect on soil arthropod diversity when compared to unburned areas. In another study from the same area, Hanula et al. (2012) found that dormant-season prescribed burning did not affect the abundance of termites (*Reticulitermes* spp.). This area of research is particularly open for additional work in the Southeast and is anticipated to yield insights into general fire effects due to the importance of soil fauna to ecosystem processes and the prevalence of prescribed fire.

Invasive Species

Invasive species represent a wide array of threats to soils and soil ecosystems across the Southeast, and these threats are complex because invaders take many forms including plants, animals, and microbes (see especially Chaps. 2, 4, and 5). Specific problems for soils and their interactions with invasive species include responses to plants that can alter the soil environment through chemical, physical, or biological channels. In many cases, these plant-mediated changes to soils result in modifications that render the soil more suitable for continued occupation or expansion by the invasive plant (a positive plant-soil feedback). There are dozens of invasive plant species in the Southeast (Miller et al. 2013), and each has potential to influence soils. Japanese stiltgrass (*Microstegium vimineum*) illustrates the kind of influence that an invasive plant can exert. This plant has invaded all 13 of the southeastern states and occurs in virtually every county in Virginia, Tennessee, Kentucky, and North Carolina (EDDMapS 2018). It has been documented to alter soil microbial (specifically root-associated fungal) communities (e.g., North and Torzilli 2017) and to alter the processing of

C and nutrients in surface soils where it invades (Strickland et al. 2010). In doing so, the plant provides itself with an advantage toward persistence or further invasion or both. Similarly, cogongrass (*Imperata cylindrica*) is a fast-spreading invasive plant with large, established populations in Florida, Alabama, Mississippi, and Georgia and is projected to spread to most of the remaining Southeast outside of the mountains (Estrada and Flory 2015). Cogongrass alters N-cycling and fire regimes such that it promotes its own growth and reproduction while native grasses and trees decline (Daneshgar and Jose 2009; Lippincott 2000).

In other cases, the invasion of an ecosystem by a plant can result in changes to soils that promote the invasion of other organisms, such as earthworms (phylum Annelida). One example of this kind of interaction is shown in the work of Lobe et al. (2014), who found that nonnative species of earthworms (suborder Lumbricina) were more abundant under thickets of invasive Chinese privet (*Ligustrum sinense*) in floodplain forests of the Piedmont. Further, these authors found that removal of Chinese privet resulted in a decrease in the abundance of the nonnative earthworms and an increase in abundance of native earthworm species.

Invasions by nonnative insects can have direct and indirect consequences for soils. One example of a direct effect is the invasion of the red imported fire ant (*Solenopsis invicta*), which has become established in most of the Southeast (USDA APHIS 2018) and has been shown to be associated with decline in native ants (and at least temporary disruption of other arthropods in the community) throughout its invaded range (LeBrun et al. 2012: Morrison 2002; Porter and Savignano 1990). One ongoing demonstration of the indirect effects that nonnative insects can have on soils is the case of eastern hemlock (*Tsuga canadensis*) decline, caused by hemlock woolly adelgid (HWA; *Adelges tsugae*) across the southern Appalachians. The death of eastern hemlock trees may alter several soil parameters with potential cascading effects through ecosystems. In one example of these ramifications, Block et al. (2013) found that HWA mortality was associated with changes in soil phosphorus cycling patterns that resulted in lower N:phosphorus stoichiometry of foliage, which in turn may influence decomposition processes and nutrient supply in these ecosystems. These authors concluded that interactions between chronic N deposition, which occurs primarily at high elevation, and HWA mortality resulted in "liberation" of phosphorus that would otherwise not be available for cycling and that this positively influenced site productivity. However, in another study conducted at lower elevation, Knoepp et al. (2011) found no detectable differences in most of the soil variables that they measured, including total C content, soil solution N concentrations, and N mineralization rates. These contrasting results from areas that are fairly close to each other but at different elevations emphasize the importance of understanding that soil responses to various disturbances depend on context.

Biodiversity

Soil represents an underappreciated reservoir of biological diversity. Indeed, in the scientific literature, soil has been referred to as "the final frontier of biodiversity" (Briones 2014). Soil biodiversity as a national resource issue is treated in Chap. 5. As home to myriad forms of life including bacteria, fungi, nearly all the invertebrate phyla, and hundreds of fossorial vertebrate species, the soil is literally teeming with life (Coleman et al. 2018). However, soil biodiversity is under threat worldwide (Orgiazzi et al. 2016), and the soils of the Southeast are no exception. Increased pressure on soils from expanding urban development, continued row-crop agriculture, sea level rise, and more intensive forestry practices all threaten the biological diversity present in the soils of the region. As one of the portions of the continental United States where native species of earthworms occur (James 1995), the Southeast has yet to be fully inventoried even for this relatively well-known group and species that have not been formally described have recently been collected across the region (Carerra-Martínez 2018). For other organisms, such as macroarthropods, we know that there are hotspots of biodiversity in the region, such as the millipede (class Diplopoda) fauna of the southern Appalachians, but even this diversity is not well documented (Snyder 2008).

Microbial biological diversity within soil represents a significant resource for pharmaceutical use, as a majority of antibiotics currently known have come from soil microbial isolates (Nesme and Simonet 2015). This resource is threatened by all of the various forces discussed throughout this assessment that have potential to change microbial communities in soil. Thus, the biological diversity of soil—bacterial, fungal, animal, and plant—represents a serious conservation concern for the Southeast.

Interactions

We have briefly discussed what we consider to be the major areas of concern for soils of the Southeast. Many of these influences on soil interact to produce outcomes that are often not intended or foreseen. For example, fire may have unintended influences on other ecosystem components, such as invasive species. Kelly and Sellers (2014) found that the red imported fire ant had invaded a relatively undisturbed habitat in North Carolina (cypress savanna dominated by *Taxodium ascendens*) and that this invasion may have been favored by the dense herbaceous understory vegetation that is promoted by frequent fires. Another example of the interactive influence of fire on soils is in the control of invasive species. Ikeda et al. (2015) found that fire may be an important tool for slowing the invasion of the earthworm *Amynthas agrestis*

by reducing the viability of the earthworm's cocoons and by reducing food availability through litter consumption by fire.

Perhaps most importantly, the human interaction with soil resources through activities such as deforestation, reforestation, urbanization, and agriculture cannot be understated. This is apparent when viewed in the context of the history of soils of the Southeast. In the mid-twentieth century, the southeastern landscape was dominated by soils degraded through deforestation and intensive agriculture. Collaborative efforts of the USDA Forest Service, the USDA Soil Conservation Service, universities, state forestry agencies, and private industry led to research breakthroughs in fertilization, tree improvement, and land management that quadrupled the productivity of pine plantations in the Southeast and transformed these lands into the "wood basket of the world" (Fox et al. 2007). Continued investment in this type of collaborative research and development, and the application of this science, is the most rational way to address the challenge of ensuring the maintenance of sustainable, resilient ecosystems, and managing the pressures on soil resources in the Southeast and beyond.

Key Findings

- The Southeast is home to highly diverse soils subject to a diverse set of challenges. Disturbances such as urbanization, floods, droughts, fire, and invasive species are among the greatest challenges facing managers of the southeastern soil resource. Nevertheless, in spite of degradations associated with the historical legacy of primitive agricultural practices over much of the region, the soils of the Southeast remain an impressive reservoir of unexplored biological diversity and are among the most productive soils in North America.
- Urbanization (conversion of land from other land uses to urban use) represents perhaps the largest threat to the southeastern soil resource, as this process necessarily results in less land area to provide equivalent or greater amounts of agricultural, forestry, ecological, and recreational products and services. Likewise, sea level rise projected for low-lying areas in the Southeast will reduce productive land area. Uncertainty surrounding climate variability and questions about the vegetation that can be sustainably supported by southeastern soils under new and different temperature and precipitation conditions are pressing concerns for researchers and land managers alike. Complex interactions between and among disturbances (e.g., drought, fire, pathogens) are likely to result in unexpected and unpredicted outcomes, relative to single disturbances considered in isolation.
- Partnerships between managers and scientists in collaborative research and development are likely to be the best way to ensure maintenance of sustainable, resilient ecosystems and expand understanding of the

capacity and limitations of the soil resources of the Southeast.

Literature Cited

Bacon AR, Richter DD, Bierman PR, Rood DH (2012) Coupling meteoric ^{10}Be with pedogenic losses of ^{9}Be to improve soil residence time estimates on an ancient North American interfluve. Geology 40(9):847–850

Block CE, Knoepp JD, Fraterrigo JM (2013) Interactive effects of disturbance and nitrogen availability on phosphorus dynamics of southern Appalachian forests. Biogeochemistry 112:329–342

Blood ER, Anderson P, Smith PA et al (1991) Effects of Hurricane Hugo on coastal soil solution chemistry in South Carolina. Biotropica 23(4) Part A:348–355

Briones MJI (2014) Soil fauna and soil functions: a jigsaw puzzle. Front Environ Sci 2:7

Burt TP, Ford Miniat C, Laseter SH, Swank WT (2017) Changing patterns of daily precipitation totals at the Coweeta Hydrologic Laboratory, North Carolina, USA. Int J Climatol 38(1):94–104

Callaham MA Jr, Scott DA, O'Brien JJ, Stanturf JA (2012) Cumulative effects of fuel management on the soils of eastern U.S. watersheds. In: LaFayette R, Brooks MT, Potyondy JP et al (eds) Cumulative watershed effects of fuel management in the eastern United States. Gen. Tech. Rep. SRS-161. U.S. Department of Agriculture, Forest Service, Southern Research Station, Asheville, pp 202–228

Carrerra-Martínez R (2018) Distribution and diversity of semiaquatic Sparganophilidae earthworms in the Southern Appalachian Piedmont, USA. University of Georgia, Athens Thesis

Coleman DC, Callaham MA Jr, Crossley DA Jr (2018) Fundamentals of soil ecology, 3rd edn. Academic Press, London, 369 p

Coyle DR, Nagendra UJ, Taylor MK et al (2017) Soil fauna responses to natural disturbances, invasive species, and global climate change: current state of the science and a call to action. Soil Biol Biochem 110:116–133

Daneshgar P, Jose S (2009) *Imperata cylindrica*, an alien invasive grass, maintains control over nitrogen availability in an establishing pine forest. Plant Soil 320(1–2):209–218

De-Campos AB, Mamedov AI, Huang CH (2009) Short-term reducing conditions decrease soil aggregation. Soil Sci Soc Am J 73(2):550–559

Early Detection & Distribution Mapping System [EDDMapS] (2018) Japanese stiltgrass. University of Georgia, Center for Invasive Species and Ecosystem Health. https://www.eddmaps.org/distribution/uscounty.cfm?sub=3051. Accessed 8 June 2018

Emanuel K (2005) Increasing destructiveness of tropical cyclones over the past 30 years. Nature 436(7051):686–688

Estrada JA, Flory SL (2015) Cogongrass (*Imperata cylindrica*) invasions in the US: mechanisms, impacts, and threats to biodiversity. Glob Ecol Conserv 3:1–10

Fox TR, Jokela EJ, Allen HL (2007) The development of pine plantation silviculture in the southern United States. J For 105(7):337–347

Hanula JL, Ulyshen MD, Wade DD (2012) Impacts of prescribed fire frequency on coarse woody debris volume, decomposition and termite activity in the longleaf pine flatwoods of Florida. Forests 3(2):317–331

Hanula JL, Wade DD (2003) Influence of long-term dormant-season burning and fire exclusion on ground-dwelling arthropod populations in longleaf pine flatwoods ecosystems. For Ecol Manag 175(1–3):163–184

Hu H, Wang GG, Bauerle WL, Klos RJ (2017) Drought impact on forest regeneration in the Southeast USA. Ecosphere 8(4):e0172

Ikeda H, Callaham MA Jr, O'Brien JJ, Hornsby BS, Wenk ES (2015) Can the invasive earthworm, *Amynthas agrestis*, be controlled with prescribed fire? Soil Biol Biochem 82:21–27

Iverson LR, Prasad AM, Matthews SN, Peters M (2008) Estimating potential habitat for 134 eastern US tree species under six climate scenarios. For Ecol Manag 254(3):390–406

Jackson CR, Martin JK, Leigh DS, West LT (2005) A southeastern Piedmont watershed sediment budget: evidence for a multi-millennial agricultural legacy. J Soil Water Conserv 60(6):298–310

James SW (1995) Systematics, biogeography, and ecology of Nearctic earthworms from eastern, central and southwestern United States. In: Hendrix PF (ed) Ecology and biogeography of earthworms in North America. Lewis Publishers, Boca Raton, pp 29–52

Jenny H (1980) The soil resource: origin and behaviour, Ecological studies, vol 37. Springer-Verlag, New York, 392 p

Kelly L, Sellers J (2014) Abundance and distribution of the invasive ant, *Solenopsis invicta* (Hymenoptera, Formicidae), in cypress savannas of North Carolina. Ann Entomol Soc Am 107:1072–1080

Knoepp JD, Vose JM, Clinton BD, Hunter MD (2011) Hemlock infestation and mortality: impacts on nutrient pools and cycling in Appalachian forests. Soil Sci Soc Am J 75:1935–1945

Kossin JP (2008) Is the North Atlantic hurricane season getting longer? Geophys Res Lett 35(23):23705

Kunkel KE, Stevens LE, Stevens SE et al (2013) Regional climate trends and scenarios for the U.S. National Climate Assessment Part 2. Climate of the southeast U.S. NOAA

Tech. Rep. NESDIS 142-2. National Oceanic and Atmospheric Administration, National Environmental Satellite, Data, and Information Service, Washington, DC, 94 p

LeBrun EG, Plowes RM, Gilbert LE (2012) Imported fire ants near the edge of their range: disturbance and moisture determine the prevalence and impact of an invasive social insect. J Anim Ecol 81:884–895

Li L, Li W, Deng Y (2013) Summer rainfall variability over the southeastern United States and its intensification in the 21st century as assessed by CMIP5 models. J Geophys Res Atmos 118(2):340–354

Lippincott CL (2000) Effects of *Imperata cylindrica* (L.) Beauv. (cogongrass) invasion on fire regime in Florida Sandhill (USA). Nat Areas J 20(2):140–149

Lobe JW, Callaham MA Jr, Hendrix PF, Hanula JL (2014) Removal of an invasive shrub (Chinese privet: *Ligustrum sinense* Lour) reduces exotic earthworm abundance and promotes recovery of native North American earthworms. Appl Soil Ecol 83:133–139

Melillo JM, Richmond TC, Yohe GW (eds) (2014) Climate change impacts in the United States: the Third National Climate Assessment. U.S. Global Change Research Program, Washington, DC, 841 p

Melvin M (2012) 2012 National prescribed fire use survey report. Tech. Rep. 01-12. Coalition of Prescribed Fire Councils, Inc, Newton 26 p

Miller JH, Manning ST, Enloe SF (2013) (slightly revised 2013 and 2015). A management guide for invasive plants in southern forests. Gen. Tech. Rep. SRS-131. U.S. Department of Agriculture, Forest Service, Southern Research Station, Asheville. 120 p

Mitchell RJ, Liu Y, O'Brien JJ et al (2014) Future climate and fire interactions in the southeastern region of the United States. For Ecol Manag 327:316–326

Moody JA, Shakesby RA, Robichaud PR et al (2013) Current research issues related to post-wildfire runoff and erosion processes. Earth Sci Rev 122:10–37

Morrison LW (2002) Long-term impacts of an arthropod-community invasion by the imported fire ant, *Solenopsis invicta*. Ecology 83(8):2337–2345

National Atmospheric and Space Administration [NASA] (n.d.). Moderate Resolution Imaging Spectroradiometer data. Available at https://modis.gsfc.nasa.gov/

National Interagency Coordination Center [NICC] (n.d.) Homepage. National Interagency Fire Center, National Interagency Coordination Center, Boise https://www.nifc.gov/nicc/

Nesme J, Simonet P (2015) The soil resistome: a critical review on antibiotic resistance origins, ecology and dissemination potential in telluric bacteria. Environ Microbiol 17:913–930

North BA, Torzilli AP (2017) Characterization of the root and soil mycobiome associated with invasive *Microstegium vimineum* in the presence and absence of a native plant community. Botany 95:513–520

O'Brien JJ, Hiers JK, Mitchell RJ et al (2010) Acute physiological stress and mortality following fire in a long-unburned longleaf pine ecosystem. Fire Ecol 6:1–12

Orgiazzi A, Panagos P, Yigini Y et al (2016) A knowledge-based approach to estimating the magnitude and spatial patterns of potential threats to soil biodiversity. Sci Total Environ 545:11–20

Porter SD, Savignano DA (1990) Invasion of polygyne fire ants decimates native ants and disrupts arthropod community. Ecology 71(6):2095–2106

Reich PB, Oleksyn J, Modrzynski J et al (2005) Linking litter calcium, earthworms and soil properties: a common garden test with 14 tree species. Ecol Lett 8(8):811–818

Scharlemann JP, Tanner EV, Hiederer R, Kapos V (2014) Global soil carbon: understanding and managing the largest terrestrial carbon pool. Carbon Manag 5(1):81–91

Snyder BA (2008) A preliminary checklist of the millipedes (Diplopoda) of the Great Smoky Mountains National Park, USA. Zootaxa 1856:16–32

Soil Survey Staff (2014) Keys to soil taxonomy, 12th edn. U.S. Department of Agriculture, Natural Resources Conservation Service, Washington, DC

Steiner JL, Briske DD, Brown DP, Rottler CM (2018) Vulnerability of Southern Plains agriculture to climate change. Clim Chang 146(1–2):201–218

Strickland MA, DeVore JL, Maerz JC, Bradford MA (2010) Grass invasion of a hardwood forest is associated with declines in belowground carbon pools. Glob Chang Biol 16:1338–1350

Trimble SW (1974) Man-induced soil erosion on the southern Piedmont 1700–1970. Soil and Water Conservation Society, Ankeny

Unger IM, Kennedy AC, Muzika R-M (2009) Flooding effects on soil microbial communities. Appl Soil Ecol 42(1):1–8

USDA Animal and Plant Health Inspection Service [USDA APHIS] (2018) Imported fire ants quarantined areas. https://www.aphis.usda.gov/aphis/maps/plant-health/ifa-quarantine-mapping/. Accessed 12 Sept 2018

Wear DN, Greis JG (2013) The southern forest futures project: technical report. Gen. Tech. Rep. SRS-178. U.S. Department of Agriculture, Forest Service, Southern Research Station, Asheville, 542 p

Weed AS, Ayres MP, Hicke JA (2013) Consequences of climate change for biotic disturbances in North American forests. Ecol Monogr 83(4):441–470

West LT (2000) Soils and landscapes in the Southern Region. In: Scott HD (ed) Water and chemical transport in soils of the southeastern USA. Southern Cooperative Series

Bulletin no. 395. Available at http://soilphysics.okstate.edu/S257/index.html. Accessed 10 Apr 2019

Caribbean

Grizelle González, Erika Marín-Spiotta, and Manuel Matos

The Setting

The United States Caribbean is an insular territory composed of a rich mix of the principal inhabited islands of Puerto Rico (including Vieques and Culebra) and the United States Virgin Islands (St. Croix, St. Thomas, St. John, and Water Island), over 800 smaller islands, and cays. It is a region of tropical humid and semiarid mountains, valleys, and coastal plains with a climate strongly influenced by the surrounding ocean. The rainfall distribution pattern on the islands surrounding the Caribbean is much more even than on most land areas within the Tropics (Ewel and Whitmore 1973). Yet precipitation in the United States Caribbean is mostly bimodal with an initial maximum around May, a relative minimum in June–August, and a second peak in September–October (Chen and Taylor 2002; Giannini et al. 2000; Rudloff 1981). Most of the precipitation in the region is orographic, given the diverse topography of the larger islands.

There are six ecological life zones in Puerto Rico and the United States Virgin Islands, ranging from subtropical dry through rain forest in the basal or sea level belt, and wet plus rain forest in the lower montane altitudinal belt (Ewel and Whitmore 1973). In general, the subtropical lower montane rainforest life zone occupies the smallest area, accounting for only 0.1% of the study region. Though subtropical moist forest is the dominant life zone, covering more than 58% of the area (Ewel and Whitmore 1973), Puerto Rico's topography results in a wide range of climatic conditions (from <850 to >5000 mm mean annual precipitation; Daly et al. 2003; Murphy et al. 2017) and ecosystem types (González et al. 2013b; Gould et al. 2006; Weaver and Gould 2013). A complex geologic history, which has given rise to alluvial, limestone, volcanic, and ultramafic soil parent materials, can be found in Puerto Rico and the United States Virgin Islands (Miller and Lugo 2009; Weaver 2006).

The ecological and geological diversity of Puerto Rico is also reflected in the diversity of its soils. Ten of the 12 soil orders established by the USDA Soil Taxonomy, the official system of soil classification of the National Cooperative Soil Survey, are present in Puerto Rico (Beinroth et al. 1996; Muñoz et al. 2018). Therefore, much of the diversity and predispositions typical of tropical soils around the world can be found within the United States Caribbean region.

The world's soils store 2–3 times more carbon (C) than the atmosphere and all terrestrial plant biomass combined (Houghton 2007). Soils play a fundamental role in the exchange of greenhouse gases with the atmosphere and in the cycling of biologically important elements (Trumbore 2009). Identifying controls on soil C storage and cycling across environmental gradients is crucial for improving predictions of feedbacks between the terrestrial biosphere and climate change. The US Caribbean has a long history of soils research, with a particular focus on understanding the role of different state factors in soil properties, biology, and ecological processes—as highlighted in the following sections.

History of Soil Surveys

The United States Government under the National Cooperative Soil Survey started soil surveys in the Caribbean region right after the United States acquired Puerto Rico during the Hispano-American war of 1898. The first soil survey in Puerto Rico was led by Clarence W. Dorsey and published in 1902. Since then, 16 formal initial and updated soil surveys have been completed covering Puerto Rico and the United States Virgin Islands. In 1928, the USDA Division of Soil Survey and the University of Puerto Rico Agricultural Experiment Station began in-depth surveys of the soils of Puerto Rico, which were published in 1942. Between 1965 and 2008, soil surveys at a scale of 1:20,000 for all of Puerto Rico were published as soil survey area reports (Table A1). Updated taxonomic classifications of the soils of Puerto Rico were later published by Beinroth et al. (2003) and Muñoz et al. (2018).

Historically, soil survey information was published in soil survey reports. Since 2005 this information has been available to the public in a digital format through the interactive application called Web Soil Survey. This digital format has substantially increased the demand for soils information at a regional scale. In addition, soil survey areas formerly were developed following political boundaries, mapped as islands, with different ages and survey crews. Recognizing the need to improve soil survey data and correct inconsistencies across soil survey areas, the Soil Science Division established major land resource areas. These areas are geographically associated land resource units. Today, the Caribbean Area National Cooperative Soil Survey, led by the USDA Natural Resources Conservation Service (USDA NRCS), manages eight Soil Survey Areas and four distinct major land resource areas. Land resource regions are a group of geographically associated major land resource areas. Identification of these large areas is important in statewide agricultural planning and has value in interstate, regional, and national planning (USDA NRCS 2006).

In 2012, the USDA NRCS implemented the Major Land Resource Areas Soil Data Join Recorrelation to address the need to improve soil survey data and reduce inconsistencies across soil survey areas. The Soil Data Join Recorrelation focused on the evaluation of map units to create a continuous coverage within the attribute database. This initiative reduced the number of map units and components in the database and improved soil properties and improved the accuracy of interpretations. This

Table A1 Soil surveys completed in Puerto Rico and US Virgin Islands

Soil survey area	Year published
Arecibo to Ponce Reconnaissance Survey	1902
St. Croix Island Reconnaissance Survey	1932
Puerto Rico Soil Survey	1942
Soil Survey of Lajas Valley	1965
Virgin Islands of the United States	1970
Soil Survey of Mayaguez	1975
Soil Survey of Humacao	1977
Soil Survey of San Juan	1978
Soil Survey of Ponce	1979
Soil Survey of Arecibo	1982
Camp Santiago and Fort Allen	2000
Soil Survey of United States Virgin Islands	2002
Caribbean National Forest and Luquillo Experimental	2002
Soil Survey of San Germán Area	2008
Soil Survey of El Yunque National Forest	2012
Soil Data Evaluation and Major Land Resource Area Soil Survey Updates	2012– Present

provided an opportunity to document decisions and identify future needs and projects[1]. In 2017, the Soil Data Join Recorrelation initiative evolved into the major land resource areas soil map-unit evaluations and major land resource areas-scale soil survey updates. New tools and data available such as geographic information systems and digital models are used to accelerate the evaluation and update process. All the changes in soil survey data become available to the public through the Web Soil Survey annual refresh each October.

The approximate total area for the distribution of the soil orders in Puerto Rico is 898,324.43 ha. Of the ten soil orders currently recognized in Puerto Rico, Inceptisols cover 29%, Ultisols 20%, Mollisols 15%, Oxisols 7.5%, Alfisols 5%, Vertisols 4%, Entisols 2%, Aridisols 1.5%, Histosols 0.5%, and Spodosols 0.2% of the total land area; the remaining 15.3% of land area is grouped in the miscellaneous soils category (Fig. A7).

For the United States Virgin Islands, the approximate total area for the distribution of the soil orders is 35,075.7 ha. Of the five soil orders currently recognized in these islands, Inceptisols cover 24.9%, Mollisols 63.6%, Alfisols 4.7%, Vertisols 3.1%, and Entisols 3.7% of the total land area (Fig. A8).

Soil Carbon Inventories in the United States Caribbean

Globally, 1576 petagrams (Pg) of C is stored in soils, of which about 506 Pg (32%) is found in soils of the Tropics. It is also estimated that about 40% of the C in soils of the

Tropics is in forested soils (Eswaran et al. 1993). In the United States Caribbean, several efforts have quantified or mapped soil C. Beinroth et al. (1992) used the modern soil survey of Puerto Rico to estimate that overall, soils on the island contained $80\ 931 \times 10^6$ kg or 80.931 teragrams (Tg) organic C in the top 1 m of soil. Of the nine soil orders recognized at the time in Puerto Rico, Ultisols contributed 28.1% of total C, Inceptisols 25.5%, Mollisols 15.6%, Oxisols 13.1%, Histosols 6.4%, Entisols 4.8%, Alfisols 3.5%, Vertisols 2.8%, and Spodosols 0.2%. Average soil organic C (SOC) content by soil order ranged from 153.7 kg m^{-2} in Histosols or wetland soils to 6.5 kg m^{-2} in Spodosols. In between, soil orders did not vary greatly in soil C content, despite temperate-biased expectations that highly weathered soil orders like Oxisols and Ultisols would have very low C content. Soil OC content did not vary predictably by soil moisture regime (with the exception of Histosols), though clayey and silty soils appeared to contain more C than loamy and sandy soils (Beinroth et al. 1992).

In an assessment of SOC across the United States, Johnson and Kern (2003) used data from the National Soil Characterization Database and STATSGO (1994) from 229 forested sites in Puerto Rico representing 7 soil orders to map SOC in mineral and organic forested soils at depths of 0–100 cm. This national analysis aggregated all forest types in Puerto Rico into a Caribbean forest-type group. Mean SOC content for mineral soils was reported as 11.8 kg m^{-2}, compared to 10.7 kg m^{-2} in Beinroth et al. (1992). These values are comparable and on the high end compared to those summarized for nontropical forest types in the contiguous United States for the top 1 m (Johnson and Kern 2003). Highly weathered tropical soils are characterized typically by deep soil profiles, and these values would be consistent with overall greater SOC storage in mineral soils of the Tropics compared to mid-latitude mineral soils.

Effect of State Factors on Soil Carbon

Much research has been conducted in Puerto Rico to better understand the role of soil-forming factors, or state factors, on SOC dynamics. A state factor approach (sensu Jenny 1941) provides a useful framework for identifying the strength of environmental predictors—climate, biota (including vegetation and human activity), parent material, topography, and time—on the processes that contribute to SOC storage to improve estimates of SOC storage and geographic distribution at different spatial scales. For example, Silver et al. (2003) reviewed data from 29 studies of tropical forest soils of the United States (including Hawaiʻi) for a total of 108 soil profiles. For the Puerto Rican soils, they reported a strong relationship between SOC and mean annual precipitation in wet and moist forests ($\geq$2500 mm yr^{-1}). Mean annual temperature explained a low proportion of variability in SOC for the same sites. Including sites on the low end of the pre-

---

[1]Matos, M.; Rios, S.; Santan, A.; Anderson, D. 2016. Soil Data Join Recorrelation in the Caribbean area. [Presentation]. A healthy soil the key for a healthy environment; Southern Regional Cooperative Soil Survey conference; June 20–23, 2016; Rincón, PR.

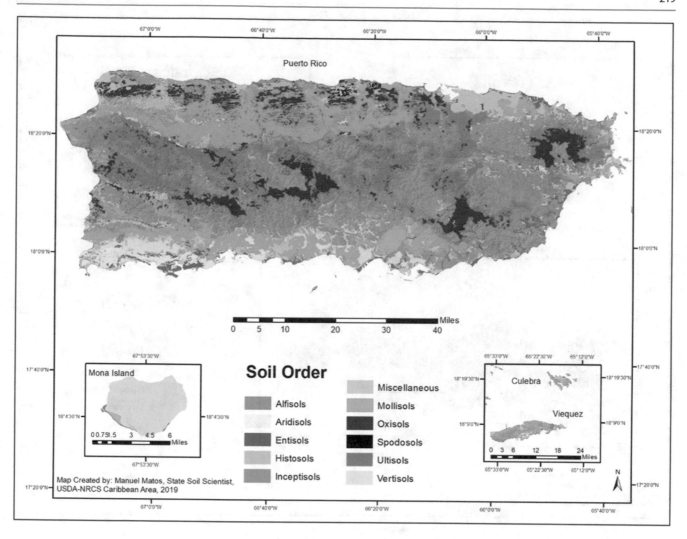

Fig. A7 Map of soil orders described for the islands of Puerto Rico

cipitation gradient weakened climatic relationships with SOC. Beinroth et al. (1996) demonstrated that the effects of climate and land cover on SOC were mediated by the role of soil mineralogy, highlighting the importance of parent material and weathering stage on C content and its response to environmental factors at the landscape surface.

More recently, the USDA NRCS surveyed 30 sites representing common soil series in Puerto Rico and the US Virgin Islands as part of the national Rapid Carbon Assessment project. Traditional predictors of soil C storage such as clay content and climate were poor predictors of regional soil C trends (Vaughan 2016). Soil OC stocks were not correlated with mean annual precipitation or mean annual temperature. Soil order and land cover were marginally significant predictors of SOC for a subset of the data with sufficient field site replication to test for the effect of these variables. Recognition of the heterogeneity of geologic substrates and weathering gradients in tropical regions and incorporation of a more

mechanistic understanding can improve soil C modeling and land management decisions.

Different state factors become important at different spatial scales. In a study of 216 soil profiles in Puerto Rico's Luquillo Mountains differing in climate, topography, parent material, and forest type, only forest type and topographic position were significant predictors of soil C content in the top 80 cm (Johnson et al. 2015). These soils contained about 70% more C than the global mean for tropical forests down to a 1 m depth. Soil C storage in sandy, low-clay Inceptisols did not differ from highly weathered clay-rich Oxisols. Differences in soil C among forest types (colorado [*Cyrilla racemiflora*] > palm [*Prestoea montana*] > tabonuco [*Dacryodes excelsa*]) were attributed to differences in the ratio of C to N (C:N) of plant litter inputs and the accumulation of C in valley soils to depressed decomposition rates due to low oxygen (Johnson et al. 2015). In contrast, Johnson et al. (2011) reported greater SOC in hilltops than in valleys

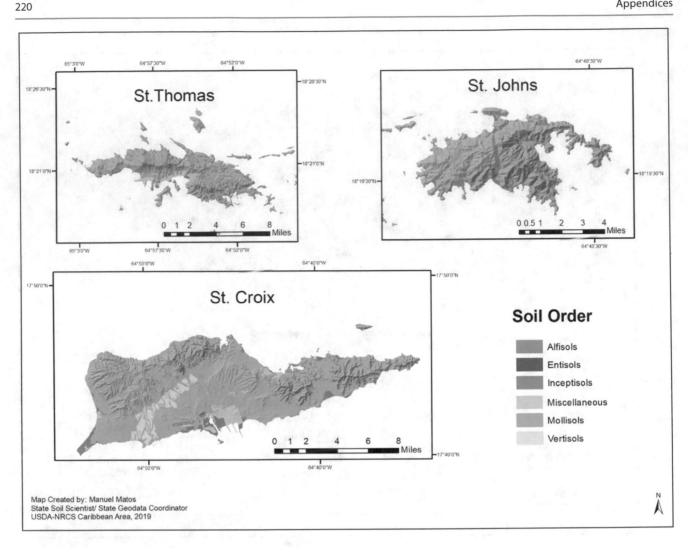

Fig. A8 Map of soil orders described for the US Virgin Islands

along steep forest soils across slopes. This atypical spatial distribution was best explained by differences in soil thickness and the concentration of iron and aluminum fractions.

Topography, climate, and parent material interact to influence organomineral assemblages and their role in SOC storage. Across a gradient of reducing soil conditions, decreasing litterfall inputs could not explain accumulation of SOC; instead, SOC stocks were correlated with concentrations of reduced iron (Hall and Silver 2015). This and other work (Hall and Silver 2013) conducted through the National Science Foundation-funded Critical Zone Observatory in Puerto Rico have focused on the role of redox in controlling SOC dynamics. These studies indicate that SOC in wet tropical soils is sensitive to fluctuations in moisture and reducing conditions, suggesting that future changes in precipitation under a warming climate have the potential to affect SOC storage.

Ping et al. (2013) studied soil properties, C distribution, and nutrient distribution along an elevation gradient in eight distinct forest types in eastern Puerto Rico. They found SOC and N stocks followed the elevation gradient: from 26.7 kg OC m^{-2} and 1.4 kg N m^{-2} in soils of the colder and wetter mountain tops to 12.3 kg OC m^{-2} and 1.0 kg N m^{-2} in soils of the lower elevation dry forests. Soil C:N ratio decreased from 20 to 11 as elevation decreased along the gradient. In addition, Ping et al. (2013) found that landscape movement on uplands through landslides, slumps, and fluvial and alluvial processes had a significant effect on variation of SOC stores, emphasizing the need to consider geomorphic processes when estimating C stores by landscape units. Coastal wetlands had exceptionally high SOC stores (>90 kg m^{-2}) due to their water-saturated and reduced environment. Soil OC content showed an inverse relation with soil bulk density and played a controlling role in cation exchange capacity, and nutrient distribution (Ping et al. 2013). Thus, Ping et al. (2013) concluded that elevation, through its influence on precipitation and temperature, exerts strong influence on the quantity and quality of terrestrial OC stores and on the depth-distribution pattern of C, N, and other nutrients.

Consistent with Gould et al. (2006), and Ping et al. (2013) showed total soil C was highest in saturated soils from upper and lower ends of the elevation gradient in northeastern Puerto Rico, that is, the roble de sierra-guayabota de sierra (*Tabebuia rigida-Eugenia borinquensis*; elfin forest) and palo de pollo or dragonsblood tree-golden leather fern (*Pterocarpus officinalis-Acrostichum aureum*) communities. Some of this C is in the form of calcium carbonate, and calcium levels are high in both the lowland dry and flooded communities. In addition, Gould et al. (2006) found that within the lowland and montane moist and wet forest communities, elevation was positively correlated with mean annual precipitation, the number of plant endemic species, total soil C, N, sulfur, C:N ratio, and organic matter content. Using a modeling approach, Dialynas et al. (2016) estimated different landslide erosion effects on SOC redistribution in two catchments with differing lithologies. This modeling study also found an effect of forest type on SOC replacement in eroding landscapes based on different net primary production rates.

Land Use Change and Management Effects

Determining management effects on soil C content can be challenging due to large background pools of organic matter and spatial heterogeneity across the landscape and with soil depth, in addition to the difficulty of identifying useful reference soils that have not been disturbed. Lugo-López (1992) provided a comprehensive review of the prior 50 years of research on soil organic matter in Puerto Rico from an agronomic perspective. He concluded that most soils maintained relatively consistent and high organic matter content despite cultivation and that when losses were observed, SOC accumulation rates post-disturbance could be very large. The review reported positive benefits of organic residue addition as a soil conservation practice to reduce erosion on steep slopes. In the lowlands, much research focused on soil improvement through remediation of alkalinity, salinity, and aluminum toxicity in soils affected by sugarcane cultivation and the rum distillery industry. In addition, past research on Puerto Rican soils identified the role of inorganic soil particles in the stabilization of soil aggregates and the protection of organic matter from microbial decomposition (Lugo-López 1992).

In a study of land use in Puerto Rico and the United States Virgin Islands, Brown and Lugo (1990) reported lower SOC in cultivated sites in wet and moist life zones than in the dry life zone relative to reference mature forest sites in each life zone. As reported for tropical soils globally, conversion to cropland resulted in greater losses of SOC than conversion to pasture. Recovery rates of SOC during postagricultural abandonment were faster in the wet and moist forests than in drier sites.

Lugo et al. (1986) in the 1980s resampled sites surveyed in Puerto Rico in the 1940s and 1960s to estimate rates of change with land use trajectories. The 1940s survey did not reveal differences in SOC (0–18 cm depth) by life zone or soil group, but according to the 1960s survey, moist and wet forests contained more SOC. The response of SOC to changes in land use in the moist life zone was more variable in gains and losses than for drier sites, yet all forest types exhibited general increases in SOC with reduced agricultural intensity. Recovery of SOC with reforestation of agricultural lands also occurred in all life zones. In addition, this study revealed the extent of urban development on former agricultural soils. Expansion occurred preferentially on the most fertile soils, resulting in greater SOC stocks in urban soils than in residual agricultural lands.

In a study of Puerto Rican and Hawaiian tree plantations established on former pastures or sugarcane (*Saccharum officinarum*) fields, more SOC accumulated under N-fixing trees than under eucalyptus (*Eucalyptus* spp.) (Resh et al. 2002). Using stable C isotopes, the authors attributed these results to greater retention of residual C derived from the past land cover, in addition to greater inputs of new tree-derived C under N-fixers; these trends were related to soil N levels in the plots. The Puerto Rican sites showed greater SOC stocks in the clayey Vertisols compared to the sandy Entisols. Soil C stocks in grasslands establishing after forest clearing contained only 70% of the soil C in surface soils (0–10 cm depth) of nearby secondary forests, regardless of whether reference forests were dominated by native or non-native tree species (Cusack et al. 2015).

Land management can affect not only the amount of SOC but also its bioavailability (Sotomayor-Ramírez et al. 2009). Conversion of sugarcane to eucalyptus or leadtree (*Leucaena* spp.) forests, pasture, and agricultural crops in a Vertisol in Puerto Rico produced differences in SOC, total N, and C cycling rates as measured by stable isotopes after a shift from warm-season to cool-season plant cover. In particular, the cropland sites had reduced SOC and microbial biomass, greater amounts of C respired during a short-term incubation, and the lowest enzyme activities.

Historical agricultural abandonment in the United States Caribbean has led to widespread forest regrowth, with potential for SOC sequestration (Silver et al. 2000). By the 1930s, much of the land area had been deforested for sugarcane, cattle, or other forms of agriculture. In Puerto Rico, up to 90% of forests had been cut down (Dietz 1986), yet by 1991 the area of wet forest had recovered by about 42% due to agricultural abandonment (Helmer 2004). The island of St. Croix, United States Virgin Islands, has experienced similar trends, with forest regrowth after sugarcane abandonment (Chakroff 2010; Daley 2010; Weaver 2006). Forest succession has the potential to restore soil ecosystem processes that were altered during deforestation (Powers and Marín-Spiotta

2017). The initial soil C content of a site and soil mineralogy can strongly influence the direction and magnitude of response of soil C to land use change (van Straaten et al. 2015).

Across Puerto Rico, secondary forests in the moist life zone contained more SOC (0–23 cm depth) than wet secondary forests (Weaver et al. 1987). At smaller scales, the effects of climate were mediated by parent material and forest successional stage. For example, on granitic soils, wet forests contained more SOC than moist forests. Topography also played an important role in SOC content, especially under coffee cultivation.

In a successional chronosequence of subtropical moist secondary forests on highly weathered Oxisols in Puerto Rico, select aboveground and belowground ecosystem components followed different successional trajectories over time, whereas others recovered in the same timeframe. Aboveground biomass C pools increased with secondary forest age and peaked in the oldest secondary forests (Marín-Spiotta et al. 2007), yet soil C storage appeared relatively stable with succession (Marín-Spiotta et al. 2009). More detailed analyses showed that certain soil organic matter pools were more sensitive to changes in land cover. Specifically, the free light fraction or particulate organic matter not associated with soil aggregates or mineral surfaces was depleted in the pastures and became replenished with reforestation, reaching levels of undisturbed forests in as little as 20 years, as did radiocarbon-based mean residence times (Marín-Spiotta et al. 2008). Stable C isotopes revealed that pasture-derived C was replaced by forest-derived C during reforestation (Marín-Spiotta et al. 2009). The greatest differences in microbial biomass, functional composition, and enzyme activities also occurred during the initial two decades of forest regrowth (Smith et al. 2014, 2015). Relative basal area and tree species richness matched those of primary forests in as little as 20 years (Marín-Spiotta et al. 2007). These data suggest that the recovery of ecosystem function to levels measured in reference forests may be rapid in postagricultural forests.

Yet into the future, it will be important to determine how the changing soil characteristics in the mountainous region of the island will intersect with management practices in the agricultural sector in the delivery of ecosystem services. Puerto Rico has been in economic crisis during the past decade and the government has decided to promote the agricultural sector. Despite the prevailing trend of reforestation for the island, significant areas of forest land are being converted for agriculture and pasture (Gao and Yu 2014). The net effect of deforestation in the central mountains of Puerto Rico may be a net increase in the water supply downstream but also a rise in large sediment discharges into streams and the ocean during large episodic rain events (Gao and Yu 2017). Ramos-Scharrón and Figueroa-Sánchez (2017)

argued the combination of a topographically abrupt wet-tropical setting with the high level of soil exposure that typifies many sun-grown coffee (*Coffea* spp.) farms in Puerto Rico represents optimal conditions for high soil erosion rates, consistent with historical research (Lugo-López 1992).

Accelerated soil loss due to human land use is still one of the most critical environmental problems in tropical mountainous regions, as it can degrade soil function and downstream resources (Ramos-Scharrón 2018). Erosion in the Insular Caribbean can have detrimental effects on soils, nearshore coral reefs, and associated ecosystem services (Ramos-Scharrón 2018). In the island of St. John, United States Virgin Islands, geomorphic evidence indicates that plantation agriculture during the eighteenth and nineteenth centuries did not cause severe erosion. However, rapid growth in roads due to increasing tourism and second-home development since the 1950s has caused at least a fourfold increase in island-wide sediment yields; unpaved roads have been the primary determining factor (McDonald et al. 1997). Similarly, in a dry tropical area of Puerto Rico, unpaved road surfaces have the potential to generate runoff 2–3.5 times more frequently than under natural conditions and can produce sediment at rates 6–200 times greater than background (Ramos-Scharrón 2018). Thus, there is increasing evidence from United States Caribbean islands that an integrated approach to tourism, urban planning, and management requires the cohesive protection of soils and coastal habitats (Hernández-Delgado et al. 2012; Ramos-Scharrón 2018).

Hurricane Effects

Hurricanes can alter C dynamics and other biogeochemical processes through effects on tree mortality, increased deposition of woody debris and litterfall to the forest floor, and changes to soil microenvironmental conditions. For example, Hurricanes Irma and Maria in 2017 deposited a pulse of litter deposition equivalent to or more than the total annual litterfall (fallen leaves and fine wood) input with at least twice the typical fraction of woody materials across four forests in Puerto Rico (Liu et al. 2018). These enormous changes in quantity and quality of litter inputs to the forest floor are likely to alter soil food webs and biogeochemical processes, although the effects on nutrient cycling may be temporally variable and dependent on storm intensity and time since the last disturbance. For example, after Hurricane Hugo crossed Puerto Rico in September 1989, it took 60 months for the total litterfall to return to the prehurricane level in a tabonuco forest of the Bisley Experimental Watersheds (Scatena et al. 1996). After Hurricane Georges (1998), forest floor standing stocks and nutrient content increased in response to the large amounts of litter deposition, but levels returned to prehurricane values within 2–10 months (Ostertag et al. 2003). This latter response to hurricane differed by forest type. Upper elevation palm forest received the lowest litter inputs, but because of the slower decomposition at this site, forest floor recovered in the same amount of time as moist forest and tabonuco forest, which had greater amounts of litter and also greater decomposition rates.

The Luquillo Long-Term Ecological Research Network Canopy Trimming Experiment (hereafter, CTE) has been conducted since 2002 in Puerto Rico to disentangle effects of debris deposition and canopy opening on ecological and biogeochemical processes. This experiment was designed to simulate a hurricane and to separate the effects of changes in temperature, humidity, throughfall, and light (canopy opening) from debris deposition (changes in nutrient levels and the physical structure of the forest floor), in a study on the effects of an increasing frequency of storms, as a possible consequence of climate change (Shiels and González 2014; Shiels et al. 2014, 2015). In the CTE, a shift in dominance in fungal decomposers from basidiomycete macrofungi to microfungi was associated with increases in fungivore specialist groups: mites (order Acari), springtails (class Collembola), and booklice (order Psocoptera) (González et al. 2014; Richardson et al. 2010; Shiels et al. 2015). Furthermore, reductions in macroarthropod decomposers of litter and basidiomycete decomposer fungi that degrade lignin together were very likely responsible for reduction in rates of leaf decomposition in plots where the canopy was opened (González et al. 2014; Lodge et al. 2014; Richardson et al. 2010). Reduction of basidiomycete fungi was associated with reduced accumulation of phosphorus via translocation by fungal root-like structures, which could have contributed to slowing of leaf decomposition in plots where the canopy was opened (Lodge et al. 2014). In addition, González et al. (2014) found a negative correlation between the Margalef index of diversity of the litter arthropods and the percentage of mass remaining of mixed species of litter, suggesting that functional complexity is an important determinant of decay in the Luquillo Experimental Forest. Further, Prather et al. (2018) reported that although canopy presence did not alter consumers' effects in the CTE, focal organisms had unexpected influences on decomposition. Decomposition was not altered by litter snails (*Megalomastoma croceum*), but herbivorous walking sticks (*Lamponius portoricensis*) reduced leaf decomposition by about 50% through reductions in high-quality litter abundance and, consequently, lower bacterial richness and abundance. This relatively unexplored but potentially important link between tropical herbivores, detritus, and litter microbes in this forest demonstrates the need to consider autotrophic influences when examining rainforest ecosystem processes (Prather et al. 2018).

Also in the CTE, Gutiérrez del Arroyo and Silver (2018) found that 10 years after an experiment simulating the addition of hurricane debris, soil C and N were elevated relative to control plots in both surface and deep soils, and both light fraction organic matter and organic molecules complexed with minerals in the heavy soil fractions were elevated. Meanwhile, results from Liu et al. (2018) suggest that hurricane disturbance can accelerate the cycling of soil light OC on a timescale of less than 2 years but can elevate soil microbial biomass C for a longer period in this tropical wet forest.

Large quantities of coarse woody debris are generated periodically during tropical storms and hurricanes in the United States Caribbean. Among many ecosystem services provided, this dead wood serves as a temporary sink for atmospheric C and a source of soil organic matter (Harmon and Hua 1991; Torres 1994). Two recently published reviews have synthesized much of the current ecological research on the interacting factors of dead wood, soil biota, and nutrient dynamics in Puerto Rico's forests (González 2016; González and Lodge 2017). In the subtropical wet forests of Puerto Rico, decaying wood contributes to the spatial heterogeneity of soil properties, through its effect on soil organic matter and nutrient dynamics, further affecting the process of soil formation and nutrient cycling (Lodge et al. 2016; Zalamea et al. 2007, 2016). Zalamea et al. (2007) studied logs with contrasting wood properties, tabonuco and Honduras mahogany (*Swietenia macrophylla*), and at two different decay stages (6 and 15 years after falling). Soil under and 50 cm away from the decaying logs was sampled for soil organic matter fractions. They found decaying logs did influence properties of the underlying soil. The effects differed by species; more sodium hydroxide-extractable C was found in the soil associated with tabonuco logs, and more water-extractable organic matter was found in the soil associated with Honduras mahogany older logs. A higher degree of condensation of water-soluble fulvic acids and other related polyaromatic residues occurred in the soil associated with the youngest logs. More divalent cations were available in the soil influenced by younger logs; availability decreased as decomposition increased (Zalamea et al. 2007).

Consistent with Gutiérrez del Arroyo Silver' (2018) results in the CTE, work by Lodge et al. (2016) in the Luquillo Mountains showed that decomposing logs from two hurricanes spaced 9 years apart had a significant signature on the underlying soil as early as 6 months after the trees fell. Further, results from Lodge et al. (2016) indicate that 20% of randomly placed soil cores at a subtropical wet forest may fall on C- and N-rich hotspots that are the legacy of decomposed previously coarse woody debris. Thus, detailed studies within a variety of tropical forest types are important to better understand the complexity and uncertainty associated with global C pools, particularly given the long-term influence of both natural and anthropogenic disturbances on the functioning of these forested ecosystems (González and Luce 2013).

Climate Change Effects on Soils

Henareh Khalyani et al. (2016) assessed different general circulation models and greenhouse gas emissions scenarios of downscaled climate projections to inform future projections of climate and its potential impacts on the United States Caribbean. From that exercise, projections indicate a reduction in precipitation and increased warming from the 1960–1990 period to the 2071–2099 period of 4–9 °C air temperature (depending on the scenario and location) in the Insular Caribbean. Consequently, they projected a high likelihood of increased energy use for cooling and shifts in ecological life zones to drier conditions. The combination of decreased rainfall, increasing variability of rainfall, and higher air temperatures would lead to reduction of soil moisture and changes in soil organic matter dynamics. Soils in the Luquillo Mountains will be affected by the changing climate through increased variability in the decay of organic matter, changes to the patterns of soil oxygen concentrations, and changes in the availability of soil nutrients to plants (González et al. 2013a). In the Luquillo Mountains, soil oxygen content decreases with increasing mean annual precipitation (Silver et al. 1999). Aerobic soils support more plant biomass but less SOC and nutrient availability, whereas anaerobic soils become a net source of methane (Silver 1998). Methane consumption increases significantly during drought, but high methane fluxes post-drought offset the sink after 7 weeks (O'Connell et al. 2018). Thus, whether tropical forests will become a source or sink of C in a warmer world remains highly uncertain (Wood et al. 2012).

Climate change can affect SOC through changes in inputs and losses via alterations to net primary production and decomposition rates through changes in moisture and temperature, as well as shifts in plant and microbial species composition. Wet tropical soils are more likely to respond to changes in soil moisture extremes (drying, flooding) than to temperature (Cusack and Marín-Spiotta in press). In an experimental study in a high-elevation wet tropical soil, a 29% reduction in soil moisture after 3 months of soil drying led to a 35% increase in soil respiration (Wood et al. 2013). In contrast, Wood and Silver (2012) concluded that decreased rainfall in humid tropical forests may cause a negative feedback to climate through lower soil CO_2 emission and greater methane and nitrous oxide consumption.

Research in the mountains of the Luquillo Experimental Forest suggests that forests along elevation gradients will respond differently to changes in climate. For example, indirect effects of temperature and precipitation (McGroddy and Silver 2000) could explain modeled results of SOC losses (up to 4.5 Mg ha^{-1}) from low to high elevation and small increases (up to 2.3 Mg ha^{-1}) at middle elevations (Wang et al. 2002). In a laboratory incubation, cooler, upper elevation rainforest soils were more sensitive to temperature than a warmer, lower elevation forest, although soil respiration in both forest types responded positively to warming (Cusack et al. 2010). In a field soil translocation experiment, Chen et al. (2017) studied the impacts of decreasing temperature but increasing moisture on SOC and respiration along an elevation gradient in northeastern Puerto Rico. They found that soils translocated from low elevation to high elevation showed an increased respiration rate with decreased SOC content at the end of the experiment, which indicated that the increased soil moisture and altered soil microbes may affect respiration rates. Further, soils translocated from high elevation to low elevation also showed an increased respiration rate with reduced SOC at the end of the experiment, indicating that increased temperature at low elevation enhanced decomposition rates. Thus, these tropical soils at high elevations may be at risk of releasing sequestered C into the atmosphere given a warming climate in the Caribbean (Chen et al. 2017). An ongoing warming experiment, the Tropical Responses to Altered Climate Experiment (TRACE), which began in 2016 in Puerto Rican montane forests on Oxisols and is sponsored by the USDA Forest Service and the Department of Energy, will provide insights into the response of aboveground and belowground ecosystem components to a warming climate (Cavaleri et al. 2015; Kimball et al. 2018).

Nitrogen Deposition

Puerto Rican soils in general have high N content (Lugo-López 1992); hence, N additions are not expected to increase aboveground C but can have contrasting effects on SOC dynamics. Research in wet tropical forest soils indicates that SOC responds positively to N additions through biotic and abiotic mechanisms (Cusack et al. 2016). In an experimental N fertilization study in montane rainforests in Puerto Rico, N fertilization enhanced soil C after 5 years via decreased microbial decomposer activity and an accumulation of C in mineral-associated pools (Cusack et al. 2011). Similarly, continuous N additions in another study enhanced mineral-associated C with a negligible effect on total SOC pool (Li et al. 2006). The effects of N deposition on soil nutrients, which can affect processes that control C accumulation and loss from soils, are variable, with different responses to soil acidification based on soil properties (Cusack and Marín-Spiotta in press). More work quantifying the effects of increasing N deposition on soils with variable nutrient content will improve predictions of the response of tropical ecosystems and their C cycling to multiple global change factors.

Soil Biology and Ecosystem Processes

Given the high rate of forest conversion and persistence of deforestation in the Tropics, it is important to study the diversity of the region's fauna and assess how global changes will affect the links between soil biota and ecosystem function (González and Barberena-Arias 2017). Understanding how environmental variation affects the dynamics of different soil microbial and faunal assemblages, and how variation in the composition of such assemblages controls decomposition processes and nutrient cycling, is critical for long-term sustainability and management of ecosystems that are subject to global change (González and Lodge 2017). Many ecological investigations in Puerto Rico are focused on the characterization of the edaphic fauna, and how they influence ecosystem processes (e.g., see review of literature in González 2016; González and Barberena-Arias 2017; González and Lodge 2017). Results from studies in Puerto Rico indicate soil organisms (and the interactions of soil fauna and microbes) are important regulating factors of litter decomposition. In Puerto Rico, more than half the decay of litter material can be explained by the effects of soil fauna alone (González and Seastedt 2001). Yet there is still a need for comprehensive and manipulative field studies that try to tease apart the distinct effect of fauna and microorganisms. This is not an easy task as the interactions of the abundance, diversity, activity, and functionality of soil organisms are at play (González 2002).

Research in Puerto Rico has revealed soil microbiome sensitivity to changes in soil moisture, which provides insight into how soil function may be altered by climate change. In a reforestation chronosequence, forest floor and surface mineral soil microbial community composition and enzyme activity were more sensitive to seasonal fluctuations in soil moisture than to changes in plant communities with forest succession (Smith et al. 2015). Soil microbial composition varied more among soil organic matter pools (macroaggregates, microaggregates, silt and clay fractions) at the fine scale than between forests of different ages (Smith et al. 2014). A throughfall exclusion study to simulate drought in the Luquillo Experimental Forest found that soil microbial diversity decreased in response to rainfall reduction and that the soil microbiome composition was sensitive even to small changes in soil water potential (Bouskill et al. 2013). Prolonged drought resulted in shifts in functional gene capacity of the microbial community as well as changes in extracellular enzyme activity (Bouskill et al. 2016). Pre-exposure to drought conditions buffered the soil microbiome response.

In the Luquillo Experimental Forest, soil fungal biovolume was found to vary directly with soil moisture (Lodge 1993). Consistent with this earlier work, Li and González (2008) reported significant decreases in total and active fungal and bacterial biomass in the drier season compared to the wetter season. While canopy trimming in the CTE decreased litter moisture, soil moisture increased due to reduction of evapotranspiration (Richardson et al. 2010; Shiels and González 2014; Shiels et al. 2015). It is therefore not surprising that Cantrell et al. (2014) found no effects of canopy trimming and debris deposition treatments in soil microbial communities using fatty acid methyl ester and terminal restriction fragment length polymorphism analyses, but did find differences attributable to drought in the control plots between years.

Conclusions

The land area of the United States Caribbean is small relative to the continental United States. However, it contains higher species diversity than all non-Caribbean national forests combined and is a globally important reservoir of C. Its level of aboveground biological complexity combined with a rich mixture of geology, climate, life zones, and land use practices is reflected in a diverse set of soil-forming factors and soil landscape. The steep gradients in elevation and climate found in Puerto Rico over short distances represent an ideal setting to study the long-term effects of global climate changes on a diverse and dynamic landscape. Most work has focused on the larger island, Puerto Rico. Supporting more soils research in the other islands, which have unique geographic and social histories, will be important for tailoring management to local environments and needs. The United States Caribbean is dynamic in time and space due to its history of natural and anthropogenic disturbances. Within the context of natural disturbance (e.g., droughts and hurricanes), we are still learning how biotic changes and interactions in the detrital food webs affect nutrient and C cycling.

Key Findings

- Identifying controls on soil C storage and turnover across environmental gradients is crucial for improving predictions of feedbacks between the terrestrial biosphere and climate change. Evaluating the long-term effects of prior land use history and management on soils is important for quantifying the C sequestration potential of human-altered landscapes. Despite the importance of land use history, this variable is often absent from regional and global assessments of soil C, although Puerto Rico leads the way in this aspect.

- The availability of historical data (aerial photographs, the Soil Survey Geographic [SSURGO] database, land tenure documents, economic and demographic data, and land cover maps) makes Puerto Rico one of the best places in the Tropics to study land use. The land use history of Puerto Rico illustrates strong feedbacks between social and ecological processes with important implications not only for C sequestration but also for food security and biodiversity conservation.

- Studies along environmental gradients and experimental manipulations in the field, such as those conducted in the Luquillo Experimental Forest, will continue to provide insights into the response of ecological and biogeochemical processes to a changing climate. Urban expansion and a growing need for diversified local agricultural production impose a different set of environmental change factors on tropical soils. Given its dynamic history, the US

Caribbean stands to provide many lessons for other areas of the world.

Acknowledgments

All research at the USDA Forest Service International Institute of Tropical Forestry is done in collaboration with the University of Puerto Rico. González was supported by the Luquillo Critical Zone Observatory (National Science Foundation grant EAR-1331841) and the Luquillo Long-Term Ecological Research Site (National Science Foundation grant DEB-1239764). Marín-Spiotta was supported by a National Science Foundation CAREER award BCS-1349952 and cooperative agreement No. 68-7482-12-525 from the USDA Natural Resources Conservation Service's National Soil Survey Center.

Literature Cited

Beinroth FH, Hernández PJ, Esnard AM et al (1992) Organic carbon content of the soils of Puerto Rico. In: Beinroth FH (ed) Organic carbon sequestration in the soils of Puerto Rico: a case study of a tropical environment. University of Puerto Rico Mayagüez Campus and USDA Soil Conservation Service, San Juan, pp 33–56

Beinroth FH, Vázquez MA, Snyder VA et al (1996) Factors controlling carbon sequestration in tropical soils: a case study of Puerto Rico. University of Puerto Rico at Mayaguez, Department of Agronomy and Soils, and USDA Natural Resources Conservation Service World Soil Resources and Caribbean Area Office, San Juan, 35 p

Beinroth FH, Engel RJ, Lugo JL et al (2003) Updated taxonomic classification of the soils of Puerto Rico, 2002. Bull. 303. University of Puerto Rico, Agricultural Experiment Station, Rio Piedras

Bouskill NJ, Lim HC, Borglin S et al (2013) Pre-exposure to drought increases the resistance of tropical forest soil bacterial communities to extended drought. ISME J 7:384–394

Bouskill NJ, Wood TE, Baran R et al (2016) Belowground response to drought in a tropical forest soil. I. Changes in microbial functional potential and metabolism. Front Microbiol 7:525

Brown S, Lugo AE (1990) Effects of forest clearing and succession on the carbon and nitrogen content of soils in Puerto Rico and US Virgin Islands. Plant Soil 124:53–64

Cantrell SA, Molina M, Lodge DJ et al (2014) Effects of a simulated hurricane disturbance on forest floor microbial communities. For Ecol Manag 332:22–31

Cavaleri MA, Reed SC, Smith K, Wood TE (2015) Urgent need for warming experiments in tropical forests. Glob Chang Biol 21:2111–2121

Chakroff M (2010) U.S. Virgin Islands forest resources assessment and strategies: a comprehensive analysis of forest-related conditions, trends, threats and opportunities. Virgin Islands Department of Agriculture, Kingshill, 100 p

Chen AA, Taylor M (2002) Investigating the link between early season Caribbean rainfall and the El Niño + 1 year. Int J Climatol 22:87–106

Chen D, Yu M, González G et al (2017) Climate impacts on soil carbon processes along an elevation gradient in the tropical Luquillo Experimental Forest. Forests 8(3):90

Cusack DF, Torn MS, McDowell WH, Silver WL (2010) The response of heterotrophic activity and carbon cycling to nitrogen additions and warming in two tropical soils. Glob Chang Biol 16:2555–2572

Cusack DF, Silver WL, Torn MS, McDowell WH (2011) Effects of nitrogen additions on above- and belowground carbon dynamics in two tropical forests. Biogeochemistry 104:203–225

Cusack DF, Lee J, McCleery T, LeCroy C (2015) Exotic grasses and nitrate enrichment alter soil carbon cycling along an urban-rural tropical forest gradient. Glob Chang Biol 21:4481–4496

Cusack DF, Karpman J, Ashdown D et al (2016) Global change effects on humid tropical forests: evidence for biogeochemical and biodiversity shifts at an ecosystem scale. Rev Geophys 54(3):523–610

Cusack DF, Marín-Spiotta E (In press) Tropical wet forests. In: Page-Dumroese D, Morris D, Giardina C, Busse M (eds) Global change and forest soils: demands and adaptations of a finite resource, Cultivating stewardship, vol 36. Elsevier, New York. Chapter 8

Daley BF (2010) Neotropical dry forests of the Caribbean: secondary forest dynamics and restoration in St. Croix, US Virgin Islands. University of Florida, Gainesville, 107 p. Ph.D. dissertation

Daly C, Helmer EH, Quiñones M (2003) Mapping the climate of Puerto Rico, Vieques and Culebra. Int J Climatol 23:1359–1381

Dialynas YG, Bastola S, Bras RL et al (2016) Impact of hydrologically driven hillslope erosion and landslide occurrence on soil organic carbon dynamics in tropical watersheds. Water Resour Res 52:8895–8919

Dietz JL (1986) Economic history of Puerto Rico: institutional change and capitalist development. Princeton University Press, Princeton

Eswaran H, Van Den Berg E, Reich P (1993) Organic carbon in soils of the world. Soil Sci Soc Am J 57:192–194

Ewel JJ, Whitmore JL (1973) The ecological life zones of Puerto Rico and the U.S. Virgin Islands. Res. Pap. IT-018. U.S. Department of Agriculture, Forest Service, Institute of Tropical Forestry, San Juan

Gao Q, Yu M (2014) Discerning fragmentation dynamics of tropical forest and wetland during reforestation, urban sprawl, and policy shifts. PLoS One 9(11):e113140

Gao Q, Yu M (2017) Reforestation-induced changes of landscape composition and configuration modulate freshwater supply and flooding risk of tropical watersheds. PLoS One 12(7):e0181315

Giannini A, Kushnir Y, Cane MA (2000) Interannual variability of Caribbean rainfall, ENSO, and the Atlantic Ocean. J Clim 13:297–311

González G (2002) Soil organisms and litter decomposition. In: Ambasht RS, Ambasht NK (eds) Modern trends in applied terrestrial ecology. Kluwer Academic/Plenum Publishers, London, pp 315–329

González G (2016) Deadwood, soil biota and nutrient dynamics in tropical forests: a review of case studies from Puerto Rico. In: Proceedings of the 112th Annual Meeting of the American Wood Protection Association. American Wood Protection Association, Birmingham, pp 206–208

González G, Barberena-Arias MF (2017) Ecology of soil arthropod fauna in tropical forests: a review of studies from Puerto Rico. J Agric Univ Puerto Rico 101(2):185–201

González G, Lodge DJ (2017) Soil biology research across latitude, elevation and disturbance gradients: a review of forest studies from Puerto Rico during the past 25 years. Forests 8(6):178

González G, Luce MM (2013) Woody debris characterization along an elevation gradient in northeastern Puerto Rico. Ecol Bull 54:181–194

González G, Seastedt TR (2001) Soil fauna and plant litter decomposition in tropical and subalpine forests. Ecology 82(4):955–964

González G, Waide RB, Willig MR (2013a) Advancements in the understanding of spatiotemporal gradients in tropical landscapes: a Luquillo focus and global perspective. Ecol Bull 54:245–250

González G, Willig MR, Waide RB (2013b) Ecological gradient analyses in a tropical landscape: multiple perspectives and emerging themes. Ecol Bull 54:13–20

González G, Lodge DJ, Richardson BA, Richardson MJ (2014) A canopy trimming experiment in Puerto Rico: the response of litter decomposition and nutrient release to canopy opening and debris deposition in a subtropical wet forest. For Ecol Manag 332:32–46

Gould WA, González G, Carrero Rivera G (2006) Structure and composition of vegetation along an elevational gradient in Puerto Rico. J Veg Sci 17:653–664

Gutiérrez Del Arroyo O, Silver WL (2018) Disentangling the long-term effects of disturbance on soil biogeochemistry in a wet tropical forest ecosystem. Glob Chang Biol 24:1673–1684

Hall SJ, Silver WL (2013) Iron oxidation stimulates organic matter decomposition in humid tropical forest soils. Glob Chang Biol 19:2804–2813

Hall SJ, Silver WL (2015) Reducing conditions, reactive metals, and their interactions can explain spatial patterns of surface soil carbon in a humid tropical forest. Biogeochemistry 125:149–165

Harmon M, Hua C (1991) Coarse woody debris dynamics in two old-growth ecosystems. Bioscience 41:604–610

Helmer EH (2004) Forest conservation and land development in Puerto Rico. Landsc Ecol 19:29–40

Henareh Khalyani A, Gould WA, Harmsen E et al (2016) Climate change implications for tropical islands: interpolating and interpreting statistically downscaled GCM projections for management and planning. J Appl Meteorol Climatol 55(2):265–282

Hernández-Delgado EA, Ramos-Scharrón CE, Guerrero-Pérez CR (et al) (2012) Long-term impacts of non-sustainable tourisms and urban development in small tropical islands coastal habitats in a changing climate: lessons from Puerto Rico. In: Kasimoglu M (ed) Visions for global tourism industry. IntechOpen, pp 357–398. https://doi.org/10.5772/38140

Houghton RA (2007) Balancing the global carbon budget. Annu Rev Earth Planet Sci 35:313–347

Jenny H (1941) Factors of soil formation: a system of quantitative pedology. McGraw-Hill, New York, 281 p

Johnson MG, Kern JS (2003) Quantifying the organic carbon held in forested soils of the United States and Puerto Rico. In: Kimble JM, Heath LS, Birdsey RA, Lal R (eds) The potential of US forest soils to sequester carbon and mitigate the greenhouse effect. CRC Press, Boca Raton, pp 47–72

Johnson KD, Scatena FN, Silver WL (2011) Atypical soil carbon distribution across a tropical steepland forest catena. Catena 87:391–397

Johnson AH, Xing HX, Scatena FN (2015) Controls on soil carbon stocks in El Yunque National Forest, Puerto Rico. Soil Sci Soc Am J 79(1):294–304

Kimball BA, Alonso-Rodríguez AM, Cavaleri MA et al (2018) Infrared heater system for warming tropical forest understory plants and soils. Ecol Evol 8(4):1932–1944

Li Y, González G (2008) Soil fungi and macrofauna in the Neotropics. In: Myster RW (ed) Post-agricultural succession in the Neotropics. Springer, Berlin, pp 93–114

Li Y, Xu M, Zou X (2006) Effects of nutrient additions on ecosystem carbon cycle in a Puerto Rican tropical wet forest. Glob Chang Biol 12:284–293

Liu X, Zeng X, Zou X et al (2018) Litterfall production prior to and during Hurricanes Irma and Maria in four Puerto Rican forests. Forests 9(6):367

Lodge DJ (1993) Nutrient cycling by fungi in wet tropical forests, British Mycological Society Symposium Series 19. Cambridge University Press, Cambridge, pp 37–37

Lodge DJ, Cantrell SA, González G (2014) Effects of canopy opening and debris deposition on fungal connectivity, phosphorus movement between litter cohorts and mass loss. For Ecol Manag 332:11–21

Lodge DJ, Winter D, González G, Clum N (2016) Effects of hurricane-felled tree trunks on soil carbon, nitrogen, microbial biomass, and root length in a wet tropical forest. Forests 7:264

Lugo AE, Sanchez MJ, Brown S (1986) Land use and organic carbon content of some subtropical soils. Plant Soil 96:185–196

Lugo-López MA (1992) Review of soil organic matter research in Puerto Rico. In: Beinroth FH (ed) Organic carbon sequestration in the soils of Puerto Rico: a case study of a tropical environment. University of Puerto Rico Mayagüez Campus and USDA Soil Conservation Service, Puerto Rico, pp 33–56

Marín-Spiotta E, Ostertag R, Silver WL (2007) Long-term patterns in tropical reforestation: plant community changes and aboveground biomass accumulation. Ecol Appl 17:828–839

Marín-Spiotta E, Swanston CW, Torn MS et al (2008) Chemical and mineral control of soil carbon turnover in abandoned tropical pastures. Geoderma 143:49–62

Marín-Spiotta E, Silver WL, Swanston CW, Ostertag R (2009) Soil organic matter dynamics during 80 years of reforestation of tropical pastures. Glob Chang Biol 15:1584–1597

McDonald LH, Anderson DM, Dietrich WE (1997) Paradise threatened: land use and erosion on St. John, US Virgin Islands. Environ Manag 21(6):851–863

McGroddy M, Silver WL (2000) Variations in belowground carbon storage and soil CO_2 flux rates along a wet tropical climate gradient. Biotropica 32(4a):614–624

Miller GL, Lugo AE (2009) Guide to the ecological systems of Puerto Rico. Gen. Tech. Rep. IITF-GTR-35. U.S. Department of Agriculture, Forest Service, and International Institute of Tropical Forestry, San Juan, p 437

Muñoz MA, Lugo WI, Santiago C et al (2018) Taxonomic classification of the soils of Puerto Rico. Bull. 313. University of Puerto Rico Mayagüez Campus, College of Agricultural Sciences, Agricultural Experiment Station, Mayagüez, 73 p

Murphy SF, Stallard RF, Scholl MA et al (2017) Reassessing rainfall in the Luquillo Mountains, Puerto Rico: local and global ecohydrological implications. PLoS One 12(7):e0180987

O'Connell CS, Ruan L, Silver WL (2018) Drought drives rapid shifts in tropical rainfall soil biogeochemistry and greenhouse gas emissions. Nat Commun 9:1348

Ostertag R, Scatena FN, Silver WL (2003) Forest floor decomposition following hurricane litter inputs in several Puerto Rican forests. Ecosystems 6:261–273

Ping CL, Michaelson GJ, Stiles CA, González G (2013) Soil characteristics, carbon stores, and nutrient distribution in eight forest types along an elevation gradient, eastern Puerto Rico. Ecol Bull 54:67–86

Powers JS, Marín-Spiotta E (2017) Ecosystem processes and biogeochemical cycles during secondary tropical forest succession. Annu Rev Ecol Evol Syst 48:497–519

Prather CM, Belovsky GE, Cantrell SA, González G (2018) Tropical herbivorous phasmids, but not litter snails, alter decomposition rates by modifying litter bacteria. Ecology 99(4):782–791

Ramos-Scharrón CE (2018) Land disturbance effects of roads in runoff and sediment production on dry-tropical settings. Geoderma 310:107–119

Ramos-Scharrón CE, Figueroa-Sánchez Y (2017) Plot-, farm-, and watershed-scale effects of coffee cultivation in runoff and sediment production in western Puerto Rico. J Environ Manag 202:126–136

Resh SC, Binkley D, Parrotta JA (2002) Greater soil carbon sequestration under nitrogen-fixing trees compared with eucalyptus species. Ecosystems 5:217–231

Richardson BA, Richardson MJ, González G et al (2010) A canopy trimming experiment in Puerto Rico: the response of litter invertebrate communities to canopy loss and debris deposition in a tropical forest subject to hurricanes. Ecosystems 13:286–301

Rudloff W (1981) World-Climates, with tables of climatic data and practical suggestions. Wissenschaftliche Verlagsgesellschaft, Stuttgart

Scatena F, Moya S, Estrada C, Chinea J (1996) The first five years in the reorganization of aboveground biomass and nutrient use following Hurricane Hugo in the Bisley Experimental Watersheds, Luquillo Experimental Forest, Puerto Rico. Biotropica 28(4a):424–440

Shiels AB, González G (2014) Understanding the key mechanisms of tropical forest responses to canopy loss and biomass deposition from experimental hurricane effects. For Ecol Manag 332:1–10

Shiels AB, González G, Willig MR (2014) Responses to canopy loss and biomass deposition in a tropical forest ecosystem: synthesis from an experimental manipulation simulating effects of hurricane disturbance. For Ecol Manag 332:124–133

Shiels AB, González G, Lodge DJ et al (2015) Cascading effects of canopy opening and debris deposition from a large-scale hurricane experiment in a tropical rain forest. Bioscience 65:871–881

Silver WL (1998) The potential effects of elevated CO_2 and climate change on tropical forest soils and biogeochemical cycling. Clim Chang 39:337–361

Silver WL, Lugo AE, Keller M (1999) Soil oxygen availability and biogeochemistry along rainfall and topographic gradients in upland wet tropical forest soils. Biogeochemistry 44(3):301–328

Silver WL, Ostertag R, Lugo AE (2000) The potential for carbon sequestration through reforestation of abandoned tropical agricultural and pasture lands. Restor Ecol 8:394–407

Silver WL, Lugo AE, Farmer D (2003) Soil organic carbon in tropical forests of the United States of America. In: Kimble JM, Heath LS, Birdsey RA, Lal R (eds) The potential of U.S. forest soils to sequester carbon and mitigate the greenhouse effect. CRC Press, Boca Raton

Smith AP, Marín-Spiotta E, de Graaff MA, Balser TC (2014) Microbial community structure varies across soil organic matter aggregate pools during tropical land cover change. Soil Biol Biochem 77:292–303

Smith AP, Marín-Spiotta E, Balser T (2015) Successional and seasonal variations in soil and litter microbial community structure and function during tropical post-agricultural forest regeneration: a multi-year study. Glob Chang Biol 21(9):3532–3547

Sotomayor-Ramírez D, Espinoza Y, Acosta-Martinez V (2009) Land use effects on microbial biomass C, beta-glucosidase and beta-glucosaminidase activities, and availability, storage, and age of organic C in soil. Biol Fertil Soils 45:487–497

Statsgo (1994) State Soil Geographic (STATSGO) database: Data use information. Vol. 1492. Tech. Rep

Torres JA (1994) Wood decomposition of *Cyrilla racemiflora* in a tropical montane forest. Biotropica 26:124–140

Trumbore S (2009) Radiocarbon and soil carbon dynamics. Annu Rev Earth Planet Sci 37:47–66

USDA Natural Resources Conservation Service [USDA NRCS] (2006) Land resource regions and major land resource areas of the United States, the Caribbean, and the Pacific Basin. USDA Handbook 296. 672 p.

van Straaten O, Corre MD, Wolf K et al (2015) Conversion of lowland tropical forests to tree cash crop plantations loses up to one-half of stored soil organic carbon. Proc Natl Acad Sci 112(32):9956–9960

Vaughan E (2016) Factors controlling soil carbon and nitrogen storage in Puerto Rico and the U.S. Virgin Islands. University of Wisconsin, Madison, Madison, 67 p. Master's thesis

Wang H, Cornell JD, Hall CAS, Marley DP (2002) Spatial and seasonal dynamics of surface soil carbon in the Luquillo Experimental Forest, Puerto Rico. Ecol Model 147(2):105–122

Weaver P (2006) Experimental Forest, St. Croix, U.S. Virgin Islands: research history and potential. Gen. Tech. Rep.

IITF–30. U.S. Department of Agriculture, Forest Service, International Institute of Tropical Forestry, San Juan. 62 p

Weaver PL, Gould WA (2013) Forest vegetation along environmental gradients in northeastern Puerto Rico. Ecol Bull 54:43–66

Weaver PL, Birdsey RA, Lugo AE (1987) Soil organic matter in secondary forests of Puerto Rico. Biotropica 19:17–23

Wood TE, Silver WL (2012) Strong spatial variability in trace gas dynamics following experimental drought in a humid tropical forest. Glob Biogeochem Cycles 26:GB3005

Wood TE, Cavaleri MA, Reed SC (2012) Tropical forest carbon balance in a warmer world: a critical review spanning microbial- to ecosystem-scale processes. Biol Rev 87:912–927

Wood TE, Detto M, Silver WL (2013) Sensitivity of soil respiration to variability in soil moisture and temperature in a humid tropical forest. PLoS One 8(12):e80965

Zalamea M, González G, Ping CL, Michaelson G (2007) Soil organic matter dynamics under decaying wood in a subtropical wet forest: effect of tree species and decay stage. Plant Soil 296:173–185

Zalamea M, González G, Lodge DJ (2016) Physical, chemical, and biological properties of soil under decaying wood in a tropical wet forest in Puerto Rico. Forests 7(8):168

Midwest

Lucas E. Nave and Chris W. Swanston

Introduction

The Midwest is a geographically diverse region. Its eight states (Illinois, Indiana, Iowa, Michigan, Minnesota, Missouri, Ohio, Wisconsin) span wide ranges in climate, vegetation, soil, land use, and history, from the distant geologic past up through early human settlement to the present. From the cold winters and glacial past of the forested north, to the long, humid summers and productive soils of the more agricultural southern Midwest states, the issues relevant to soil health and vulnerability are as varied as the lands in this region. The geographic factors that affect life for Midwesterners do not follow state lines, so any resource assessment—soils included—is more meaningful when based on a map drawn with different boundaries. In this regional summary, we use the USDA Forest Service ECOMAP (Cleland et al. 1997), which divides the United States into ecoregions of similar climate, vegetation, and other geographic factors, as our organizational structure. We highlight key soil health and vulnerability issues for the Midwest's five ecoregional provinces (Fig. A9). Two of these provinces—the Laurentian Mixed Forest Province and the Midwest Broadleaf Forest Province—occur exclusively within the Midwest; the other three extend into adjacent

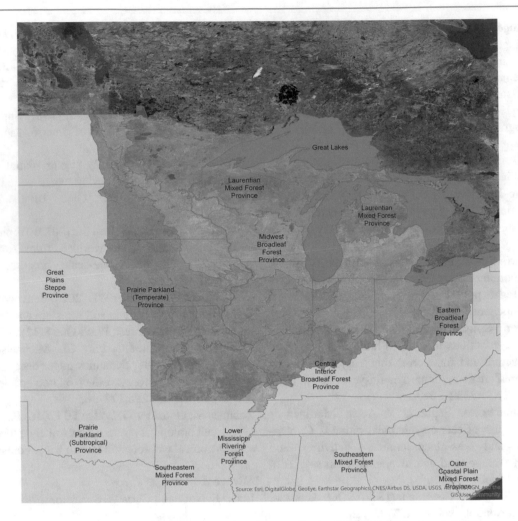

Fig. A9 Ecoregional provinces of the Midwest, as defined by the USDA Forest Service ECOMAP (Cleland et al. 1997)

regions, where concerns will be much the same as described here.

Midwestern lands and ecosystems are globally important providers of food, fiber, and fuel. Employing millions and generating hundreds of billions of dollars annually for the United States economy (Oppedahl 2018), agriculture and forestry in the Midwest are utterly dependent on the health of its soils and ecosystems. These outputs, and the industries and people who depend on them, are no accident—they are due to a fortunate geologic history and favorable present climate. The soils of the Midwest, which are the foundation for the productivity of its lands, are generally younger than most other United States soils, thanks to the glaciers that moved across much of the region until about 15,000 years ago. These glaciers ground up the bedrock over which they passed during their trips south, leaving behind deep deposits of fresh, mineral-rich sediments that are today some of the most productive soils on Earth. The hospitable present-day climate of the region brings out the best in these productive soils, providing plentiful moisture and a reasonably long growing season for forests, rangelands, and crops.

The Midwest has been experiencing accelerating climate changes in recent decades, however, and these changes will increasingly affect ecosystems, industries, and people. Climate change impacts are likely to magnify primary existing soil health issues, such as land use change and invasive species. Among other changes, average temperature of the region has increased 0.83 °C since 1900; increases of 2.8–5.0 °C are expected by 2100 (Pryor et al. 2014). Past climate variability and change have contributed to historic disruptions of ecosystems, organisms, and civilizations. Examples range from the Dust Bowl of the 1930s, which led to significant demographic shifts and economic disruptions in the United States (McLeman et al. 2013), to the collapse of the Maya civilization during drought-induced crop failures more than 1,000 years ago (Douglas et al. 2016). The extent and severity of the effects of past climate change raise concerns about the impacts of projected climate change. The combination of new and amplified problems associated with accelerated climate change elevates it to the most significant threat to soil health in the Midwest. In the following sections, we report the key vulnerabilities, their impacts, and proposed mitigation or adaptation strategies for sustaining the health of soils across the Midwest's ecoregions.

Laurentian Mixed Forest Province

The Laurentian Mixed Forest Province is the proverbial "Northwoods" of the Midwest. It is a mostly forested ecoregion, abundant in waterways and wetlands, and most of its agriculture is relatively low input, such as grazing and forage production. These land uses reflect the fairly cool, short growing seasons and soils that are in many areas either too wet or too sandy to support more demanding crops or uses. However, production of fruit crops (e.g., orchards and berries) and nontimber forest products (e.g., maple syrup) is significant (USDA NASS 2012), and any lack of agricultural economy is compensated for by tourism to the ecoregion for its aesthetic and recreational opportunities. The ecoregion's sizable forestry industry is driven by regeneration harvesting of aspen-birch-mixed conifer forests (*Populus* spp., *Betula papyrifera*, *Pinus* spp., *Abies balsamea*, *Picea* spp.) mostly for pulp and bioenergy, long-lived hardwoods such as sugar maple (*Acer saccharum*) and red oak (*Quercus rubra*) for veneer and other high-value products, and plantation pine (e.g., red and jack; *P. resinosa* and *P. banksiana*) for a combination of pulpwood, fiber, and dimensional lumber. In one or more ways, each of these land uses, from agriculture to recreation to forestry, depends on the health of the soils in this ecoregion.

The most critical forest soil health issues are (1) vulnerability of wetland soils to climate change and altered hydrology; (2) storm-, drought-, and fire-driven decreases in the productivity of dry soils; and (3) direct and indirect invasive species impacts. The first of these issues may be the farthest-reaching, not only because wetland soils are very extensive, but also because they provide many ecosystem services that people value. The deep, organic matter-rich soils of northern wetlands have formed over thousands of years where waterlogging promotes accumulation, rather than decomposition, of plant and animal remains (Trettin and Jurgensen 2003). However, carbon and other elements that are associated with it, such as toxic mercury, are released to adjacent aquatic ecosystems in runoff and groundwater (Kolka et al. 1999). Climate change is increasing the frequency and intensity of large runoff events, as well as prolonged dry periods, which draw down water levels, aerate the soil profile, and increase the export of carbon, mercury, and other elements to the atmosphere (Haynes et al. 2017) and to aquatic ecosystems (Heathcote et al. 2015). In addition to accelerating the pace of climate change (i.e., due to higher greenhouse gas emissions from soils), these changes are likely to increase the toxicity of already impaired fisheries in the thousands of lakes that receive mercury inputs from wetlands across the ecoregion (Perlinger et al. 2018).

The second forest soil health issue—interacting drought and wildfire—could profoundly change forestry across the ecoregion. The extensive forests growing on drought-prone (sandy or rocky) soils in this ecoregion are subject to increasingly frequent severe storms and tree dieback, which have increased flammable fuel density. When critical fire weather aligns with these conditions, extreme fires can occur, such as the Pagami Creek Fire in northern Minnesota in 2011 or the Duck Lake Fire in Michigan's Upper Peninsula in 2012, which together burned well over 40 000 ha. In addition to their many other ecological and resource impacts, severe fires burn away the all-important soil organic matter that sustains productivity of forests on water- and nutrient-poor soils (Nave et al. 2011).

The consequences of increased wildfire are likely to be exacerbated by climate change impacts on tree species, because the species best suited to dry soils, red pine and jack pine, are climate change "losers" already at the southern limits of their range in this ecoregion (Swanston and Handler 2012). With these most appropriate species—which are very important to plantation forestry—expected to decline by 2100, fuel reduction, fire management (including prescribed fire), and reforestation will be critical for maintaining productive forests on dry sites. Ecosystems currently supporting these and other northern xeric species may provide suitable future habitat for species that today occur in mixed-conifer forests and oak woodlands to the south (e.g., black oak [*Quercus velutina*], eastern redcedar [*Juniperus virginiana*]). These new arrivals may be managed by using similar approaches, but disruptions to soil properties and health are nonetheless likely during the uncertain transition to future forest composition.

The third major forest soil health issue stems from the many impacts that invasive species have on soils (Ehrenfeld 2010). These range from the direct consumption of soil organic matter, tree roots, and seeds by earthworms (suborder Lumbricina; all of which are nonnative, having been introduced during the past century or so (Bohlen et al. 2004)) to indirect impacts from the loss of tree species due to species-specific pests, such as the emerald ash borer (*Agrilus planipennis*) (Slesak et al. 2014). Indirect impacts are as many and varied as the invasive species that are disturbing forests throughout the ecoregion. These impacts include changes to soil microclimate that affect tree regeneration, feedbacks between soil organic matter loss and the cycling of nutrients, and synergies with other disturbances, such as fires that occur after insect-driven tree mortality.

Key soil health issues for rangeland and agricultural uses include (1) increasing agricultural use of marginal soils and (2) the potential for more frequent soil freezing to damage tree roots and disrupt nutrient cycling. The first of these issues may counteract decades of ecosystem recovery following the historical logging, burning, and failed attempts at agriculture that occurred across the ecoregion in the late 1800s to early 1900s. Essentially the entire ecoregion was cleared of its native forest during a 50-year period, with no attempt at reforestation, and the wildfires that followed decreased the productivity of its soils. Those who attempted to farm the poor soils and overcome the short growing seasons had largely learned the hard lessons by the 1930s, and many degraded lands reverted to state or Federal ownership. Where agriculture persisted, it has emphasized grazing or favored the limited areas of soils that can be cultivated to grow more-demanding crops. But increasing pressure for agricultural production—especially as heat and drought grow worse in the traditional breadbasket areas to the south—may once again drive agriculture into vulnerable soils that are best left to continue their slow recovery from past abuse.

The issue of soil freezing and its impact on roots and nutrient cycling may seem at odds with a warming climate. The explanation lies not in an overall increase in temperature, but in the reduction of snow cover. Across the region, the seasonally persistent snowpack has become shorter and more sporadic (Demaria et al. 2016). When snow melts during winter thaws, its insulating cover over the soil is diminished. Freezing of the soil physically damages tree roots from the outside and ruptures the internal cells and tissues of roots and soil-dwelling organisms. While some level of root death is normal—a tree's root network is continuously growing, dying, and replacing itself—elevated root mortality contributes to health decline, especially if stresses are compounded, such as by a drought or insect infestation after a winter with frozen ground. The consequences of these physical disruptions to tree roots and soil structure include decreased production of agricultural commodities (e.g., fruits, maple sap) (Pryor et al. 2014), and the leaching of nutrients from soils into groundwater and surface water, which can exacerbate water quality problems (Fitzhugh et al. 2001).

The final soil health issue of note stems from the major role that this ecoregion plays in providing recreational opportunities for people from the Midwest and beyond. The natural resources and aesthetics of Northwoods ecosystems are the foundation of this ecoregion's tourism economy. They are also important because in becoming connected to these ecosystems, people gain awareness of threats facing them. Unfortunately, many of the most iconic sites in the ecoregion's recreational areas also have the most sensitive soils, topography, or vegetation. Thus, increasing visitation rates and more intensive recreational activities such as motorized uses bring with them the prospect of more-detrimental site impacts, such as devegetation, erosion, and water quality degradation.

Midwest Broadleaf Forest Province

The Midwest Broadleaf Forest Province begins in the United States as a narrow zone in northwestern Minnesota and then extends southeast across the Midwest as an ever-broadening transition area between the Laurentian Mixed Forest and the warmer, drier (Prairie Parkland), or wetter (Central Interior and Eastern Broadleaf Forest) ecoregions. In general, its soils are more productive than those of the latter two provinces; its topography is gentler because it was more recently glaciated, and its soil parent materials more fertile. Agriculture, including cultivated crops, forage production, and rangeland for grazing, is much more extensive than in the north. The ecoregion has a lower percentage of forest cover than all other provinces in the Midwest except the Prairie Parkland. Those forests that do occur in this province are highly fragmented, and because of its fairly long history of agricultural and other human impacts, very few of its forest ecosystems or soils support fully intact ecosystem services. Most forest soils of the ecoregion are being impacted by invasive organisms (especially plants and insects), past or present grazing, and inputs of atmospheric pollutants, agricultural fertilizers, and pesticides (Knox 2001; Weathers et al. 2001). Forest soils of the Midwest Broadleaf Forest Province are critically important perhaps because they are so limited in extent and subject to so many disturbances. Across this heavily impacted ecoregion, forest soils are hotspots that disproportionately provide important ecosystem services, although their ability to continue providing these services in the face of climate change may require revisions to how they, and adjacent (e.g., agricultural) soils, are managed. Forest fragments follow riparian or wetland areas in many landscapes, making them a "line of defense" between agricultural land uses and surface waters, which are susceptible to water quality problems due to nutrient runoff (Bharati et al. 2002). In such settings, forest soils remove nutrients and pesticides from runoff and groundwater, storing them in soil organic matter (Reichenberger et al. 2007). Similarly, forest ecosystems and their soils are hotspots for biodiversity in agriculture- and human-dominated landscapes, providing habitat for everything from earthworm-feeding songbirds to soil bacteria (Andren 1994; Buckley and Schmidt 2001).

Similar to the Laurentian Mixed Forest, climate change is the cause of several soil health issues in the Midwest Broadleaf Forest; conversely, because this ecoregion has a longer history of more active land use, it has several unique concerns. Climate change will impact all soils, regardless of land use, in a variety of ways. Increases in the frequency and severity of storms (particularly rainfall) have already occurred due to climate change (Pryor et al. 2014). Coupled with more frequent droughts, which can cause soils to become water repellent, surface runoff and soil erosion are likely to become increasingly important problems in the future. Even as these processes contribute to pollution and nutrient loading in surface water and groundwater, they also represent the loss of soils and their nutrients. These soils formed over very long time periods and would be better managed if retained in place to support productive farmlands and forest lands. In similar fashion, commercial and residential development remove productive soils from agricultural production and decrease the disturbance-mitigating capacity of forest soils. The loss of agricultural soils is to the obvious detriment of food production and the agricultural economy.

Where forest soils are allowed to persist in urban areas, they are important providers of ecosystem services such as water quality protection. Overall, however, urbanization has negative impacts on soil properties, including water infiltration and nutrient retention (Lorenz and Lal 2009). Forest soils are increasingly impacted by invasive species. Though impacts vary across species, generally common problems resulting from species invasions are declines in biodiversity due to the loss of native species (e.g., plants, soil microorganisms) (Nuzzo et al. 2009; Wolfe et al. 2008), diminished organic matter or nutrient retention (Ashton et al. 2005), and the leakage of nutrients into adjacent waterways (Costello and Lamberti 2008).

Prairie Parkland Province

The Prairie Parkland Province is the least forested and driest ecoregion of the Midwest. This ecoregion also has the widest temperature extremes (particularly in the north). Its climate results from its location in the center of the continent, far removed from the moist, temperature-moderating influences of oceans and the Great Lakes. Land use is overwhelmingly agricultural (row crops and rangelands). Although dry in comparison to the other ecoregions of the Midwest, it is moister than the provinces to its west and therefore supports more extensive rain-fed production of moisture-demanding crops (e.g., corn; *Zea mays*) than to the west, where crops better suited to dry summers and rangeland grazing become increasingly important. Most of the ecoregion's limited forest land is either the result of deliberate planting or outward expansion of forest from moist bottomlands that were subject to less frequent wildfires. This tension between forest and grassland is the hallmark of the ecoregion not only in terms of its past but also for its present and future. It is an ecoregion of transition.

Within the last 10,000 years, prairies spread eastward (well into Michigan and Ohio), becoming established as the dominant vegetation for thousands of years and persisting in large areas until Euro-American settlement in the 1800s (Transeau 1935). This expansion left behind areas of fertile, organic matter-rich soils that, in combination with the current climate, are today some of the most productive land on Earth. However, the climate that drove the eastward expansion of grassland was decidedly different. Much warmer and drier, the period of prairie expansion also supported the northward expansion of more drought-tolerant tree species (e.g., oaks and pines) into the Laurentian Mixed Forest ecoregion, and wildfires were much more common and widespread across the entire Midwest (Clark et al. 2001). Recent climate changes, and projections of continued change, bear some similarity to this period from the recent geologic past and to a much better known period from the recent historic past: the Dust Bowl of the 1930s. In this regard, the critical soil health and vulnerability issues are well known; our responses to and the consequences of rapid climate changes are less certain (Alley et al. 2003).

The critical soil health issue for the Prairie Parkland Province is the loss of forest cover from areas where it has recently established, along with degradation of grasslands and wetlands, as the climate becomes warmer and drier. Loss of vegetation density and cover, coupled with the windy climate of the region, will lead to soil erosion (Cook et al. 2009). If the past is a lesson, these changes to soils may have severe ecological, economic, and social repercussions (McLeman et al. 2013). Where irrigation is intensified or expanded in order to maintain agricultural and rangeland uses, groundwater withdrawals will cause subsidence. Wetland ecosystems and soils already under direct threat from a warmer, drier climate (Johnson et al. 2005) will be further impacted by diversions of groundwater and surface water intended to support increasingly water-limited agricultural soils. Direct negative impacts to wetlands, such as the loss of waterfowl breeding habitat, will be magnified by problems in aquatic ecosystems that depend on functioning wetlands (Covich et al. 1998). Last, where natural forests are most extensive, in the far north of the ecoregion, they are largely dominated by species at the southern edges of their range. For example, in the aspen parklands of northern Minnesota, trees will be under increasing threat from moisture stress, compounded by insect pests (Frelich and Reich 2010). On these landscapes, reforestation—including use of more drought-tolerant southern species—may be required to maintain forest soils and the ecosystem services that they provide. The alternative may well be a direct incursion of grassland into the Laurentian Mixed Forest ecoregion, which has occurred in the postglacial past, and a resulting loss of forest area and forest soils (Hogg and Hurdle 1995).

Central Interior and Eastern Broadleaf Forest Provinces

The Central Interior Broadleaf Forest Province and the Eastern Broadleaf Forest Province occupy a smaller area of the Midwest, along its southeastern margins. Although they have unique properties, including their more distant glacial past, older soils, and more rugged topography, these provinces may provide a glimpse of future soil health issues for more northern ecoregions. Specifically, the vegetation and climate of these ecoregions are similar to what is projected for more northern portions of the Midwest by 2100. By that time, average temperatures are projected to increase by at least 2.5 °C under moderate emissions scenarios and the suitable habitat for many of the dominant tree species is expected to shift northward by 500 km or more (Iverson et al. 2008).

The key current and future soil health issues in these provinces include (1) invasive species impacts, (2) drought- and fire-induced losses in forest cover in the west, and (3) increased runoff and erosion due to more frequent intense rainfall. The first of these issues is already a major challenge in these ecoregions, where invasive plants such as kudzu (*Puereria montana* var. *lobata*) and privet (*Ligustrum vulgare*) are dramatically altering soil microbial communities, nutrient cycling, and nutrient retention (Bradley et al. 2010; Forseth and Innis 2004). In addition to the feedbacks that these changes have to other forest health issues, such as the regeneration of economically and ecologically important tree species, invasive species decrease biodiversity by eliminating native species—even in the soil—that have closely coevolved with the native vegetation (Kourtev et al. 2002). In the western portions of these ecoregions, where they border the Prairie Parkland Province, a warmer, drier climate may cause declines in forest health and area (Handler et al. 2012).

Wildfires and their subsequent soil impacts (losses of organic matter and nutrients, runoff, and erosion) can be expected as tree mortality becomes more widespread. Such changes are more likely than they might have been if not for a long period of fire suppression (nearly 100 years) in these ecoregions. Historically, frequent low-severity fires kept fuel levels low and favored tree species (e.g., oaks) that are adapted to fires and drought (Abrams 1992). As a consequence of this fire suppression, species that require more moisture and are not tolerant of fire (e.g., maples) have expanded (Nowacki and Abrams 2008). These more maple-dominated forests are accordingly at greater risk of decline or outright loss in the future climate than if oak-dominated forests had been maintained all along (Lenihan et al. 2008). Fire-damaged ecosystems and soils are more vulnerable to erosion, but even intact, undisturbed forest soils are likely to suffer increased runoff and erosion due to the more frequent, intense rainfall events that are occurring with climate change. Though this soil health issue is present across all ecoregions of the Midwest, it may be most important in the Central Interior and Eastern Broadleaf Forest Provinces because of their already higher precipitation, steeper topography, and older, more weathered soils. Soils of these southernmost ecoregions of the Midwest were glaciated much longer ago; as a result, they not only lack geologically recent inputs of fresh mineral-rich sediment but have had hundreds of thousands more years to leach away their nutrients. Thus, the climate change-driven precipitation patterns expected for these ecoregions (Pryor et al. 2014) are likely to exacerbate issues associated with sustaining their soil productivity for forest, rangeland, and limited agricultural use.

Key Findings

- The Midwest is a region of diverse soils, ecosystems, climates, and land uses.
- Society depends on continued soil-based ecosystem services to maintain food, fiber, and fuel resources and to provide recreational and economic opportunities.
- Climate change is the most widespread and critical threat to the continuation of soil-based ecosystem services across the region.
- Climate change will continue to interact with other stressors and sources of soil vulnerability, exacerbating problems but in some cases allowing a single solution to fix multiple problems.
- Climate impacts and mitigation strategies vary across the region according to fundamental soil properties, geographic factors, and societal constraints.

Literature Cited

Abrams MD (1992) Fire and the development of oak forests. Bioscience 42:346–353

Alley RB, Marotzke J, Nordhaus WD et al (2003) Abrupt climate change. Science 299:2005–2010

Andren H (1994) Effects of habitat fragmentation on birds and mammals in landscapes with different proportions of suitable habitat—a review. Oikos 71(3):355–366

Ashton IW, Hyatt LA, Howe KM et al (2005) Invasive species accelerate decomposition and litter nitrogen loss in a mixed deciduous forest. Ecol Appl 15(4):1263–1272

Bharati L, Lee KH, Isenhart TM, Schultz RC (2002) Soil-water infiltration under crops, pasture, and established riparian buffer in Midwestern USA. Agrofor Syst 56(3):249–257

Bohlen PJ, Scheu S, Hale CM et al (2004) Nonnative invasive earthworms as agents of change in northern temperate forests. Front Ecol Environ 2(8):427–435

Bradley BA, Wilcove DS, Oppenheimer M (2010) Climate change increases risk of plant invasion in the eastern United States. Biol Invasions 12(6):1855–1872

Buckley DH, Schmidt TM (2001) The structure of microbial communities in soil and the lasting impact of cultivation. Microb Ecol 42(1):11–21

Clark JS, Grimm EC, Lynch J, Mueller PG (2001) Effects of Holocene climate change on the C-4 grassland/woodland boundary in the Northern Plains, USA. Ecology 82(3):620–636

Cleland DT, Avers PE, McNab WH et al (1997) National hierarchical framework of ecological units. Ecosyst Manag Appl Sustain For Wildl Resourc 20:181–200

Cook BI, Miller RL, Seager R (2009) Amplification of the North American "Dust Bowl" drought through human-induced land degradation. Proc Natl Acad Sci 106(13):4997–5001

Costello D, Lamberti G (2008) Nonnative earthworms in riparian soils increase nitrogen flux into adjacent aquatic ecosystems. Oecologia 158(3):499–510

Covich AP, Fritz SC, Lamb PJ et al (1998) Potential effects of climate change on aquatic ecosystems of the Great Plains of North America. Hydrol Process 11(8):993–1021

Demaria EMC, Roundy JK, Wi S, Palmer RN (2016) The effects of climate change on seasonal snowpack and the hydrology of the northeastern and upper Midwest United States. J Clim 29(18):6527–6541

Douglas P, Demarest AA, Brenner M, Marcello A (2016) Impacts of climate change on the collapse of lowland Maya civilization. Annu Rev Earth Planet Sci 44:613–645

Ehrenfeld JG (2010) Ecosystem consequences of biological invasions. Annu Rev Ecol Evol Syst 41:59–80

Fitzhugh RD, Driscoll CT, Groffman PM et al (2001) Effects of soil freezing disturbance on soil solution nitrogen, phosphorus, and carbon chemistry in a northern hardwood ecosystem. Biogeochemistry 56(2):215–238

Forseth IN, Innis AF (2004) Kudzu (*Pueraria montana*): history, physiology, and ecology combine to make a major ecosystem threat. Crit Rev Plant Sci 23(5):401–413

Frelich LE, Reich PB (2010) Will environmental changes reinforce the impact of global warming on the prairie–forest border of central North America? Front Ecol Environ 8(7):371–378

Handler SD, Swanston CW, Butler PR et al (2012) Climate change vulnerabilities within the forestry sector for the midwestern United States. In: Winkler JA, Andresen JA, Hatfield JL et al (eds) Climate change in the Midwest: a synthesis report for the National Climate Assessment. Island Press, Washington, DC, pp 114–151

Haynes KM, Kane ES, Potvin L et al (2017) Gaseous mercury fluxes in peatlands and the potential influence of climate change. Atmos Environ 154:247–259

Heathcote AJ, Anderson NJ, Prairie T et al (2015) Large increases in carbon burial in northern lakes during the Anthropocene. Nat Commun 6:10016

Hogg EH, Hurdle PA (1995) The aspen parkland in western Canada: a dry-climate analogue for the future boreal forest? Water Air Soil Pollut 82(1–2):391–400

Iverson LR, Prasad AM, Matthews SN, Peters M (2008) Estimating potential habitat for 134 eastern U.S. tree species under six climate scenarios. For Ecol Manag 254(3):390–406

Johnson WC, Millett BV, Gilmanov T et al (2005) Vulnerability of northern prairie wetlands to climate change. Bioscience 55(10):863–872

Knox JC (2001) Agricultural influence on landscape sensitivity in the Upper Mississippi River Valley. Catena 42(2–4):193–224

Kolka RK, Grigal DF, Verry ES, Nater EA (1999) Mercury and organic carbon relationships in streams draining forested upland peatland watersheds. J Environ Qual 28(3):766–775

Kourtev PS, Ehrenfeld JG, Haggblom M (2002) Exotic plant species alter the microbial community structure and function in the soil. Ecology 83(11):3152–3166

Lenihan JM, Bachelet D, Neilson RP, Drapek R (2008) Simulated response of conterminous United States ecosystems to climate change at different levels of fire suppression, CO_2 emission rate, and growth response to CO_2. Glob Planet Chang 64(1–2):16–25

Lorenz K, Lal R (2009) Biogeochemical C and N cycles in urban soils. Environ Int 35(1):1–8

McLeman RA, Dupre J, Ford LB et al (2013) What we learned from the Dust Bowl: lessons in science, policy, and adaptation. Popul Environ 35(4):417–440

Nave LE, Vance ED, Swanston CW, Curtis PS (2011) Fire effects on temperate forest soil C and N storage. Ecol Appl 21(4):1189–1201

Nowacki GJ, Abrams MD (2008) The demise of fire and "mesophication" of forests in the eastern United States. Bioscience 58(2):123–128

Nuzzo VA, Maerz JC, Blossey B (2009) Earthworm invasion as the driving force behind plant invasion and community change in northeastern North American forests. Conserv Biol 23(4):966–974

Oppedahl DB (2018) Midwest agriculture's ties to the global economy. Chicago Fed Letter No. 393. Federal Reserve Bank of Chicago, Chicago. https://www.chicagofed.org/publications/chicago-fed-letter/2018/393. Accessed 11 Apr 2019

Perlinger JA, Urban NR, Giang A et al (2018) Responses of deposition and bioaccumulation in the Great Lakes region to policy and other large-scale drivers of mercury emissions. Environ Sci — Process Impacts 20(1):195–209

Pryor SC, Scavia D, Downer C et al (2014) Midwest. In: Melillo JM, Richmond TC, Yohe GW (eds) Climate change impacts in the United States: the Third National Climate Assessment. U.S. Global Change Research Program, Washington, DC, pp 418–440. Chapter 18

Reichenberger S, Bach M, Skitschak A, Frede HG (2007) Mitigation strategies to reduce pesticide inputs into ground- and surface water and their effectiveness; a review. Sci Total Environ 384(1–3):1–35

Slesak RA, Lenhart CF, Brooks KN et al (2014) Water table response to harvesting and simulated emerald ash borer mortality in black ash wetlands in Minnesota, USA. Can J For Res 44(8):961–968

Swanston CW, Handler SD (2012) Regional summary: midwest. In: Vose JM, Peterson DL, Patel-Weynand T (eds)
Effects of climatic variability and change on forest ecosystems: a comprehensive science synthesis for the U.S.
forest sector. Gen. Tech. Rep. PNW-GTR-870.
U.S. Department of Agriculture, Forest Service, Pacific
Northwest Research Station, Portland, pp 227–230
Transeau EN (1935) The prairie peninsula. Ecology
16:324–437
Trettin CC, Jurgensen MF (2003) Carbon cycling in wetland
forest soils. In: Kimble JM, Heath LS, Birdsey RA, Lal R
(eds) The potential of U.S. forest soils to sequester carbon
and mitigate the greenhouse effect. Lewis Publishers,
CRC Press, Boca Raton, pp 311–332. Chapter 19
USDA National Agricultural Statistics Service [USDA
NASS] (2012) Crop production 2011 summary. 95 p.
Available at https://usda.library.cornell.edu/concern/pub-
lications/k3569432s?locale=en. Accessed 11 Apr 2019
Weathers KC, Cadenasso ML, Pickett STA (2001) Forest
edges as nutrient and pollutant concentrators: potential
synergisms between fragmentation, forest canopies, and
the atmosphere. Conserv Biol 15(6):1506–1514
Wolfe BE, Rodgers VL, Stinson KA, Pringle A (2008) The
invasive plant *Alliaria petiolata* (garlic mustard) inhibits
ectomycorrhizal fungi in its introduced range. J Ecol
96(4):777–783

Great Plains

Charles H. (Hobie) Perry, Brian A. Tangen, and Sheel Bansal

Introduction

In this regional summary, the Great Plains are identified as a region spanning eight states: Montana, Wyoming, North Dakota, South Dakota, Nebraska, Kansas, Oklahoma, and Texas (note that Oklahoma and Texas are also discussed in the Southeast regional summary). This is an incredibly diverse region, ranging from the Canadian to Mexican borders and from the Continental Divide to the Gulf of Mexico. Here we review concerns impacting soil health across the Great Plains generally while also paying special attention to the Prairie Pothole Region. We give special attention to the Prairie Pothole Region because of the particular impacts that healthy potholes have on carbon (C) sequestration (Euliss et al. 2006), as well as the significant additional ecosystem services that are at risk (Gascoigne et al. 2011). Historically, prairie potholes are the most productive breeding ground for North American waterfowl, and expected local droughts associated with climate change would reduce their productivity (Johnson et al. 2005; Sorenson et al. 1998).

Agencies in the United States Department of Agriculture have prepared at least two ways of understanding the landscape: ecoregions and major land resource areas. Ecological regions are mapped by the Forest Service (Cleland et al. 2007; McNab et al. 2007) using the National Hierarchical Framework of Ecological Units (Cleland et al. 1997). In this framework, large areas of relatively homogeneous physical and biological components are mapped together (Fig. A10). The fundamental criteria include elevation, temperature, soils and geology, and potential natural vegetation. Map units define unique ecological characteristics and potentials. A strong east-to-west gradient is clear in the Great Plains, and ecological regions range from coastal plain forests to semidesert in the south and from prairie parkland to dry steppe and coniferous forest in the north.

Alternatively, lands may be mapped as a function of their current land use activities (USDA NRCS 2006). In this framework, soil units are integrated with climate, water resource, and land use information to create land resource units several thousand acres (hectares) in size. These units are aggregated into major land resource areas and then into land resource regions (LRRs). Though LRRs are similar to ecological regions, the inclusion of land use is a notable distinguishing characteristic. As such, this framework delineates major cropping, forage, and forest regions of the Great Plains (Fig. A11). The Great Plains are characterized primarily by agricultural and grazing land uses.

As the reader might expect, forest communities vary significantly across a landscape of this size. The grasslands, rangeland, and cropping of the Great Plains separate the forests of the eastern United States from those of the western United States (Fig. A12). Eastern forests intrude most notably in the southern part of the region with oak/hickory and loblolly/shortleaf pine forest-type groups grading from east to west into pinyon/juniper and woodlands (Table A2; see table for scientific names of trees mentioned in this paragraph). Farther north, the Douglas-fir, fir/spruce/mountain hemlock, and lodgepole pine forest-type groups of the Rocky Mountains dominate the landscape until giving way to the ponderosa pine forests of the Black Hills in South Dakota. Across the Great Plains, several states—Texas, Oklahoma, Kansas, and Nebraska—are using statewide digital soil maps to facilitate fine-resolution mapping of ecological systems relevant at local scales.

Concerns

Concerns linked to soil management vary as widely as the land use and forest cover found across the Great Plains region. Here, we address several items identified in the chapters and other regional summaries of this document as well as through conversations with foresters and soil scientists across the region. These include interactions between C, wildfire and prescribed fire, and invasive pests, as well as impacts of changing water levels in the rivers and the soil.

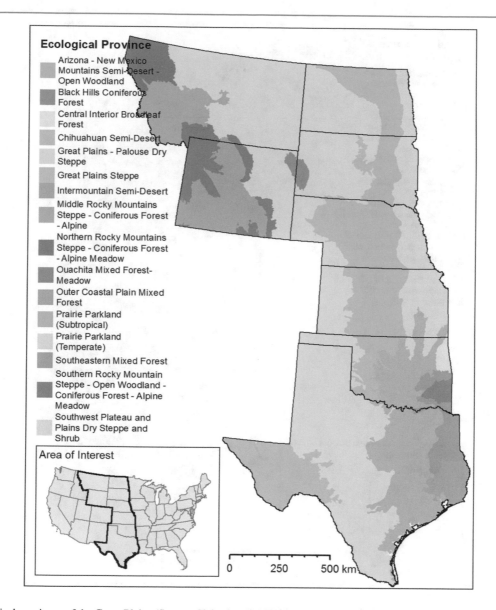

Ecological Province

Arizona - New Mexico Mountains Semi-Desert - Open Woodland

Black Hills Coniferous Forest

Central Interior Broadleaf Forest

Chihuahuan Semi-Desert

Great Plains - Palouse Dry Steppe

Great Plains Steppe

Intermountain Semi-Desert

Middle Rocky Mountains Steppe - Coniferous Forest - Alpine

Northern Rocky Mountains Steppe - Coniferous Forest - Alpine Meadow

Ouachita Mixed Forest-Meadow

Outer Coastal Plain Mixed Forest

Prairie Parkland (Subtropical)

Prairie Parkland (Temperate)

Southeastern Mixed Forest

Southern Rocky Mountain Steppe - Open Woodland - Coniferous Forest - Alpine Meadow

Southwest Plateau and Plains Dry Steppe and Shrub

Area of Interest

0   250   500 km

Fig. A10 Ecological provinces of the Great Plains (Source: Cleland et al. 2007)

Wildland Fire

Wildland fires are increasing in number and severity (Dennison et al. 2014), but there is some disagreement about the role of lightning-caused versus anthropogenic fires in the Great Plains (Changnon et al. 2002; Higgins 1986). Regardless of fire origin, restoration of diverse tallgrass prairie mosaics can be achieved through management of the interactions between fire and grazing (Fuhlendorf and Engle 2004). The frequency of fire can have a direct influence on soil organic carbon (SOC) in the soil food web; annual burning yields greater flow of C derived from roots and litter than infrequent burning (Shaw et al. 2016). However, fires may also remove the C stored in native tallgrass prairie between prescribed burns (Suyker and Verma 2001). In forests of the Greater Yellowstone Ecosystem in

Wyoming, Montana, and Idaho, crown fires reduce recruitment of lodgepole pine seedlings, and severely burned areas have delayed canopy recovery (Turner et al. 2000). Mountain pine beetle (*Dendroctonus ponderosae*) outbreaks may reduce the probability of crown fires by thinning forest canopies (Simard et al. 2011). Even so, the fire regime of the Greater Yellowstone Ecosystem is changing in a way that may fundamentally alter the landscape; conifer species currently dominant are likely to be replaced by lower montane woodland or nonforest vegetation (Westerling et al. 2011). Fire management is more complicated in the ponderosa pine forests of the Black Hills, where there is a history of both severe- and low-intensity fires, depending on climate, topography, slope, and exposure (Shinneman and Baker 1997).

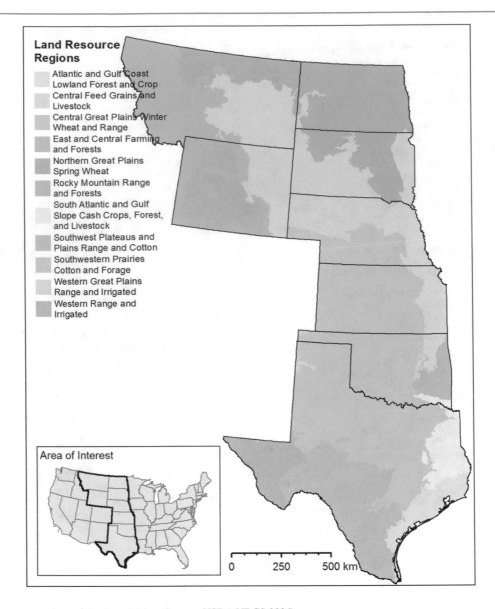

Land Resource Regions

   Atlantic and Gulf Coast Lowland Forest and Crop

   Central Feed Grains and Livestock

   Central Great Plains Winter Wheat and Range

   East and Central Farming and Forests

   Northern Great Plains Spring Wheat

   Rocky Mountain Range and Forests

   South Atlantic and Gulf Slope Cash Crops, Forest, and Livestock

   Southwest Plateaus and Plains Range and Cotton

   Southwestern Prairies Cotton and Forage

   Western Great Plains Range and Irrigated

   Western Range and Irrigated

Area of Interest

0   250   500 km

Fig. A11 Land resource regions of the Great Plains (Source: USDA NRCS 2006)

Invasive Species

Several invasive species are impacting important tree species with subsequent risks to forest C stocks and other ecological services. Emerald ash borer (*Agrilus planipennis*) is present along the eastern fringe of the Great Plains (see USDA FS and Michigan State University n.d.). Ash mortality results in lost regional forest productivity and forest C, at least until the growth of non-ash species partly compensates (Flower et al. 2013); this pest may result in other species becoming dominant in these forests. The impacts could be particularly significant in urban landscapes where ash dominates the population of street trees (Ball et al. 2007). Similarly, American elm (*Ulmus americana*) was present in the Great Plains (Little 1971) and is well represented as young trees in some stands in Kansas, but its potential to influence soils and store C as a

dominant species is limited by Dutch elm disease (caused by the fungus *Ophiostoma ulmi*) (Abrams 1986). Large-scale ponderosa pine mortality following beetle outbreaks can lead to organic C export in streams (Vik et al. 2017).

Trees may also be invasive species. Eastern redcedar (*Juniperus virginiana*) is increasing in abundance on grasslands (Briggs et al. 2002; Jones and Bowles 2016) and forests, where it is associated with decreases in tree species diversity (Hanberry et al. 2014; Meneguzzo and Liknes 2015). Invasion of grasslands by eastern redcedar alters soil physical and chemical properties. This species may shape the soil microbial community to its advantage over invaded oak upland forests by altering the dominant type of mycorrhizal fungi (Williams et al. 2013). Some studies document reduced organic matter quality and slower decomposition

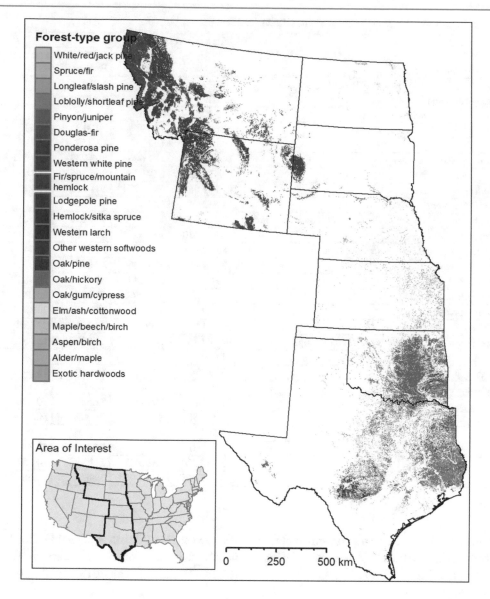

Fig. A12 Forest-type groups of the Great Plains (Source: Ruefenacht et al. 2008)

rates (Norris et al. 2001a), but significant C accumulates in invaded soils (McKinley and Blair 2008) and associated aboveground biomass (Norris et al. 2001b).

Restoration of areas dominated by eastern redcedar is difficult and expensive, particularly as stand density increases (Bidwell and Weir 2002). Prescribed fire, herbicides, and cutting are common strategies to manage encroachment by this species (Ortmann et al. 1998). However, these trees have hydrophobic leaf litter (Wine et al. 2012). This characteristic raises concerns about the potential for fires to increase soil water repellency, which inhibits soil microbial activity, seed germination, and plant growth (Fernelius et al. 2017). Even after the soil water repellency dissipates, the high soil nitrogen (N) present after these fires in pinyon-juniper woodlands may promote weed invasion (Fernelius et al. 2017).

Earthworms (phylum Annelida) are another threat to soil development and stability. Soils develop over time under the influence of climate, parent material, topographic relief, and biological organisms (Jenny 1941). Invasive earthworms (suborder Lumbricina) are a significant threat to soil development and native plant communities because they alter the manner in which leaf litter is deposited and stored on the soil surface, leading to fundamental changes in soil habitat and structure (Bohlen et al. 2004). A well-documented problem in northern temperate forests with unknown impacts on successional trajectory (Frelich et al. 2006), invasive earthworms have also been observed in northern Great Plains prairies (Henshue et al. 2018). Invasive earthworms in Oklahoma are more commonly associated with tallgrass prairie than with oak woodlands (Loss et al. 2017). Managing these invaders will prove challenging;

Table A2 Estimates of area of forest land in the Great Plains (USDA FS 2018)

Forest-type group (scientific name)	Area of forest land (ha)				
	Total	National Forest	Other Federal	State and local	Private
White/red/jack pine (*Pinus strobus*/*P. resinosa*/*P. banksiana*)	6 679	–	1 484	–	5 195
Spruce/fir (*Picea* spp./*Abies* spp.)	29 295	20 056	4 815	–	4 425
Longleaf/slash pine (*Pinus palustris*/*P. elliottii*)	52 365	8 076	–	–	44 289
Loblolly/shortleaf pine (*Pinus taeda*/*P. echinata*)	2 646 099	269 866	29 312	47 113	2 299 808
Other eastern softwoods	465 051	1 663	13 355	26 198	423 835
Pinyon/juniper (*Pinus* spp./*Juniperus* spp.)	4 994 915	117 646	502 933	240 711	4 133 625
Douglas-fir (*Pseudotsuga menziesii*)	3 419 157	2 049 767	266 365	148 672	954 353
Ponderosa pine (*Pinus ponderosa*)	1 971 987	651 927	144 728	174 726	1 000 606
Western white pine (*Pinus monticola*)	4 512	2 588	–	629	1 294
Fir/spruce/mountain hemlock (*Abies* spp./*Picea* spp./*Tsuga mertensiana*)	3 320 372	2 737 842	364 234	33 501	184 795
Lodgepole pine (*Pinus contorta*)	2 650 610	1 903 479	522 925	41 201	183 005
Hemlock/Sitka spruce (*Tsuga* spp./*Picea sitchensis*)	88 122	74 336	2 504	5 044	6 237
Western larch (*Larix occidentalis*)	372 999	259 735	27 053	27 032	59 179
Other western softwoods	594 229	438 244	63 831	14 985	77 169
Oak/pine (*Quercus* spp./*Pinus* spp.)	1 193 421	45 289	43 409	35 233	1 069 490
Oak/hickory (*Quercus* spp./*Carya* spp.)	8 869 606	87 561	228 077	200 757	8 353 210
Oak/gum/cypress (*Quercus* spp./*Nyssa* spp./*Taxodium* spp.	951 131	6 826	79 261	22 328	842 716
Elm/ash/cottonwood (*Ulmus* spp./*Fraxinus* spp./*Populus* spp.)	2 407 858	26 191	137 288	158 518	2 085 862
Maple/beech/birch (*Acer* spp./*Fagus* spp./*Betula* spp.)	17 603	–	–	–	17 603
Aspen/birch (*Populus* spp./*Betula* spp.)	530 934	305 893	51 069	15 541	158 431
Alder/maple (*Alnus* spp./*Acer* spp.)	6 975	4 458	–	2 517	–
Other hardwoods	328 760	20 644	8 652	10 276	289 187
Woodland hardwoods	9 711 505	8 457	74 002	502 908	9 126 138
Tropical hardwoods	2 564	–	–	2, 564	–
Exotic hardwoods	192 350	–	5 539	13 387	173 423
Nonstocked	3 148 520	531 382	231 189	144 575	2 241 375
Total	47 977 617	9 571 924	2 802 025	1 868 416	33 735 251

nonnative earthworms are more common in unburned plots in the Flint Hills in central Kansas (Callaham and Blair 1999) and burned plots in Oklahoma (Loss et al. 2017).

Oil and Gas Development

Oil and gas development is increasing in the region with the potential to impact soil properties and associated ecosystem services. Coal bed methane development may make the landscape more susceptible to invasion by nonnative species (Bergquist et al. 2007). The location of the well pad influences erosion and water quality; moving the well pad as little as 15 m from an intermittent stream significantly reduces erosion, N losses, and phosphorus losses from the site (McBroom et al. 2012). Fine-scale soil heterogeneity and soil texture are significant predictors of soil organic matter in abandoned well sites from the prerestoration era (33–90 years old), whereas disturbance is not (Avirmed et al. 2014). These results should not be generalized to the current reclaimed areas, where soil removal and stockpiling occur as part of well development (Avirmed et al. 2014). Indeed, soil organic matter is mineralized or released when well pad development and restoration include stripping, stockpiling,

and respreading (Mason et al. 2011). Road development has persistent impacts on soil properties, even following road removal (Matthees et al. 2018).

Grazing

Proper grazing management and use of grazing systems are important to soil health. Grazing can promote regrowth of grazing-tolerant plants and root exudation of C with positive feedbacks on N cycling and photosynthesis (Frank and Groffman 1998; Hamilton and Frank 2001), but heavy grazing—reducing standing biomass to 10% of average aboveground peak biomass—leads to declines in N cycling (Biondini et al. 1998). The impact of grazing on native species richness or invasion appears to be minimal at landscape scales (Lyseng et al. 2018); instead, soil characteristics, climate, and other disturbances drive these impacts (Stohlgren et al. 1999). The underlying causes of invasion by introduced species—whether soil, climate, or management history—need to be better understood. This understanding could inform an adaptive management approach for mitigation of the effects of these species and for grassland restoration (Grant et al. 2009).

The Prairie Pothole Region: A Focus on Wetlands

Prairie pothole wetlands formed across the northern Great Plains following the Wisconsin glaciation. These wetlands have complex hydrology that yields freshwater ponds or brackish and saline ponds, both dominated by direct precipitation (Pennock et al. 2014, van der Kamp and Hayashi 2009). Wetlands are being lost to row crop expansion at the rate of 5203–6223 ha yr^{-1}); this land use change is concentrated in those areas most conducive to agriculture (Johnston 2013). Techniques that use the historical record can delineate persisting, permanently lost, and temporarily lost wetlands, assisting with prioritization of restoration activities (Waz and Creed 2017). While wetlands in the Prairie Pothole Region, which extends far into Canada, currently store over 200 × 10^9 kg or 200 teragrams (Tg) C in the United States, restoration could yield sequestration of an additional 115 Tg C in the United States over a 10-year period (Euliss et al. 2006). Here we focus on the unique features, ecosystem services, and management challenges of this region.

The Prairie Pothole Region encompasses approximately 770 000 km^2 (boundaries vary slightly by source) of north-central North America and includes portions of Alberta, Saskatchewan, and Manitoba in Canada and Montana, North Dakota, South Dakota, Minnesota, and Iowa in the United States (Badiou et al. 2011; Dahl 2014; Euliss et al. 2006; Goldhaber et al. 2014). Before nineteenth-century Euro-American settlement, wetlands may have covered nearly 70 000 km^2 of mixed-grass and tallgrass prairies in the Prairie Pothole Region (Dahl 2014). Currently (circa 2009), the Prairie Pothole Region is characterized by millions of wetland basins, which cover roughly 26 000 km^2 and are dispersed throughout a landscape consisting primarily of croplands and grasslands (Dahl 2014; Pennock et al. 2010; Tangen et al. 2015). Pothole wetland densities and sizes (circa 2009) averaged nearly 7 basins km^{-2} and 1.3 ha, respectively. Wetland densities ranged between 12 basins km^{-2} in North Dakota and 1–2 basins km^{-2} in Minnesota and Iowa, with a maximum density of 57 basins km^{-2} in parts of North Dakota (Dahl 2014).

The Prairie Pothole Region was shaped around 12,000 years ago when glaciers receded during the Pleistocene Epoch. The Prairie Pothole Region generally is partitioned into three overarching physiographic regions that vary by local relief (i.e., maximum elevation difference within 93 km^2): the Glaciated Plains, Missouri Coteau, and Prairie Coteau. The Glaciated Plains region is a gently sloping, rolling landscape; the Missouri and Prairie Coteaus are hummocky plains of superglacial sediment associated with dead-ice moraines (Bluemle 2000; Gleason et al. 2008; Kantrud et al. 1989a). Over time, ice masses within the glacial till melted, resulting in the formation of closed depressions underlain by low-permeability till (Johnson et al. 2008). As these shallow basins collected water, they devel-

oped into what now are called pothole wetlands or prairie potholes. Due to their vast expanse and ecological value, prairie potholes have been the subject of a wide variety of research and reviews examining general ecology, biotic communities, soils, hydrogeochemistry, hydrology, and status (Dahl 2014; Goldhaber et al. 2011; Hayashi et al. 2016; Kantrud et al. 1989a; Richardson et al. 1994; van der Valk 1989).

Classification of Prairie Pothole Region Wetlands

Although there are several wetland classification systems (e.g., Brinson 1993; Cowardin et al. 1979; Wells and Zoltai 1985), pothole wetlands generally are classified according to the United States national system of Cowardin et al. (1979) or the region-specific system of Stewart and Kantrud (1971). Under Cowardin et al.'s (1979) system, potholes generally are classified as palustrine emergent wetlands with hydrology subclasses ranging from temporarily to permanently ponded. The regional system of Stewart and Kantrud (1971) classifies potholes based on vegetation zones and water permanence (e.g., temporary, seasonal, semipermanent), with subclass modifiers for characteristics such as salinity. In a natural grassland setting, wetland vegetation zones span a range from peripheral low-prairie or wet-meadow zones to deep-marsh emergent vegetation or open-water zones (Stewart and Kantrud 1971). On the basis of the regional wetland classes, mean surface areas of potholes are 0.4–9.5 ha (Dahl 2014); water depths range from centimeters to a few meters, although depths of most wetlands do not exceed 2 m.

Mineral soils of Prairie Pothole Region wetlands generally are classified within the Mollisol order; Aquoll, Udoll, or Ustoll suborders; and various finer classifications (e.g., groups, families, series). Soils of Prairie Pothole Region wetlands formed over time under the influence of climate, topography (i.e., closed depressions), fluctuating water tables, saturated flow, and intermittent anoxic conditions. Thus, soil profile properties (e.g., chemistry, color, texture) can vary among wetlands that differ in their relation to the groundwater (e.g., recharge, discharge). For example, seasonally ponded recharge wetlands often are characterized by soil profiles where dissolved solids and clay have been leached or translocated from the upper portion of the profile due to the downward movement of water over time. Conversely, groundwater discharge wetlands, where high water tables often persist, typically are characterized by saline, calcareous, and gypsiferous soils that are more homogeneous than recharge wetlands (Arndt and Richardson 1988; Richardson et al. 1994). Pothole wetlands can have high rates of primary production that, when combined with typical anoxic conditions and slow decomposition rates, result in the accumulation of organic matter (e.g., C sequestration), especially in the deeper, more permanent zones. Soil biogeochemical processes, such as methanogenesis and denitrification, can be highly variable depending on various factors such as microbial community composition, chemistry, and hydrology (Badiou et al. 2011: Bedard-Haughn et al. 2006; Dalcin Martins et al. 2017; Pennock et al. 2010; Tangen et al. 2015).

Hydrology and Water Chemistry
The water balance of Prairie Pothole Region wetlands is dominated by inputs from direct precipitation and precipitation runoff (including snowmelt); water losses largely are attributed to evapotranspiration and overland flow when inputs exceed capacity. Depending on a wetland's location in the landscape, as well as factors such as groundwater levels, potholes generally function as groundwater recharge, flow-through, or discharge sites (Winter 1989; Winter and Rosenberry 1995). Climate of the Prairie Pothole Region varies along a northwest-to-southeast gradient, with precipitation and temperature increasing toward the southeast. Though some pothole wetlands contain water throughout most years, a large proportion of these systems dry up seasonally because annual evapotranspiration and infiltration exceed precipitation. Water chemistry (e.g., ionic composition, salinity) of Prairie Pothole Region wetlands is highly variable, with salinities spanning a gradient from fresh to hypersaline. Much of the variability in water chemistry is associated with factors such as relation to groundwater (e.g., recharge, discharge), time of year (e.g., dilution, evapoconcentration), and climate (wet and dry cycles) (Goldhaber et al. 2014; LaBaugh 1989; Mushet et al. 2015; Stewart and Kantrud 1972; Swanson et al. 1988; Tangen et al. 2013).

Biotic Communities
Biotic communities of Prairie Pothole Region wetlands are generally composed of relatively resilient taxa capable of persisting in harsh and variable environments (Euliss et al. 1999; Kantrud et al. 1989a, b; Stewart and Kantrud 1972; Tangen et al. 2003). Within a pothole, composition of plant, aquatic invertebrate, and amphibian communities largely are determined by factors such as water depth and permanence, salinity, climate, biotic interactions, and anthropogenic disturbance (Euliss et al. 1999; Hanson et al. 2005; Kantrud et al. 1989a, b; Stewart and Kantrud 1972). Likewise, use of potholes by migratory waterfowl and waterbirds during any given year depends on the aforementioned habitat conditions, as well as the makeup of the other wetland biotic communities and the availability of adjacent upland habitats (Igl et al. 2017).

Ecosystem Services
Wetlands of the Prairie Pothole Region are recognized for providing an array of ecosystem services. Most notably, the Prairie Pothole Region provides breeding, brood-rearing,

and migration habitats for most of North America's migratory waterfowl (Batt et al. 1989). Potholes also provide critical habitats for a range of other wildlife such as mammals, game and nongame birds, reptiles, amphibians, and pollinators (Igl et al. 2017; Kantrud et al. 1989a; Otto et al. 2016; Smart et al. 2017). In the western portions of the Prairie Pothole Region, potholes can be an important water source for cattle (*Bos taurus*). Potholes also provide services related to conserving biological diversity, flood mitigation, filtration of pollutants, groundwater recharge, C sequestration, nutrient retention, and recreational opportunities (Badiou et al. 2011; Euliss et al. 2006; Gleason et al. 2008, 2011; Knutsen and Euliss 2001; Winter and Rosenberry 1995).

Anthropogenic and Climate Effects
The multitude of functions and ecosystem services provided by Prairie Pothole Region wetlands, along with their soils and biotic communities, can be affected by a host of anthropogenic disturbances, as well as by climate change (Anteau 2012; Johnson et al. 2010; Mushet et al. 2014; Niemuth et al. 2014; Tangen and Finocchiaro 2017). Agricultural drainage has been the primary driver of wetland losses in the United States (Dahl 1990, 2014; Dahl and Johnson 1991; Johnston 2013). By the mid-1980s, wetland losses in the Prairie Pothole Region were as high as 90% in Iowa, 80% in Minnesota, 49% in North Dakota, and 35% in South Dakota; statewide losses for Montana were approximately 27% (Dahl 1990, 2014). The Prairie Pothole Region lost an additional 4% of wetlands between 1997 and 2009 (Dahl 2014). Recently (circa 2012), increased crop demand has resulted in the rapid spread of subsurface drainage systems into the North Dakota and South Dakota portions of the Prairie Pothole Region (Johnston 2013; Tangen and Finocchiaro 2017; Werner et al. 2016). Consolidation drainage, the draining of small wetlands into larger wetlands, has also been identified as a practice that has negative ecological effects on the region's wetland habitats and associated biota (Anteau 2012; McCauley et al. 2015; Wiltermuth and Anteau 2016).

Remaining wetlands, especially those in a cropland setting, are susceptible to accelerated sedimentation rates and burial of upper soil horizons (Gleason and Euliss 1998; Gleason et al. 2003; Martin and Hartman 1987; Tangen and Gleason 2008), inputs of agricultural chemicals and nutrients (Grue et al. 1989; McMurry et al. 2016; Neely and Baker 1989), unnatural variance in water-level fluctuations (Euliss and Mushet 1996; van der Kamp et al. 1999, 2003), and altered vegetative communities (Kantrud and Newton 1996; Kantrud et al. 1989b; Laubhan and Gleason 2008). Additionally, when basins dry seasonally or during drought, they often are physically disturbed through burning or tillage.

Portions of the western Prairie Pothole Region overlie the Williston Basin, a large sedimentary basin with significant oil reserves that straddles the border of the United States and Canada. The Williston Basin has experienced a recent boom in oil production resulting from advances in hydraulic fracturing techniques and high demand for domestic petroleum production. Pothole wetlands and associated biota in this area are at risk from physical disturbances and habitat fragmentation caused by well pad development, infrastructure construction, building of roads and pipelines, and increased water demand (Post van der Burg et al. 2017). Most notable, however, is the potential contamination of wetlands by produced waters (brines) or oil spills (Gleason and Tangen 2014; Hossack et al. 2017; Preston and Chesley-Preston 2015).

Potholes are highly vulnerable to climate change owing to the dominant role that precipitation and evapotranspiration play in their water balance (Larson 1995). Long-term monitoring across wet and dry climate cycles has demonstrated considerable variability in wetland surface-water expression (e.g., depth, ponded surface area), water chemistry, and the composition of bird, plant, invertebrate, and amphibian communities (Beeri and Phillips 2007; Dahl 2014; Euliss and Mushet 2011; Euliss et al. 2004; Johnson et al. 2004; Larson 1995; Mushet et al. 2015; Niemuth et al. 2010). Moreover, modeling efforts examining potential effects of climate change suggest reduced water volumes and periods of inundation, stabilized vegetation communities, and, ultimately, diminished waterfowl habitats (Johnson and Poiani 2016; Johnson et al. 2005, 2010; Millet et al. 2009; Werner et al. 2013).

Wetland catchments of the Prairie Pothole Region are highly valued for their ecosystem services and diverse biotic communities and abiotic environments. The biotic and abiotic diversity of Prairie Pothole Region wetlands is a reflection of highly variable land uses and wide-ranging climatic, hydrologic, edaphic, and geographic conditions. The fact that the Prairie Pothole Region is one of the most intensely managed wetland ecosystems in North America, combined with concerns over climate change, emphasizes the need for multidisciplinary research to enhance our understanding of short- and long-term effects of climate and land use change on these wetlands and the services that they provide to society.

Key Findings

- Wildland fires are increasing in number and severity. Their impact varies across the diverse ecosystems of the Great Plains.
- Several invasive pests are impacting important tree species with subsequent risks to forest C stocks and other ecological services.
- Eastern redcedar is increasing in abundance on grasslands and forests, where it is associated with decreases in tree species diversity.

- Oil and gas development is increasing in the region with the potential to impact soil properties and related ecosystem services.
- Proper grazing management and use of grazing systems are important to soil health.
- The multitude of functions and ecosystem services provided by Prairie Pothole Region wetlands, along with their soils and biotic communities, can be affected by a host of anthropogenic disturbances, as well as by climate change.

Literature Cited

Abrams MD (1986) Historical development of gallery forests in northeast Kansas. Vegetatio 65(1):29–37

Anteau MJ (2012) Do interactions of land use and climate affect productivity of waterbirds and prairie-pothole wetlands? Wetlands 32:1–9

Arndt JL, Richardson JL (1988) Hydrology, salinity and hydric soil development in a North Dakota prairie-pothole wetland system. Wetlands 8:93–108

Avirmed O, Burke IC, Mobley ML et al (2014) Natural recovery of soil organic matter in 30–90-year-old abandoned oil and gas wells in sagebrush steppe. Ecosphere. 5(3):art24

Badiou P, McDougal R, Pennock D, Clark B (2011) Greenhouse gas emissions and carbon sequestration potential in restored wetlands of the Canadian Prairie Pothole Region. Wetl Ecol Manag 19(3):237–256

Ball J, Mason S, Kiesz A et al (2007) Assessing the hazard of emerald ash borer and other exotic stressors to community forests. Arboricult Urban Forest 33(5):350–359

Batt BDJ, Anderson MG, Anderson CD, Caswell FD (1989) The use of prairie potholes by North American ducks. In: van der Valk AG (ed) Northern prairie wetlands. Iowa State University Press, Ames, pp 204–227

Bedard-Haughn A, Matson AL, Pennock DJ (2006) Land use effects on gross nitrogen mineralization, nitrification, and N₂O emissions in ephemeral wetlands. Soil Biol Biochem 38:3398–3406

Beeri O, Phillips RL (2007) Tracking palustrine water seasonal and annual variability in agricultural wetland landscapes using Landsat from 1997 to 2005. Glob Chang Biol 13:897–912

Bergquist E, Evangelista P, Stohlgren TJ, Alley N (2007) Invasive species and coal bed methane development in the Powder River Basin, Wyoming. Environ Monit Assess 128(1):381–394

Bidwell TG, Weir JR (2002) Eastern redcedar management and control: best management practices to restore Oklahoma's ecosystems. F-2876. Oklahoma State University, Division of Agricultural Sciences and Natural Resources, Stillwater, 4 p. Available at https://www.extension.okstate.edu/fact-sheets/eastern-redcedar-control-and-management-best-management-practices-to-restore-oklahomas-ecosystems.html. Accessed 29 May 2020

Biondini ME, Patton BD, Nyren PE (1998) Grazing intensity and ecosystem processes in a northern mixed-grass prairie, USA. Ecol Appl 8(2):469–479

Bluemle JP (2000) The face of North Dakota—the geologic story, Educational series 26, 3rd edn. North Dakota Department of Mineral Resources, Geological Survey, Bismarck, 210 p

Bohlen PJ, Scheu S, Hale CM et al (2004) Non-native invasive earthworms as agents of change in northern temperate forests. Front Ecol Environ 2(8):427–435

Briggs JM, Hoch GA, Johnson LC (2002) Assessing the rate, mechanisms, and consequences of the conversion of tallgrass prairie to *Juniperus virginiana* forest. Ecosystems 5(6):578–586

Brinson MM (1993) A hydrogeomorphic classification for wetlands. Wetlands Research Program Tech. Rep. WRP-DE-4. U.S. Army Corps of Engineers, Vicksburg, 79 p

Callaham MA Jr, Blair JM (1999) Influence of differing land management on the invasion of North American tallgrass prairie soils by European earthworms. Pedobiologia 43(6):507–512

Changnon SA, Kunkel KE, Winstanley D (2002) Climate factors that caused the unique tall grass prairie in the central United States. Phys Geogr 23(4):259–280

Cleland DT, Avers PE, McNab WH et al (1997) National hierarchical framework of ecological units. In: Boyce MS, Haney AW (eds) Ecosystem management: applications for sustainable forest and wildlife resources. Yale University Press, New Haven, pp 181–200. Chapter 9

Cleland DT, Freeouf JA, Keys JE et al (2007) Ecological subregions: sections and subsections for the conterminous United States. Gen. Tech. Rep. GTR-WO-76D [Map on CD-ROM] (Sloan, A.M., cartog.). U.S. Department of Agriculture, Forest Service, Washington Office, Washington, DC. [1:3,500,000]

Cowardin LM, Carter V, Golet FC, LaRoe ET (1979) Classification of wetlands and deepwater habitats of the United States. FWS/OBS-79/31. U.S. Department of the Interior, Fish and Wildlife Service, Office of Biological Services, Washington, DC. 131 p. Available at https://www.fws.gov/wetlands/Documents/classwet/index.html. Accessed 12 Apr 2019

Dahl TE (1990) Wetland losses in the United States 1780's to 1980's. U.S. Department of the Interior, Fish and Wildlife Service, Washington, DC. 13 p. Available at http://www.fws.gov/wetlands/Status-And-Trends/index.html. Accessed 9 Mar 2019

Dahl TE (2014) Status and trends of prairie wetlands in the United States 1997 to 2009. U.S. Department of the Interior, Fish and Wildlife Service, Washington, DC. 67 p. Available at http://www.fws.gov/wetlands/Status-And-Trends/index.html. Accessed 9 Mar 2019

Dahl TE, Johnson CE (1991) Wetlands status and trends in the conterminous United States, mid-1970s to mid-1980s. U.S. Department of the Interior, Fish and Wildlife Service, Washington, DC. 28 p. Available at http://www.fws.gov/wetlands/Status-And-Trends/index.html. Accessed 9 Mar 2019

Dalcin Martins P, Hoyt DW, Bansal S et al (2017) Abundant carbon substrates drive extremely high sulfate reduction rates and methane fluxes in Prairie Pothole wetlands. Glob Chang Biol 23(8):3107–3120

Dennison PE, Brewer SC, Arnold JD, Moritz MA (2014) Large wildfire trends in the western United States, 1984–2011. Geophys Res Lett 41(8):2928–2933

Euliss NH Jr, Mushet DM (1996) Water-level fluctuation in wetlands as a function of landscape condition in the Prairie Pothole Region. Wetlands 16(4):587–593

Euliss NH Jr, Mushet DM (2011) A multi-year comparison of IPCI scores for prairie pothole wetlands: implications of temporal and spatial variation. Wetlands 31:713–723

Euliss NH Jr, Wrubleski DA, Mushet DM (1999) Wetlands of the Prairie Pothole Region: invertebrate species composition, ecology, and management. In: Batzer DP, Rader RB, Wissinger SA (eds) Invertebrates in freshwater wetlands of North America: ecology and management. John Wiley & Sons, New York, pp 471–514

Euliss NH Jr, LaBaugh JW, Fredrickson LH et al (2004) The wetland continuum: a conceptual framework for interpreting biological studies. Wetlands 24(2):448–458

Euliss NH, Gleason RA, Olness A et al (2006) North American prairie wetlands are important nonforested land-based carbon storage sites. Sci Total Environ 361(1):179–188

Fernelius KJ, Madsen MD, Hopkins BG et al (2017) Post-fire interactions between soil water repellency, soil fertility and plant growth in soil collected from a burned piñon-juniper woodland. J Arid Environ 144:98–109

Flower CE, Knight KS, Gonzalez-Meler MA (2013) Impacts of the emerald ash borer (*Agrilus planipennis* Fairmaire) induced ash (*Fraxinus* spp.) mortality on forest carbon cycling and successional dynamics in the eastern United States. Biol Invasions 15(4):931–944

Frank DA, Groffman PM (1998) Ungulate vs. landscape control of soil C and N processes in grasslands of Yellowstone National Park. Ecology 79(7):2229–2241

Frelich LE, Hale CM, Scheu S et al (2006) Earthworm invasion into previously earthworm-free temperate and boreal forests. Biol Invasions 8(6):1235–1245

Fuhlendorf SD, Engle DM (2004) Application of the fire-grazing interaction to restore a shifting mosaic on tallgrass prairie. J Appl Ecol 41(4):604–614

Gascoigne WR, Hoag D, Koontz L et al (2011) Valuing ecosystem and economic services across land-use scenarios in the Prairie Pothole Region of the Dakotas, USA. Ecol Econ 70(10):1715–1725

Gleason RA, Euliss NH Jr (1998) Sedimentation of prairie wetlands. Great Plains Res 8:97–112

Gleason RA, Euliss NH Jr, Hubbard DE, Duffy WG (2003) Effects of sediment load on emergence of aquatic invertebrates and plants from wetland soil egg and seed banks. Wetlands 23(1):26–34

Gleason RA, Tangen BA (eds) (2014) Brine contamination to aquatic resources from oil and gas development in the Williston Basin, United States. Scientific Investigations Rep. 2014-5017. U.S. Department of the Interior, Geological Survey, Reston 127 p

Gleason RA, Laubhan MK, Euliss NH Jr (eds) (2008) Ecosystem services derived from wetland conservation practices in the United States Prairie Pothole Region with an emphasis on the U.S. Department of Agriculture Conservation Reserve and Wetlands Reserve Programs. Prof. Pap. 1745. U.S. Department of the Interior, Geological Survey, Reston. 58 p. Available at https://pubs.usgs.gov/pp/1745/. Accessed 12 Apr 2019

Gleason RA, Euliss NH Jr, Tangen BA et al (2011) USDA conservation program and practice effects on wetland ecosystem services in the Prairie Pothole Region. Ecol Appl. 21(supplement):S65–S81

Goldhaber MB, Mills C, Stricker CA, Morrison JM (2011) The role of critical zone processes in the evolution of the Prairie Pothole Region wetlands. Appl Geochem 26:S32–S35

Goldhaber MB, Mills CT, Morrison JM et al (2014) Hydrogeochemistry of prairie pothole region wetlands: role of long-term critical zone processes. Chem Geol 387:170–183

Grant TA, Flanders-Wanner B, Shaffer TL et al (2009) An emerging crisis across northern prairie refuges: prevalence of invasive plants and a plan for adaptive management. Ecol Restor 27(1):58–65

Grue CE, Tome MW, Messmer TA et al (1989) Agricultural chemicals and prairie pothole wetlands: meeting the needs of the resource and the farmer—U.S. perspective. Trans N Am Wildl Nat Resour Conf 54:43–58

Hamilton EW III, Frank DA (2001) Can plants stimulate soil microbes and their own nutrient supply? evidence from a grazing tolerant grass. Ecology 82(9):2397–2402

Hanberry BB, Kabrick JM, He HS (2014) Changing tree composition by life history strategy in a grassland-forest landscape. Ecosphere 5(3):1–16

Hanson MA, Zimmer KD, Butler MG et al (2005) Biotic interactions as determinants of ecosystem structure in prairie wetlands: an example using fish. Wetlands 25(3):764–775

Hayashi M, van der Kamp G, Rosenberry DO (2016) Hydrology of prairie wetlands: understanding the integrated surface-water and groundwater processes. Wetlands 36(supplement 2):S237–S254

Henshue N, Mordhorst C, Perkins L (2018) Invasive earthworms in a Northern Great Plains prairie fragment. Biol Invasions 20(1):29–32

Higgins KF (1986) Interpretation and compendium of historical fire accounts in the Northern Great Plains. Resour. Publ. 161. U.S. Department of the Interior, Fish and Wildlife Service, Washington, DC, 39 p

Hossack BR, Puglis HJ, Battaglin WA et al (2017) Widespread legacy brine contamination from oil production reduces survival of chorus frog larvae. Environ Pollut 231:742–751

Igl LD, Shaffer JA, Johnson DH, Buhl DA (2017) The influence of local- and landscape-level factors on wetland breeding birds in the Prairie Pothole Region of North and South Dakota. Open-File Rep. 2017-1096. U.S. Department of the Interior, Geological Survey, Reston, 65 p

Jenny H (1941) Factors of soil formation: a system of quantitative pedology. McGraw-Hill, New York, 281 p

Johnson WC, Poiani KA (2016) Climate change effects on prairie pothole wetlands: findings from a twenty-five year numerical modeling project. Wetlands 36(supplement):273–285

Johnson WC, Boettcher SE, Poiani KA, Guntenspergen G (2004) Influence of weather extremes on the water levels of glaciated prairie wetlands. Wetlands 24(2):385–398

Johnson WC, Millett BV, Gilmanov T et al (2005) Vulnerability of northern prairie wetlands to climate change. Bioscience 55:863–872

Johnson RR, Oslund FT, Hertel DR (2008) The past, present, and future of prairie potholes in the United States. J Soil Water Conserv 63:84A–87A

Johnson WC, Werner BA, Guntenspergen GR et al (2010) Prairie wetland complexes as landscape functional units in a changing climate. Bioscience 60(2):128–140

Johnston CA (2013) Wetland losses due to row crop expansion in the Dakota Prairie Pothole Region. Wetlands 33(1):175–182

Jones MD, Bowles ML (2016) Eastern redcedar dendrochronology links hill prairie decline with decoupling from climatic control of fire regime and reduced fire frequency. J Torrey Bot Soc 143(3):239–253

Kantrud HA, Krapu GL, Swanson GA (1989a) Prairie basin wetlands of the Dakotas— a community profile. Biol. Rep. 85(7.28). U.S. Department of the Interior, Fish and Wildlife Service, Washington, DC, 116 p

Kantrud HA, Millar JB, van der Valk AG (1989b) Vegetation of wetlands of the Prairie Pothole Region. In: van der Valk

AG (ed) Northern prairie wetlands. Iowa State University Press, Ames, pp 132–187

Kantrud HA, Newton WE (1996) A test of vegetation-related indicators of wetland quality in the prairie pothole region. J Aquat Ecosyst Health 5:177–191

Knutsen GA, Euliss NH Jr (2001) Wetland restoration in the Prairie Pothole Region of North America–a literature review. Biol. Sci. Rep. USGS/BRD/BSR 2001-0006. U.S. Department of the Interior, Geological Survey, Reston, 54 p

LaBaugh JW (1989) Chemical characteristics of water in northern prairie wetlands. In: van der Valk AG (ed) Northern prairie wetlands. Iowa State University Press, Ames, pp 56–90

Larson DL (1995) Effects of climate on numbers of northern prairie wetlands. Clim Chang 30:169–180

Laubhan MK, Gleason RA (2008) Plant community quality and richness. In: Gleason RA, Laubhan MK, Euliss NH Jr (eds) Ecosystem services derived from wetland conservation practices in the United States Prairie Pothole Region with an emphasis on the U.S. Department of Agriculture Conservation Reserve and Wetlands Reserve Programs. Prof. Pap. 1745. U.S. Department of the Interior, Geological Survey, Reston, pp 15–22

Little EL (1971) Atlas of United States trees. Vol. 1, Conifers and important hardwoods. Misc. Publ. 1146. U.S. Department of Agriculture, Forest Service, Washington, DC, 9 p.; 200 maps

Loss SR, Paudel S, Laughlin CM, Zou C (2017) Local-scale correlates of native and nonnative earthworm distributions in juniper-encroached tallgrass prairie. Biol Invasions 19(5):1621–1635

Lyseng MP, Bork EW, Hewins DB et al (2018) Long-term grazing impacts on vegetation diversity, composition, and exotic species presence across an aridity gradient in northern temperate grasslands. Plant Ecol 219(6):649–663

Martin DB, Hartman WA (1987) The effect of cultivation on sediment and deposition in prairie pothole wetlands. Water Air Soil Pollut 34:45–53

Mason A, Driessen C, Norton J, Strom C (2011) First year soil impacts on well-pad development and reclamation on Wyoming's sagebrush steppe. Nat Resourc Environ Issues 17:Art. 5

Matthees HL, Hopkins DG, Casey FXM (2018) Soil property distribution following oil well access road removal in North Dakota, USA. Can J Soil Sci 98(2):369–380

McBroom M, Thomas T, Zhang Y (2012) Soil erosion and surface water quality impacts of natural gas development in East Texas, USA. Water 4(4):944

McCauley LA, Anteau MJ, Post van der Burg MP, Wiltermuth MT (2015) Land use and wetland drainage affect water levels and dynamics of remaining wetlands. Ecosphere 6(6):1–22

McKinley DC, Blair JM (2008) Woody plant encroachment by *Juniperus virginiana* in a mesic native grassland promotes rapid carbon and nitrogen accrual. Ecosystems 11(3):454–468

McMurry ST, Belden JB, Smith LM et al (2016) Land use effects on pesticides in sediments of prairie pothole wetlands in North and South Dakota. Sci Total Environ 565:682–689

McNab WH, Cleland DT, Freeouf JA et al (2007) Description of ecological subregions: sections of the conterminous United States. Gen. Tech. Rep. GTR-WO-76B. Washington, DC: U.S. Department of Agriculture, Forest Service, Washington Office, 80 p

Meneguzzo DM, Liknes GC (2015) Status and trends of eastern redcedar (*Juniperus virginiana*) in the central United States: analyses and observations based on Forest Inventory and Analysis data. J For 113(3):325–334

Millet B, Johnson WC, Guntenspergen G (2009) Climate trends of the North American prairie pothole region 1906–2000. Clim Chang 93:243–267

Mushet DM, Neau JL, Euliss NH Jr (2014) Modeling effects of conservation grassland losses on amphibian habitat. Biol Conserv 174:93–100

Mushet DM, Goldhaber MB, Mills CT et al (2015) Chemical and biotic characteristics of prairie lakes and large wetlands in south-central North Dakota—effects of a changing climate. Scientific Investigations Rep. 2015-5126. U.S. Department of the Interior, Geological Survey, Reston, 55 p

Neely RK, Baker JL (1989) Nitrogen and phosphorus dynamics and the fate of agricultural runoff. In: van der Valk AG (ed) Northern prairie wetlands. Iowa State University Press, Ames, pp 92–131

Niemuth ND, Wangler B, Reynolds RE (2010) Spatial and temporal variation in wet area of wetlands in the Prairie Pothole Region of North Dakota and South Dakota. Wetlands 30(6):1053–1064

Niemuth ND, Fleming KK, Reynolds RE (2014) Waterfowl conservation in the U.S. Prairie Pothole Region: confronting the complexities of climate change. PLoS One 9:e100034

Norris MD, Blair JM, Johnson LC (2001a) Land cover change in eastern Kansas: litter dynamics of closed-canopy eastern redcedar forests in tallgrass prairie. Can J Bot 79(2):214–222

Norris MD, Blair JM, Johnson LC, McKane RB (2001b) Assessing changes in biomass, productivity, and C and N stores following *Juniperus virginiana* forest expansion into tallgrass prairie. Can J For Res 31(11):1940–1946

Ortmann J, Stubbendieck J, Masters RA et al (1998) Efficacy and costs of controlling eastern redcedar. J Range Manag 51(2):158–163

Otto CRV, Roth CL, Carlson BL, Smart MD (2016) Land-use change reduces habitat suitability for supporting managed honey bee colonies in the Northern Great Plains. Proc Natl Acad Sci 113:10430–10435

Pennock D, Yates T, Bedard-Haughn A et al (2010) Landscape controls on N₂O and CH₄ emissions from freshwater mineral soil wetlands of the Canadian Prairie Pothole Region. Geoderma 155(3–4):308–319

Pennock D, Bedard-Haughn A, Kiss J, van der Kamp G (2014) Application of hydropedology to predictive mapping of wetland soils in the Canadian Prairie Pothole Region. Geoderma 235–236:199–211

Post van der Burg M, Symstad AJ, Igl LD et al (2017) Potential effects of energy development on environmental resources of the Williston Basin in Montana, North Dakota, and South Dakota—species of conservation concern. Scientific Investigations Report 2017-5070-D. U.S. Department of the Interior, Geological Survey, Reston, 41 p

Preston TM, Chesley-Preston TL (2015) Risk assessment of brine contamination to aquatic resources from energy development in glacial drift deposits: Williston Basin, USA. Sci Total Environ 508:534–545

Richardson JL, Arndt JL, Freeland J (1994) Wetland soils of the prairie potholes. Adv Agron 52:121–171

Ruefenacht B, Finco MV, Nelson MD et al (2008) Conterminous U.S. and Alaska forest type mapping using Forest Inventory and Analysis data. Photogramm Eng Remote Sens 74(11):1379–1388

Shaw EA, Denef K, Milano de Tomasel C et al (2016) Fire affects root decomposition, soil food web structure, and carbon flow in tallgrass prairie. Soil 2:199–210

Shinneman DJ, Baker WL (1997) Nonequilibrium dynamics between catastrophic disturbances and old-growth forests in ponderosa pine landscapes of the Black Hills. Conserv Biol 11(6):1276–1288

Simard M, Romme WH, Griffin JM, Turner MG (2011) Do mountain pine beetle outbreaks change the probability of active crown fire in lodgepole pine forests? Ecol Monogr 81(1):3–24

Smart MD, Cornman RS, Iwanowicz DD et al (2017) A comparison of honey bee-collected pollen from working agricultural lands using light microscopy and ITS metabarcoding. Environ Entomol 46(1):38–49

Sorenson LG, Goldberg R, Root TL, Anderson MG (1998) Potential effects of global warming on waterfowl populations breeding in the Northern Great Plains. Clim Chang 40(2):343–369

Stewart RE, Kantrud HA (1971) Classification of natural ponds and lakes in the glaciated prairie region. Resour. Publ. 92. U.S. Department of the Interior, Fish and Wildlife Service, Washington, DC, 57 p

Stewart RE, Kantrud HA (1972) Vegetation of prairie potholes, North Dakota, in relation to quality of water and other environmental factors. Prof. Pap. 585-D. U.S. Department of the Interior, Geological Survey, Washington, DC, 36 p

Stohlgren TJ, Schell LD, Vanden Heuvel B (1999) How grazing and soil quality affect native and exotic plant diversity in Rocky Mountain grasslands. Ecol Appl 9(1):45–64

Suyker AE, Verma SB (2001) Year-round observations of the net ecosystem exchange of carbon dioxide in a native tall-grass prairie. Glob Chang Biol 7(3):279–289

Swanson GA, Winter TC, Adomaitis VA, LaBaugh JW (1988) Chemical characteristics of prairie lakes in south-central North Dakota—their potential for influencing use by fish and wildlife. Tech. Rep. 18. U.S. Department of the Interior, Fish and Wildlife Service, Washington, DC, 44 p

Tangen BA, Gleason RA (2008) Reduction of sedimentation and nutrient loading. In: Gleason RA, Laubhan MK, Euliss NH Jr (eds) Ecosystem services derived from wetland conservation practices in the United States Prairie Pothole Region with an emphasis on the U.S. Department of Agriculture Conservation Reserve and Wetlands Reserve Programs. Prof. Pap. 1745. U.S. Department of the Interior, Geological Survey, Reston, pp 38–44

Tangen BA, Butler MG, Ell MJ (2003) Weak correspondence between macroinvertebrate assemblages and land use in Prairie Pothole Region wetlands, USA. Wetlands 23:104–115

Tangen BA, Finocchiaro RG, Gleason RA et al (2013) Assessment of water-quality data from Long Lake National Wildlife Refuge, North Dakota—2008 through 2012. Scientific Investigations Rep. 2013-5183. U.S. Department of the Interior, Geological Survey, Reston, 28 p

Tangen BA, Finocchiaro RG, Gleason RA (2015) Effects of land use on greenhouse gas fluxes and soil properties of wetland catchments in the Prairie Pothole Region of North America. Sci Total Environ 533:391–409

Tangen BA, Finocchiaro RG (2017) A case study examining the efficacy of drainage setbacks for limiting effects to wetlands in the Prairie Pothole Region, USA. J Fish Wildl Manag 8:513–529

Turner MG, Romme WH, Gardner RH (2000) Prefire heterogeneity, fire severity, and early postfire plant reestablishment in subalpine forests of Yellowstone National Park, Wyoming. Int J Wildland Fire 9(1):21–36

USDA Forest Service [USDA FS] (2018) Forest Inventory EVALIDator web-application ver. 1.7.0.01. https://apps.fs.usda.gov/EVALIDator/evalidator.jsp. Accessed 29 May 2020

USDA Forest Service [USDA FS] and Michigan State University. (n.d.) Emerald Ash Borer Information Network. https://www.emeraldashborer.info. Accessed 6 Apr 2019

USDA Natural Resources Conservation Service [USDA NRCS] (2006) Land resource regions and major land resource areas of the United States, the Caribbean, and the Pacific Basin. Agric. Handb. 296. USDA, Washington, DC, 672 p

van der Kamp G, Hayashi M (2009) Groundwater-wetland ecosystem interaction in the semiarid glaciated plains of North America. Hydrogeol J 17(1):203–214

van der Kamp G, Stolte WJ, Clark RG (1999) Drying out of small prairie wetlands after conversion of their catchments from cultivation to permanent brome grass. Hydrol Sci J 44:387–397

van der Kamp G, Hayashi M, Gallén D (2003) Comparing the hydrology of grassed and cultivated catchments in the semi-arid Canadian prairies. Hydrol Process 17:559–575

van der Valk A (ed) (1989) Northern prairie wetlands. Iowa State University Press, Ames, 400 p

Vik ES, Sieverding HL, Punsal JJ et al (2017) Timing of organic carbon release from mountain pine beetle impacted ponderosa pine forests in South Dakota. Water Environ J 31(3):375–379

Waz A, Creed IF (2017) Automated techniques to identify lost and restorable wetlands in the Prairie Pothole Region. Wetlands 37(6):1079–1091

Wells ED, Zoltai S (1985) The Canadian system of wetland classification and its application to circumboreal wetlands. Aquil Ser Bot 21:45–52

Werner BA, Johnson WC, Guntenspergen GR (2013) Evidence for 20th century climate warming and wetland drying in the North American Prairie Pothole Region. Ecol Evol 3:3471–3482

Werner B, Tracy J, Johnson WC et al (2016) Modeling the effects of tile drain placement on the hydrologic function of farmed prairie wetlands. J Am Water Resour Assoc 52(6):1482–1492

Westerling AL, Turner MG, Smithwick EAH et al (2011) Continued warming could transform Greater Yellowstone fire regimes by mid-21st century. Proc Natl Acad Sci 108(32):13165–13170

Williams RJ, Hallgren SW, Wilson GWT, Palmer MW (2013) *Juniperus virginiana* encroachment into upland oak forests alters arbuscular mycorrhizal abundance and litter chemistry. Appl Soil Ecol 65:23–30

Wiltermuth MT, Anteau MJ (2016) Is consolidation drainage an indirect mechanism for increased abundance of cattail in northern prairie wetlands? Wetl Ecol Manag 24(5):533–544

Wine ML, Ochsner TE, Sutradhar A, Pepin R (2012) Effects of eastern redcedar encroachment on soil hydraulic properties along Oklahoma's grassland-forest ecotone. Hydrol Process 26(11):1720–1728

Winter TC (1989) Hydrologic studies of wetlands in the northern prairie. In: van der Valk AG (ed) Northern prairie wetlands. Iowa State University Press, Ames, pp 16–54

Winter TC, Rosenberry DO (1995) The interaction of ground water with prairie pothole wetlands in the Cottonwood Lake area, east-central North Dakota, 1979–1990. Wetlands 15(3):193–211

Southwest

Robert Vaughan and Erin Berryman

Introduction

The landscapes of the Southwest region (California, Nevada, Utah, Colorado, New Mexico, and Arizona; Fig. A13) can best be characterized by their extensive range and magnitude of physiography (e.g., precipitation, temperature, elevation, and vegetation). From the arid Chihuahuan, Sonoran, and Mojave Deserts to the cool and moist montane forests and alpine environments of the Sierra Nevada, one can observe every soil order with the exception of Oxisols.

Forest and rangeland soils of the Southwest are susceptible to the effects of several interrelated influences. The most pressing issues stem from long-term changes in climate, which are having direct effects on soil organic carbon (SOC), water availability, wildfire frequency and intensity, and invasive species.

Soil Carbon

Drought, insect pest outbreaks, wildfire, and invasive species are among the factors that pose critical management challenges in forests and rangelands of the southwestern United States and are likely to have consequences for SOC storage if effects are not mitigated or preventive actions are not taken. Climate change models project increasing severity of droughts in the southwestern United States, potentially exceeding what the region has experienced in the last millennia (Williams et al. 2012). In the Southwest, forest managers are primarily concerned with maintaining forest health and reducing risk of wildfire through forest thinning and fuels management rather than managing for timber harvest. Over the past decade, bark beetle (subfamily Scolytinae) outbreaks and drought-induced forest mortality have severely reduced live standing biomass in forests across the region, with the greatest losses in Colorado (Hicke et al. 2016). These losses are the result of a combination of factors: climate change effects on pests and trees; a lack of active forest management and history of fire suppression, which have led to overstocking in dry forests; and past stand-replacing disturbances in high altitude forests, which are now ideal settings for pest outbreaks.

In response, managers may take action to improve forest health in overstocked stands by thinning, which can mitigate drought-induced tree mortality (Bradford and Bell 2017). A reduction in tree mortality maintains inputs of organic matter to soil, thus building SOC stocks. However, there is the potential for negative impacts on SOC with any harvest operation. Site management during and after forest operations should prioritize leaving residues in place where possible to promote accumulation of SOC provided it does not hinder the regeneration process (e.g., mineral soil exposure required for light-seeded spe-

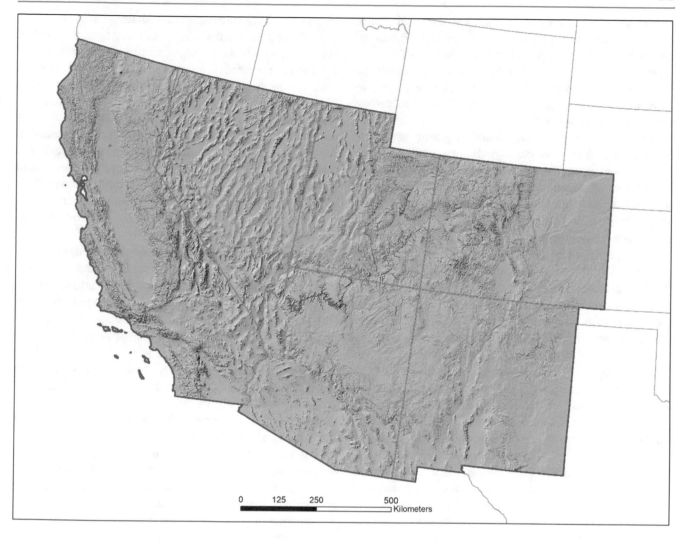

Fig. A13 Map of the Southwest region

cies). Repeated low-intensity prescribed burning, such as that used for surface fuel reduction, also appears to have benefits for SOC.

In Arizona, California, Nevada, New Mexico, and Utah, wildfire has killed as many or more trees than have bark beetles over the past few decades (Hicke et al. 2016). Area burned by wildfire in the southwestern United States has been on the rise since the 1980s (Westerling 2016). Therefore, forest management has largely focused on how to mitigate wildfire hazard in forests where fire has been suppressed and fuel loadings are high enough to support rapid growth and spread of crown fire. Repeated prescribed fires can help build stable forms of SOC, such as pyrogenic C. As with any decision on site removal of C, however, the potential loss of organic matter inputs during controlled burning must be weighed against the risk that the forest would burn down in a wildfire. Postwildfire management to reduce erosion, especially with wood mulch, may help maintain SOC at forest sites and rebuild SOC through enhanced plant regeneration and mulch decomposition.

A decline in snowpack has been occurring in mountain ranges in the southwestern United States over the past few decades, and it is expected to continue under future climate change (Mote et al. 2005), with implications for SOC stores. The changes in hydrology will reduce SOC through reductions in plant productivity and plant inputs to SOC.

Nitrogen (N) deposition can also affect SOC. On the east side of the Continental Divide in the southern Rocky Mountains, forests have been undergoing N deposition derived from the greater Denver-Front Range metropolitan area in Colorado (Baron et al. 2000). High altitude forests have displayed a shift in soil microbial activity due to increased N deposition, although effects on SOC are not consistent. In one subalpine forest in Rocky Mountain National Park in Colorado, SOC levels declined after chronic N deposition (Boot et al. 2016).

Invasive plants alter soil physical and chemical properties, including SOC storage. Invasive annual grasses are dominant on many rangeland and forest sites. Restoration of areas dominated by invasive annual grasses is difficult, but prescribed fire, biochar, and native forb and shrub islands are all methods that may help to increase SOC and reduce the spread of invasive species.

Responses of SOC to large mammal grazing are mixed and depend on the type of plant species (e.g., cool-season vs. warm-season plants, shortgrass vs. tallgrass communities) and amount of grazing. Generally, mid-grass and tallgrass communities have the potential to build larger SOC pools than shortgrass communities. In addition, the amount of fine roots produced is likely to alter SOC and its distribution.

Proper grazing management and use of grazing systems are important for decreasing the extent and spread of invasive plants and increasing SOC storage. Grazing management actions for invasive species mitigation can involve holistic approaches such as minimizing disturbance, reducing hoof action to protect soil surface and avoid release of excess nutrients for invasive plants to exploit, rotating livestock, and cleaning equipment between pastures. Mitigating for nonnative species simultaneously improves soil health, decreases soil disturbance, and promotes increased SOC storage through a variety of processes. Belowground and aboveground net primary productivity are increased as the mass of roots and green leafy matter increases. Natural ecosystem functions such as fire are allowed to resume. Rather than manage for homogeneous forage grasses, managers can promote heterogeneous native vegetation. Minimizing overgrazing and breakup of biological soil crusts helps to decrease the frequency of large flushes of nutrients, which invasive plants find easy to exploit; seedbanks of nonnative species are thus suppressed.

Water

Serving over 66 million people and 3400 communities, lands managed by the USDA Forest Service (hereafter, Forest Service) are the single largest supplier of municipal water in the country (USDA FS n.d.). Beyond providing valuable drinking water, freshwater resources provide local communities a host of other related benefits such as recreation and tourism. Forests play a vital role in regulating water and energy cycles (Ellison et al. 2017). Long-term drought and decreased winter snowpack in the Southwest coupled with elevated temperatures are having substantial impacts on water quantity in southwestern forest ecosystems. Wetlands and riparian areas that provide key wildlife habitat are drying up. Long-term soil moisture deficits may lead to the increased spread of invasive plant species and replacement of native species (Archer and Predick 2008).

Other Management Concerns

Land and resource managers in the Southwest continue to be challenged with persistent drought and elevated tempera-

tures, which have led to accelerated desertification, increased fire risk, and the rapid spread of invasive species. Arid and semiarid intermountain ecosystems may be more at risk for climate change impacts because of the decreased inputs of water and increased risks for wildfire (Chambers and Pellant 2008). Additional research is needed to understand how woodland and shrubland plant composition and diversity have been altered by land use and to inform development of appropriate restoration strategies. Managers have sought to combat these issues by managing for increased ecosystem resistance and resilience. They use adaptive management strategies that rely on long-term monitoring to identify successful approaches and emerging issues (McCollum et al. 2017).

Increasing desertification is a growing threat due to accelerated wind and water erosion, habitat loss and fragmentation, and poor grazing management (Sivakumar 2006). Woody plant encroachment and loss of native grasses resulting from changes in moisture, overgrazing, and decreased wildfire are altering landscapes at a rapid pace (Brunelle et al. 2014).

A lack of wildfire in the region due to fire suppression efforts has led to changes in forest composition and structure (e.g., increased density of shade-tolerant and fire-prone species), as well as an increase in wildfire size, severity, and intensity—and risk (Reynolds et al. 2013). Wildfire in the Southwest and Great Plains (that part of the Great Plains region within the Rockies) has made it difficult to predict the spatial extent of water repellent soils (those that are hydrophobic in response to wildfire) (Zvirzdin et al. 2017). A better understanding of where they might be on the landscape would help guide postfire amelioration and rehabilitation efforts to minimize erosion (Zvirzdin et al. 2017).

The rapid spread and establishment of animal, insect, and plant invasive species is a perennial management concern and threat to natural resources in the Southwest. Invasive species cause both economic and environmental harm when native species are removed from ecosystems (Lilleskov et al. 2010). Invasive species can severely alter belowground soil nutrient dynamics and aboveground soil surface erosion rates leading to long-term soil degradation (Lilleskov et al. 2010). Insect and disease outbreaks have wide-ranging implications for long-term soil quality and health such as bare ground and SOC loss. Increased soil moisture and N concentrations have been observed in forests affected by mountain pine beetle (*Dendroctonus ponderosae*) outbreaks (Clow et al. 2011). In addition, surface leaf litter decay has increased due to higher air temperatures, leading to higher total N and phosphorus concentrations in aquatic habitats (Clow et al. 2011).

Many riparian areas in the Southwest need stream channel and vegetative restoration because of factors such as overgrazing, logging, invasive species, and fire (Dwire and Kauffman 2003). Cost-effective, rapid assessment tools are needed to identify the status of habitats for restoration or protection.

Legacy anthropogenic ecosystem impacts from poor road construction and lack of maintenance pose substantial challenges. The thousands of abandoned mine lands that have no chemical toxicity problems could be restored. Though many areas can be returned to some level of productivity rather quickly (e.g., by using soil amendments), more research is needed to determine the best soil restoration or soil-building methods and plant species mixes for these sites. Further, increased energy exploration and development have caused soil degradation. Whether wind, gas, oil, or solar, all forms of mass energy production have impacts on the landscape, with concomitant effects on soil health.

Mapping

The Southwest has a rich history of Terrestrial Ecological Unit Inventory (formerly known as Terrestrial Ecosystem Survey) and participation in the USDA Natural Resources Conservation Service's (NRCS's) National Cooperative Soil Survey. In Arizona and New Mexico, the Forest Service maintains a large regional Terrestrial Ecological Unit Inventory mapping program that has almost finished mapping all Forest Service lands. However, sizable proportions of federal lands outside of California and southwest Colorado lack any modern NRCS Soil Survey Geographic (SSURGO) soil survey. Several cooperative efforts between the Forest Service and NRCS over the last decade have begun filling gaps in SSURGO coverage on Forest Service lands. Interagency cooperative agreements such as these continue to be a valuable mechanism for forest managers to obtain soil resource information and interpretations important for land management activities.

Key Findings

- Forest soil loss increases by orders of magnitude following wildfires and especially in high-severity burned areas. As wildfire size and severity continue to grow, forest soils are eroding at historically peak rates and sediments are collecting in reservoirs. There has been no regional-scale assessment of postfire soil erosion risk.
- Erosion by wind, water, and gravity is prevalent at multiple spatial scales and is driven by both well-known and novel disturbances that are becoming less predictable. The loss of soil vegetative cover coupled with drought, other extreme climatic events, and uncharacteristic disturbances (e.g., wildfire, energy development) is a leading cause of accelerated erosional processes that jeopardize resource sustainability.
- Sheet and rill erosion are recognized, understood, and modeled. Gullies and larger fluvial eroded waterways are growing and responding more dramatically to upslope disturbances. These processes are less understood and documented.

- Dust in the form of fugitive dust (particulates not emitted from a vent or stack) through wind erosion has become a substantial, highly acknowledged stressor in the western United States. Its impact on human health and safety, in addition to environmental effects (e.g., dust on snow), is rapidly becoming a critical focus area for urban and rural communities. Spatial data and better forecasting mechanisms are needed.
- Debris flows affecting multiple watersheds are becoming more evident following large-scale disturbances. The debris is not only establishing new geomorphic benchmarks but is also impacting infrastructure below. More training and better documentation of occurrences can inform new approaches to analyzing landscapes vulnerable to debris flows before disturbances take place.
- The pressures and dynamics associated with uncontrolled mechanical impacts are soil stressors that lead to compaction, sealing, displacement, and loss of nutrients. Operation of mechanized equipment (e.g., mastication) has improved as effects on soil are better understood. Local training is needed, however, to understand the level of controlled impacts that ecosystems can resist and to minimize uncontrollable effects.
- Burning influences soil biological diversity and SOC, which are essential to site productivity. A secondary effect is soil hydrophobicity, or water repellency, which promotes soil erosion. Research results are available to optimize implementation of prescribed fire for the desired biodiversity and SOC outcomes. But incident burnouts, backfires, and treatment of slash piles can pose operational problems. Risk management needs to incorporate the predicted and acceptable environmental effects in addition to human safety into planning and implementation.

Literature Cited

Archer SR, Predick KI (2008) Climate change and ecosystems of the southwestern United States. Rangelands 30(3):23–28

Baron JS, Rueth HM, Wolfe AM et al (2000) Ecosystem responses to nitrogen deposition in the Colorado Front Range. Ecosystems 3(4):352–368

Boot CM, Hall EK, Denef K, Baron JS (2016) Long-term reactive nitrogen loading alters soil carbon and microbial community properties in a subalpine forest ecosystem. Soil Biol Biochem 92:211–220

Bradford JB, Bell DM (2017) A window of opportunity for climate-change adaptation: easing tree mortality by reducing forest basal area. Front Ecol Environ 15(1):11–17

Brunelle A, Minckley TA, Delgadillo J, Blissett S (2014) A long-term perspective on woody plant encroachment in the desert southwest, New Mexico, USA. J Veg Sci 25(3):829–838

Chambers JC, Pellant M (2008) Climate change impacts on northwestern and intermountain United States range-lands. Rangelands 30(3):29–33

Clow DW, Rhoades CC, Briggs J et al (2011) Responses of soil and water chemistry to mountain pine beetle induced tree mortality in Grand County, Colorado, USA, Appl Geochem 26:S174–S178

Dwire KA, Kauffman JB (2003) Fire and riparian ecosystems in landscapes of the western USA. For Ecol Manag 178:61–74

Ellison D, Morris CE, Locatelli B et al (2017) Trees, forests and water: cool insights for a hot world. Glob Environ Chang 43:51–61

Hicke JA, Meddens AJH, Kolden CA (2016) Recent tree mortality in the western United states from bark beetles and forest fires. For Sci 62(2):141–153

Lilleskov E, Callaham MA Jr, Pouyat R et al (2010) Invasive soil organisms and their effects on belowground pro-cesses. In: Dix ME, Britton K (eds) A dynamic invasive species research vision: opportunities and priorities 2009–29. Gen. Tech. Rep. WO-79/83. U.S. Department of Agriculture, Forest Service, Research and Development, Washington, DC, pp 67–83

McCollum DW, Tanaka JA, Morgan JA et al (2017) Climate change effects on rangelands and rangeland management: affirming the need for monitoring. Ecosyst Health Sustain 3(3):e01264

Mote PW, Hamlet AF, Clark MP, Lettenmaier DP (2005) Declining mountain snowpack in western North America. Bull Am Meteorol Soc 86:39–49

Reynolds RT, Sánchez Meador AJ, Youtz JA et al (2013) Restoring composition and structure in Southwestern fre-quent-fire forests: a science-based framework for improv-ing ecosystem resiliency. Gen. Tech. Rep. RMRS-GTR-310. U.S. Department of Agriculture, Forest Service, Rocky Mountain Research Station, Fort Collins, 76 p

Sivakumar MVK (2006) Dissemination and communication of agrometeorological information—global perspectives. Meteorological Applications, 13(S1): 21-30

USDA Forest Service [USDA FS] (n.d.) Water facts. https://www.fs.fed.us/managing-land/national-forests-grass-lands/water-facts. Accessed 11 June 2018

Westerling AL (2016) Increasing western US forest wildfire activity: sensitivity to changes in the timing of spring. Philos Trans R Soc Lond Ser B Biol Sci 371(1696):20150178

Williams AP, Allen CD, Macalady AK et al (2012) Temperature as a potent driver of regional forest drought stress and tree mortality. Nat Clim Chang 3(3):292–297

Zvirzdin DL, Roundy BA, Barney NS et al (2017) Postfire soil water repellency in piñon-juniper woodlands: extent, severity, and thickness relative to ecological site charac-teristics and climate. Ecol Evol 7(13):4630–4639

Northwest

Deborah S. Page-Dumroese, Brian Gardner, Paul McDaniel, and Steve Campbell

Introduction

"Essentially, all life depends upon the soil … There can be no life without soil and no soil without life; they have evolved together." —Charles E. Kellogg (USDA 1938, p. 864).

Forests in the Northwest region (Washington, Oregon, and Idaho) represent a complex array of composition and structure. These complexities indicate the diverse range of soils and the physical environment which affect soil and plant interactions. In this region the textural classes range from gravel to heavy clay, many are skeletal (>35% rocks), and many are derived from volcanic materials (Isaac and Hopkins 1937). Today, these forests and soils are being impacted by warmer, drier years often associated with El Niño or the Pacific Decadal Oscillation (or both), below-average snowpack and streamflow, below-average forest growth, and an above-average risk of fire (Mote et al. 2005).

This regional summary describes the two distinct areas of the northwestern United States: the Pacific (coastal) Northwest and the Inland Northwest. It also provides an overview of some of the management challenges.

Description of the Region

The Northwest is characterized by a diversity of soil, cli-matic regimes, parent material, surficial deposits, and vege-tation. Land uses and opportunities for soil conservation and ecosystem services from soil vary throughout this area. Proximity to the Pacific Ocean and large elevational gradi-ents throughout the region contribute to unique soil and veg-etation relationships. The region contains two distinct areas: (1) the wet, ocean-moderated coastal zone on the west side of the Cascade Mountains and (2) the Inland Northwest zone, which includes drier forests with long summer drought periods (Fig. A14).

The Pacific Northwest

The coastal forests of the Pacific Northwest are highly pro-ductive, a result of favorable climate, geology, and soil. The dominant commercial tree species is Douglas-fir (*Pseudotsuga menziesii*). Old-growth forests are dominated by Sitka spruce (*Picea sitchensis*) along the exceedingly wet coastline, by western hemlock (*Tsuga heterophylla*) and western redcedar (*Thuja plicata*) at mid-elevations, and by Engelmann spruce (*Picea engelmannii*), true firs (*Abies* spp.), and mountain hemlock (*Tsuga mertensiana*) at higher

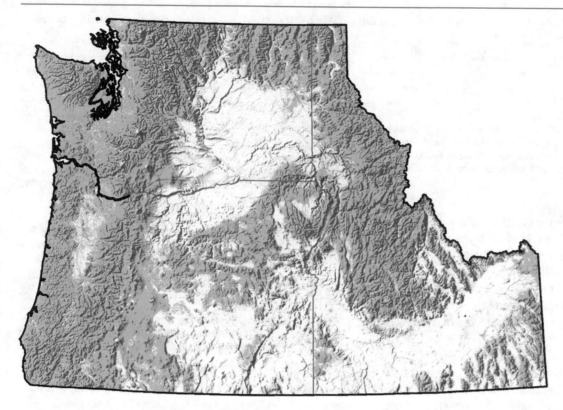

Fig. A14 Map of the Pacific Northwest and Inland Northwest

elevations. These forests grow primarily on steep mountainous areas, which might suggest young, thin soils (Waring and Franklin 1979); however, the soils are at least as deep and fertile as other temperate forests. This area experiences wet, mild winters, and warm, dry summers (Shumway 1979), which allow many conifers to flourish.

The dominant soils of the Cascade Range and Olympic Mountains are Andisols and Spodosols (Bockheim 2017). Almost 35% of the soil particle-size families are ashy or medial (Table A3), reflecting the influence of volcanic ash parent materials and the weathering processes associated with andic soils. Spodosols occur at the higher elevations, which have a greater amount of effective precipitation. On the drier east side of the Cascades, Inceptisols are extensive (Southard et al. 2017).

Soils in the Pacific Northwest include all textural classes—from heavy clay to gravel—but the most common textural class is loam. In the southern part of the coastal area, soils are primarily derived from basalt (Mangum 1913), and in the northern portion, many soils are derived from sedimentary gravels, sands, and silts from past glacial activity (Isaac and Hopkins 1937). In addition, many soils of this region have a high coarse-fragment content, which has been shown to have a strong negative effect on tree growth (Table A4). Skeletal soils have at least 35% rock-fragment content on a volume basis, contributing to reduced water holding capacity and lower rooting volume compared to nonskeletal soils.

Table A3 Relative abundance of soil particle-size classes in the Pacific (coastal) Northwest (Soil Survey Staff 2017, 2018)

Soil textural class[a]	Abundance (%)
Ashy or medial	34.9
Loamy skeletal	23.5
Loamy or silty	21.4
Clayey, fine, or very fine	15.3
Sandy or sandy skeletal	2.6
Cindery or pumiceous	2.1

[a]Textural classes are derived from the textural family either in its entirety or from the first element of multipart textural families. Approximately 720 000 ha were evaluated, and no data were excluded

Table A4 Relative abundance of soils with or without a skeletal feature in textural control section in the Pacific (coastal) Northwest (Soil Survey Staff 2017, 2018)

Soil feature	Abundance (%)
All skeletal[a]	43.8
All nonskeletal	56.2

[a]Skeletal soils have >35% rock fragments in some part of the particle size control section (1 m depth). Approximately 720 000 ha were evaluated, and no data were excluded

On the west side of the Cascades, the soils are primarily porous and most precipitation infiltrates rapidly as long as the site has not been impacted by compaction, rutting, or displacement of the forest floor (Heninger et al. 2011). More

Table A5 Occurrence of soil depth classes in the Pacific (coastal) Northwest (Soil Survey Staff 2017, 2018)

Soil feature[a]	Abundance (%)
Very deep to restriction	43.7
Deep to restriction	17.9
Moderately deep to restriction	33.5
Very shallow and shallow	1.8

[a]Depth classes are very shallow and shallow <50 cm, moderately deep 50–99 cm, deep 100–150 cm, very deep >150 cm. Approximately 720 000 ha were evaluated, and no data were excluded. Percentages do not add up to 100 because of the occurrence of rock outcrops or water features

than 95% of the soils in the Pacific Northwest are moderately deep to very deep to any kind of layer that restricts root growth or the movement of water (Table A5).

Because rapid regrowth generally occurs on the coastal Northwest, there is very little bare soil after management activities. Therefore, the dominant erosion process here is landslides, which move large masses of the soil mantle downslope.

The Inland Northwest

Inland Northwest soils are much drier because of the rain shadow effect of the Cascades. The bulk of the precipitation is dropped by westerlies as they approach the mountain crests along the eastern boundaries. Although the mountains of the Okanogan Highlands in northeastern Washington rise as high as the Selkirk and Cabinet Mountains, which lie successively east of them, they are less massive and considerably drier. In the Okanogan Highlands, forests and soils are typically xerophytic, even near the mountain summits. In the Selkirk and Cabinet, however, typical vegetation is mesophytic to the valley floor (Daubenmire and Daubenmire 1968). Parent materials consist of igneous, sedimentary, and metamorphic rocks. Glacial till and outwash deposits are common north of the Columbia River and Spokane-Rathdrum Prairies. An extensive mantle of mixed volcanic ash and loess overlies the local tills, colluvium, and residuum.

The dominant forest soils of the Inland Northwest are Andisols, Inceptisols, and Alfisols. Andisols and andic subgroups of Inceptisols and Alfisols account for 37% of the particle-size classes in the region (Table A6), underscoring the importance of volcanic ash as a parent material. Spodosols are found at higher elevations in northeastern Washington, northern Idaho, and northwestern Montana, where colder temperatures, greater snowpack, and abundant organic compounds derived from coniferous litter promote the podzolization process (Valerio et al. 2016). Mollisols are transitional soils associated with drier forest communities.

Geologic history and climate have led to strong mechanical weathering of the mountain slopes where most forest lands occur. This has produced extensive areas of soils high in rock fragments, especially at higher elevations (Table A7). Deeply weathered soils with fewer fragments do occur, how-

Table A6 Relative abundance of soil particle-size classes in the Inland Northwest (Soil Survey Staff 2017, 2018)

Soil texture[a]	Abundance (%)
Ashy or medial	37.1
Loamy skeletal	29.9
Loamy or silty	19.0
Clayey, fine, or very fine	9.0
Sandy or sandy skeletal	4.9
Cindery or pumiceous	0.1

[a]Textural classes are derived from the textural family either in its entirety or from the first element of multipart textural families. Approximately 11 600 000 ha were evaluated, and no data were excluded

Table A7 Relative abundance of soils with or without a skeletal feature in textural control section in the Inland Northwest (Soil Survey Staff 2017, 2018)

Soil feature[a]	Abundance (%)
All skeletal	53.7
All nonskeletal	46.3

[a]Skeletal soils have >35% rock fragments in some part of the particle size control section (1 m depth). Approximately 11 600 000 ha were evaluated, and no data were excluded.

ever. These are most common in the unglaciated low mountains and foothills south of Lake Coeur D'Alene and north of the Idaho Batholith.

Skeletal soils are abundant across the region, accounting for more than half of all soils in the Inland Northwest (Table A7). Because of this, the mantle of volcanic ash greatly enhances the plant-available water holding capacity of many forest soils. A 40-cm-thick layer of weathered volcanic ash can typically provide 12–16 cm of plant available water, 2–3 times as much as underlying horizons (McDaniel et al. 2005). This difference can be critical to forest community composition and productivity in the seasonally dry landscapes of the region. Although most soils of the Inland Northwest are deep or very deep (Table A8), many of the benefits associated with deep soils may be offset by the abundance of skeletal soils. Rock fragments within skeletal soils occupy a significant volume of mineral soil, making less room for root growth and moisture storage.

Impacts on the Soil Resource

Harvesting and Site Preparation

Stand harvesting and site preparation alter the inputs of soil organic matter into the soil. Soil organic matter is critical for maintaining site productivity because of its role in supporting nutrient availability, gas exchange, and water supply (Blake and Ruark 1992; Henderson 1995; Jurgensen et al. 1997; Powers et al. 1990). Organic matter is also essential to microflora and macrofauna active in nutrient cycling, soil aggregation, and disease incidence or prevention (Harvey et al. 1987). Recent changes in land management now com-

Table A8 Occurrence of soil depth classes in the Inland Northwest (Soil Survey Staff 2017, 2018)

Soil feature[a]	Abundance (%)
Very deep to restriction	46.7
Deep to restriction	17.3
Moderately deep to restriction	26.0
Very shallow and shallow	10.0

[a]Depth classes are very shallow and shallow <50 cm, moderately deep 50–99 cm, deep 100–150 cm, very deep >150 cm. Approximately 11 600 000 ha were evaluated, and no data were excluded.

monly favor restoration of fire regimes or watershed health and precommercial thinnings to restore ecological function. Loss of soil organic matter from periodic stand disturbances (e.g., repeated thinning) could have negligible short-term impacts (Sanchez et al. 2006) or more significant impacts, depending on soil type, tree species, ecosystem, or climatic regime (Grigal and Vance 2000; Henderson 1995). Removal of logging slash for bioenergy, rather than leaving the harvest residues on site, can change nutrient availability (Sinclair 1992), soil temperature, water availability, and biological activity (Covington 1981; Harvey et al. 1976). Conversely, excess biomass left after stand harvesting or thinning may negatively impact soil quality by providing fuel for uncharacteristically severe wildfire (Page-Dumroese et al. 2010) or impair natural regeneration potential (Jurgensen et al. 1997). Traditional harvest methods (e.g., feller bunchers, harvesters, skyline logging, forwarders), when used with care, can cause less than 10% soil displacement and compaction (McIver et al. 2003). Newer logging systems (e.g., tethered logging) have yet to be tested for their soil impacts. For example, tethered logging has been used in New Zealand on steep terrain (>40% slope) and is now being used in a few locations in the Pacific Northwest and Inland Northwest. It is unclear how much compaction or soil displacement may be generated from this type of logging and how that may alter site productivity.

Fire

Current forest structure in the Northwest is a result of a successful fire suppression program that began after the 1910 fires (Pyne 1982). Since that time, vegetation and surface fuels have been slowly accumulating, and we are faced with many ecological issues associated with overstocked forests (Keane 2008; Keane et al. 2002). When wildfires burn today, they do so with greater intensity and severity during a longer fire season (Jolly et al. 2015). Wildfires have the potential to shift lifeform dominance across large areas (Moreira et al. 2011; Odion et al. 2010), which can result in shifting microbial populations, carbon stocks, and nutrients. Soil carbon is critical for maintaining soil physical properties, supporting hydrologic function, providing clean water, and circulating greenhouse gases, but these are at risk after high-severity wildfires or prescribed burns (Busse et al. 2014). In addition

to affecting soil carbon and other nutrients, wildfire impacts the soil fungal community structure. The impacts are mixed and are dependent on site or fire severity (or both). Usually, there are greater impacts in the surface horizons than deeper in the profile, and they are more pronounced in areas of frequent burning (Cairney and Bastias 2007). These mixed results of wildfire or prescribed burning underscore the importance of understanding individual soil and vegetation relationships as well as changes at the landscape scale.

Although wildfire is a concern on many Pacific and Inland Northwest sites, another management concern is the threat of a reburn, where dead trees remain and eventually fall to create "jackstraw" conditions. These intersecting points of downed logs can create hotspots when the site reburns (Passovoy and Fulé 2006). Reburns can be difficult to suppress, have the potential to kill any regeneration (Kemp 1967), and can impact soil carbon stocks (Page-Dumroese et al. 2015). For example, in the Scapegoat Wilderness in northwestern Montana, soil carbon in the 0–30 cm depth declined 9–16%; the change depended on the number of reburns and time since the last fire (Page-Dumroese et al. 2015).

It is important to understand the management options for areas with many dead or dying trees. One method to lower the fuel load is prescribed burning. This tool can manipulate the amount of woody residues on the soil surface and can control the amount of damage to plants and soil (Brown et al. 2003). Salvage logging is another option on sites with numerous snags. This method can reduce the amount of coarse wood on the soil surface and reduce insect populations (Simon et al. 1994). However, Page-Dumroese et al. (2006) note that salvage logging can produce a considerable amount of soil disturbance (>15% of the harvest unit) unless care is taken to limit logging to winter or dry soils and equipment is matched to site conditions.

Climate Change

Changes in twenty-first-century climate are projected to increase wildfire in many ecosystems (Flannigan et al. 2009), which may adversely affect the terrestrial carbon sink and cascade to affect soil biodiversity. Both the Pacific Northwest and Inland Northwest have many ecosystems that are tightly coupled with fire. These ecosystems have been sensitive to past climatic shifts because of their strong seasonal warm-wet winters and hot-dry summers. These seasonal conditions are projected to become amplified in most climate models (Mote and Salathé 2010; Rogers et al. 2011). Currently, the global soil carbon (C) pool is 3.3 times that of the atmospheric pool (Lal 2004). It is predicted that semiarid biomes in the Pacific and Inland Northwest could have increases in biomass production and C sequestration despite changing climate and fire regimes because of increased productivity during nonsummer months (Rogers et al. 2011). However, forests in the Pacific Northwest may become more vulnerable to summer drought and fire, resulting in less soil C (Rogers et al. 2011).

Another consequence of a changing climate is the decline in winter snowpack and snow water equivalent, particularly below 1800 m elevation (Mote 2003). Seasonal snow cover has decreased by 7% between 1973 and 1998 and is correlated with increasing air temperatures (Edwards et al. 2007). Reduced snow cover leads to a decline in soil C stocks (Neilson et al. 2001); nitrogen (N) leaching from increased snowmelt (Hazlett et al. 1992); increased freeze-thaw cycles, which result in more damaged roots (Weih and Karlsson 2002); and changing soil biota (Sjursen et al. 2005).

Soil Restoration

There are many methods for reducing the risk of wildfire on forest and range wildlands. These methods include restoration activities such as prescribed fire and mechanical practices (e.g., thinning, mastication, or chipping). They are used to reduce the wildfire hazard, limit the spread of fire, protect life and property, and restore forest or rangeland soil and site health (Busse et al. 2014). Each site treatment method is not without soil impacts, however, and managers will need to determine what ecosystem services are most important to maintain and then monitor the effectiveness of these treatments (Page-Dumroese et al. 2009). For example, restoration fire and thinning in ponderosa pine (*Pinus ponderosa*) resulted in similar litter depths and mycorrhizal fungi after 15 years as compared to the unharvested control, indicating that some forest sites are resilient (Hart et al. 2018). This resilience provides management flexibility to restore forest sites. However, repeated stand entries may produce a considerable amount of soil disturbance or compaction; the degree and extent of this may impact microbial and vegetative biodiversity, C sequestration, water quantity or quality, and nutrient cycling (Page-Dumroese et al. 2010). Some ecosystem services may not be compatible with restoration activities and therefore managers will have to balance maintaining soil productivity with managing for other site properties or desired outcomes (Burger et al. 2010).

The increased use of prescribed fire for restoration of rangeland sites may increase runoff and erosion in both the short term and the long term, leading to damage to soil and water resources. Increased runoff and erosion are a result of the loss of ground cover, aggregate stability, and soil roughness, and an increase in soil hydrophobicity (water repellency) (Pierson et al. 2011). Among long-term responses to burning in sagebrush-grassland ecosystems was a rebound of microbial community structure to preburn levels after 7 years (Dangi et al. 2010). Frequent rangeland fires, fuel accumulation from invasive species (e.g., cheatgrass; *Bromus tectorum*), and postfire annual weed dominance coupled with a low decomposition rate of plant materials (Knapp 1996) promote the alien grass-fire cycle (Brooks et al. 2004) that perpetuates nonnative species and can often lead to intensifying efforts to restore belowground processes.

Problems and Limitations

Management of forests and rangelands in the Pacific Northwest and Inland Northwest will continue to get more complex as rural communities seek to harvest timber or conduct restoration activities while conservationists push to protect wildlands. These issues are exacerbated when combined with a changing climate. Many policies, Federal agencies, and Congress have noted a forest health crisis in the Pacific and Inland Northwest that is arising mostly in overstory vegetation. Impaired health is usually related to dense overstories caused by fire suppression, numerous dead trees caused by drought or insect and disease outbreaks, or even-aged stand compositions that lack resistance to insect and disease invasions (DellaSalla et al. 1995; O'Laughlin et al. 1993).

Current solutions for these problems are to increase harvesting by thinning stands or salvage logging dead trees. However, there is a great need to expand the scope of the discussions about forest health to include whole ecosystem processes that involve changes to belowground properties as they relate to management and restoration goals. Understanding the role of forest and rangeland soil properties in hydrologic function, nutrient cycling, C sequestration, and resilience to climate change is critical to both ongoing harvesting and restoration activities. In areas where forest productivity has declined, it is often because of soil displacement, compaction, puddling (smearing of surface pores), and losses of organic matter (Heninger et al. 1997). In addition, taking land out of production due to roads and harvest landings contributes to the decline in forest productivity potential.

Controlling the amount of disturbance relies on integrated decision-making processes that guide harvest methods. This process requires monitoring to determine soil disturbance after management and relating that information to best management practices and operational ratings (Heninger et al. 1997; Page-Dumroese et al. 2010). Greater collaboration to remove barriers among agencies, universities, scientists, and land managers is critical for continuing to move forward on training efforts, monitoring protocols, and joint research efforts to restore or maintain soil processes and hydrologic function.

Key Findings

- Since the early twentieth century, wildfires in the Northwest have burned with greater intensity and severity. These catastrophic fires may shift forest community composition and structure, which in turn can lead to shifts in C stocks, soil nutrient availability, and microbial populations. There is a need for documentation of wildfire effects on belowground processes and microbial community structure and function.

- Vegetation and surface fuels have been slowly accumulating in the Northwest during the last century, increasing the risk of wildfire and reburns. We need long-term data on the impacts of fuel reduction treatments in low-elevation semiarid forests, including the impacts of using biochar, biosolids, or mastication to build surface and mineral soil organic matter.

- Forests in the Pacific Northwest are vulnerable to summer drought and fire. Areas undergoing aridification (long-term drought) need to be identified and monitored so that we can define the soil changes that lead to vegetation shifts.

- Ongoing harvesting and restoration activities affect forest and rangeland soil properties that impact C sequestration. Frequent rangeland fires, fuel accumulation from invasive plant species, and postfire annual weed dominance contribute to a cycle that perpetuates nonnative species and may require intensified efforts to restore belowground processes. We need a better understanding of how to protect and promote forest and rangeland soil and vegetation productivity to encourage C sequestration and limit invasive species.

- Careful use (e.g., logging on dry soils, limited off-trail travel) of traditional harvest methods such as feller bunchers and forwarders can have limited soil impacts, but data are needed on the impacts of new logging systems (e.g., tethered logging) on soil productivity, C sequestration, and stand sustainability.

- Land management is increasingly focusing on restoration of ecological function. We need to determine the long-term impacts of forest, rangeland, riparian area, and mine site development, as well as energy exploration and development, on our ability to restore soil processes. Applications of this research include helping managers (1) determine how a changing climate impacts restoration objectives, (2) use monitoring techniques to ensure objectives are met, and (3) understand whether and how ecosystems are progressing toward the desired restoration goal (active vs. passive restoration).

Literature Cited

Blake JI, Ruark GA (1992) Soil organic matter as a measure of forest productivity: some critical questions. In: Proceedings of the soil quality standards symposium. Soil Science Society of America meeting. WO-WSA-2. U.S. Department of Agriculture, Forest Service, Watershed and Air Management, Washington, DC, pp 28–40

Bockheim JG (2017) Soils of the USA: the broad perspective. In: West L, Singer M, Hartemink A (eds) The soils of the USA, World soils book series. Springer International Publishing, Cham, pp 43–75

Brooks ML, D'Antonio CM, Richardson DM et al (2004) Effects of invasive alien plants on fire regimes. Bioscience 54(7):677–688

Brown JK, Reinhardt ED, Kramer KA (2003) Coarse woody debris: managing benefits and fire hazard in the recovering forest. Gen. Tech. Rep. RMRS-GTR-105. U.S. Department of Agriculture, Forest Service, Rocky Mountain Research Station, Ogden, 16 p

Burger JA, Gray G, Scott DA (2010) Using soil quality indicators for monitoring sustainable forest management. Proceedings RMRS-P-59. U.S. Department of Agriculture, Forest Service, Rocky Mountain Research Station, Fort Collins, pp 12–31

Busse MD, Hubbert KR, Moghaddas EEY (2014) Fuel reduction practices and their effects on soil quality. Gen. Tech. Rep. PSW-GTR-241. U.S. Department of Agriculture, Forest Service, Pacific Southwest Research Station, Albany, 156 p

Cairney JWG, Bastias BA (2007) Influences of fire on forest soil fungal communities. Can J For Res 37:207–215

Covington WW (1981) Changes in forest floor organic matter and nutrient content following clear cutting in northern hardwoods. Ecology 62:41–48

Dangi SR, Stahl PD, Pendall E et al (2010) Recovery of soil microbial community structure after fire in a sagebrush-grassland ecosystem. Land Degrad Dev 21(5):423–432

Daubenmire R, Daubenmire JB (1968) Forest vegetation of eastern Washington and northern Idaho. Tech. Bull. 60. Washington State University, Washington Agricultural Experiment Station, Pullman, 106 p

DellaSalla DA, Olson DM, Barth SE et al (1995) Forest health: moving beyond rhetoric to restore healthy landscapes in the Inland Northwest. Wildl Soc Bull 23:346–356

Edwards AC, Scalenghe R, Freppaz M (2007) Changes in the seasonal snow cover of alpine regions and its effect on soil processes: review. Quat Int 162-163:172–181

Flannigan MD, Krawchuk MA, deGroot WJ et al (2009) Implications of changing climate for global wildland fire. Int J Wildland Fire 18(5):483–507

Grigal DF, Vance ED (2000) Influence of soil organic matter on forest productivity. N Z J For 30:169–205

Hart BTN, Smith JE, Luoma DL, Hatten JA (2018) Recovery of ectomycorrhizal fungus communities fifteen years after fuels reduction treatments in ponderosa pine forest of the Blue Mountains, Oregon. For Ecol Manag 422:11–22

Harvey AE, Jurgensen MF, Larsen MJ (1976) Intensive fiber utilization and prescribed fire: effects on the microbial ecology of forests. Gen. Tech. Rep. INT-GTR-28. U.S. Department of Agriculture, Forest Service, Intermountain Research Station, Ogden, 46 p

Harvey AE, Jurgensen MF, Larsen MJ, Graham RT (1987) Relationships among soil microsite, ectomycorrhizae, and natural conifer regeneration of old-growth forests in western Montana. Can J For Res 17:58–62

Hazlett PW, English MC, Foster NW (1992) Ion enrichment of snowmelt water by processes within a podzolic soil. J Environ Qual 21:102–109

Henderson GS (1995) Soil organic matter: a link between forest management and productivity. In: McFee WW, Kelley JM (eds) Carbon forms and functions in forest soils. Soil Science Society of America, Madison, pp 419–535

Heninger RL, Terry TA, Dobkowski A, Scott W (1997) Managing for sustainable site productivity: Weyerhaeuser's forestry perspective. Biomass Bioenergy 13(4–5):255–267

Heninger RL, Scott W, Dobkowski A, Terry TA (2011) Managing soil disturbance. In: Angima SD, Terry TA (eds) Best management practices for maintaining soil productivity in the Douglas-fir region. Oregon State University, Extension Service, Corvallis, pp 18–32. Chapter 5

Isaac LA, Hopkins HG (1937) The forest soils of the Douglas-fir region, and the changes wrought upon it by logging and slash burning. Ecology 18:264–279

Jolly WM, Cochrane MA, Freeborn PH et al (2015) Climate-induced variations in global wildfire danger from 1979 to 2013. Nat Commun 6:7537

Jurgensen MF, Harvey AE, Graham RT et al (1997) Impacts of timber harvesting on soil organic matter, nitrogen, productivity, and health of Inland Northwest forests. For Sci 43(2):234–251

Keane RE (2008) Biophysical controls on surface fuel litterfall and decomposition in the northern Rocky Mountains, USA. Can J For Res 38:1431–1445

Keane RE, Ryan KC, Veblen TT et al (2002) Cascading effects of fire exclusion in the Rocky Mountain ecosystems: a literature review. Gen. Tech. Rep. RMRS-GTR-91. U.S. Department of Agriculture, Forest Service, Rocky Mountain Research Station, Fort Collins, 24 p

Kemp JL (1967) Epitaph for the giants: the story of the Tillamook Burn. Touchstone Press, Portland

Knapp PA (1996) Cheatgrass (Bromus tectorum L.) dominance in the Great Basin desert. Glob Environ Chang 6:37–52

Lal R (2004) Soil carbon sequestration impacts on global climate change and food security. Science 11:1623–1627

Mangum AW (1913) Reconnaissance survey of southwestern Washington. Advance sheets— field operations of the Bureau of Soils, 1911. U.S. Department of Agriculture, Washington, DC, 137 p

McDaniel PA, Wilson MA, Burt R et al (2005) Andic soils of the Inland Pacific Northwest, USA: properties and ecological significance, Soil Sci 170:300–311

McIver JD, Adams PW, Doyal JA et al (2003) Environmental effects and economics of mechanized logging for fuel reduction in northeastern Oregon mixed-conifer stands. West J Appl For 18(4):238–249

Moreira F, Viedma O, Arianoutsou M et al (2011) Landscape–wildfire interactions in southern Europe: implications for management. J Environ Manag 92(10):2389–2402

Mote PW (2003) Trends in snow water equivalent in the Pacific Northwest and their climatic causes. Geophys Res Lett 30(12):1601

Mote PW, Salathé EP (2010) Future climate in the Pacific Northwest. Climate Change 102:29–50

Neilson CB, Groffman PM, Hamburg SP et al (2001) Freezing effects on carbon and nitrogen cycling in northern hardwood forest soils. Soil Sci Soc Am J 65(6):1723–1730

Odion DC, Moritz MA, DellaSalla DA (2010) Alternative community states maintained by fire in the Klamath Mountains, USA. J Ecol 98:96–105

O'Laughlin J, MacCracken AY, Adams DL et al (1993) Forest health conditions in Idaho. Executive summary. University of Idaho: Wildlife and Range Policy Analysis Group, Moscow, 37 p

Page-Dumroese D, Jurgensen M, Abbott A, et al (2006) Monitoring changes in soil quality from post-fire logging in the Inland Northwest. In Andrews PL, Butler BW (comps) Fuels management—how to measure success: conference proceedings. Proceedings RMRS-P-41. U.S. Department of Agriculture, Forest Service, Rocky Mountain Research Station, Fort Collins, pp 605–614

Page-Dumroese DS, Abbott AM, Rice TM (2009) Forest soil disturbance monitoring protocol. Vol II: supplementary methods, statistics, and data collection. WO-GTR-82b. U.S. Department of Agriculture, Forest Service, Washington, DC, 64 p

Page-Dumroese DS, Jurgensen M, Terry T (2010) Maintaining soil productivity during forest or biomass-to-energy thinning harvests in the western United States. West J Appl For 25:5–12

Page-Dumroese DS, Jain TB, Sandquist JE et al (2015) Reburns and their impact on carbon pools, site productivity, and recovery. In: Potter KM, Conkling BL (eds) Forest Health Monitoring: national status, trends, and analysis 2014. Gen. Tech. Rep. SRS-209. U.S. Department of Agriculture, Forest Service, Southern Research Station, Asheville, pp 143–150

Passovoy MD, Fulé PZ (2006) Snag and woody debris dynamics following severe wildfires in northern Arizona ponderosa pine forests. For Ecol Manag 223:237–246

Pierson FB, Williams CJ, Hardegree SP et al (2011) Fire, plant invasions, and erosion events on western rangelands. Rangel Ecol Manag 64(5):439–449

Powers RF, Alban DH, Miller RE et al (1990) Sustaining site productivity in North American forests: problems and prospects. In: Gessel SP, Lacate DS, Weetman GF, Powers RF (eds) Proceedings of the seventh North American forest soils conference on sustained productivity of forest soils. University of British Columbia, Faculty of Forestry, Vancouver, pp 49–79

Pyne SJ (1982) Fire in America: a cultural history of wildland and rural fire. Princeton University Press, Princeton, 630 p

Rogers BM, Neilson RP, Drapek R et al (2011) Impacts of climate change on fire regimes and carbon stocks of the U.S. Pacific Northwest. J Geophys Res Lett 116:G03037

Sanchez FG, Scott DA, Ludovici KH (2006) Negligible effects of severe organic matter removal and soil compaction on loblolly pine growth over 10 years. For Ecol Manag 227(1–2):145–154

Shumway SE (1979) Climate. In: Heilman PE, Anderson HW, Baumgartner DM (eds) Forest soils of the Douglas-fir region. Washington State University, Cooperative Extension Service, Pullman, pp 87–93

Simon J, Christy S, Vessels J (1994) Clover-mist fire recovery: a forest management response. J For 92:41–44

Sinclair TR (1992) Mineral nutrition and plant growth response to climate change. J Exp Bot 43:1141–1146

Sjursen H, Michelsen A, Holmstrup M (2005) Effects of freeze-thaw cycles on microarthropods and nutrient availability in a sub-arctic soil. Appl Soil Ecol 28:79–93

Soil Survey Staff (2017) Web soil survey. U.S. Department of Agriculture, Natural Resources Conservation Service. Available at https://websoilsurvey.sc.egov.usda.gov/App/HomePage.htm. Accessed April 2018

Soil Survey Staff (2018) Gridded Soil Survey Geographic (gSSURGO) Database for the conterminous United States [FY 2018 official release]. U.S. Department of Agriculture, Natural Resources Conservation Service. Available at https://gdg.sc.egov.usda.gov/. Accessed 7 Apr 2019

Southard SB, Southard RJ, Burns S (2017) Soils of the Pacific Coast Region: LRRs A and C. In: West LT, Singer MJ, Hartemink AE (eds) The soils of the USA, World soils book series. Springer International Publishing, Cham, pp 77–100

U.S. Department of Agriculture (1938) Soils and men. Yearbook of Agriculture 1938, 1249 p

Valerio MW, McDaniel PA, Gessler PE (2016) Distribution and properties of podzolized soils in the northern Rocky Mountains. Soil Sci Soc Am J 80:1308–1316

Waring RH, Franklin JF (1979) Evergreen coniferous forests of the Pacific Northwest. Science 204:1380–1386

Weih M, Karlsson PS (2002) Low winter soil temperature effects summertime nutrient uptake capacity and growth rate of mountain birch seedlings in the subarctic, Swedish Lapland. Arct Antarct Alp Res 34:434–439

Alaska
David V. D'Amore

Introduction
Alaska is by far the largest state in the United States but is often discussed within the same context as other states in reviews of natural resources. Regardless of the size of a state or region, it is a difficult task to disseminate all available information that will adequately address many aspects of a soil resource evaluation in one overview. This summary provides key references to soil resource inventory, soil vulnerability, and basic soil landscape information for obtaining additional soils information needed for specific uses. An overview of Alaska soil resources is presented within the context of several major functional groupings including permafrost, fire, and landscape position.

Current Condition of Alaska Soils
The areal coverage and diversity of soils in Alaska are extremely large. Alaska soils represent 7 of the 12 recognized soil orders (Ping et al. 2017; Soil Survey Staff 2014) and a large component of endemic soil taxonomic units (Carpenter et al. 2014). Alaska contains approximately 18% (1 717 900 km^2) of the total land and water area of the United States, covering a wide range of soil ecosystems (Fig. A15). Alaska soils can be broadly characterized by three key metrics: presence or absence of permafrost, presence or absence of fire, and landscape position as an upland or lowland. These attributes are the primary controls on dominant soil-forming factors that are related to the soil geography of the state. Permafrost manifests itself regionally with more prevalence from north to south, and fire dominates the boreal forest in the central part of the state. Uplands and lowlands are general landscape descriptors that distinguish areas of freely flowing matrix water (uplands) and areas of water accumulation (lowlands).

Alaska soils are subject to current climate disturbance as northern latitudes are expected to have more dramatic climate change than lower latitudes (Bekryaev et al. 2010). Alaska regions are also subject to extreme climate normal. Temperature oscillations include seasonal temperature changes of 48 °C in interior Alaska. Precipitation is extreme in the coastal forest region, where total rainfall can reach 5 m annually. The state contains a disproportionate area susceptible to change, including wetlands and permafrost. Changes in soils should be of great concern in Alaska due to the influence of soils on the stock and flow of carbon (C), stability of structures, and both animal and human habitat, including

Fig. A15 Map of the state of Alaska

supply of water and food. Changing climate normals, including widening climate oscillations, increasing temperatures, and more variable rainfall, have the potential to destabilize existing soil functions.

One of the more compelling soil attributes in Alaska is the presence of permafrost. Permafrost soils have frozen soil layers that are sustained over a 2-year period (IPA n.d.). Permafrost can be continuous and cover the entire portion of certain landscapes or discontinuous and co-occur with non-permafrost soils. Permafrost soils have interactions with many aspects of soil functions including C cycling and storage, water processing, wetlands, and rural and urban infrastructure. These functions are particularly vulnerable due to the susceptibility of permafrost soils to thaw with warming temperatures (Grosse et al. 2010).

The thawing of permafrost has both physical and chemical implications. Physical vulnerabilities occur when permafrost is lost below structures such as buildings and roads (Larsen et al. 2008). The loss of the permafrost due to incursion of heat at depth leads to subsidence and collapse of physical structures. In addition, deep thawing and then refreezing leads to frost heave, which can buckle pavement on roadways. Permafrost melt below forests is common and leads to a feature of "jackstrawed" trees with impaired hydrologic functions. The loss of deeper permafrost leads to collapse of soil horizons and changes to water flow (Liljedahl et al. 2016). Polygons form within melt pathways in arctic landscapes that create vectors for erosion with drainage water in tundra soils (Nitze and Grosse 2016). Carbon that was previously protected by the low temperatures and physical occlusion in ice are subject to microbial processing, leading to increased soil organic matter oxidation and loss of C under aerobic, thawed conditions (Schuur et al. 2008). A

major concern across the boreal forest is the redistribution of soil C due to warming. Projections for interior zones of both continuous and discontinuous permafrost indicate that soil C will increase due to increases in vegetation growth (Genet et al. 2018). However, there is a great deal of uncertainty about the stability of older C stored deeper in soil profiles that may be exported by lateral fluxes of C as dissolved organic C.

Fire is prevalent in the boreal forest and has a great impact on surface soils, permafrost persistence, and vegetation feedbacks to soil. Larger fires have become more common in Alaska in recent decades (Barrett et al. 2011; Kasischke et al. 2002), which could lead to a higher risk for soils. Fire is more frequent in well-drained uplands, making these soil landscapes more vulnerable to change with either more frequent or more severe fire (Genet et al. 2013). Fire can alter vegetation and the quantity of organic matter in the forest floor (Barrett et al. 2011). Intense fires, which lead to the loss of insulating vegetation, can accelerate permafrost degradation. High-frequency and larger fires can alter both the vegetation cover and the soil conditions by leading to the loss of permafrost across large areas and a conversion of permafrost soil to nonpermafrost conditions. Therefore, increasing fire severity will dictate the future fate of large areas of the boreal forest.

Alaska soils can be broadly divided into two topographic categories: uplands and lowlands. This distinction is based on topography, but several key functional attributes follow the geographic distinction. As mentioned earlier, fire is more prevalent on well-drained uplands (Genet et al. 2013). Water flow is dominated by recharge and the large supply of freshwater flowing from snow. The physiographic distinction also facilitates assessments of soil C in Alaska. Soil C accumulates in lowlands where saturated soil and low temperatures form deep accumulations of organic matter. Histosols are common features, especially in the coastal forest region (Neiland 1971). The coastal forest region is one of the densest C reservoirs in the world with average soil C densities of 300 Mg ha⁻¹ (Leighty et al. 2006). Lowlands are undergoing change as wetlands over permafrost are susceptible to loss as permafrost thaws (Avis et al. 2011). The result is not only a loss of wetland area and habitat but also potential erosion of C stocks and mobilization to aquatic ecosystems (O'Donnell et al. 2014).

Uplands

Enhanced vegetation production under higher atmospheric CO_2 could increase litterfall rates and alter surface soil C and nutrient dynamics. These changes may be either beneficial or deleterious depending on the loading of debris and the subsequent processing of the material. Increased biomass loading could result in the risk of more severe wildfires in certain portions of the landscape. Rapid turnover of biomass could have a positive impact on vegetation productivity and terrestrial C sequestration, including increases in soil C storage.

Lowlands (Wetlands)

The risk of fire is much lower on lowlands, or wetlands, in Alaska. The main vulnerability of wetlands is loss of the characteristic soil moisture that maintains wetland functions. Increasing temperatures in interior Alaska have resulted in drying out of wetlands and the loss and conversion of wetlands to uplands. The loss of wetlands causes shifts in plant communities, reduction in animal habitat, and loss of water storage leading to drought, and decreased baseflow to freshwater streams. Wetlands are also the primary sources of dissolved material flowing from terrestrial to aquatic systems. Large quantities of organic acids are exported from soils and carried via inland waters to the coastal ocean (D'Amore et al. 2015; Stackpoole et al. 2017). Alterations to both soil moisture and soil C mineralization could change the trajectory of concentrations and delivery of dissolved organic carbon (DOC) with consequences from either increased or decreased loading of this material in the coastal ocean.

Soil Carbon

Alaska has some of the largest stores of soil C in the United States. Estimates of soil C storage for the state range from 31 to 72 petagrams (Pg) of C over the period 1950–2009 (McGuire et al. 2016). Interior Alaska permafrost stocks are part of the circumpolar C reserve, which is nearly twice the amount in the atmosphere (Hugelius et al. 2014), and the coastal forest region has some of the densest C stocks in the world (Johnson et al. 2011; Leighty et al. 2006; McNicol et al. 2019; Vitharana et al. 2017). The stock of C in permafrost regions interacts with climate and fire, where the thaw-

ing of permafrost becomes a critical factor in the stability of the soil C pool (Kasischke and Turetsky 2006). Predicting the trajectory of soil C will be closely tied to the flow of biomass and its interaction with fire dynamics. The boreal regime will have a dynamic interaction with plant growth balanced by C oxidation by fire, while the coastal forest region is expected to continue to accumulate C (Genet et al. 2018).

Understanding soil change, especially with respect to soil C, is important in Alaska given the potential for the magnitude of shifts in the terrestrial C stock to both aquatic and atmospheric pools. Opportunities for advancing knowledge of terrestrial soil resources are growing through both ground inventory (USDA FS 2018) and combinations of remote sensing technologies through satellite technology focused on both C (NASA 2018a) and soil moisture (NASA 2018b). Increased data on soil resources coupled with process models for C flux within and across pools will enhance our ability to predict changes and anticipate future needs across the diverse Alaska landscape.

Soil Mapping and Assessment

Soil mapping in Alaska is a paradox: a great number of soil pedon descriptions along with extensive laboratory data for the state, but a huge geographic area to characterize with that database. New tools such as digital soil maps are leveraging the existing data and environmental variables to fill gaps and provide new insights into soil types and soil functions. An evaluation of soil C prediction throughout the state revealed that the confidence in prediction for the coastal forest region was actually very high given the number of pedon observations, but the interior region of the state will need many more samples to achieve a satisfactory level of confidence (Mishra and Riley 2015). While large areas of the state remain unmapped, technological advances in remote sensing provide a means to address some soil evaluation needs and meet some of the mapping objectives of the Soil2026 initiative[2]. Alaska can serve as a robust proving ground for prediction of soil properties and processes and therefore plays a valuable role in the future efforts of mapping soils and soil function across diverse and mountainous terrain.

Key Findings

- Recent advances in knowledge of permafrost thaw and soil change have expanded the ability to inform decisions about soil change related to soil moisture, soil C, and human infrastructure. Enhanced mapping of perma-

---

[2]Lindbo, D.L.; Thomas, P. 2016. Shifting paradigms - Soil Survey 2026. [Presentation and poster]. Resilience emerging from scarcity and abundance; meeting of the American Society of Agronomy–Crop Science Society of America–Soil Science Society of America; Nov. 6–9, 2016; Phoenix, AZ.

frost, along with identification of vulnerable areas that have the potential for rapid change in the presence of permafrost, provides a framework for risk evaluation and mitigation of permafrost change. There remains a need for increased knowledge about soil change in permafrost regions.

- Alaska contains a large portion of the worldwide soil C stock. This C store ranges across frozen ground, forest, and deep organic soils. Recent soil C inventories have improved accounting of these stocks for integration with C cycle initiatives such as C trading and greenhouse gas mitigation strategies. The fate of the large soil C stock is highly dependent on fire dynamics, vegetation change, and soil temperature and moisture relationships. All of these are key points of information needs for future research.

- Alaska is benefiting from enhanced soil inventory, mapping, and prediction facilitated by remote sensing and machine learning techniques. Extrapolation of soil properties based on numerous field pedon sample data has increased the power of prediction across both mapped and unmapped landscapes. However, Alaska continues to present a challenge for predicting soil properties in remote terrain, where natural change is occurring more rapidly than in most other areas.

Literature Cited

Avis CA, Weaver AJ, Meissner KJ (2011) Reduction in areal extent of high-latitude wetlands in response to permafrost thaw. Nat Geosci 4:444–448

Barrett K, McGuire AD, Hoy EE, Kasischke EE (2011) Potential shifts in dominant forest cover in interior Alaska driven by variations in fire severity. Ecol Appl 21(7):2380–2396

Bekryaev RV, Polyakov IV, Alexeev VA (2010) The role of polar amplification in long-term surface air temperature variations and modern arctic warming. J Clim 23:3888–3906

Carpenter DN, Bockheim JG, Reich PF (2014) Soils of temperate rainforests of the North American Pacific Coast. Geoderma 230–231:250–264

D'Amore DV, Edwards RT, Biles FE (2015) Landscape controls on dissolved organic carbon export from watersheds in the coastal temperate rainforest. Aquat Sci 78:381–393

Genet H, McGuire AD, Barrett K et al (2013) Modeling the effects of fire severity and climate warming on active layer thickness and soil carbon storage of black spruce forests across the landscape in interior Alaska. Environ Res Lett 8(4):045016

Genet H, He Y, Lyu Z et al (2018) The role of driving factors in historical and projected carbon dynamics of upland ecosystems in Alaska. Ecol Appl 28:5–27

Grosse G, Romanovsky V, Nelson FE et al (2010) Why permafrost is thawing, not melting. EOS Trans Am Geophys Union. 91(2):87

Hugelius G, Strauss J, Zubrzycki S et al (2014) Estimated stocks of circumpolar permafrost carbon with quantified uncertainty ranges and identified data gaps. Biogeosciences 11:6573–6593

International Permafrost Association [IPA] (n.d.) What is permafrost? https://ipa.arcticportal.org/publications/occasional-publications/what-is-permafrost. Accessed 8 June 2018

Johnson KD, Harden J, McGuire AD et al (2011) Soil carbon distribution in Alaska in relation to soil-forming factors. Geoderma 167–168:71–84

Kasischke ES, Turetsky MR (2006) Recent changes in the fire regime across the North American boreal region—spatial and temporal patterns of burning across Canada and Alaska. Geophys Res Lett 33(9):09703

Kasischke ES, Williams D, Barry D (2002) Analysis of the patterns of large fires in the boreal forest region of Alaska. Int J Wildland Fire 11(2):131–144

Larsen PH, Goldsmith S, Smith O et al (2008) Estimating future costs for Alaska public infrastructure at risk from climate change. Glob Environ Chang 18(3):442–457

Leighty WW, Hamburg SP, Caouette J (2006) Effects of management on carbon sequestration in forest biomass in southeast Alaska. Ecosystems 9(7):1051–1065

Liljedahl AK, Boike J, Daanen RP et al (2016) Pan-Arctic ice-wedge degradation in warming permafrost and its influence on tundra hydrology. Nat Geosci 9(4):312–318

McGuire AD, Genet H, He Y, Stackpoole SM, D'Amore D, Rupp TS, Wylie BK, Zhou X, Zhu Z (2016) In: Zhu Z, McGuire AD (eds) Baseline and projected future carbon storage and greenhouse gas fluxes in ecosystems of Alaska. Prof. Pap. 1826. U.S. Department of the Interior, Geological Survey, Reston, pp 189–196. Chapter 9. https://doi.org/10.3133/pp1826

McNicol G, Bulmer C, D'Amore DV et al (2019) Large, climate-sensitive soil carbon stocks mapped with pedology-informed machine learning in the North Pacific coastal temperate rainforest. Environ Res Lett 14(1):014004

Mishra U, Riley WJ (2015) Scaling impacts on environmental controls and spatial heterogeneity of soil organic carbon stocks. Biogeosciences 12(13):3993–4004

National Aeronautics and Space Administration [NASA] (2018a) GeoCarb: a new view of carbon over the Americas. https://www.nasa.gov/feature/jpl/geocarb-a-new-view-of-carbon-over-the-americas. Accessed 13 Sept 2018

National Aeronautics and Space Administration [NASA] (2018b) SMAP: soil moisture active passive. Homepage. https://smap.jpl.nasa.gov. Accessed 13 Sept 2018

Neiland BJ (1971) The forest-bog complex of southeast Alaska. Vegetatio 22:1–64

Nitze I, Grosse G (2016) Robust trends of landscape dynamics in the Arctic Lena Delta with temporally dense Landsat time-series stacks, with links to GeoTIFFs. https://doi.org/10.1594/PANGAEA.854640

O'Donnell JA, Aiken GR, Walvoord MA et al (2014) Using dissolved organic matter age and composition to detect permafrost thaw in boreal watersheds of interior Alaska. J Geophys Res 119(11):2155–2170

Ping CL, Clark MH, D'Amore DV et al (2017) Soils of Alaska: LRRs W1, W2, X1, X2, and Y. In: West L, Singer M, Hartemink A (eds) The soils of the USA. World Soils book Series. Springer International Publishing, Cham, pp 329–350. Chapter 17

Schuur EA, Bockheim J, Canadell JG et al (2008) Vulnerability of permafrost carbon to climate change: implications for the global carbon cycle. Bioscience 58(8):701–714

Soil Survey Staff (2014) Keys to soil taxonomy, 12th edn. U.S. Department of Agriculture, Natural Resources Conservation Service, Washington, DC

Stackpoole SM, Butman DE, Clow DW et al (2017) Inland waters and their role in the carbon cycle of Alaska. Ecol Appl 27(5):1403–1420

USDA Forest Service [USDA FS] (2018) 2014 Interior Alaska highlights: forests of the Tanana Valley State Forest and Tetlin National Wildlife Refuge Alaska. https://www.fs.fed.us/pnw/rma/local-resources/documents/TananaHighlights2-8-18.pdf. Accessed 13 Sept 2018

Vitharana UWA, Mishra U, Jastrow JD et al (2017) Observational needs for estimating Alaskan soil carbon stocks under current and future climate. J Geophys Res Biogeosci 122(2):415–429

Hawaii and U.S. Affiliated Pacific Islands

Christian P. Giardina, Richard A. MacKenzie, and Jonathan L. Deenik

Introduction

This regional summary addresses the effects of climate change on soils of seven of the eight main Hawaiian islands (Kaua'i, O'ahu, Moloka'i, Lāna'i, Kaho'olawe, Maui, and Hawai'i), the territories of Guam and American Samoa, the Commonwealth of the Northern Mariana Islands, and the freely associated Republic of Palau, Federated States of Micronesia, and Republic of the Marshall Islands. The distribution of soil types in Hawai'i and the US Affiliated Pacific Islands (USAPI), and their sensitivity to climate change, is determined as elsewhere by soil-forming factors and their interactions (Jenny 1941; Vitousek 2004). These influences include geologic substrate age and type, temperature and precipitation, topographic features that affect hydrology and

movement of sediments, vegetation occupying the site, and human impacts. For wetlands, hydric soils are shaped by available moisture, which can be dynamic, especially where water is diverted from natural wetlands for agriculture or development purposes, or in coastal areas subject to tidal inputs, storm surges, and baseline sea level rise. This understanding of climate-hydrology-soil relationships suggests that many soils will experience some change as a result of Hawai'i's changing climate.

Hawaiian Islands

The Hawaiian Islands represent a chronosequence of parent material, ranging from literally seconds-old cooling lava on the southeastern district of Puna on Hawai'i Island to the more than 4 million-year-old substrate on Kaua'i Island and of course all ages in between. In Papahānaumokuākea, or the Northwestern Hawaiian Islands, jurisdictionally a part of the state of Hawai'i but not thoroughly discussed here, coral-covered surface substrate dates back tens of millions of years. This geologically driven diversity in substrate age across the Hawaiian Islands is orthogonally married to remarkably steep gradients of temperature and moisture, from lowland deserts to montane shrublands, to alpine grasslands, to montane wet forests, and to lowland rainforests. Nearly every combination of substrate age, temperature condition, and moisture environment can be found on the Hawaiian Islands—resulting in a complex array of soil types across very short distances (Vitousek 2004).

The geographic area of the Hawaiian Islands covers 16 587 km^2 distributed across the islands as follows: Hawai'i Island (10 432 km^2), Maui (1884 km^2), O'ahu (1545 km^2), Kaua'i (1430 km^2), Moloka'i (637 km^2), Lāna'i (364 km^2), and Kaho'olawe (115 km^2)[3]. Ten of the 12 soil orders in the USDA Soil Taxonomy are found in the Hawaiian Archipelago (Fig. A16) (Soil Survey Staff 2014). Andisols are the most extensive soil across Hawai'i. Most of these soils occur on the Island of Hawai'i, the youngest and largest of the Hawaiian Islands. Much of the remaining Andisols (40 163 ha) are found on Maui Island and much smaller isolated areas on Moloka'i, O'ahu, and Kaua'i. Andisols are soils formed in volcanic ash characterized by high concentrations of short-range-order clay minerals (e.g., allophane and ferrihydrite), low bulk density, and high organic matter. We note, however, that steep rainfall gradients produce three broad categories of Andisols in Hawai'i: Udands (wet, highly weathered, low fertility Andisols), Ustands (wet/dry, moderately weathered, high fertility Andisols), and Torrands (dry, weakly weathered Andisols). The influence of rainfall on soil weathering and fertility is exemplified in the Kohala District of Hawai'i Island (Fig. A17). As described in Chadwick

---

[3]The eighth main Hawaiian Island, Ni'ihau, is private and lacks soil survey data.

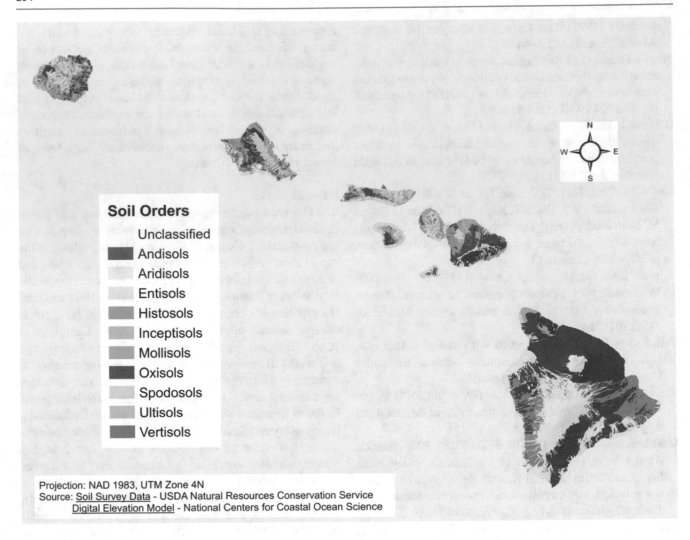

Soil Orders

Unclassified
Andisols
Aridisols
Entisols
Histosols
Inceptisols
Mollisols
Oxisols
Spodosols
Ultisols
Vertisols

Projection: NAD 1983, UTM Zone 4N
Source: <u>Soil Survey Data</u> - USDA Natural Resources Conservation Service
<u>Digital Elevation Model</u> - National Centers for Coastal Ocean Science

Fig. A16 Soils map of the main Hawaiian Islands (Source: University of Hawaii 2014)

et al. (2003), increasing rainfall has a dramatic effect on soil properties, producing three Andisols with distinctly different fertility capabilities. In the Torrands, found at the dry low elevations where evapotranspiration exceeds precipitation, rock weathering has proceeded slowly, creating a soil dominated by coarse materials (80% >2 mm diameter) with low clay content (<13%). The soil is rich in weatherable carbonates, which result in its alkaline pH and high base saturation, but is low in organic carbon (C) (Table A9). In the moderate weathering regime associated with moderate levels of bimodally distributed rainfall (750–1500 mm yr^{-1}), the Ustand has accumulated clay (30–40%) and substantial amounts of organic C, resulting in a near neutral soil with exceptionally high cation exchange capacity (CEC) and high base saturation. At the wet summit of Kohala Mountain, intense leaching and weathering have produced a very acidic, nutrient-depleted soil with minimal CEC and substantial amounts of soluble Al^{3+}. Torrands are also found at low elevations on the dry leeward (with respect to the prevailing

trade winds) slopes of Mauna Kea and Haleakalā on Maui, while Ustands dominate the mid-elevations between Kohala and Mauna Kea and the leeward slopes of Haleakalā. The Udands are common on the wet windward slopes of Mauna Kea and the Hana region of Maui (Table A10).

A closer look at Kaua'i, with its ten soil orders, shows how the five soil-forming factors have interacted to create tremendous soil diversity on such a small island (Fig. A18). Rocky, poorly developed Entisols cover large areas of steep mountainous land in central Kaua'i and some sandy soils on leeward coastal flats. Highly weathered, acidic Oxisols with low fertility dominate the old volcanic slopes of windward Kaua'i. High rainfall over millennia has leached these soils of base cations, leaving behind low-CEC oxidic clays. Their productivity lies in an organic matter-rich surface horizon developed over millennia from tropical wet forest cover. After deforestation and cultivation, however, the organic matter is susceptible to rapid oxidation, and the loss of organic matter quickly renders these soils dependent on

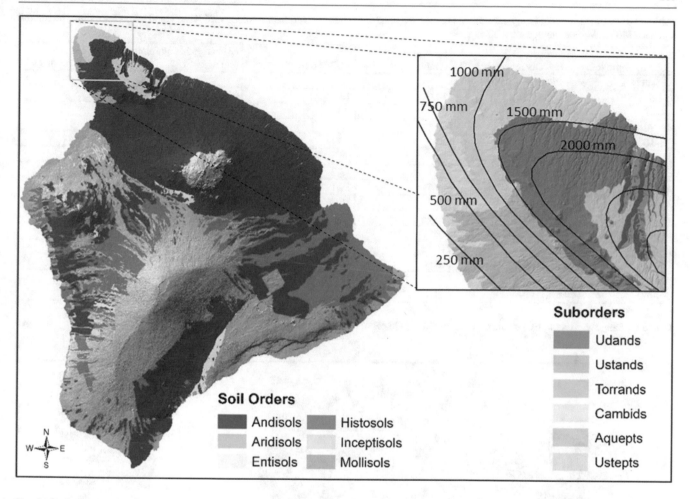

Fig. A17 Soil orders of Hawaiʻi Island with inset illustrating the effect of rainfall (contours) on differentiating dry (Torrands), wet/dry (Ustands), and wet (Udands) Andisols in Kohala District. Maps generated in ArcMap (ver. 10.5.1; Esri®, Redlands, CA) using SSURGO data

heavy nutrient inputs to remain productive. On the other hand, large areas of fertile Mollisols cover the lowlands of leeward Kauaʻi. These soils have developed under a drier climate, which slows weathering and preserves the presence of high activity clays with ample CEC.

Three representative soils—a highly weathered Oxisol, a moderately weathered Mollisol, and a cool, wet Spodosol from the summit of Waiʻaleʻale in the center of the island—illustrate some key fertility differences among Kauaʻi soils. The Oxisol is strongly acidic, characterized by negligible CEC at field pH; relatively high exchangeable Al, suggesting potential toxicity; and very low base status. Together, these characteristics put this soil at the far end of the spectrum of low fertility soils. Any semblance of fertility is restricted to the surface horizon where organic matter has accumulated. The much larger value of CEC_{pH7} compared with effective cation exchange capacity (ECEC) indicates the dominance of variable-charge clay minerals, specifically iron (Fe) oxides. The Mollisol, on the other hand, is a fertile soil with high CEC throughout the profile and a high base saturation,

Table A9 Selected physical and chemical properties[a] for three Andisols found on Kohala Mountain, Hawaiʻi Island

Soil property (0–30 cm)	Torrands (Hapuna series)	Ustands (Waimea series)	Udands (Kehena series)
Texture	Cobbly silt loam	Silt loam	Silty clay loam
pH	8.0	7.0	4.5
OC (%)	2.1	9.6	4.54
Ca^{2+} ($cmol_c$ kg^{-1})	21.9	59.0	3.0
Mg^{2+} ($cmol_c$ kg^{-1})	15.5	12.7	1.7
K^+ ($cmol_c$ kg^{-1})	9.7	1.1	0.3
ECEC ($cmol_c$ kg^{-1})	44.8	74.0	7.3
BS (%)	80	100	10
Al^{3+} ($cmol_c$ kg^{-1})	0	0	2.2
Al^{sat} (%)[b]	0	0	29.7

Source: USDA NCSS n.d.

[a]*OC* organic carbon, *Ca* calcium, *Mg* magnesium, *K* potassium, *ECEC* effective cation exchange capacity, *BS* base saturation, *Al* aluminum
[b]Al^{sat} = (Exchangeable Al/ECEC) × 100%

Table A10 Selected chemical properties for an Oxisol from windward Kaua'i, a Mollisol from leeward Kaua'i, and a Spodosol from the summit region of Mount Wai'ale'ale in central Kaua'i

Soil and depth (cm)	pH	Organic carbon (%)	CEC$_{pH7}$ (cmol$_c$ kg^{-1})	ECEC^a (cmol$_c$ kg^{-1})	Exchangeable aluminum (cmol$_c$ kg^{-1})	Base saturation (%)
Oxisol (Pooku series), rainfall = 3005 mm yr^{-1}						
0–38	5.2	3.88	14.6	1.6	0.3	8.9
38–48	4.7	2.43	9.6	0.3	0.1	2.0
48–76	5.0	1.58	4.8	0.7	0.1	12.5
76–100	4.8	1.10	2.6	0.1	0.1	0
Mollisol (Kekaha series), rainfall = 584 mm yr^{-1}						
0–18	7.2	2.18	44.1	43.2	0	72.0
18–35	7.0	1.72	46.7	39.1	0	84.0
35–70	6.9	1.36	44.2	36.4	0	84.0
70–110	7.0	0.69	26.7	22.0	0	83.0
Spodosol (Wai'ale'ale series), rainfall = 4445 mm yr^{-1}						
0–22	3.7	68.4	161.2	27.2	11.1	10.5
23–43	3.7	63.0	221.0	100.5	96.6	1.5
43–50	4.2	1.86	56.6	18.3	17.6	1.0
50–100	4.5	0.90	29.0	7.3	6.6	2.5

Source: USDA NCSS n.d.

^{a}Cation exchange capacity (CEC) or the CEC measured at field pH

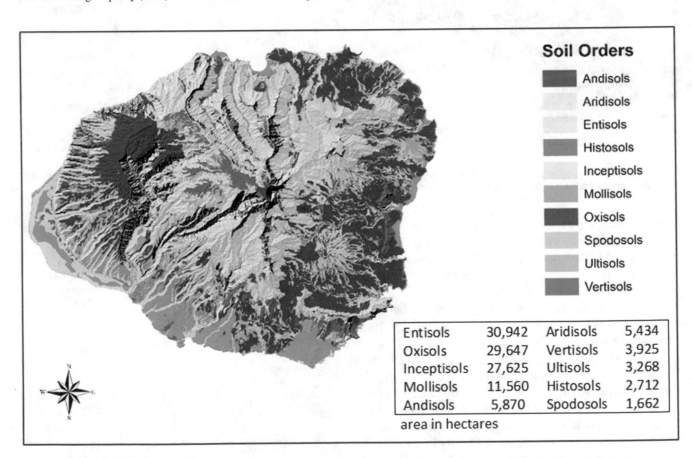

Fig. A18 The distribution of soil orders on Kaua'i and their approximate area. Map generated in ArcMap (ver. 10.5.1) using SSURGO data

indicating ample quantities of essential plant nutrients: calcium (Ca), magnesium (Mg), and potassium (K). The small differences between CEC$_{pH7}$ and ECEC indicate the dominance of permanently charged high activity clays. This Mollisol is a prime agricultural soil known for its high productivity. The Spodosol, found in one of the rainiest locations in the world, is a poorly drained organic matter-rich, infertile soil with low base content and toxic levels of soluble

Al. Despite its very low fertility, it supports a diverse forest of rare and endemic species.

United States Affiliated Pacific Islands

The United States Affiliated Pacific Islands (USAPI) is also highly diverse with respect to soils, climate, and other characteristics but geographically much more dispersed, and the region's pedology and climate are much less studied than those of the Hawaiian Islands. The region is made up of some 2,000 separate islands and thousands more islets extending across an area larger than the continental United States. The jurisdictions of the USAPI contain an enormous variety of geologies from low-lying coral atoll islands to high volcanic islands of various ages. Because of the location of these island systems in both the northwestern and southwestern Pacific regions, with most USAPI jurisdictions north of the equator and American Samoa south of the equator, they are affected by different climate systems.

Furthermore, these jurisdictions span enormous areas of ocean, so there is no common weather pattern for the islands beyond the exposure of all jurisdictions to warm tropical temperatures and moderate to high levels of precipitation. Weather patterns, drought periods and severity, and timing and expression of large events such as El Niño vary dramatically across the USAPI. Similarly, changes in response to climate warming and drying will vary across these jurisdictions. There also are important certainties with respect to future climate-related impacts to the region. Ocean temperatures, air temperatures, and sea levels are all expected to rise. There is also a strong scientific basis for projections that cyclone events will strengthen because of a warmer, moister atmosphere, which stores more energy. Some evidence indicates that droughts in the region may be getting more severe (Polhemus 2017).

Given the geologic and climatic variation across the USAPI, soils show great diversity. Differences in parent material, degree of weathering, and landscape position are the main controllers of variability. Soils on Guam cover a single main island and several very small islets, which add up to 549 km² of land. These soils are composed of diverse

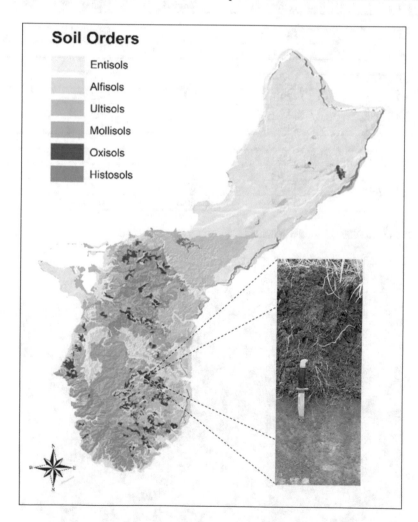

Fig. A19 The distribution of soil orders on Guam with a picture of an acidic, infertile Oxisol commonly found in the uplands of southern Guam (Photo credit: Robert Gavenda, USDA NRCS)

geologies that lead to highly contrasting soils over short distances (Fig. A19). Very shallow Entisols that developed on the limestone plateaus of northern Guam are the dominant soil (27 737 ha). The soils are neutral to slightly alkaline in pH, rich in organic matter, and very high in exchangeable Ca but have accumulated gibbsite, an Al-oxide clay with high phosphorus fixation (Table A11). In stark contrast, we find Ultisols (7364 ha) and Oxisols (1728 ha) in the highly dissected and steep mountainous areas of southern Guam, which formed on old volcanic parent material now heavily weathered into acidic soils with low nutrient availability and high erosion potential. The fertility of these soils is tied to organic matter accumulation in the surface horizons (Table A11). A distinct increase in extractable Al in the subsoil results in Al saturation of 50%, which makes these soils Al toxic for many plants.

Much of southern Guam has been heavily impacted by land use conversion from forests to pasture, frequent (often arson caused) fires and hurricanes, and past military activities including bombing (Fig. A20). The result is large areas of degraded pastures and badlands from ridges to valley bottoms where the topsoil has eroded, leaving a nutrient-depleted and Al-toxic subsoil. More fertile, but very shallow, Mollisols formed in marine-deposited tuff are found in the uplands of southern Guam (6622 ha). The Northern Mariana Islands north of Guam include 474 km^2 of land over 14 islands and are geologically similar to Guam. These islands have dissected limestone plateaus and complex topography with steep and, in some places, eroded soils of volcanic origin.

Palau, located approximately 1300 km southwest of Guam, is geologically complex. It includes a mixture of older volcanic and metamorphic substrates on the large

Table A11 Selected chemical properties for an Entisol from northern Guam and an Ultisol/Oxisol and Mollisol from the volcanic uplands of southern Guam

Soil and depth (cm)	pH	Organic carbon (%)	CEC$_{pH7}$ (cmol$_c$ kg^{-1})	ECEC^a (cmol$_c$ kg^{-1})	Exchangeable aluminum (cmol$_c$ kg^{-1})	Base saturation (%)
Entisol (Guam series)						
0–6	7.1	8.57	42.7	–	0	75
6–20	7.5	5.56	26.3	–	0	76
Ultisol/Oxisol (Akina series)						
0–10	5.0	5.04	28.9	12.4	1.9	29
10–20	4.9	2.81	20.7	8.4	4.3	15
20–60	5.0	1.08	13.5	7.3	3.6	17.5
60–110	5.3	0.31	19.5	11.6	1.5	41
Mollisol (Agfayan series)						
0–10	6.4	2.66	60.3	52.4	0	87
10–20	6.7	1.36	69.7	61.5	0	88
20–50	6.7	0.61	73.5	62.0	0	87
35–50	6.9	0.85	74.2	66.5	0	90

Source: USDA NCSS n.d.

^{a}Cation exchange capacity (CEC) or the CEC measured at field pH

Fig. A20 Severe erosion is common in the uplands of southern Guam. The exposed subsoil is incapable of supporting vegetation (Photo credit: Robert Gavenda, USDA NRCS)

island of Babeldaob (20–40 million years before the present [Ma]), limestone parent material capping volcanic core material on smaller islands to the south, and hundreds of small, steep limestone islands farther south. Palau also includes atoll islands at its northern and southern extremes. The resulting soils are a mixture of highly weathered and acidic soils on the forested uplands and poorly drained organic matter-rich soils in the lowlands of Babadoab, minimally developed thin soils on the karst Rock Islands, and deeper and more fertile soils on the islands where substrates mix (Fig. A21). Historically, the acidic, infertile Oxisols (such as the Aimeliik series) of the Babeldoab uplands were forested and not cultivated. These soils' high acidity and excessive concentrations of soluble Al (Table A12) limit their use for agriculture. As with the Akina series on Guam, the minimal fertility of these soils resides in the organic matter-rich surface layer, and loss of this important layer leaves behind a severely degraded soil. The increasing incidence of wildfires across Babeldoab has played a strong role in denud-ing forested uplands, causing topsoil loss to erosion and the creation of severely degraded soils that remain exposed and difficult to revegetate. The Histosols, on the other hand, are base cation-rich organic soils that have been continuously planted to wetland taro (*Alocasia* spp.) for as long as 1,000 years. Adjacent, poorly drained organic matter-rich Inceptisols are also important agricultural soils.

The Federated States of Micronesia represents some 600 islands covering 700 km^2 of land, ranging from high island volcanic islands to low-lying coral atolls. The easternmost islands of Kosrae are the youngest at just under 1 million years. The geology becomes progressively older over the 2700 km to the westernmost islands of Yap (about 14 million years old). The island of Pohnpei shows a range of soil types, from mucky sandy soils supporting dense mangrove (*Avicennia* spp.) forests that circle the shoreline to patches of highly weathered, infertile Oxisols on 7 Ma volcanic substrate (Fig. A22). Fertile Inceptisols and Alfisols have developed on more recent lava flows, which support a diverse and

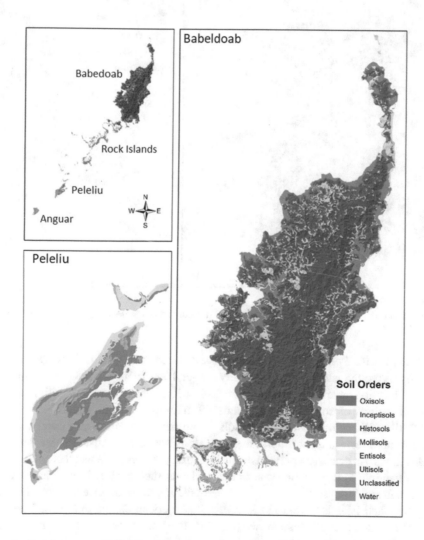

Fig. A21 The distribution of soil orders across the Palau archipelago

Table A12 Selected soil chemical properties for an Oxisol and Histosol from Babedoap Island, Palau

Soil and depth (cm)	pH	Organic C (%)	CEC$_{pH7}$ (cmol$_c$ kg^{-1})	ECEC^a (cmol$_c$ kg^{-1})	Exchangeable aluminum (cmol$_c$ kg^{-1})	Aluminum saturation^b (%)
Oxisol (Aimeliik series)						
0–4	6.2	35.1	101	–	0	0
4–14	5.4	4.4	22.2	10.5	2.3	22
14–35	5.2	2.1	16.3	7.9	6.0	76
35–90	5.5	0.1	17.2	8.8	6.2	70
Histosol (Mesei series)						
0–12	5.5	39.8	69.8	31.7	0.2	0.6
12–40	5.0	14.9	36.6	9.1	2.0	22
40–86	5.1	15.8	31.9	7.0	0.2	3
86–150	3.8	5.8	26.1	20.9	3.1	15

Source: USDA NCSS n.d.

^{a}Cation exchange capacity (CEC) or the CEC measured at field pH.

^{b}%Aluminum (Al) saturation = (Exchangeable Al/ECEC) × 100%.

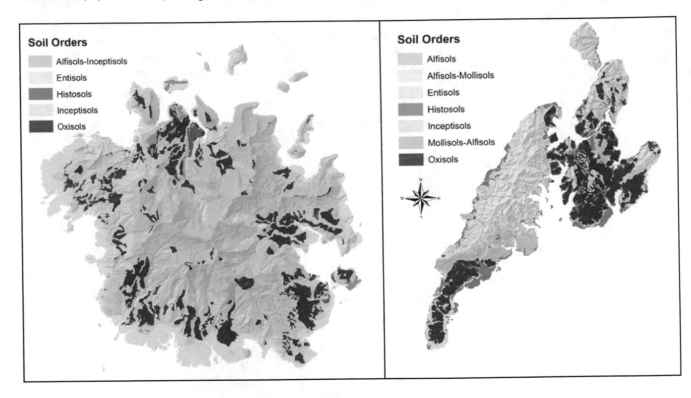

Fig. A22 The distribution of soil orders on the main islands of Pohnpei (L) and Yap (R) in the Federated States of Micronesia

productive agroforest system. Like Pohnpei, mangrove forests on mucky sandy soils fringe the island, whereas highly weathered Oxisols, low in fertility, have formed on volcanic deposits on flat bench formations of southern and eastern portions of Yap. More fertile Alfisols mixed with Mollisols have formed on older, but more resistant, metamorphic formations at the higher elevations. As discussed previously, the Oxisols are especially susceptible to degradation due to the erosional loss of organic matter-rich topsoil.

The Republic of the Marshall Islands covers 181 km^2 of land spread across some 1 500 000 km^2 of ocean in two parallel north-south aligned chains of islands, consisting of 5 inhabited main atoll islands, 29 low-lying atoll islands, and over 1000 islets. The highest point in the entire country is just 10 m above sea level, with most of the country less than 2 m above sea level. Natural soils of this country are therefore derived from young coral material. Soils of the coral atolls are sandy Entisols characterized by high pH and low fertility. As with the weathered Oxisols of the high volcanic islands, their fertility is associated with organic matter accumulation in the surface horizon (Deenik and Yost 2006).

American Samoa in the southern hemisphere is made up of five main volcanic islands and two coral atoll islands covering 197 km^2. The contrasting parent material and complex

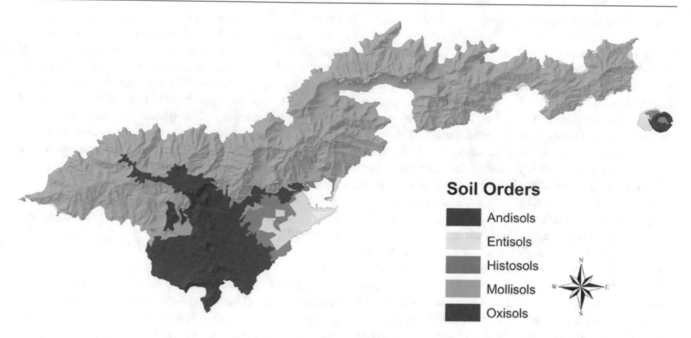

Fig. A23 The distribution of soil orders on the islands of Tutuila and Aunu'u, American Samoa

topography result in diverse soils. Fertile Mollisols dominate the steep forested mountains (10 515 ha) from the east to the west end of Tutuila (the main populated island) (Fig. A23), whereas fertile Andisols are the primary soil on the plains of southwest Tutuila (5396 ha). The Andisols formed from volcanic ash deposits laid down during posterosional volcanism. They are well-drained soils where continuous agriculture is practiced. Histosols and Entisols cover much smaller areas and are primarily in urban use.

Coastal Wetlands

We provide more detail here on coastal wetland soils because we anticipate that climate change via sea level rise will have the biggest effects on these soils. Coastal wetland soils are very deep, are very poorly drained soils, and are generally very high in organic matter. In the western Caroline Islands (e.g., Palau and Yap), mangrove soils are dominated by Histosols (Illachetomel series) (64%) composed of organic deposits derived from decomposing mangrove roots and leaf litter. The other soil series are loamy and mucky Entisols (Naniak series) (19%), composed of organic deposits and alluvium derived from volcanic material or limestone and the peaty and mucky Histosols (Chia series) (18%) composed of organic matter and coralline sand and gravel (USDA SCS 1983b; USDA NRCS 2009). In the eastern Carolines (e.g., Kosrae, Pohnpei), mangrove soils are dominated by Naniak Entisols (65%) formed from alluvium in this location-derived igneous rock. Other soils include the sandy and peaty Entisols of the Insak series (20%) and Chia Histosol soils (<15%) described earlier. Freshwater peatland forests domi-

nated by ka (*Terminalia carolensis*) are commonly found behind mangrove forests in Micronesia, but the largest intact forest is found on the island of Kosrae. Soil types here are dominated by deep and poorly drained loamy Inceptisols of the Nansepsep series (45%) and Entisols of the Inkosr series (40%), both formed from alluvial deposits of igneous origin (USDA SCS 1982, 1983a). Soils of the Insak series are also found along coasts of American Samoa, although the predominant soil type found in mangroves are Histosols of the Ngerungor series, which is a mixture of peat, basalt, and calcareous sand (Gilman et al. 2007; USDA 1984).

In Hawai'i, coastal wetlands typically develop on soil types composed of alluvium. On Kaua'i and Oahu, these soils include the Hanalei series, an Inceptisol that is a silty clay, silty loam, or peaty silty clay alluvium derived from igneous rock. On the younger islands, wetlands form Inceptisols from mixed alluvial soils. These Tropaquepts consist of alluvium washed from soils that are formed in volcanic ash and that have a surface layer of mucky silt loam overlying a silty clay loam (USDA SCS 1972, 1973).

The organic deposits derived from decomposing roots and litter combined with the poorly drained anoxic soils and the high root to shoot ratio of these coastal forested systems has resulted in relatively high soil C content (Krauss et al. 2014; Lovelock 2008). Percent soil C content from mangroves ranges from 2.8 to 39.5% with an average of 16.4 ± 6.7% (Donato et al. 2011, 2012; Gilman et al. 2007; R.A. MacKenzie et al., unpublished data of the coauthor) and from 19.9 to 47.0% in ka peat swamps sampled on Kosrae (Chimner and Ewel 2004).

Major Threats and Impacts to Soils

Changing Climate

Hawai'i is projected to become both warmer and generally drier, especially for leeward areas of each of the larger islands (Keener et al. 2012a, b), which have experienced significant warming (Giambelluca et al. 2008) and drying over the past century (Frazier and Giambelluca 2017). It is also expected that impacts of climate change will be strong, though varied, for the USAPI. While future air temperatures are predicted to increase, shifts in the Pacific Decadal Oscillation will result in areas with more intense storms and significantly more rainfall (e.g., Palau, Yap), whereas other areas will have longer drought periods between rain events and overall decreased rainfall (e.g., the Marshall Islands) (Polhemus 2017). We expect that climate change effects may be very large in Hawai'i. The vulnerability of soils to these changes is driven by (1) the fast rate at which climate is changing in Hawai'i; (2) the diversity of interacting, climate-related threats including precipitation change, invasive species, and expanding fire regimes; and (3) limited financial and technological resources to direct to adaptation or mitigation projects. These concerns will also drive impacts in the USAPI. Across the region, coastal areas of all islands are expected to be one of the most at-risk areas for increased rates of sea level rise (McLeod and Salm 2006).

There is compelling evidence that in the past few decades, increases in mean annual air temperature for Hawai'i's native ecosystems have exceeded the global average rate of increase (Giambelluca et al. 2008). There is also strong evidence for drying over the past century (Frazier and Giambelluca 2017). Below-average precipitation has had an impact on watershed hydrology (Keener et al. 2012a); streamflows have decreased by 10% in the past half-century compared to the previous 50 years (Oki 2004), in line with model projections for moisture responses to warming (Safeeq and Fares 2011). Dynamical downscaling for the Hawaiian Islands projects further warming and drying. These changes will be accompanied by increased periodicity of a smaller total rainfall amount, hence more intense rain events separated by longer periods of dry weather (Chu and Chen 2005; Chu et al. 2010; Norton et al. 2011). Forecasting further indicates stronger periodicity and stronger reductions in total rainfall in the drier leeward areas of Hawai'i.

Fire

Most direct effects of climate change on soils will take millennia to be manifested as soil development and responses to change are typically very slow, but there is a high likelihood that soils may change more rapidly in response to interactions between climate change and other threats. For example, novel fire regimes are expanding into new areas because of invasion by fire-prone exotic grass and shrub species, including fountain grass (*Cenchrus setaceus*), common gorse (*Ulex europaeus*), and guinea grass (*Urochloa maxima*). Such invasive species have dramatically magnified fire danger not only in Hawai'i (D'Antonio and Vitousek 1992; Trauernicht et al. 2015) but also across portions of the USAPI, especially Guam, Palau, and parts of the Federated States of Micronesia (Fig. A24). These accelerating invasions are also complicat-

Fig. A24 Fire-prone nonnative grasses and shrubs, like those shown here in Hawai'i Volcanoes National Park, cover 25% of Hawai'i's land area, and with expanding cover of these weeds, a warming and drying climate is exacerbating the wildfire threat across the Pacific region, especially Hawai'i, Guam, the Northern Mariana Islands, and the Federated States of Micronesia (Photo courtesy of the National Park Service)

ing fire management, a situation that is further exacerbated by warmer and drier conditions.

The combined effects of these threats have been devastating to native dry and mesic forests of Hawai'i and mesic to wet forests in the more strongly seasonal areas of the USAPI. In Hawai'i, these forests are converted by fire to nonnative grasslands, with important impacts to soil. For example, introduced fire regimes result in fire events that kill native trees and their roots (Cabin et al. 2002); consume organic matter on the soil surface; consume organic matter in the top few centimeters of mineral soil, releasing large amounts of nutrients (Giardina et al. 2000); and kill soil flora and fauna (Døckersmith et al. 1999). Combined, these fire-related impacts reduce soil organic matter, arguably the most important component of tropical soils, and render them vulnerable to rapid degradation.

In the USAPI, fires have converted forests to savanna ecosystems that are dominated primarily by native fire-adapted species. In contrast to Hawai'i, savanna of the USAPI will be colonized by woody tree species and, with sufficient time (decades) between fires, savanna will return to native forest (Donnegan et al. 2007). In Hawai'i, where fire return intervals are long, nonnative woody species will typically invade nonnative grasslands (Cordell et al. 2002). Immediately following fire, freshly exposed soils are the most vulnerable to rain events, which can cause significant erosion of the unprotected soils. Because of the compact nature of watersheds in Hawai'i and throughout the USAPI, rain-on-fire events can quickly move soil from upper watershed areas into coastal or nearshore areas. Critically, once invaded and burned, nonnative grasses in Hawai'i or native savanna in the USAPI will occupy a site and drive a novel largely anthropogenic fire cycle that will repeatedly burn an area, with each burn further degrading soils.

Soil Moisture Changes

Though it is widely known that nearly all native-dominated dry ecosystems in Hawai'i are invaded or have been converted to grasslands by fire, it is less appreciated that most mesic and wet forests in Hawai'i have also been invaded by nonnative species, and in large areas, invasions are very dense (Asner et al. 2005, 2008). Remnant long-lived trees such as ohia (*Metrosideros polymorpha*) can persist in the overstory for many decades, thereby delaying complete conversion of a site to nonnative condition. But when an individual dies, the complete absence of native species recruits results in full occupation of gaps by nonnative species. Thus, wet forest soils are subject to both climate change and climate change interactions with invasive plants. Specifically, plant invasions increase plant and stand water use (Cavaleri and Sack 2010; Strauch et al. 2017), with increases in water use being magnified under warmer and drier conditions (Strauch et al. 2017). The combined effects of climate

change-induced drying and plant invasions reduce soil moisture and hinder potential reestablishment of native plant communities.

On longer timesteps, climate-accelerated plant invasions can displace native vegetation within a few decades, fundamentally changing forest structure, rainfall interactions with soils, sediment delivery to streams, and overall watershed hydrology. According to the census of Hawai'i and forests recently completed by the USDA Forest Service's Forest Inventory and Analysis program (USDA FS 2018), the nonnative strawberry guava (*Psidium cattleianum*) is now the most common tree in Hawai'i. Total estimated stem numbers exceed that of Hawai'i's most common native tree, ohia. Guava has a wide variety of hydrological effects on soils, including decreased soil moisture and very different litter inputs that are likely to affect soil organisms and both biological and physical processes.

Sea Level Rise

By far the biggest climate-related impacts to the region's soils will be in low-lying areas affected by sea level rise, more intense storms, and altered hydrology due to altered rainfall patterns and warmer temperatures. Conservation and restoration of USAPI coastal forests are especially important given the large role that they play in many islands' subsistence economies. Conservation of these systems is also important for mitigation of impacts from climate change and increased nutrient and sediment runoff. For example, the high C content in coastal forested wetland soils, coupled with their high productivity and waterlogged anaerobic condition, results in massive amounts of relatively permanent C stocks. Carbon stocks from USAPI coastal forested wetlands range from 700 to 870 Mg C ha^{-1} with an average of 780 ± 320 Mg C ha^{-1}. These forests represent about 10% of forest cover in the USAPI but contribute up to 30% of total island C stocks (Donato et al. 2012). This is largely due to soil C, which makes up more than 80% of total C stocks in these coastal forests. This C can provide funding through C markets for the conservation and restoration of coastal forested wetlands and the many ecosystem services that they provide.

Conservation of coastal wetlands is also important as these ecosystems reduce sediment and nutrient loading to valuable nearshore ecosystems, such as seagrass beds and coral reefs. Coastal forests can reduce sediment input to coral reef ecosystems by as much as 30% (Golbuu et al. 2003; Victor et al. 2004). Soils in coastal wetlands can also reduce phosphorus (P) and nitrogen (N) loads that can result in eutrophication of nearshore waters. Sorption values for P from coastal wetland soils around Hawai'i were as great or greater than values reported from intact wetlands in the Midwest and on the east coast (Bruland and De Ment 2009). These wetlands also appear to play an important role in removal of N loads (Bruland and MacKenzie 2010).

Forest, Wetland, and Agricultural Products

Compared to elsewhere in the Asia Pacific region, coastal wetlands in the USAPI remain relatively intact. Selective harvesting of trees occurs in mangroves for cooking fuel or for house timber and in ka forests for canoes (Balick 2009). Fish and crabs are also harvested from mangroves, and the organic rich soils of freshwater peat swamps support banana (*Musa* spp.), sugarcane (*Saccharum officinarum*), and soft taro (*Colocasia esculenta*) agroforestry. Combined, coastal wetlands provide up to 60% of the median household income on the island of Kosrae (Conroy et al. 2011; Drew et al. 2005; Naylor and Drew 1998). In Hawai'i, coastal wetlands were traditionally converted into loko i'a kalo, irrigated agriculture plots with raised islands for growing taro, a dominant form of starch consumed by Hawaiian peoples. Interspersed among these islands were ponded areas that were also used for capturing and raising aholehole (*Kuhlia sandvicensis*) and 'o'opu fish (*Eleotris sandwicencis, Awaous stamineus*) (Kikuchi 1976).

Higher air temperatures and lower rainfall (Frazier and Giambelluca 2017; Giambelluca et al. 2008; Keener et al. 2012b) will have adverse effects on coastal wetland soils of Hawai'i and the USAPI through greater soil salinity. Increased salinity will result from rising sea level and will be exacerbated by lower water tables and freshwater inputs due to increased drought frequency (Scavia et al. 2002). Rising salinity levels will severely limit food productivity because the key food crops (e.g., taro, banana) grown in the coastal lowlands are typically sensitive to soil salinity (Day et al. 2008; Field 1995). Warmer and drier conditions will also expose coastal soils to aerobic conditions that increase decomposition as well as nutrient loss through oxidation and subsequent reflooding (Burkett and Kusler 2000; Chimner and Ewel 2004; Morrissey et al. 2014). Increased decomposition of peat soils results in subsidence in coastal ecosystems, making these areas increasingly vulnerable to seawater inundation (Krauss et al. 2014; Rogers et al. 2014). Alternatively, increased precipitation can increase soil productivity by reducing salinity and osmotic stress of plants and trees (Clough et al. 1982; Ellison 2000). Increased precipitation can also increase peat and sediment accumulation in coastal ecosystems through increased delivery of sediments from flooding, peat production through root growth, and organic matter accumulation from litter inputs (Rogers et al. 2014; Sanders et al. 2016; Scavia et al. 2002). This process has allowed mangroves and salt marshes to keep up with sea level rise in the past (Ellison and Stoddart 1991; Rogers et al. 2014) as well as migrate inland (Eslami-Andargoli et al. 2009). We note, however, that prolonged flooding can also result in the loss of wetland plants and thus belowground productivity.

Invasive plants are likely to have a negative impact on coastal wetland soils of Hawai'i and the USAPI, although little research has been conducted on this topic. Vegetation from 40 wetlands sampled across the five main Hawaiian Islands was dominated by nonnative species, representing more than 60% of the species sampled (Bantilan-Smith et al. 2009). In continental United States wetlands, this kind of invasion had a negative impact on soil properties (Zedler and Kercher 2004), yet no studies have examined these relationships in Hawai'i and USAPI coastal systems. Preliminary results from Hawai'i have shown that the invasion by mangroves can significantly increase C content of coastal soils (R.A. MacKenzie, unpublished data of the coauthor).

Sea level rise is the factor expected to have the greatest negative impact on coastal ecosystems. Mangroves are dynamic systems that play a critical role in protecting island ecosystems from a rising sea level by (1) maintaining the forest floor relative to sea level through root growth and sediment trapping and (2) migrating with a rising sea level to more inland locations suitable for soil formation and forest growth (Krauss et al. 2014). However, there appear to be strong geologically driven differences across the USAPI in resilience to sea level rise. Stratigraphic records from the USAPI revealed that high-island mangrove ecosystems were able to keep up with past rates of sea level rise of 4.5 mm yr^{-1}, whereas low-island mangroves were not able to keep up with rates of 1.2 mm yr^{-1} (Ellison and Stoddart 1991). Thus, high-island mangrove systems are thought to be more resilient to current rates of sea level rise (2.0–4.0 mm yr^{-1}) than low-island mangroves. This difference may be related to (1) greater sediment inputs from high-island watersheds, which allow these coastal forests to trap larger amounts of sediment and maintain their elevation and (2) greater high-island diversity of inland sites for mangrove migration.

We anticipate that similar patterns will emerge with current high rates of sea level rise for low atoll islands of the USAPI but also of Papahānaumokuākea (the Northwestern Hawaiian Islands). We expect low islands to be more sensitive to sea level rise than high islands, adding burdens and further reducing the socioecological resilience of these low islands and their soils to sea level rise. Further, a rising sea level will result in longer salt water inundation periods for soils, salinity shifts, and increased coastal erosion. While many wetland plants are adapted to live in extreme conditions that occur from inundation, such as waterlogged and anoxic hydric soils, many plants cannot tolerate inundation, especially with higher salinity water, for extended periods of time. Permanent flooding of coastal areas will also alter soils by decreasing oxygen availability and increasing the amount of reduced elements (e.g., Fe, manganese) that can lead to plant stress and decreased plant growth, reproductive output, and productivity of some plant species (Pezeshki and DeLaune 2012). These conditions can also result in the release of P and N bound to wetland soils, potentially increasing nutrient loading to nearshore coral reefs or seagrass beds

(Bruland and De Ment 2009). Loss of vegetation from flooding can result in peat collapse through the decomposition of root material or leaching of organic matter, lowering the elevation of the wetland, and creating flooded conditions again (Day et al. 2008).

Increased flooding from sea level rise and from storm surges of increased severity and frequency can also impact the elevation of coastal soils through the deposition of new sediments or erosion of coastal edges. The latter can lead to the redistribution and deposition of sediments or vegetation wrack elsewhere in the systems (Fagherazzi et al. 2012; Morton and Barras 2011). Either can have positive or negative effects. Small amounts of sediment benefit coastal wetlands by contributing to vertical soil building through deposition on the marsh or mangrove surface. In contrast, large quantities of sediment delivered to low-saline marshes can smother and kill vegetation (Guntenspergen et al. 1995). Quantifying how fast the elevation of coastal ecosystems is rising relative to sea level rise is an important research need. Research and monitoring can identify areas that may be resilient to sea level rise. These areas may be sequestering C at a faster rate than less productive systems and thus require special conservation designation to protect the many ecosystem services that are provided by coastal soils. Less productive areas that are not keeping up with sea level rise and are sequestering C at a slower rate may require a more proactive restoration and management action (reforestation, hydrologic restoration) in order to protect them.

Key Findings

- Soils of Hawaii and the USAPI represent a remarkable gradient in soil-forming factors and, hence, soil types, complicating efforts to develop management solutions for global change-related challenges to soil health. There remains a need for improved knowledge about how synergistic effects of rising sea level, changing climate, expanding invasive species cover, and increased exposure to novel fire regimes all affect soil resources of the Pacific.
- Recent advances in knowledge of invasive species, drought, and fire have expanded capacity to inform decisions regarding soil changes that accompany novel fires—the frequency, severity, and extent of which are increasing as climate warms, rainfall becomes more variable, and the cover of fire-promoting invasive species increases. While understanding of both the drivers of these disturbances and their often multiplicative effects on soils has greatly increased, strategies are needed for successfully reducing fires and their impacts, especially during periods of high fire danger.
- Hawai'i and the USAPI are home to a number of the Earth's plant and animal biodiversity hotspots, including a remarkable diversity of nearshore coral habitats, as well as an enormous diversity of cultural systems strongly connected to soils via traditional management systems. These are all affected by impacts to watershed condition and soil erosional processes and by sea level rise. Likely future changes for the region include (1) elevated fire impacts to watersheds; (2) expanded agricultural land use, which elevates erosional processes; and (3) continued rise in sea level and salt water intrusion events, which drastically alter soil tilth and compromise ecosystem health. Preventing human-caused fires across the region and enhancing coastal forest cover on low-lying islands remain important areas that require enhanced management attention.
- Soils and soil-building processes are critical in maintaining and protecting coastal habitats that support indigenous populations of flora and fauna and humans.

Literature Cited

Asner GP, Elmore AJ, Hughes RF et al (2005) Ecosystem structure along bioclimatic gradients in Hawai'i from imaging spectroscopy. Remote Sens Environ 96(3–4):497–508

Asner GP, Hughes RF, Vitousek PM et al (2008) Invasive plants transform the three-dimensional structure of rain forests. Proc Natl Acad Sci 105(11):4519–4523

Balick MJ (ed) (2009) Ethnobotany of Pohnpei: plants, people, and island culture. New York Botanical Garden/University of Hawaii Press, New York

Bantilan-Smith M, Bruland GL, MacKenzie RA et al (2009) A quantitative assessment of the vegetation and soils of coastal lowland wetlands in Hawai`i. Wetlands 29:1023–1036

Bruland GL, De Ment G (2009) Phosphorus sorption dynamics of Hawaii's coastal wetlands. Estuar Coasts 32:844–854

Bruland GL, MacKenzie RA (2010) Nitrogen source tracking with $\delta^{15}N$ content of coastal wetland plants in Hawaii. J Environ Qual 39:409–419

Burkett V, Kusler J (2000) Climate change: potential impacts and interactions in wetlands of the United States. J Am Water Resour Assoc 36:313–320

Cabin RJ, Weller SG, Lorence DH et al (2002) Effects of light, alien grass, and native species additions on Hawaiian dry forest restoration. Ecol Appl 12(6):1595–1610

Cavaleri MA, Sack L (2010) Comparative water use of native and invasive plants at multiple scales: a global meta-analysis. Ecology 91:2705–2715

Chadwick OA, Gavenda RT, Kelly EF et al (2003) The impact of climate on the biogeochemical functioning of volcanic soils. Chem Geol 202:195–223

Chimner RA, Ewel KC (2004) Differences in carbon fluxes between forested and cultivated Micronesian tropical peatlands. Wetl Ecol Manag 12:419–427

Chu P-S, Chen H (2005) Interannual and interdecadal rainfall variations in the Hawaiian Islands. J Clim 18:4796–4813

Chu P-S, Chen YR, Schroeder TA (2010) Changes in precipitation extremes in the Hawaiian Islands in a warming climate. J Clim 23:4881–4900

Clough BF, Andrews TJ, Cowan IR (1982) Physiological processes in mangroves. In: Clough BF (ed) Mangrove ecosystems in Australia: structure, function and management. Australian National University Press, Canberra, pp 193–210

Conroy NK, Fares A, Ewel KC et al (2011) A snapshot of agroforestry in *Terminalia carolinensis* wetlands in Kosrae, Federated States of Micronesia. Micronesica 41(2):177–195

Cordell S, Cabin R, Hadway L (2002) Physiological ecology of native and alien dry forest shrubs in Hawaii. Biol Invasions 4(4):387–396

D'Antonio CM, Vitousek PM (1992) Biological invasions by exotic grasses, the grass/fire cycle, and global change. Annu Rev Ecol Syst 23:63–87

Day JW, Christian RR, Boesch DM et al (2008) Consequences of climate change on the ecogeomorphology of coastal wetlands. Estuar Coasts 31:477–491

Deenik JL, Yost RS (2006) Chemical properties of atoll soils in the Marshall Islands and constraints to crop production. Geoderma 136(3–4):666–681

Døckersmith IC, Giardina CP, Sanford RL Jr (1999) Persistence of tree related patterns in soil nutrients following slash-and-burn disturbance in the tropics. Plant Soil 209(1):137–157

Donato DC, Kauffman JB, Murdiyarso D et al (2011) Mangroves among the most carbon-rich forests in the tropics. Nat Geosci 4(5):293–297

Donato DC, Kauffman JB, MacKenzie RA et al (2012) Whole-island carbon stocks in the tropical Pacific: implications for mangrove conservation and upland restoration. J Environ Manag 97:89–96

Donnegan JA, Butler SL, Kuegler O et al (2007) Palau's forest resources, 2003. Resour. Bull. PNW-RB-252. U.S. Department of Agriculture, Forest Service, Pacific Northwest Research Station, Portland, 52 p

Drew WM, Ewel KC, Naylor RL, Sigrah A (2005) A tropical freshwater wetland: III. Direct use values and other goods and services. Wetl Ecol Manag 13(6):685–693

Ellison JC (2000) How South Pacific mangroves may respond to predicted climate change and sea-level rise. In: Gillespie A, Burns WCG (eds) Climate change in the South Pacific: impacts and responses in Australia, New Zealand, and small island states. Springer, Dordrecht, pp 289–300

Ellison AM, Stoddart DR (1991) Mangrove ecosystem collapse during predicted sea level rise: holocene analogues and implications. J Coast Res 7:151–165

Eslami-Andargoli L, Dale P, Sipe N, Chaseling J (2009) Mangrove expansion and rainfall patterns in Moreton Bay, Southeast Queensland, Australia. Estuar Coast Shelf Sci 85:292–298

Fagherazzi S, Kirwan ML, Mudd SM et al (2012) Numerical models of salt marsh evolution: ecological, geomorphic, and climatic factors. Rev Geophys 50:RG1002

Field CD (1995) Impact of expected climate change on mangroves. Hydrobiologia 295:75–81

Frazier AG, Giambelluca T (2017) Spatial trend analysis of Hawaiian rainfall from 1920 to 2012. Int J Climatol 37(5):2522–2531

Giambelluca TW, Diaz HF, Luke SA (2008) Secular temperature changes in Hawai'i. Geophys Res Lett 35:L12702

Giardina CP, Sanford RL, Døckersmith IC, Jaramillo V (2000) The effects of slash burning on ecosystem nutrients during the land preparation phase of shifting cultivation. Plant Soil 220(1–2):247–260

Gilman E, Ellison J, Coleman R (2007) Assessment of mangrove response to projected relative sea-level rise and recent historical reconstruction of shoreline position. Environ Monit Assess 124:105–130

Golbuu Y, Victor S, Wolanski E, Richmond RH (2003) Trapping of fine sediment in a semi-enclosed bay, Palau, Micronesia. Estuar Coast Shelf Sci 57:941–949

Guntenspergen G, Cahoon D, Grace J et al (1995) Disturbance and recovery of the Louisiana coastal marsh landscape from the impacts of Hurricane Andrew. J Coast Res Spec Issue 21:324–339

Jenny H (1941) Factors of soil formation: a system of quantitative pedology. McGraw-Hill, New York, 281 p

Keener VW, Izuka SK, Anthony S (2012a) Freshwater and drought on Pacific Islands. In: Keener VW, Marra JJ, Finucane ML et al (eds) Climate change and Pacific Islands: indicators and impacts. Report for the 2012 Pacific Islands Regional Climate Assessment (PIRCA). Island Press, Washington, DC. Chapter 2

Keener VW, Marra JJ, Finucane ML et al (eds) (2012b) Climate change and Pacific Islands: indicators and impacts. Report for the 2012 Pacific Islands Regional Climate Assessment (PIRCA). Island Press, Washington, DC

Kikuchi WK (1976) Prehistoric Hawaiian fishponds. Science 193:295–299

Krauss KW, McKee KL, Lovelock CE et al (2014) How mangrove forests adjust to rising sea level. New Phytol 202:19–34

Lovelock CE (2008) Soil respiration and belowground carbon allocation in mangrove forests. Ecosystems 11:342–354

McLeod E, Salm RV (2006) Managing mangroves for resilience to climate change. The International Union for the Conservation of Nature and Natural Resources, Gland

Morrissey EM, Gillespie JL, Morina JC, Franklin RB (2014) Salinity affects microbial activity and soil organic matter content in tidal wetlands. Glob Chang Biol 20:1351–1362

Morton RA, Barras JA (2011) Hurricane impacts on coastal wetlands: a half-century record of storm-generated features from southern Louisiana. J Coast Res 27:27–43

Naylor R, Drew M (1998) Valuing mangrove resources in Kosrae, Micronesia. Environ Dev Econ 3:471–490

Norton CW, Chu PS, Schroeder TA (2011) Projecting changes in future heavy rainfall events for Oahu, Hawaii: a statistical downscaling approach. J Geophys Res 116:D17

Oki DS (2004) Trends in streamflow characteristics at long-term gaging stations, Hawaii. Scientific Investigations Rep. 2004-5080. U.S. Department of the Interior, Geological Survey. Available at https://pubs.er.usgs.gov/publication/sir20045080. Accessed 2 May 2020

Pezeshki S, DeLaune R (2012) Soil oxidation-reduction in wetlands and its impact on plant functioning. Biology 1:196–221

Polhemus DA (2017) Drought in the U.S. – affiliated Pacific Islands: a multi-level assessment. Report prepared for the Pacific Islands Climate Science Center. 67 p. Available at https://www.sciencebase.gov/catalog/item/598dbb58e4b09fa1cb13ef85

Rogers K, Saintilan N, Woodroffe CD (2014) Surface elevation change and vegetation distribution dynamics in a subtropical coastal wetland: implications for coastal wetland response to climate change. Estuar Coast Shelf Sci 149:46–56

Safeeq M, Fares A (2011) Hydrologic response of a Hawaiian watershed to future climate change scenarios. Hydrol Process 26(18):2745–2764

Sanders CJ, Maher DT, Tait DR et al (2016) Are global mangrove carbon stocks driven by rainfall? J Geophys Res Biogeosci 121:2600–2609

Scavia D, Field JC, Boesch DF et al (2002) Climate change impacts on U.S. coastal and marine ecosystems. Estuaries 25(2):149–164

Soil Survey Staff (2014) Keys to soil taxonomy, 12th edn. U.S. Department of Agriculture, Natural Resources Conservation Service, Washington, DC

Strauch AM, Giardina CP, MacKenzie RA et al (2017) Modeled effects of climate change and plant invasion on watershed function across a steep tropical rainfall gradient. Ecosystems 20(3):583–600

Trauernicht C, Pickett E, Giardina CP et al (2015) The contemporary scale and context of wildfire in Hawai'i. Pac Sci 69(4):427–444

University of Hawaii (2014) Hawaii soil atlas. University of Hawaii, College of Tropical Agriculture and Human Resources. http://gis.ctahr.hawaii.edu/downloads/soilAtlas. Accessed 20 June 2018

USDA Forest Service [USDA FS] (2018) Inventory data. Available at https://www.fs.fed.us/pnw/rma/fia-topics/inventory-data/. U.S. Department of Agriculture, Forest Service, Pacific Northwest Research Station

USDA National Cooperative Soil Survey [USDA NCSS] (n.d.) National Cooperative Soil Survey soil characterization database. Available at https://ncsslabdatamart.sc.egov.usda.gov/. Accessed 17 Mar 2019

USDA Natural Resources Conservation Service [USDA NRCS] (2009) Soil survey of the Islands of Palau, Republic of Palau, 402 p. https://www.nrcs.usda.gov/Internet/FSE_MANUSCRIPTS/pacific_basin/palauPB2011/palauCD.pdf

USDA Soil Conservation Service [USDA SCS] (1972) Islands of Kauai, Oahu, Maui, Molokai, and Lanai, State of Hawaii, 249 p. https://www.nrcs.usda.gov/Internet/FSE_MANUSCRIPTS/hawaii/islandsHI1972/Five_islands_SS.pdf

USDA Soil Conservation Service [USDA SCS] (1973) Island of Hawaii, State of Hawaii, 118 p. https://www.nrcs.usda.gov/Internet/FSE_MANUSCRIPTS/hawaii/HI801/0/hawaii.pdf

USDA Soil Conservation Service [USDA SCS] (1982) Soil survey of Island of Ponape, Federated States of Micronesia, 81 p. https://www.nrcs.usda.gov/Internet/FSE_MANUSCRIPTS/pacific_basin/PB932/0/ponape.pdf

USDA Soil Conservation Service [USDA SCS] (1983a) Soil survey of Island of Kosrae, Federated States of Micronesia, 67 p. https://www.nrcs.usda.gov/Internet/FSE_MANUSCRIPTS/pacific_basin/PB931/0/Kosrae.pdf

USDA Soil Conservation Service [USDA SCS] (1983b) Soil survey of Islands of Yap, Federated States of Micronesia, 90 p. https://www.nrcs.usda.gov/Internet/FSE_MANUSCRIPTS/pacific_basin/PB934/0/yap.pdf

USDA Soil Conservation Service [USDA SCS] (1984) Soil survey of American Samoa, 97 p. https://www.nrcs.usda.gov/Internet/FSE_MANUSCRIPTS/pacific_basin/PB630/0/american_samoa.pdf

Victor S, Golbuu Y, Wolanksi E, Richmond RH (2004) Fine sediment trapping in two mangrove-fringed estuaries exposed to contrasting land-use intensity, Palau, Micronesia. Wetl Ecol Manag 12:277–283

Vitousek PM (2004) Nutrient cycling and limitation: Hawai'i as a model system. Princeton University Press, Princeton, 232 p

Zedler JB, Kercher S (2004) Causes and consequences of invasive plants in wetlands: opportunities, opportunists, and outcomes. Crit Rev Plant Sci 23:431–452

Appendix B: Soils Networks and Resources

Yvonne Shih

Table B1 Soils monitoring and long-term experimental networks

Name	Chapter	Sponsor (s)[a]	Scope	Purpose	Parameters, measured, modeled, mapped, or recorded in database	Start year	Web address
Climate Reference Network (CRN)	3	NOAA National Centers for Environmental Information	National, multi-media	"To provide a continuous series of climate observations for monitoring trends in the Nation's climate and supporting climate-impact research."	Measure temperature, precipitation, wind speed, soil conditions, and more	2008	https://www.ncdc.noaa.gov/crn/
Forest Soil Disturbance Monitoring Protocol (FSDMP)	9	USDA Forest Service	National, soils only	"To monitor forest sites before and after ground-disturbing management activities for physical attributes that could influence site resilience and long-term sustainability."	Monitor the attributes of surface cover, ruts, compaction, and platy structure	2009	https://www.fs.usda.gov/treesearch/pubs/34427
Forest Inventory and Analysis/Forest Health Monitoring (FIA/FHM) Soil Condition Indicators	1, 2, 9	USDA Forest Service	National, multi-media	"To collect data about erosion, compaction, and important physical and chemical soil properties."	Measure chemical and physical properties, including nutrient information, such as the amount of exchangeable cations (e.g., calcium, magnesium, potassium, and phosphorus); pH level; carbon and nitrogen; toxics such as heavy metals; and the bulk density (weight of soil per unit volume)	1930s	https://www.fia.fs.fed.us/library/brochures/docs/Forest_Health_Indicators.pdf
Integrated Forest Study (IFS)	4	Electric Power Research Institute	International, multi-media	To evaluate the effects of atmospheric deposition on nutrient cycling in forest ecosystems	Monitor deposition and nutrient cycling, especially significant inputs, outputs, and internal fluxes	1970	N/A
Long-Term Ecological Research (LTER) Network	9	National Science Foundation (NSF)	National	"To provide the scientific community, policymakers, and society with the knowledge and predictive understanding necessary to conserve, protect, and manage the nation's ecosystems, their biodiversity, and the services they provide."	In these living laboratories, Forest Service scientists not only make discoveries but also demonstrate research results for cooperators and stakeholders	1980	https://lternet.edu/
Long-Term Soil Productivity (LTSP) Experiment	1, 2, 5, 8	USDA Forest Service	National, multi-media	"Centers on core experiments."	"Manipulate site organic matter, soil porosity, and the complexity of the plant community."	1989	https://www.fs.fed.us/psw/topics/forest_mgmt/ltsp/

Name	Region	Organization	Scale	Objective	Description	Year	URL
Luquillo Long-Term Ecological Research Network Canopy Trimming Experiment (CTE)	Caribbean	NSF, the University of Puerto Rico's Department of Environmental Sciences, and the USDA Forest Service's International Institute of Tropical Forestry	International, multi-media	"To study the long-term effects of natural and human disturbances on tropical forests and streams in the Luquillo Mountains."	"Luquillo scientists from varied disciplinary backgrounds carry out long-term measurements of the forest's physical environment, living organisms, and chemical cycles."	1980	https://luq.lter.network/
National Atmospheric Deposition Program (NADP)		US state Agricultural Experiment Stations	National, multi-media	To measure atmospheric deposition and study its effects on the environment.	Monitor concentrations of acids, nutrients, mercury, and base cations in wet and dry atmospheric deposition	1977	http://nadp.slh.wisc.edu/
National Ecological Observatory Network (NEON)	1, 4, 9	NSF and Battelle	National, multi-media	"To collect and provide open data that characterize and quantify complex, rapidly changing ecological processes across the US."	Collect and process "data that characterize plant, animals, soil, nutrients, freshwater and atmosphere."	2011	https://www.neonscience.org/
National Park Service Inventory and Monitoring Network	9	National Park Service	National, multi-media	To understand "the range of natural resources in and around parks" and "how these resources are doing over the long term."	Gather and analyze information on specific park natural resources – the plants, animals, and ecosystems that can indicate the overall biological health of parks	1990	https://www.nps.gov/im/index.htm
Natural Resources Conservation Service [NRCS] Natural Resources Inventory	9	USDA NRCS	National, multi-media	To collect current information on the status, conditions, and trends of soil, water, and related natural resources.	Obtain current information on the status, condition and trends of soil, water, and related natural resources, including expanded ecological concerns; substrate level inference, temperature, specific humidity, wind components, radiation and surface pressure	1977	https://www.nrcs.usda.gov/wps/portal/nrcs/main/national/technical/nra/nri/
National Soil Moisture Network	3	USDA, NOAA, NIDIS, USGS, The Ohio State University, Texas A&M University	National, soils only	"To utilize in situ measurements of soil moisture and NRCS SSURGO soil characteristics and PRISM data to develop high-resolution gridded soil moisture products."	"Collect daily soil moisture observations from the NASMD stations."	2013	http://nationalsoilmoisture.com/
Remote Automated Weather Stations (RAWS)	3	USDA Forest Service and Bureau of Land Management	National, multi-media	To monitor the weather and fire danger so fire managers can "use this data to predict fire behavior and monitor fuels; resource managers can use the data to monitor environmental conditions."	Collect, store, and forward weather data "such as monitoring air quality, rating fire danger, and providing information for research applications."	Mid-1970s	https://raws.nifc.gov/

(continued)

Table B1 (continued)

Name	Chapter	Sponsor (s)[a]	Scope	Purpose	Parameters, measured, modeled, mapped, or recorded in database	Start year	Web address
Soil Climate Analysis Network (SCAN)	3	USDA NRCS and National Water Climate Center	National, soils only	To provide data to support natural resource assessments and conservation activities. The SCAN system focuses on agricultural areas.	Monitor soil moisture content at several depths, air temperature, relative humidity, solar radiation, wind speed and direction, liquid precipitation, and barometric pressure	1991	https://www.wcc.nrcs.usda.gov/scan/
Soil Moisture Active Passive (SMAP)	3, Alaska	National Aeronautics and Space Administration (NASA)	National, soils only	"To enhance understanding of processes that link the water, energy, and carbon cycles, and to extend the capabilities of weather and climate prediction models."	Measure the land surface soil moisture and freeze-thaw state as well as how much water is in the top layer of soil	2015	https://smap.jpl.nasa.gov/
Snow Telemetry (SNOTEL)	3	USDA NRCS and National Water Climate Center	National, multi-media	To forecast yearly water supplies, predict floods, and for general climate research.	Measure snow water content, accumulated precipitation, and air temperature. Some sites also measure snow depth, soil moisture and temperature, wind speed, solar radiation, humidity, and atmospheric pressure.	Mid-1960s	https://www.wcc.nrcs.usda.gov/snow/
Tropical Responses to Altered Climate Experiment (TRACE)	Caribbean	USDA Forest Service's International Institute of Tropical Forestry, US Department of Energy, USGS, NSF National Program Critical Zone Observatories, Michigan Tech, Luquillo Long-Term Ecological Research Network	Regional, multi-media	"To assess the effects of increased temperatures on tropical forest ecosystems."	"Measure the potential impacts of climate change— particularly temperature increase- on soil structure, carbon cycling, and plant physiology."	2016	https://www.forestwarming.org/
Web Soil Survey (WSS)	1, 9, Caribbean	USDA NRCS	National, soils only	To provide "soil data and information produced by the National Cooperative Soil Survey."	Measure soil tabular data, soil boundary color and thickness, soil label size, and provides thematic maps.	2005	https://websoilsurvey.sc.egov.usda.gov/App/HomePage.htm

[a]Other acronyms: *NASMD* North American Soil Moisture Database, *NIDIS* National Integrated Drought Information System, *NOAA* National Oceanic and Atmospheric Administration, *USGS* US Geological Survey

Table B2 Soils mapping, modeling, and database archiving and access

Name	Chapter	Sponsor(s)	Scope	Purpose	Parameters measured, model, mapped or stored in a database	Start year	Web address
Digital Soil Mapping (DSM)	9	USDA Natural Resources Conservation Service (NRCS)	International, soils only	To use geospatial techniques for mapping soils and in the prediction of soil classes or properties from point data using a statistical algorithm.	Estimate the spatial distribution of soil classes (e.g., soil series) and soil properties (e.g., soil organic matter).	Late 1900s	https://www.nrcs.usda.gov/wps/portal/nrcs/detail/soils/survey/geo/%3Fcid=stelprdb1254424
Disaggregation and harmonization of Soil Map units through Resampled classification Trees (DSMART)	9	Research Division of Soils and Landscapes New Zealand	International, soils only	To estimate the probability of occurrence of the individual soil classes.	Sample the polygons of a legacy soil map and uses classification trees to generate a number of realizations of the potential soil class distribution	2014	https://meetingorganizer.copernicus.org/EGU2015/EGU2015-10962-1.pdf
ESRI ArcGIS Soil Inference Engine (ArcSIE®)	9	USDA NRCS	National, soils only	Designed for field soil scientists to implement knowledge-based raster soil mapping.	Implement "rule-based reasoning and case-based reasoning to facilitate construction of soil-landscape models and perform automatic fuzzy soil inference."	2010	http://www.arcsie.com/index.htm
Forest Service Experimental Forest System	9	USDA Forest Service	National, multi-media	To provide a "wealth of records and knowledge of environmental change in natural and managed forest and rangeland ecosystems across the United States."	Measure and record climate, forest dynamics, hydrology, ecosystem components, and forest and stream ecosystems.	1908	https://www.fs.fed.us/research/efr/
International Soil Carbon Network (ISCN)	1, 2, 9	USDA (Forest Service, NRCS, National Institute of Food and Agriculture), US Geological Survey, Lawrence Berkeley National Laboratory	International, soils only	"Science-based network that facilitates data sharing, assembles databases, identifies and fills gaps in data coverage," and enables "spatially explicit assessments of soil carbon in context of landscape, climate, land use, and biotic variables."	Measure soil carbon-interrelated processes to further investigate "the capacity for soil to provide food, fiber, nutrients, water and a balance among land, atmosphere, and terrestrial waters."	2012	https://iscn.fluxdata.org/

(continued)

Table B2 (continued)

Name	Chapter	Sponsor(s)	Scope	Purpose	Parameters measured, model, mapped or stored in a database	Start year	Web address
Gridded Soil Survey Geographical (gSSURGO) Database	2, 9	USDA Natural Resources Conservation Service	National, soils only	"Gridded SURGO (gSSURGO) is similar to the standard USDA-NRCS Soil Survey Geographic (SSURGO) Database product but in the format of an Environmental Systems Research Institute, Inc. (ESRI®) file geodatabase. It is possible to offer these data in statewide or even Conterminous United States (CONUS) tiles. gSSURGO contains all of the original soil attribute tables in SSURGO."	"The tabular data represent the soil attributes and are derived from properties and characteristics stored in the National Soil Information System (NASIS)." Some of the data measures are soil organic carbon, available water storage, National Commodity Crop Productivity Index, root-zone depth of commodity crops, available water storage within the root-zone depth, drought-vulnerable soil landscapes, and potential wetland soil landscapes.	N/A	https://www.nrcs.usda.gov/wps/portal/nrcs/detail/soils/survey/geo/?cid=nrcs142p2_053628
Major Land Resource Area Soil Data Join Recorrelation (SDJR) initiative	Caribbean	USDA NRCS	National, soils only	To create a continuous and joined coverage within the attribute database through a process of data harmonization.	Use the current soils information and data (SSURGO) and to use the current vector depiction of soil distribution for disaggregation.	2012	https://efotg.sc.egov.usda.gov/references/public/WI/SDJR_Harmonizing_Info_Sheet.pdf
National Hierarchical Framework of Ecological Units	9, Great Plains	USDA Forest Service	National, multi-media	It "is a regionalization, classification, and mapping system for stratifying the Earth into progressively smaller areas of increasingly uniform ecological potentials."	Measure "associations of those biotic and environmental factors that directly affect or indirectly express energy, moisture, and nutrient gradients which regulate the structure and function of ecosystems. These factors include climate, physiography, water, soils, air, hydrology, and potential natural communities."	1993	https://www.fs.fed.us/land/pubs/ecoregions/intro.html
National Soil Information System (NASIS)	N/A	USDA NRCS	National, soils only	To provide a "dynamic resource of soils information for a wide range of needs."	Collect soil pedon descriptions and soil samples to be analyzed for soil physical and chemical properties; and create and maintain map unit information.	2017	https://www.nrcs.usda.gov/wps/portal/nrcs/detail/soils/survey/tools/?cid=nrcs142p2_053552
North American Land Data Assimilation System (NLDAS-2)	3	National Center for Atmospheric Research, National Science Foundation	National, multi-media	"To construct quality-controlled, and spatially and temporally consistent, land-surface model (LSM) datasets from the best available observations and model output to support modeling activities."	Datasets are available on hourly, daily, monthly and climatological time scales that measure total precipitation, potential evaporation, total convection precipitation and convective available potential energy.	1970	https://climatedataguide.ucar.edu/climate-data/nldas-north-american-land-data-assimilation-system

Name		Source	Scope	Purpose	Description	Year	URL
Rapid Carbon Assessment (RaCA)	2	USDA NRCS Soil Science Division	National, soils only	"To develop statistically reliable quantitative estimates of amounts and distribution of carbon stocks for US soils under various land covers and to the extent possible, differing agricultural management; to provide data to support model simulations of soil carbon change related to land use change, agricultural management, conservation practices, and climate change; to provide a scientifically and statistically defensible inventory of soil carbon stocks for the US"	"Soil morphology and landscape characteristics were measured at each site, limited vegetation and agricultural management information was collected from each location." Amounts and distributions of carbon stocks and soil carbon change were estimated.	2010	https://www.nrcs.usda.gov/wps/portal/nrcs/detail/soils/survey/?cid=nrcs142p2_054164
Riparian Ecosystem Management Model (REMM)	6	USDA Agricultural Research Service	National, multi-media	"A modeling tool that can help to quantify the water quality benefits of riparian buffers under varying site conditions."	Measure "surface and subsurface hydrology; sediment transport and deposition; carbon, nitrogen, and phosphorus transport, removal, and cycling; and vegetation growth."	1999	https://www.ars.usda.gov/IS/np/RiparianEcosystem/REMMpub.pdf
Soil Data Viewer (SDV)	9	USDA NRCS	National, soils only	To allow "a user to create soil-based thematic maps."	Measure soil tabular and spatial data collected from the Web Soil Survey (WSS).	N/A	https://www.nrcs.usda.gov/wps/portal/nrcs/detailfull/soils/survey/geo/?cid=nrcs142p2_053620
Soil Survey Geographic (SSURGO) Database	1, 3, 9, Great Plains	USDA NRCS	National, soils only	To provide "information about soil as collected by the National Cooperative Soil Survey." "The information can be displayed in tables or as maps."	Walk over the land and observe the soil. The map units describe soils and other components that have unique properties, interpretations, and productivity.	1990s	https://www.nrcs.usda.gov/wps/portal/nrcs/detail/soils/survey/?cid=nrcs142p2_053627
Soil Water Assessment Tool (SWAT)	6	USDA Agricultural Research Service and Texas A&M AgriLife Research	International, soils only	"SWAT is a small watershed to river basin-scale model to simulate the quality and quantity of surface and ground water and predict the environmental impact of land use, land management practices, and climate change."	Measure weather, surface runoff, return flow, percolation, evapotranspiration, transmission losses, pond and reservoir storage, crop growth and irrigation, groundwater flow, reach routing, nutrient and pesticide loading, and water transfer.	Early 1990s	https://swat.tamu.edu/
SoilGrids1km	9	ISRIC[a]–World Soil Information and the Africa Soil Information Service (AfSIS) project	International, soils only	To make spatial predictions for a selection of soil properties.	Measure soil organic carbon, soil pH, sand, silt and clay fractions, bulk density, cation exchange capacity, coarse fragments, soil organic carbon stock, depth to bedrock, World Reference Base soil groups, and USDA Soil Taxonomy suborders.	2014	http://soilgrids.org

(continued)

Table B2 (continued)

Name	Chapter	Sponsor(s)	Scope	Purpose	Parameters measured, model, mapped or stored in a database	Start year	Web address
State Soil Geographic (STATSGO and STATSGO2) database	1, 9	USDA NRCS	National, soils only	"The level of mapping is designed for broad planning and management uses covering state, regional, and multistate areas."	Measure geology, topography, vegetation, and climate of soil and nonsoil areas.	1994	https://www.nrcs.usda.gov/wps/portal/nrcs/detail/soils/survey/geo/?cid=nrcs142p2_053629
Terrestrial Ecological Unit Inventory (TEUI)	9, Southwest	USDA Forest Service	National, multi-media	"A system to classify ecosystem types and map ecological units at different spatial scales."	Measure differences across "land areas that differ in important ecological factors, such as geology, climate, soils, hydrology, and vegetation."	2005	https://www.fs.fed.us/soils/teui.shtml
The Soil2026 Initiative	2, Alaska	USDA NRCS	National, soils only	Data soil mapping in terms of the Soil Survey Program's goal "'to keep the soil survey relevant to ever-changing needs'"	Measure soil-forming factors (climate, organisms, parent material, relief and time) and direct measurements based on a soil landscape approach.	N/A	https://www.nrcs.usda.gov/wps/portal/nrcs/detail/soils/research/?cid=nrcseprd1321708
US Global Change Research Program (USGCRP)	1	National Science and Technology Council's Committee on Environment, Natural Resources, and Sustainability	International, multi-media	"To assist the Nation and the world to understand, assess, predict, and respond to human-induced and natural processes of global change."	Observe and measure short- and long-term changes in climate, the ozone layer, and land cover.	1989	https://www.globalchange.gov/about

aInternational Soil Reference and Information Centre

Table B3 Soils education and outreach networks for students, communities, and policymakers

Name	Chapter	Sponsor(s)	Scope	Purpose	Parameters measured, modeled, mapped, or stored in a database	Start year	Website
The "4 per 1000" Initiative	2	CGIAR[a] System Organization	International, multi-media	"To demonstrate that agriculture, and in particular agricultural soils, can play a crucial role where food security and climate change are concerned."	Encourage stakeholders to implement actions on soil carbon storage through "practices adapted to local environmental, social and economic conditions, such as agro-ecology, agro-forestry, conservation agriculture or landscape management."	2015	https://www.4p1000.org/
The Edible Schoolyard Project	7	Martin Luther King, Jr. Middle School in Berkeley, CA	Local, multi-media	"To build and share a national edible education curriculum for pre-kindergarten through high school."	Demonstrate "the edible education curriculum and pedagogy to the whole school community" by connecting "a one-acre teaching garden and a dedicated kitchen classroom to the science and humanities curricula taught to all students."	1995	https://edibleschoolyard.org/berkeley
Hundred Dollar Bill Project	7	Fundred	National, multi-media	Collective art project with the goal of bringing awareness to the dangers of soil lead and the importance of investing in a solution to improve both human health and soil.	Measure success through individual drawings; children and families can create bills to call for a lead-free future.	2006	https://fundred.org/
National Cooperative Soil Survey (NCSS)	1, 2, 9, Caribbean, Southwest, Hawaii	USDA Natural Resources Conservation Service	National, soils only	The program "works to cooperatively investigate, inventory, document, classify, interpret, disseminate, and publish information about soils."	Relate to the technology for collecting, managing, and presenting information about the properties, patterns, and responses of soils. Also to measure ecological site inventory, training, coordinated research, and operations.	1896	https://www.nrcs.usda.gov/wps/portal/nrcs/main/soils/survey/partnership/ncss/
Sisters Area Fuels Reduction Project (SAFRP)	8	USDA Forest Service	Local, multi-media	"To improve forest health, provide safe escape routes throughout the area, reduce risk to homes and structures in the area, reduce the risk of uncharacteristic wildfire on forest ecosystem components, improve the sustainability of conifer stands to withstand frequent fire, and increase firefighter safety."	Measure the effects that thinning trees, mechanically treating brush, and prescribed burning have on forest health and fire safety.	2008	https://data.ecosystem-management.org/nepaweb/nepa_project_exp.php?project=7800

(continued)

Table B3 (continued)

Name	Chapter	Sponsor(s)	Scope	Purpose	Parameters measured, modeled, mapped, or stored in a database	Start year	Website
Soil and Water Integrated Model (SWIM)	6	Potsdam Institute for Climate Impact Research	International, soils only	"To investigate climate and land use change impacts at the regional scale, where the impacts are manifested and adaptation measures take place."	Measure interlinked processes at the mesoscale such as runoff generation, plant and crop growth, nutrient and carbon cycling, and erosion. It provides numerous model outputs such as river discharge, crop yield, and nutrient concentrations and loads.	2000	https://www.pik-potsdam.de/research/climate-impacts-and-vulnerabilities/models/swim/swim-description
Soil Health Institute	2	Samuel Roberts Nobel Foundation and the Farm Foundation	National, soils only	"Safeguard and enhance the vitality and productivity of soil through scientific research and advancement"	Measure "key soil processes to increase productivity, resilience and environmental quality"; and "identify research and adoption gaps."	2013	https://soilhealthinstitute.org/
Southeast Tree Research and Education Site (SETRES)	1	USDA Forest Service, North Carolina State University, Duke University, and North Carolina State Forest Nutrition cooperative member companies	Regional, multi-media	"To understand and model the mechanisms that permit loblolly pine to respond to 'global change.'"	Measure nutrient and water availability in a randomized experimental design, in which the split-plot effect is CO_2 fertilization.	N/A	https://www.srs.fs.usda.gov/sifg/research/productivity/setres.html
Southern Forest Futures Project	Southeast	USDA Forest Service Southern Research Station	Regional, multi-media	To forecast changes in southern forests between 2010 and 2060 under various scenarios.	Measure concerns that the USDA Forest Service identified in public meetings held throughout the 13 southern states.	2011	https://www.srs.fs.usda.gov/futures/

[a]Consultative Group on International Agricultural Research

Appendix C: Summary of Research Questions

To help fill information gaps identified in this assessment, the following questions are proposed to address challenges and opportunities. They are not intended as recommendations or to be prescriptive in any way, but are intended to encourage thinking and action to enhance soil conditions and their management

Soil Carbon and Soil Organic Matter

- How do carbon dynamics differ between O horizons and mineral soil horizons, where most of the organic matter resides?
- How will climate change-driven increases in precipitation and temperature variability impact soil carbon vulnerability and greenhouse gas fluxes?
- How do rangeland and forest management actions interact with global change phenomena (e.g., shifting plant communities, altered fire regimes) to influence soil carbon vulnerability?
- How does soil carbon change over time and how can uncertainties in carbon estimates related to soil heterogeneity and rock content be reduced?
- How can technological, modeling, and statistical solutions for dealing with uncertainty be standardized to better constrain future soil organic carbon predictions?
- What is the contribution of deeper roots to organic matter inputs and how much organic matter is occurring in surface (0–30 cm) soil and soil deeper than 30 cm (preferably studied to a depth of at least 1 m)?
- How much carbon is associated with canopy soils (e.g., temperate rainforests of PNW, cloud forests, etc) and, if present, what and how do adventitious roots, mycorrhizas, and soil biota contribute to pools and processes of carbon and nutrient cycling in those canopy soils?

Soil Water

- To what spatial extent does grazing impact water movement and quality?
- What are the cumulative impacts of trails, roads, and recreational activities on forest and rangeland soils, particularly relative to water quality and quantity?
- What are the consequences to water quality of making snow and how are these consequences affected by the use of additives?
- What are the effects of natural gas development on soils, and how do they link with surface water quantity and quality in forest and rangeland soils at local and cumulative scales?
- How can hydrologic models be improved to predict soil water dynamics at watershed scales?

- How does soil moisture change with depth and how do moisture profiles vary among soils?
- How can water storage capacity and water dynamics be accurately quantified at a variety of scales?
- What is the isolated effect of soil water repellency on infiltration and runoff and how can that effect be quantified at the watershed scale given the large temporal and spatial variability of soil water repellency?

Soil Fertility and Nutrient Cycling

- How can mineral weathering, a major source of soil nutrients, be quantified? New conceptual models and interdisciplinary research teams, new and transformative instrumentation, and new uses of long-term forest research sites will be needed.
- What are the best techniques for ameliorating soil pollution beyond organic matter amendments and how can these techniques be made operational?
- What are the risks from soil pollution in forests and rangelands and what is the spatial extent of those risks?
- What are the impacts of repeated fire and what are the spatially explicit effects on soil properties and soil biota?

Soil Biota, Biodiversity, and Invasive Species

- What is the taxonomic identity of bacteria, archaea, fungi, nematodes, insects, and other groups of organisms in soils?
- What invasive soil organisms are present and how much have they spread? Some monitoring efforts are currently underway for invasive earthworms and flatworms, but there is a need for sustained, systematic monitoring.
- How do plant, animal, insect, and pathogen invasive species impact soil carbon and how can managers effectively change land management to alter invasive species effects on ecosystem services?
- How will climate change impact soil biotic communities?
- How can forest and rangeland management anticipate climate change and protect or enhance soil biodiversity to promote ecosystem resistance and resilience?
- How does the biotic diversity of soils contribute to key ecosystem functions including water filtration and storage, nutrient and carbon cycling, and wildlife habitat? Combining emerging techniques such as high-throughput sequencing and stable isotope probing will deepen understanding of these areas.
- What are the long-term consequences (>5 years) of management practices on species richness, species diversity, and community composition?
- What indicators can be used to identify priority conservation areas given that linkages between plant diversity and soil biodiversity are ambiguous in US forests and rangelands?

- What is the distribution of soil organisms across ecosystems in North America and which areas are at risk from pathogens and communities that perform specific ecosystem functions well (e.g., carbon storage, plant production, water infiltration)?
- What do we need to know about the biology of late-seral soils to inform restoration of late-seral and other desired plant communities?
- What are the aboveground and belowground interactions between microbes, organic matter, decomposition, and nutrient cycles?

Wetland and Hydric Soils

- What are the effects of interactions between management regimes, climate, and stand development on the functionality of pine plantations in wetlands? How should understanding of these effects inform the design of silvicultural drainage systems that must sustain functional properties of the wetland?
- Are hydric soil processes in restored wetlands comparable to natural systems after the establishment phase, thereby regaining capacity related to ecosystem services?
- What are the soil processes of tidal freshwater wetlands, especially forests, and how is their complex hydrologic setting mediated by tidal and terrestrial processes? This information is important because these wetlands are at the interface of the changing sea level and commonly occur within rapidly urbanizing regions (e.g., southeastern United States).
- What are the long-term (>10 years) responses of soil processes in restored wetlands and forested wetlands where silvicultural drainage has been used? Opportunities should focus on established (>10 years) sites that have some prior information from studies or monitoring that could provide a foundation for further assessment, with the goal of better understanding long-term responses to ensure sustainability.
- What are the dynamics and transfer of organic matter, nutrients, and chemicals through tidal freshwater wetlands and adjoining uplands and tidal waters on a watershed scale?
- How do management regimes and extreme events interact with wetland soil processes that affect surface water quality? Large-scale manipulation experiments can be used to answer this question. Particularly important are systems sensitive to mercury methylation, nutrient loading, and burning.

Urban Soils

- How does soil formation from anthropogenic parent materials occur under urban environmental conditions?

- How can the ecosystem services of urban soils be enhanced?
- What are the soil-social interactions in urban areas?
- How can the "new heterogeneity" of soils in urban landscapes be quantified and mapped, and how do urban soil characteristics translate into soil processes and functions? The ability to connect soil characteristics with soil processes and functions is a desirable outcome from soil surveys and the interpretation of soil survey map information.
- Can some relationships be generalized for all urban soils and can characteristics measured at fine resolutions be scaled up for coarser-scaled modeling efforts?
- How can urban soils be surveyed, classified, and mapped? Can they be incorporated into current classification and mapping systems? How can this information be used for urban planning and design?
- What are the quantifiable spatial characteristics of urban soils?
- What are the quantifiable benefits and consequences of soil contaminants or pathogens in relation to human health?
- How can biosolids and other soil amendments derived from waste materials be used sustainably to maximize food production in the small-plot systems typically found in urban areas?
- How can urban soils be used to facilitate science, technology, engineering, and math education?
- What are soil conditions beneath impervious surfaces and what techniques can be used to restore highly degraded soils?
- Does the combination of soil degradation and sealed surfaces require restoration of soil functions to a higher level (i.e., on a per unit area basis) than typically found with native soils?
- How can "hyperfunctioning" urban soils be created? For example, is it possible to design soil-plant systems that can sequester soil carbon at rates that are multiple times greater than native soil-plant conditions? Can soil-plant systems be designed to improve infiltration compared to native soils?

Soil Mapping and Modeling

- How can soil surveys be downscaled using the predictive capacity of terrain-based digital soil modeling?
- How can soil variability be adequately described and parameterized in models?
- What are the temporal and spatial variations in greenhouse gas fluxes (especially carbon dioxide and methane) and how does disturbance (i.e., both land use and climate) influence those fluxes?
- How can robust mechanistic models be built that predict soil biogeochemical processes in response to changing

environmental conditions and management regimes? Such models can provide tools for assessing and protecting ecosystem services.

- How do key soil ecosystems and soil stressors interact with climate change, air pollution, disturbances such as fire, and land use change and what are the anticipated effects on soil functions and services? This knowledge gap speaks to the need for robust, process-based models to better understand uncertainties and risks so that management and protective measures can be developed.

Printed in the United States
by Baker & Taylor Publisher Services